MATHEMATICS RESEARCH DEVELOPMENTS

ADVANCED MATHEMATICS FOR ENGINEERS WITH APPLICATIONS IN STOCHASTIC PROCESSES

MATHEMATICS RESEARCH DEVELOPMENTS

Queueing Models in Industry and Business
A.M. Haghighi and D.P. Mishev
2008. ISBN: 978-1-60456-189-0

Boundary Properties and Applications of the Differentiated Poisson Integral for Different Domains
Sergo Topuria
2009. ISBN: 978-1-60692-704-5

Quasi-Invariant and Pseudo-Differentiable Measures in Banach Spaces
Sergey Ludkovsky
2009. ISBN: 978-1-60692-734-2

Operator Splittings and their Applications
Istvan Farago and Agnes Havasiy
2009. ISBN: 978-1-60741-776-7

Measure of Non-Compactness for Integral Operators in Weighted Lebesgue Spaces
Alexander Meskhi
2009. ISBN: 978-1-60692-886-8

Mathematics and Mathematical Logic: New Research
Peter Milosav and Irene Ercegovaca (Editors)
2009. ISBN: 978-1-60692-862-2

Role of Nonlinear Dynamics in Endocrine Feedback
Chinmoy K. Bose
2009. ISBN: 978-1-60741-948-8

Geometric Properties and Problems of Thick Knots
Yuanan Diao and Claus Ernst
2009. ISBN: 978-1-60741-070-6

Lie Groups: New Research
Altos B. Canterra (Editor)
2009. ISBN: 978-1-60692-389-4
2009. ISBN: 978-1-61668-164-7 (E-book)

Emerging Topics on Differential Geometry and Graph Theory
Lucas Bernard and Francois Roux (Editors)
2009. ISBN: 978-1-60741-011-9

Weighted Norm Inequalities for Integral Transforms with Product Kernals
Vakhtang Kokilashvili, Alexander Meskh and Lars-Erik Persson
2009. ISBN: 978-1-60741-591-6

Group Theory: Classes, Representation and Connections, and Applications
Charles W. Danellis (Editor)
2010. ISBN: 978-1-60876-175-3

Functional Equations, Difference Inequalities and Ulam Stability Notions (F.U.N.)
John Michael Rassias (Editor)
2010. ISBN: 978-1-60876-461-7

Control Theory and its Applications
Vito G. Massari (Editor)
2010. ISBN: 978-1-61668-384-9
2010. ISBN: 978-1-61668-447-1 (E-book)

MATHEMATICS RESEARCH DEVELOPMENTS

ADVANCED MATHEMATICS FOR ENGINEERS WITH APPLICATIONS IN STOCHASTIC PROCESSES

ALIAKBAR MONTAZER HAGHIGHI
JIAN-AO LIAN
AND
DIMITAR P. MISHEV

DEPARTMENT OF MATHEMATICS
PRAIRIE VIEW A&M UNIVERSITY
(A MEMBER OF THE TEXAS A&M UNIVERSITY SYSTEM)

Nova Science Publishers, Inc.
New York

Revised Edition 2011

LIBRARY OF CONGRESS CATALOGING-IN-PUBLICATION DATA

Haghighi, Aliakbar Montazer.
  Advanced mathematics for engineers with applications in stochastic processes / Aliakbar Montazer Haghighi, Jian-ao Lian, Dimitar P. Mishev.
     p. cm.
  Includes bibliographical references and index.
  ISBN 978-1-60876-880-6 (hardcover)
  1. Functions of several complex variables. 2. Stochastic analysis. I. Lian, Jian-ao. II. Mishev, D. P. (Dimiter P.) III. Title.
  QA331.H175 2009
  510--dc22
                        2009044328

Published by Nova Science Publishers, Inc. ✦ New York

To My Better Half, Shahin, for her long patience and support
-A. M. Haghighi

To Xiaoli, you and your beautiful smile are always my best home remedies
-J.-A. Lian

To My Mother and My Late Father
-D. P. Mishev

CONTENTS

PREFACE

Topics in advanced mathematics for engineers, probability and statistics typically span three subject areas, are addressed in three separate textbooks and taught in three different courses in as many as three semesters. Due to this arrangement, students taking these courses have had to shelf some important and fundamental engineering courses until much later than is necessary.

This practice has generally ignored some striking relations that exist between the seemingly separate areas of statistical concepts, such as moments and estimation of Poisson distribution parameters. On one hand, these concepts commonly appear in stochastic processes - for instance, in measures on effectiveness in queuing models. On the other hand, they can also be viewed as applied probability in engineering disciplines – mechanical, chemical, and electrical, as well as in engineering technology.

There is obviously, an urgent need for a textbook that recognizes the corresponding relationships between the various areas and a matching cohesive course that will see through to their fundamental engineering courses as early as possible. This book is designed to achieve just that. Its seven chapters, while retaining their individual integrity, flow from selected topics in advanced mathematics such as complex analysis and wavelets to probability, statistics, difference equations and stochastic processes. Chapter One establishes a strong foundation for more recent and complex topics in wavelet analysis and classical Fourier analysis that are included in Chapter Two.

The book is not for *all* engineers. However, it allows most engineering majors who need the necessary mathematical concepts after two semesters of calculus and a semester of ordinary differential equations to be able to complete their degree. Few instructors teach the combined subject areas together due to the difficulties associated with handling such a rigorous course with such hefty materials. Instructors can easily solve this issue by teaching the class as a multi-instructor course, as it is practiced at Prairie View A&M University in Texas.

The authors would like to recognize three other colleagues (Stephen Durham, Nancy Flournoy, and Charles Goddard) for their work, parts of which have been used sporadically throughout chapters Four, Five, and Seven. Some of the work was developed from the lecture notes of the first author (Aliakbar Montazer Haghighi) and his three colleagues starting in 1991.

We hope the book will be an interesting and useful one to both students and faculty in science, technology, engineering, and mathematics (STEM).

Chapter 1

INTRODUCTION

The first and second semesters of calculus cover differential and integral calculus of functions with only one variable. However, in reality, students have to deal with functions with two or more variables. That is exactly what the third semester calculus is all about: the differential and integral calculus of functions with two or more variables, usually called multiple variable calculus. Why bother? Calculus of functions of a single variable only provides a blurry image of a picture. Yet, many real world phenomena can only be described by functions with several variables. Calculus of functions of multiple variables perfects the picture and makes it more complete. Therefore, we introduce a few critical key points for functions of several variables in this book. Further details can always be found in a textbook for multivariable calculus.

1.1. FUNCTIONS OF SEVERAL VARIABLES

We start with a simple example, $f(x, y) = \sin(x^2 + y^2)$ as a function of *two* variables x and y, shown in Figure 1.1.1.

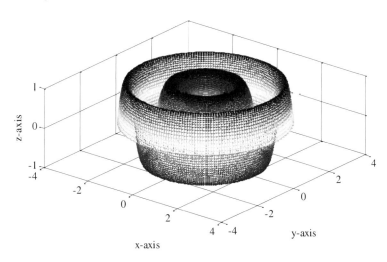

Figure 1.1.1. Graph of the two-variable function $f(x, y) = \sin(x^2 + y^2)$.

Definition 1.1.1.

A *function of several variables* is a function with *domain* being a subset of $\mathbb{R}^n$ and *range* being a subset of $\mathbb{R}$, where the positive integer n is the number of independent variables.

It is clear that explicit functions with two variables can in general be visualized or plotted, as the function shown in Figure 1.1.1.

Example 1.1.1.

Find the domain and range, and sketch the graph of the function of two variables $f(x, y) = \sqrt{1 - x^2 - y^2}$.

Solution

The domain of $f(x, y)$ can be determined from the inequality $1 - x^2 - y^2 \geq 0$, i.e., the unit disk in $\mathbb{R}^2$: $\{(x, y) \in \mathbb{R}^2 : x^2 + y^2 \leq 1\}$. If we let $z = f(x, y)$, the range is clearly $[0, 1]$, simply because $x^2 + y^2 + z^2 = 1$. Henceforward, the graph of $f(x, y)$ is the top half of the unit sphere in space, or $\{(x, y, z) \in \mathbb{R}^3 : x^2 + y^2 + z^2 = 1; 0 \leq z \leq 1\}$. See Figure 1.1.2.

Definition 1.1.2.

A *level set* of a real-valued function $f(x_1, \cdots, x_n)$ of n variables $x_1, x_2, \ldots, x_n$ with respect to an appropriate real value c, called at *level c* for short, is

$$\{(x_1, x_2, \cdots, x_n) \in \mathbb{R}^n : f(x_1, x_2, \cdots, x_n) = c\}.$$

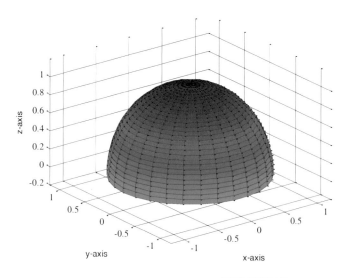

Figure 1.1.2. Graph of the two-variable function $f(x, y) = \sqrt{1 - x^2 - y^2}$, the unit hemisphere.

It is also called a *contour* of f. A *contour plot* is a plot displaying level sets for several different levels. A level set of a function f with two variables x and y at level c is called a *level curve* or a *contour curve*, i.e., $f(x, y) = c$. Level curves of the function in Figure 1.1.1 at level $c = 1$ is a collection of infinite circles

$$\{x^2 + y^2 = 2k\pi + \frac{\pi}{2} : k \in \mathbb{Z}_+\}.$$

A level set of a function f with three variables x, y, and z at level c is called a *level surface* or a *contour surface*, i.e., $f(x, y, z) = c$. The level surface of $f(x, y, z) = x^2 + y^2 + z^2$ at level c, which is nonnegative, is the sphere $x^2 + y^2 + z^2 = \sqrt{c}$.

Functions with three or more variables cannot be visualized anymore. As a simple example, the function $u = g(x, y, z) = \sin(xyz)$ is a function with three variables. Although u cannot be visualized in the four-dimensional space, we can study the function by investigating its contour plots, i.e., $\sin(xyz) = c$ for different level values c on $[-1, 1]$.

Definition 1.1.3.

A *parametric equation* or *parametric form* of a real-valued function or an equation is a method of expressing the function or the equation by using parameters.

It is known that the equation of an ellipse in standard position, i.e., $x^2/a^2 + y^2/b^2 = 1$, has its parametric form $x = a\cos\theta$, $y = b\sin\theta$, with the parameter $\theta \in [0, 2\pi]$. Similarly, a standard axis-aligned ellipsoid $x^2/a^2 + y^2/b^2 + z^2/c^2 = 1$ has a parametric form

$$x = a\cos\theta\sin\phi,$$
$$y = b\sin\theta\sin\phi,$$
$$z = c\cos\phi. \tag{1.1.1}$$

where $\theta \in [0, 2\pi]$, which is used for azimuthal angle, or longitude (denoted by λ), and $\phi \in [0, \pi]$, which is used for polar angle, or zenith angle, or colatitude (latitude $= \pi/2 - \phi$). Equations in (1.1.1) are the usual *spherical coordinates* $(r, \theta, \phi) = $ (radial, azimuthal, polar), a natural description of positions on a sphere.

A surface can sometimes also be given in terms of two parameters, or in its parametric form, like the function in Figure 1.1.1. Figure 1.1.3 is the plot of a typical seashell surface [cf., e.g., Davis & Sigmon 2005], given by a parametric form

$$x = 2\left(1 - e^{u/(6\pi)}\right)\cos u \cos^2 \frac{v}{2},$$
$$y = -2\left(1 - e^{u/(6\pi)}\right)\sin u \cos^2 \frac{v}{2},$$
$$z = 1 - e^{u/(3\pi)} - \left(1 - e^{u/(6\pi)}\right)\sin v. \tag{1.1.2}$$

where $u \in [0.6\pi]$ and $v \in [0.2\pi]$.

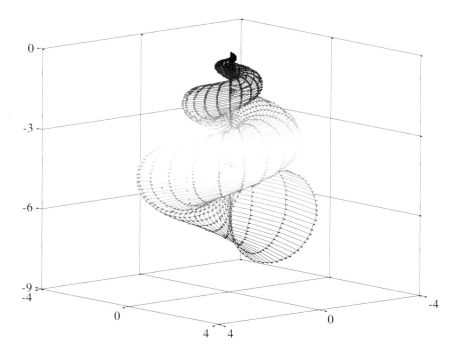

Figure 1.1.3. Graph of a seashell surface with its parametric form given by (1.1.2).

Definition 1.1.4.

For $\delta > 0$, a δ-*neighborhood* (or δ-*ball*) of a point P in $\mathbb{R}^n$ is the set of points that are less δ units away from P, i.e., if $P = (a_1, a_2, \cdots, a_n)$, the δ-neighborhood of P is the set

$$\left\{ (x_1, x_2, \cdots, x_n) \in \mathbb{R}^n : \sum_{j=1}^{n} (x_j - a_j)^2 < \delta^2 \right\}.$$

This of course coincides with the usual δ-neighborhood of a point x_1, when $n = 1$, namely, the length 2δ open interval $\{x \in \mathbb{R} : |x - x_1| < \delta\}$. When $n = 2$, the δ-neighborhood of a point (x_1, y_1) is the open circle with diameter 2δ: $\{(x, y) \in \mathbb{R}^2 : (x - x_1)^2 + (y - y_1)^2 < \delta^2\}$. Naturally, when $n = 3$, the δ-neighborhood of a point $(x_1, y_1, z_1) \in \mathbb{R}^3$ is the following open sphere with diameter 2δ: $\{(x, y, z) \in \mathbb{R}^3 : (x - x_1)^2 + (y - y_1)^2 + (z - z_1)^2 < \delta^2\}$.

Definition 1.1.5.

Let f be a function on $\mathbb{R}^n$ with n variables $x_1, x_2, \cdots, x_n$, and $P \in \mathbb{R}^n$. Then the *limit* of $f(x_1, x_2, \cdots, x_n)$ is L when $(x_1, x_2, \cdots, x_n)$ approaches P, denoted by

$$\lim_{(x_1,x_2,\cdots,x_n)\to P} f(x_1,x_2,\cdots,x_n) = L,$$

if for any $\varepsilon > 0$, there is a $\delta > 0$ such that $|f(x_1,x_2,\cdots,x_n) - L| < \varepsilon$ for all points in the δ-neighborhood of P, except possibly P itself.

By $(x_1,x_2,\cdots,x_n) \to P$, it means the point $(x_1,x_2,\cdots,x_n)$ goes to P by arbitrary or all possible ways. This indeed implies that $\lim_{(x_1,x_2,\cdots,x_n)\to P} f(x_1,x_2,\cdots,x_n) = L$ requires $f(x_1,x_2,\cdots,x_n)$ to be somehow well-defined in the δ-neighborhood of P.

Example 1.1.2.

Find: (a) $\displaystyle\lim_{(x,y)\to(0,0)} \frac{xy}{x^2 + 2y^2}$ and (b) $\displaystyle\lim_{(x,y)\to(0,0)} \frac{x^2 y}{x^2 + 2y^2}$.

Solution

Let $f(x,y) = \dfrac{xy}{x^2 + 2y^2}$. It is clear that the domain of $f(x,y)$ is all points in $\mathbb{R}^2$ but the origin, i.e., $\{(x,y) \in \mathbb{R}^2 : (x,y) \neq (0,0)\}$. Although the function $f(x,y)$ is a rational function, it is undefined at the origin. So, to find the limit of $f(x,y)$ at $(0,0)$, it is not simply substituting both x and y by 0. Observe that $f(x,kx) = \dfrac{k}{1 + 2k^2}$, which gives different values when k changes. In other words, if $(x,y) \to (0,0)$ along two lines $y = x$ and $y = 2x$, the function $f(x,y)$ goes to $1/3$ and $2/9$, respectively. Therefore, $\lim_{(x,y)\to(0,0)} \frac{xy}{x^2+2y^2}$ does not exist. Similarly, to find the limit in (b), we introduce $g(x,y) = \dfrac{x^2 y}{x^2 + 2y^2}$. Then, for all $(x,y) \neq (0,0)$,

$$|g(x,y)| = \frac{x^2|y|}{x^2 + 2y^2} \leq \frac{x^2|y|}{x^2} = |y|,$$

so that $\lim_{(x,y)\to(0,0)} |g(x,y)| = 0$ by using the fact that $\lim_{(x,y)\to(0,0)} |y| = 0$. Hence, $\lim_{(x,y)\to(0,0)} \frac{x^2 y}{x^2+2y^2} = 0$.

Definition 1.1.6.

A convenient notation called *big O notation*, describes the limiting behavior of a single variable real-valued function in terms of a simpler function when the argument of the function approaches a particular value or infinity. The first common big O notation is

$$f(x) = O(g(x)) \text{ as } x \to \infty,$$

meaning there are a sufficiently large number x_0 and a positive number C such that

$$|f(x)| \leq C|g(x)|, \quad \text{for all } x > x_0.$$

The second big O notation is

$$f(x) = O(g(x)) \text{ as } x \to a,$$

meaning there exist positive numbers δ and C such that

$$|f(x)| \leq C|g(x)|, \quad \text{whenever } |x - a| < \delta.$$

The big O notation is very useful for determining the growth rate of functions. For example, the two limits in Example 1.1.2 suggest that the orders or speed when both the numerator and denominator approaching 0 when $(x, y) \to (0, 0)$ play significant rule in determining the existence of a limit. If we let $t = x^2 + y^2$, both the numerator and denominator in (a) $= O(t)$ as $t \to 0$ but the numerator in (b) $= O\left(t^{3/2}\right)$ as $t \to 0$ and its denominator $= O(t)$ as $t \to 0$.

Definition 1.1.7.

Let f be a function of n variables $x_1, x_2, \cdots, x_n$ with domain D. The *graph* of f is the set of all points $(x_1, x_2, \cdots, x_n, z) \in \mathbb{R}^{n+1}$ such that $z = f(x_1, x_2, \cdots, x_n)$ for all $(x_1, x_2, \cdots, x_n) \in D$.

The graph of a function f with two variables x and y and domain D is the set of all points in

$$\left\{(x, y, f(x, y)) \in \mathbb{R}^3 : (x, y) \in D \subset \mathbb{R}^2\right\}.$$

For example, Figure 1.1.1 shows the graph of $f(x, y) = \sin(x^2 + y^2)$.

1.2. PARTIAL DERIVATIVES, GRADIENT, AND DIVERGENCE

To extend the derivative concept of functions with a single variable to functions with several variables, we must use a new name "partial derivative" since all independent variables have equal positions. Without loss of generality, let us concentrate on $f(x, y)$, a function with only two variables x and y.

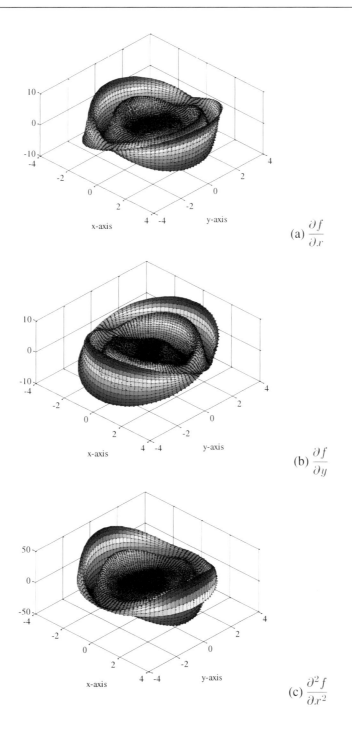

(a) $\dfrac{\partial f}{\partial x}$

(b) $\dfrac{\partial f}{\partial y}$

(c) $\dfrac{\partial^2 f}{\partial x^2}$

Figure 1.1.4.(Continues)

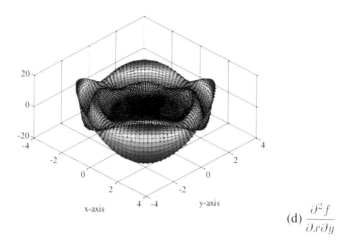

(d) $\dfrac{\partial^2 f}{\partial x \partial y}$

Figure 1.1.4. Graph of the partial derivatives $\dfrac{\partial f}{\partial x}$, $\dfrac{\partial f}{\partial y}$, $\dfrac{\partial^2 f}{\partial x^2}$, and $\dfrac{\partial^2 f}{\partial x \partial y}$ of the two variable function $f(x, y) = \sin(x^2 + y^2)$ shown in Figure 1.1.1.

Definition 1.2.1.

The *partial derivatives* of $f(x, y)$ with respect to x and y, denoted by $\dfrac{\partial f}{\partial x}$ and $\dfrac{\partial f}{\partial y}$, or f_x and f_y, respectively, are defined by

$$\frac{\partial f}{\partial x} = f_x = \lim_{h \to 0} \frac{f(x + h, y) - f(x, y)}{h} \tag{1.2.1}$$

and

$$\frac{\partial f}{\partial y} = f_y = \lim_{h \to 0} \frac{f(x, y + h) - f(x, y)}{h}. \tag{1.2.2}$$

Clearly, the partial derivative of $f(x, y)$ with respect to x is simply the usual or regular derivative of a function with one variable x, with y in $f(x, y)$ being considered as a "constant". The higher order partial derivatives can be similarly defined, e.g.,

$$\frac{\partial^2 f}{\partial y \partial x} = \frac{\partial}{\partial y}\left(\frac{\partial f}{\partial x}\right) = f_{xy}. \quad \frac{\partial^2 f}{\partial x \partial y} = \frac{\partial}{\partial x}\left(\frac{\partial f}{\partial y}\right) = f_{yx}. \tag{1.2.3}$$

$$\frac{\partial^2 f}{\partial x^2} = \frac{\partial}{\partial x}\left(\frac{\partial f}{\partial x}\right) = f_{xx}. \quad \frac{\partial^2 f}{\partial y^2} = \frac{\partial}{\partial y}\left(\frac{\partial f}{\partial y}\right) = f_{yy}. \tag{1.2.4}$$

Figure 1.1.4 gives the plots of the four partial derivatives $\dfrac{\partial f}{\partial x}, \dfrac{\partial f}{\partial y}, \dfrac{\partial^2 f}{\partial x^2}$ and $\dfrac{\partial^2 f}{\partial x \partial y}$ for the function $f(x, y)$ plotted in Figure 1.1.1.

Example 1.2.1.

Find $\dfrac{\partial f}{\partial x}, \dfrac{\partial f}{\partial y}, \dfrac{\partial^2 f}{\partial x^2}, \dfrac{\partial^2 f}{\partial y \partial x}$, and $\dfrac{\partial^2 f}{\partial y^2}$, for $f(x, y) = \sqrt{1 - x^2 - y^2}$.

Solution

For $\dfrac{\partial f}{\partial x}$, we view y as a constant in $f(x, y)$. So,

$$\frac{\partial f}{\partial x} = \frac{1}{2\sqrt{1 - x^2 - y^2}}(-2x) = -\frac{x}{\sqrt{1 - x^2 - y^2}}.$$

Similarly,

$$\frac{\partial f}{\partial y} = \frac{1}{2\sqrt{1 - x^2 - y^2}}(-2y) = -\frac{y}{\sqrt{1 - x^2 - y^2}}.$$

Next, to get $\dfrac{\partial^2 f}{\partial x^2}$, we again view all y in the expression of $\dfrac{\partial f}{\partial x}$ as constants to get:

$$\frac{\partial^2 f}{\partial x^2} = \frac{\partial}{\partial x}\left(\frac{\partial f}{\partial x}\right) = \frac{\partial}{\partial x}\left(-x \cdot \frac{1}{\sqrt{1 - x^2 - y^2}}\right)$$

$$= -\frac{1}{\sqrt{1 - x^2 - y^2}} - x\left(-\frac{1}{2}\right)(1 - x^2 - y^2)^{-3/2}(-2x)$$

$$= \frac{y^2 - 1}{(1 - x^2 - y^2)^{3/2}}.$$

Analogously,

$$\frac{\partial^2 f}{\partial y \partial x} = \frac{\partial}{\partial y}\left(\frac{\partial f}{\partial x}\right) = \frac{\partial}{\partial y}\left(-x \cdot \frac{1}{\sqrt{1 - x^2 - y^2}}\right)$$

$$= -x\left(-\frac{1}{2}\right)(1 - x^2 - y^2)^{-3/2}(-2y)$$

$$= -\frac{xy}{(1 - x^2 - y^2)^{3/2}},$$

and

$$\frac{\partial^2 f}{\partial y^2} = \frac{\partial}{\partial y}\left(\frac{\partial f}{\partial y}\right) = \frac{x^2 - 1}{(1 - x^2 - y^2)^{3/2}}.$$

For high order partial derivatives, does the order with respect to different variables matter? In other words, is $f_{xy} = f_{yx}$ for a function $f(x, y)$ in general?

Theorem 1.2.1 (Clairaut's[1] Theorem or Schwarz's[2] Theorem)

If a function f of n variables $x_1, x_2, \cdots, x_n$ has continuous second order partial derivatives at a point $(a_1, a_2, \cdots, a_n) \in \mathbb{R}^n$, then

$$\frac{\partial^2 f}{\partial x_i \partial x_j}(a_1, a_2, \cdots, a_n) = \frac{\partial^2 f}{\partial x_j \partial x_i}(a_1, a_2, \cdots, a_n), \tag{1.2.5}$$

for $i, j = 1, 2, \ldots, n$. The continuity of $\dfrac{\partial^2 f}{\partial x_i \partial x_j}$ is necessary for (1.2.5) to be true.

Example 1.2.2.

For $f(x, y) = \begin{cases} \dfrac{xy(x^2 - y^2)}{x^2 + y^2}, & \text{if } (x, y) \neq (0, 0), \\ 0, & \text{if } (x, y) = (0, 0), \end{cases}$ find $f_{xy}(0, 0)$ and $f_{yx}(0, 0)$.

Solution

First of all, $f(x, y)$ is continuous at $(0, 0)$ due to the fact that

$$|f(x, y)| = \frac{|xy|\,|x^2 - y^2|}{x^2 + 2y^2} \leq |xy| \leq \frac{1}{2}(x^2 + y^2) \to 0, \quad \text{when } (x, y) \to (0, 0).$$

Secondly, by using (1.2.1) and (1.2.2),

$$f_x(0, 0) = \lim_{h \to 0} \frac{f(h, 0) - f(0, 0)}{h} = 0, \quad f_y(0, 0) = \lim_{h \to 0} \frac{f(0, h) - f(0, 0)}{h} = 0,$$

so that

---

[1] *Alexia Clairaut*, May 3, 1713 – May 17, 1765, was a prominent French mathematician, astronomer, geophysicist, and intellectual, and was born in Paris, France. http://en.wikipedia.org/wiki/Alexis_Clairaut.

[2] *Karl Hermann Amandus Schwarz*, January 25, 1843 – November 30, 1921, was a German mathematician, known for his work in complex analysis. He was born in Hermsdorf, Silesia (now Jerzmanowa, Poland). There are a lot of mathematical phrases being named after Schwarz's name, such as Cauchy-Schwarz inequality, Schwarz lemma, Schwarz triangle, just named a few. http://en.wikipedia.org/wiki/Hermann_Schwarz.

$$f_x(x,y) = \begin{cases} \dfrac{y(x^4 + 4x^2y^2 - y^4)}{(x^2+y^2)^2}, & \text{if } (x,y) \neq (0,0), \\ 0, & \text{if } (x,y) = (0,0), \end{cases}$$

and

$$f_y(x,y) = \begin{cases} \dfrac{x(x^4 - 4x^2y^2 - y^4)}{(x^2+y^2)^2}, & \text{if } (x,y) \neq (0,0), \\ 0, & \text{if } (x,y) = (0,0). \end{cases}$$

Thirdly, by using the definition for partial derivatives,

$$f_{xy}(0,0) = \lim_{h \to 0} \frac{f_x(0,h) - f_x(0,0)}{h} = \lim_{h \to 0} \frac{\dfrac{-h^5}{h^4} - 0}{h} = -1,$$

and

$$f_{yx}(0,0) = \lim_{h \to 0} \frac{f_y(h,0) - f_x(0,0)}{h} = \lim_{h \to 0} \frac{\dfrac{h^5}{h^4} - 0}{h} = 1,$$

i.e., $f_{xy}(0,0) = -1$ and $f_{yx}(0,0) = 1$. It then follows from (1.2.3) that

$$f_{xy}(x,y) = \begin{cases} \dfrac{(x^2 - y^2)(x^4 + 10x^2y^2 + y^4)}{(x^2+y^2)^3}, & \text{if } (x,y) \neq (0,0), \\ -1, & \text{if } (x,y) = (0,0), \end{cases}$$

and

$$f_{yx}(x,y) = \begin{cases} \dfrac{(x^2 - y^2)(x^4 + 10x^2y^2 + y^4)}{(x^2+y^2)^3}, & \text{if } (x,y) \neq (0,0), \\ 1, & \text{if } (x,y) = (0,0). \end{cases}$$

Hence, both $f_{xy}(x,y)$ and $f_{yx}(x,y)$ are discontinuous at $(0,0)$, due to the fact

$$\lim_{(x,y) \to (0,0)} \frac{(x^2 - y^2)(x^4 + 10x^2y^2 + y^4)}{(x^2+y^2)^3}$$

does not exist.

Definition 1.2.2.

The *total differential* or *exact differential* of a function f of n variables $x_1, x_2, \ldots, x_n$, denoted by $df(x_1, x_2, \cdots, x_n)$, is defined by

$$df(x_1, x_2, \cdots, x_n) = \frac{\partial f}{\partial x_1}\, dx_1 + \frac{\partial f}{\partial x_2}\, dx_2 + \cdots + \frac{\partial f}{\partial x_n}\, dx_n$$
$$= f_{x_1}\, dx_1 + f_{x_2}\, dx_2 + \cdots + f_{x_n}\, dx_n.$$

So the total differential of a function f of two variables x and y is

$$df(x, y) = \frac{\partial f}{\partial x}\, dx + \frac{\partial f}{\partial y}\, dy.$$

In a typical introductory differential equation course, there is a section called "exact equations" of first order ordinary differential equations (ODE). For example, the exact ODE $x\, dx + y\, dy = 0$ has solutions $x^2 + y^2 = C$, where C is an arbitrary constant, due to $d(x^2 + y^2) = 2(x\, dx + y\, dy)$. In fact, the expression $M(x, y)\, dx + N(x, y)\, dy$ is a total differential if and only if $M_y = N_x$, provided that M, N, M_x, N_x, M_y, and N_y are all continuous. This is indeed a consequence of Theorem 1.2.1.

Example 1.2.3.

Solve the initial-value problem

$$\left(\frac{1}{1 + y^2} + \cos x - 2xy\right) \frac{dy}{dx} = y(y + \sin x), \quad y(0) = 1.$$

Solution

First, rewrite the ODE into $M(x, y)\, dx + N(x, y)\, dy = 0$ with

$$M(x, y) = y(y + \sin x), \quad N(x, y) = -\frac{1}{1 + y^2} - \cos x + 2xy.$$

It is easy to verify that $M_y = 2y + \sin x = N_x$, i.e., the new ODE, namely, $M(x, y)\, dx + N(x, y)\, dy = 0$, is exact. To find a function f so that $f_x = M$ and $f_y = N$, we can start off by either integrating M with respect to x or integrating N with respect to y, and then apply $f_y = N$ or $f_x = M$. In details,

$$f(x, y) = \int M(x, y)\, dx = \int y(y + \sin x)\, dx = xy^2 - y\cos x + g(y)$$

for an arbitrary function g of y. Then $f_y = N$ becomes

$$2xy - \cos x + g'(y) = -\frac{1}{1 + y^2} - \cos x + 2xy,$$

which leads to $g(y) = -\tan^{-1} y$. Hence, the general solution for the first order ODE is

$$xy^2 - y\cos x - \tan^{-1} y = C.$$

With the initial condition $y(0) = 1$, we have $C = -1 - \pi/4$. Hence, the solution for the initial-value problem is

$$xy^2 - y\cos x - \tan^{-1} y = -1 - \frac{\pi}{4}.$$

Definition 1.2.3.

The *gradient* or *gradient vector* is a (row) vector operator denoted by ∇,

$$\nabla = \frac{\partial}{\partial x_1}\mathbf{u}_1 + \frac{\partial}{\partial x_2}\mathbf{u}_2 + \cdots \frac{\partial}{\partial x_n}\mathbf{u}_n = \left(\frac{\partial}{\partial x_1}, \frac{\partial}{\partial x_2}, \cdots, \frac{\partial}{\partial x_n}\right),$$

where $\mathbf{u}_1$, $\mathbf{u}_2$, ..., $\mathbf{u}_n$ are the unit vectors along the positive directions of axes ox_1, ox_2, ..., ox_n. It is often applied to a real function of n variables, e.g., $f(x_1, x_2, \cdots, x_n)$, namely,

$$\nabla f(x_1, x_2, \cdots, x_n) = \text{grad}(f) = \left(\frac{\partial f}{\partial x_1}, \frac{\partial f}{\partial x_2}, \cdots, \frac{\partial f}{\partial x_n}\right). \tag{1.2.6}$$

For a real function of three variables, e.g., $f(x, y, z)$,

$$\nabla f(x, y, z) = \text{grad}(f) = \frac{\partial f}{\partial x}\mathbf{i} + \frac{\partial f}{\partial y}\mathbf{j} + \frac{\partial f}{\partial z}\mathbf{k} = \left(\frac{\partial f}{\partial x}, \frac{\partial f}{\partial y}, \frac{\partial f}{\partial z}\right). \tag{1.2.7}$$

The notion of gradient has rich geometric interpretations. It can be used to define the tangent plane and normal direction of a surface at a point.

Definition 1.2.4.

The *tangent plane* of a surface $z = f(x, y)$ at a point (x_0, y_0, z_0) is

$$z - z_0 = f_x(x_0, y_0)(x - x_0) + f_y(x_0, y_0)(y - y_0).$$

Definition 1.2.5.

The *Laplace operator* or *Laplacian,* denoted by Δ, is defined from ∇ by

$$\Delta f = \nabla^2 f = \nabla \cdot \nabla f = \frac{\partial^2 f}{\partial x_1^2} + \frac{\partial^2 f}{\partial x_2^2} + \cdots + \frac{\partial^2 f}{\partial x_n^2}. \tag{1.2.8}$$

In particular, for a function f of three variables x, y, and z,

$$\Delta f = \frac{\partial^2 f}{\partial x^2} + \frac{\partial^2 f}{\partial y^2} + \frac{\partial^2 f}{\partial z^2}. \tag{1.2.9}$$

It was named after a French mathematician Pierre-Simon Laplace[3]. The *Laplace's equation* is a partial differential equation (PDE) defined by

$$\Delta f = \frac{\partial^2 f}{\partial x^2} + \frac{\partial^2 f}{\partial y^2} + \frac{\partial^2 f}{\partial z^2} = 0. \tag{1.2.10}$$

All solutions of the Laplace's equations are called *harmonic functions*. Some simple examples of harmonic functions include $e^{ax} \cos ay$, $e^{-ax} \cos ay$, $e^{ax} \sin ay$, $x^3 - 3xy^2$, and $y^3 - 3x^2 y$. If the right-hand side is changed to a given function g, then it is called *Poisson's equation*,

$$\Delta f = \frac{\partial^2 f}{\partial x^2} + \frac{\partial^2 f}{\partial y^2} + \frac{\partial^2 f}{\partial z^2} = g. \tag{1.2.11}$$

It is named after another French mathematician Siméon-Denis Poisson[4]. Both the Laplace's equation and Poisson's equation are the simplest *elliptic* partial differential equations.

Example 1.2.4.

Given $f(x, y) = (\cos^2 x + \cos^2 y)^2$, find ∇f and Δf.

Solution
First of all, all the partial derivatives needed are

$$f_x = -2 \sin 2x (\cos^2 x + \cos^2 y),$$
$$f_y = -2 \sin 2y (\cos^2 x + \cos^2 y),$$
$$f_{xx} = -2 \left[2 \cos 2x (\cos^2 x + \cos^2 y) - \sin^2 2x \right],$$

---

[3] *Pierre-Simon, marquis de Laplace,* March 23, 1749 – March 5, 1827, was a French mathematician and astronomer, and was born in Beaumont-en-Auge, Normandy. His work was pivotal to the development of mathematical astronomy and statistics. The Bayesian interpretation of probability was mainly developed by Laplace. He formulated Laplace's equation, and pioneered the Laplace transform which appears in many branches of mathematical physics. The Laplace operator, widely used in applied mathematics, is also named after him. (http://en.wikipedia.org/wiki/Pierre-Simon_Laplace)

$$f_{yy} = -2\left[2\cos 2y(\cos^2 x + \cos^2 y) - \sin^2 2y\right].$$

Hence, it follows from (1.2.6) and (1.2.8) that

$$\nabla f = (f_x, f_y) = -2(\cos^2 x + \cos^2 y)(\sin 2x, \sin 2y),$$
$$\Delta f = f_{xx} + f_{yy} = -16(\cos^4 x + \cos^2 x \cos^2 y + \cos^4 y - \cos^2 x - \cos^2 y).$$

Definition 1.2.6.

The *divergence* of a vector field $\mathbf{F}(x, y, z) = (M(x, y, z), N(x, y, z), P(x, y, z))$ is an operator defined by

$$\nabla \cdot \mathbf{F} = \mathrm{div}(\mathbf{F}) = \frac{\partial M}{\partial x} + \frac{\partial N}{\partial y} + \frac{\partial P}{\partial z} = M_x + N_y + P_z. \qquad (1.2.12)$$

It is a scalar function and measures the magnitude of a vector field's source or sink at a point. That is why it is sometimes also called *flux density*. A vector field $\mathbf{F}(x, y, z)$ is *divergence free* (or *divergenceless*, or *solenoidal*), if $\nabla \cdot \mathbf{F} \equiv 0$.

Definition 1.2.7.

The *curl* of a vector field is a matrix operator applied to a vector-valued function $\mathbf{F}(x, y, z) = (M(x, y, z), N(x, y, z), P(x, y, z))$, defined by using the gradient operator ∇,

$$\nabla \times \mathbf{F} = \mathrm{curl}(\mathbf{F}) = \begin{vmatrix} \mathbf{i} & \mathbf{j} & \mathbf{k} \\ \dfrac{\partial}{\partial x} & \dfrac{\partial}{\partial y} & \dfrac{\partial}{\partial z} \\ M & N & P \end{vmatrix}$$

$$= \left(\frac{\partial P}{\partial y} - \frac{\partial N}{\partial z}\right)\mathbf{i} + \left(\frac{\partial M}{\partial z} - \frac{\partial P}{\partial x}\right)\mathbf{j} + \left(\frac{\partial N}{\partial x} - \frac{\partial M}{\partial y}\right)\mathbf{k}$$

$$= \begin{pmatrix} 0 & -\dfrac{\partial}{\partial z} & \dfrac{\partial}{\partial y} \\ \dfrac{\partial}{\partial z} & 0 & -\dfrac{\partial}{\partial x} \\ -\dfrac{\partial}{\partial y} & \dfrac{\partial}{\partial x} & 0 \end{pmatrix} \mathbf{F}. \qquad (1.2.13)$$

---

[4] *Siméon-Denis Poisson*, June 21, 1781 – April 25, 1840, was a French mathematician, geometer, and physicist, and was born in Pithiviers, Loiret. (http://en.wikipedia.org/wiki/Simeon_Poisson)

It measures the rotation of a vector field. So it is also sometimes called *circulation density*. The direction of the curl is determined by the right-hand rule. A vector field is *irrotational* or *conservative* if its curl is zero.

Example 1.2.5.

Given

$$\mathbf{F} = \left(\ln(x^2 + y^2), \ -\frac{2z}{x} \tan^{-1} \frac{y}{x}, \ z\sqrt{x^2 + y^2} \right),$$

find $\nabla \cdot \mathbf{F}$ and $\nabla \times \mathbf{F}$.

Solution

Let M, N, and P be the three components of $\mathbf{F}$. Then all the first order partial derivatives of $M, N,$ and P are

$$M_x = \frac{2x}{x^2 + y^2}, \quad M_y = \frac{2y}{x^2 + y^2}, \quad M_z = 0,$$

$$N_x = \frac{2z}{x^2}\left(\tan^{-1} \frac{y}{x} + \frac{xy}{x^2 + y^2} \right), \quad N_y = -\frac{2z}{x^2 + y^2}, \quad N_z = -\frac{2}{x} \tan^{-1} \frac{y}{x},$$

$$P_x = \frac{zx}{\sqrt{x^2 + y^2}}, \quad P_y = \frac{zy}{\sqrt{x^2 + y^2}}, \quad P_z = \sqrt{x^2 + y^2}.$$

Hence, it follows from (1.2.12) and (1.2.13) that

$$\nabla \cdot \mathbf{F} = \frac{2(x - z)}{x^2 + y^2} + \sqrt{x^2 + y^2},$$

and

$$\nabla \times \mathbf{F} = \left(\frac{zy}{\sqrt{x^2 + y^2}} + \frac{2}{x} \tan^{-1} \frac{y}{x}, \ -\frac{zx}{\sqrt{x^2 + y^2}}, \ \frac{2z}{x^2} \tan^{-1} \frac{y}{x} + \frac{2y(z - x)}{x(x^2 + y^2)} \right).$$

Definition 1.2.8.

The *directional derivative* of a function f of n variables $x_1, x_2, \cdots, x_n$ at a point $\mathbf{x} = (x_1, x_2, \cdots, x_n)$ along a direction $\mathbf{v} = (v_1, v_2, \cdots, v_n)$, denoted by $\nabla_{\mathbf{v}} f(\mathbf{x})$ or $D_{\mathbf{v}} f(\mathbf{x})$, is the rate at which $f(\mathbf{x})$ changes at $\mathbf{x}$ in the direction $\mathbf{v}$, defined by the limit

$$\nabla_{\mathbf{v}} f(\mathbf{x}) = D_{\mathbf{v}} f(\mathbf{x}) = \lim_{h \to 0} \frac{f(\mathbf{x} + h\mathbf{v}) - f(\mathbf{x})}{h}.$$

So for a function $f(x, y)$ with two variables, its directional derivative at a point (x_0, y_0) along a vector $v = (x_1, y_1)$, denoted simply by $D_v f(x_0, y_0)$, is

$$D_v f(x_0, y_0) = \frac{d}{dt} f(x_0 + x_1 t, y_0 + y_1 t)\Big|_{t=0},$$

and for a function $f(x, y, z)$ with three variables, its directional derivative at a point (x_0, y_0, z_0) along a vector $v = (x_1, y_1, z_1)$, denoted simply by $D_v f(x_0, y_0, z_0)$, is

$$D_v f(x_0, y_0, z_0) = \frac{d}{dt} f(x_0 + x_1 t, y_0 + y_1 t, z_0 + z_1 t)\Big|_{t=0}.$$

The directional derivatives can be simply calculated by using the gradient, as indicated in the following.

Theorem 1.2.2.

If all first order partial derivatives of $f(\mathbf{x})$ at a point $\mathbf{x}_0$ exist, then

$$D_{\mathbf{v}} f(\mathbf{x}_0) = (\nabla f(\mathbf{x}_0) \cdot \mathbf{v}. \tag{1.2.14}$$

Example 1.2.6.

Find the derivative of the function $f(x, y, z) = x^2 + 2y^2 - 3z^2$ at $P_0(1, -1, 1)$ along the direction $\mathbf{v} = -2\mathbf{i} + 3\mathbf{j} - 4\mathbf{k}$.

Solution
By applying (1.2.14),

$$\begin{aligned}
D_v f(1, -1, 1) &= (\nabla f(x, y, z))\big|_{(1, -1, 1)} \cdot \mathbf{v} \\
&= (2x, 4y, -6z)\big|_{(1, -1, 1)} \cdot (-2, 3, -4) = 8.
\end{aligned}$$

1.3. FUNCTIONS OF A COMPLEX VARIABLE

Functions of a complex variable are complex-valued functions with an independent variable being also complex-valued.

Definition 1.3.1.

A *function of a complex variable*, say, f, defined on a set S is a rule that assigns to every z in S a complex number w, called the function value of f at z, written as $w = f(z)$. The set S is called the *domain* of f and the set of all values of f is called the *range* of f.

Write $w = u + iv$, with u and v being the real and imaginary parts of w. Since w depends on $z = x + iy$, with x and y being the real and imaginary parts of z, it is clear that both u and v are real-valued functions of two real variables x and y. Hence, we may also write

$$w = f(z) = u(x, y) + iv(x, y).$$

Example 1.3.1.

Let $f(z) = (z - 2)/(z + 2)$. Find the values of $\operatorname{Re} f$ and $\operatorname{Im} f$ at $4i$.

Solution

By simply replacing z by $x + iy$, we have

$$f(z) = \frac{z - 2}{z + 2} = \frac{x + iy - 2}{x + iy + 2} = \frac{(x + iy - 2)(x - iy + 2)}{(x + 2)^2 + y^2} = \frac{x^2 + y^2 - 4 + 4iy}{(x + 2)^2 + y^2},$$

so that

$$\operatorname{Re} f(z) = u(x, y) = \frac{x^2 + y^2 - 4}{(x + 2)^2 + y^2} \quad \text{and} \quad \operatorname{Im} f(z) = v(x, y) = \frac{4y}{(x + 2)^2 + y^2}.$$

Hence, at $z = 4i$,

$$\operatorname{Re} f(4i) = u(4, 0) = \frac{3}{5} \quad \text{and} \quad \operatorname{Im} f(4i) = v(0, 4) = \frac{4}{5}.$$

However, if we are not interested in the actual expressions of both $u(x, y)$ and $v(x, y)$, $\operatorname{Re} f$ and $\operatorname{Im} f$ at $4i$ can also be easily obtained directly and quickly from $f(z)$ at $4i$:

$$f(4i) = \left(\frac{z - 2}{z + 2} \right) \bigg|_{z=4i} = \frac{4i - 2}{4i + 2} = \frac{3}{5} + i\frac{4}{5} = \operatorname{Re} f(4i) + i\operatorname{Im} f(4i).$$

The *complex natural exponential function* e^z is defined in terms of real-valued functions by

$$e^z = \exp z = e^x(\cos y + i \sin y), \quad z = x + iy. \tag{1.3.1}$$

Clearly, $e^z = e^x$ if $y = 0$, meaning it is a natural extension of the real-valued natural exponential function. When $x = 0$ we have the *Euler formula*

$$e^{iy} = \cos y + i \sin y. \tag{1.3.2}$$

It is named after the Swiss mathematician Leonhard Paul Euler[5]. By using the usual polar coordinates r and θ for x and y, i.e.,

$$x = r \cos \theta, \qquad y = r \sin \theta, \tag{1.3.3}$$

a complex number $z = x + iy$ has its *polar form*

$$z = x + iy = r(\cos \theta + i \sin \theta) = r e^{i\theta}, \tag{1.3.4}$$

Table 1.3.1. Some Elementary Complex-Valued Functions

	Functions	Period
Complex natural exponential function	$e^z = \exp z = e^x(\cos y + i \sin y)$	$2\pi i$
Complex trigonometric functions	$\cos z = \dfrac{e^{iz} + e^{-iz}}{2} = \cos x \cosh y - i \sin x \sinh y$	2π
	$\sin z = \dfrac{e^{iz} - e^{-iz}}{2i} = \sin x \cosh y + i \cos x \sinh y$	2π
	$\tan z = \dfrac{\sin z}{\cos z} = \dfrac{\sin 2x + i \sinh 2y}{\cos 2x + \cosh 2y}$	π
	$\cot z = \dfrac{\cos z}{\sin z} = -\dfrac{\sin 2x - i \sinh 2y}{\cos 2x - \cosh 2y}$	π
	$\sec z = \dfrac{1}{\cos z} = 2\dfrac{\cos x \cosh y + i \sin x \sinh y}{\cos 2x + \cosh 2y}$	2π
	$\csc z = \dfrac{1}{\sin z} = -2\dfrac{\sin x \cosh y - i \cos x \sinh y}{\cos 2x - \cosh 2y}$	2π
Complex hyperbolic functions	$\cosh z = \dfrac{e^z + e^{-z}}{2} = \cos iz = \cos y \cosh x + i \sin y \sinh x$	$2\pi i$
	$\sinh z = \dfrac{e^z - e^{-z}}{2} = -i \sin iz = \cos y \sinh x + i \sin y \cosh x$	$2\pi i$
	$\tanh z = \dfrac{\sinh z}{\cosh z} = -i \tan iz = \dfrac{\sinh 2x + i \sin 2y}{\cosh 2x + \cos 2y}$	πi
	$\coth z = \dfrac{\cosh z}{\sinh z} = i \cot iz = \dfrac{\sinh 2x - i \sin 2y}{\cosh 2x - \cos 2y}$	πi
	$\operatorname{sech} z = \dfrac{1}{\cosh z} = \sec iz = 2\dfrac{\cos y \cosh x - i \sin y \sinh x}{\cos 2y + \cosh 2x}$	$2\pi i$
	$\operatorname{csch} z = \dfrac{1}{\sinh z} = i \csc iz = -2\dfrac{\cos y \sinh x - i \sin y \cosh x}{\cos 2y - \cosh 2x}$	$2\pi i$

---

[5] *Leonhard Paul Euler,* April 15, 1707 – September 18, 1783, was a pioneering Swiss mathematician and physicist, and was born in Basel. Euler made important discoveries in calculus and graph theory. He introduced the mathematical terminology such as "function" and "mathematical analysis". A statement attributed to Pierre-Simon Laplace expresses Euler's influence on mathematics: "Read Euler, read Euler, he is the master of us all." (http://en.wikipedia.org/wiki/Leonhard_Euler)

	Functions	Period
Complex natural logarithm function	$\ln z = \mathrm{Ln}\, z \pm 2k\pi i = \ln|z| + i\mathrm{Arg}\, z \pm 2k\pi i$	n/a
Complex general exponential function	$z^c = e^{c \ln z}$, c is a complex-valued constant	n/a
Complex general power function	$a^z = e^{z \ln a}$, a is a complex-valued constant, $a \neq 1$	$\dfrac{2\pi i}{\ln a}$

where r is the *absolute value* or *modulus* of z, denoted by $|z|$, θ is the *argument* of z, denoted by $\arg z$, namely:

$$|z| = r = \sqrt{x^2 + y^2} = \sqrt{z\bar{z}}, \tag{1.3.5}$$

$$\arg z = \theta = \arctan \frac{y}{x}, \quad z \neq 0, \tag{1.3.6}$$

and the Euler formula (1.3.2) was used in the last equality of (1.3.4). Obviously, the complex natural exponential function e^z is periodic with period $2\pi i$. To define $\ln z$, we have to rewrite $\arg z$ in (1.3.6) more precisely, i.e.,

$$\arg z = \mathrm{Arg}\, z \pm 2k\pi, \quad k = 0, 1, \cdots, \tag{1.3.7}$$

where $-\pi \leq \mathrm{Arg}\, z \leq \pi$ represents the *principle value* of $\arg z$. So by defining $\ln z$ as the inverse of the complex natural exponential function e^z, we arrive at

$$\ln z = \ln|z| + i\mathrm{Arg}\, z \pm 2k\pi i = \mathrm{Ln}\, z \pm 2k\pi i, \tag{1.3.8}$$

where $\mathrm{Ln}\, z$ is the *principle value* of $\ln z$. The complex trigonometric functions and the complex hyperbolic functions can be defined naturally through

$$\cos z = \frac{e^{iz} + e^{-iz}}{2} \quad \text{and} \quad \sin z = \frac{e^{iz} - e^{-iz}}{2i}. \tag{1.3.9}$$

In summary, these and some other complex elementary functions are included in Table 1.3.1.

1.4. Power Series and their Convergent Behavior

Complex power series are a natural extension of the real power series in calculus. Although all the properties and convergent tests of complex power series are completely similar to those of the real power series, complex power series have their own distinct properties, e.g., they play a pivotal role for analytic functions in complex analysis. More precisely, complex power functions are analytic functions in complex analysis, and vise versa. For convenience, we will not separate real from complex power series, and simply omitting the word "complex" in the sequel. It is easy to distinguish from the context that whether it is real or complex. For examples, series may mean complex series, and power series may mean complex power series.

A *series* is the sum of a sequence, which can be finite or infinite. A *finite series* is a sum of finite sequences and an *infinite series* is the sum of an infinite sequence. Simple examples of series are

$$(1 + 2i) + (4 + 6i) + \cdots + (148 + 198i) \tag{1.4.1}$$

and

$$1 - \left(\frac{1}{2} - \frac{i}{3}\right) + \left(\frac{1}{2} - \frac{i}{3}\right)^2 - \left(\frac{1}{2} - \frac{i}{3}\right)^3 + \cdots, \tag{1.4.2}$$

where the former is a finite series of the *arithmetic* sequence

$$\{1 + 2i, 4 + 6i, 7 + 10i, \cdots, 148 + 198i\},$$

while the latter is an infinite series of the *geometric* sequence

$$\left\{1, -\left(\frac{1}{2} - \frac{i}{3}\right), \left(\frac{1}{2} - \frac{i}{3}\right)^2, -\left(\frac{1}{2} - \frac{i}{3}\right)^3, \cdots\right\}.$$

By using the summation notations, these two series can be simply written as

$$\sum_{n=1}^{50} (3n - 2 + (4n - 2)i) \quad \text{and} \quad \sum_{n=0}^{\infty} (-1)^n \left(\frac{1}{2} - \frac{i}{3}\right)^n, \tag{1.4.3}$$

respectively.

Definition 1.4.1.

A *power series* is an infinite series of the sum of all nonnegative integer powers of $z - z_0$, namely,

$$\sum_{n=0}^{\infty} a_n (z - z_0)^n = a_0 + a_1(z - z_0) + a_2(z - z_0)^2 + \cdots, \tag{1.4.4}$$

where z_0 is a fixed constant or the *center* of the power series, $a_0, a_1, a_2, \ldots$, are constants or the *coefficients* of the power series, and z is a variable. The nth *partial sum*, denoted by $S_n(z)$, is the sum of the first n terms of the power series:

$$S_n(z) = \sum_{m=0}^{n-1} a_m(z - z_0)^m$$

$$= a_0 + a_1(z - z_0) + a_2(z - z_0)^2 + \cdots + a_{n-1}(z - z_0)^{n-1}. \quad (1.4.5)$$

The *remainder* of the power series after the $(n - 1)^{\text{st}}$ term $a_{n-1}(z - z_0)^{n-1}$, denoted by $R_n(z)$, is also an infinite series

$$R_n(z) = \sum_{m=n}^{\infty} a_m(z - z_0)^m$$

$$= a_n(z - z_0)^n + a_{n+1}(z - z_0)^{n+1} + a_{n+2}(z - z_0)^{n+2} + \cdots. \quad (1.4.6)$$

Definition 1.4.2.

A power series *converges* at z, if its partial sum $S_n(z)$ in (1.4.5) converges at z as a sequence, i.e.,

$$S_n(z) = \sum_{m=0}^{n-1} a_m(z - z_0)^m \xrightarrow{n \to \infty} f(z), \quad (1.4.7)$$

or, the sum of the series is $f(z)$, which is simply denoted by

$$f(z) = \sum_{n=0}^{\infty} a_n(z - z_0)^n.$$

Otherwise, it *diverges*, meaning the partial sum sequence $S_n(z)$ diverges. For a convergent series, its remainder $R_n(z)$, as a series, converges to 0. In fact, $f(z) = S_n(z) + R_n(z)$. A power series always converges at its center z_0 (converges to a_0); and, in general, a power series has *radius of convergence* R, if it converges for all z in the R-neighborhood of z_0:

$$f(z) = \sum_{n=0}^{\infty} a_n(z - z_0)^n, \quad |z - z_0| < R.$$

In the complex plane, $|z - z_0| < R$ denotes an "open circular disk" centered at z_0 and with radius R. The actual value R of the radius of convergence of a power series can be calculated from either of the two formulations

$$R = \frac{1}{\lim\limits_{n \to \infty} \sqrt[n]{|a_n|}} \quad \text{and} \quad R = \frac{1}{\lim\limits_{n \to \infty} \left| \dfrac{a_{n+1}}{a_n} \right|}, \quad (1.4.8)$$

provided the limit exists. The second formulation in (1.4.8) is also called Cauchy[6]-Hadamard[7] formula.

Example 1.4.1.

Show that the *geometric series* $\sum_{n=0}^{\infty} z^n = 1 + z + z^2 + \cdots$ converges for all z satisfying $|z| < 1$.

Proof

Due to the identity $\dfrac{1 - z^n}{1 - z} = 1 + z + z^2 + \cdots + z^{n-1}$ for any positive integer n, it is obvious that the geometric series converges to $\dfrac{1}{1 - z}$ when $|z| < 1$ (it diverges when $|z| \geq 1$). It is also easy to see from (1.4.8) that the radius of convergence $R = 1$ due to the fact that the coefficients $a_n = 1$. Hence,

$$\frac{1}{1 - z} = \sum_{n=0}^{\infty} z^n = 1 + z + z^2 + \cdots , \quad |z| < 1. \tag{1.4.9}$$

Replacing z by $-z$ does not have any effect on the radius. So it is also true that

$$\frac{1}{1 + z} = \sum_{n=0}^{\infty} (-1)^n z^n = 1 - z + z^2 - \cdots , \quad |z| < 1. \tag{1.4.10}$$

Example 1.4.2.

Find the power series expression for e^z.

Solution

The most trivial but important power series in mathematics is the one for the *natural exponential* function, namely,

$$e^z = \sum_{n=0}^{\infty} \frac{z^n}{n!} = 1 + \frac{z}{1!} + \frac{z^2}{2!} + \frac{z^3}{3!} + \cdots , \tag{1.4.11}$$

which has radius of convergence as ∞ since

---

[6] *Augustin-Louis Cauchy*, August 21, 1789–May 23, 1857, was a French mathematician and was born in Paris, France. As a profound mathematician and prolific writer, Cauchy was an early pioneer of analysis, particularly complex analysis. His mathematical research spread out to basically the entire range of mathematics. He started the project of formulating and proving the theorems of calculus in a rigorous manner and was thus an early pioneer of analysis. He also gave several important theorems in complex analysis and initiated the study of permutation groups. (http://en.wikipedia.org/wiki/Augustin_Louis_Cauchy)

[7] *Jacques Salomon Hadamard*, December 8, 1865 – October 17, 1963, was a French mathematician who best known for his proof of the prime number theorem in 1896. (http://en.wikipedia.org/wiki/Jacques_Hadamard)

$$\frac{a_{n+1}}{a_n} = \frac{1}{n+1} \xrightarrow{n \to \infty} 0.$$

In other words, (1.4.11) is an identity for all complex number z.

Example 1.4.3.

Find the radius of convergence for $\displaystyle\sum_{n=0}^{\infty} \frac{(z-2)^{2n}}{3^n}$.

Solution

To find the radius of convergence for a series such as $\displaystyle\sum_{n=0}^{\infty} \frac{(z-2)^{2n}}{3^n}$, it is helpful if we apply the formulation to a generic geometric series. More precisely, the series can be viewed as

$$\sum_{n=0}^{\infty} \frac{(z-2)^{2n}}{3^n} = \sum_{n=0}^{\infty} (b_n(z))^n,$$

with $b_n(z) = \dfrac{(z-2)^2}{3}$. Hence, it converges when $|b(z)| < 1$ or $\left| \dfrac{(z-2)^2}{3} \right| < 1$, i.e.,

$$\frac{1}{1 - \dfrac{(z-2)^2}{3}} = \sum_{n=0}^{\infty} \frac{(z-2)^{2n}}{3^n}, \quad |z-2| < \sqrt{3}. \tag{1.4.12}$$

1.5. REAL-VALUED TAYLOR SERIES AND MACLAURIN SERIES

To be easily moving to the complex-valued Taylor series and Maclaurin series, let us first recall real-valued Taylor series and Maclaurin series in calculus.

Definition 1.5.1.

The *Taylor series* of a real-valued function $f(x)$ at a is defined as the power series

$$\sum_{n=0}^{\infty} \frac{f^{(n)}(a)}{n!} (x-a)^n = f(a) + f'(a)(x-a) +$$

$$\frac{f''(a)}{2!} (x-a)^2 + \frac{f^{(3)}(a)}{3!} (x-a)^3 + \cdots. \tag{1.5.1}$$

provided that its nth order derivative exists at $x = a$ for all $n = 0, 1, \cdots$. Taylor series was named in honor of the English mathematician Brook Taylor[8].

Definition 1.5.2.

The *Maclaurin series* of a real-valued function $f(x)$ is the Taylor series of $f(x)$ at $x = a = 0$, i.e.,

$$\sum_{n=0}^{\infty} \frac{f^{(n)}(0)}{n!} x^n = f(0) + f'(0)x + \frac{f''(0)}{2!}x^2 + \frac{f^{(3)}(0)}{3!}x^3 + \cdots . \tag{1.5.2}$$

The Maclaurin series was named in honor of the Scottish mathematician Colin Maclaurin[9].

Again, for the geometric series $\sum_{n=0}^{\infty} x^n = 1 + x + x^2 + \cdots$, it is clear from $f(x) = 1/(1-x)$ and $f^{(n)}(x) = n!/(1-x)^{n+1}$ that $f^{(n)}(0) = 1$ for all $n = 0, 1, \cdots$. In other words, The geometric series $\sum_{n=0}^{\infty} x^n$ is the Maclaurin series of the function $1/(1-x)$.

1.6. POWER SERIES REPRESENTATION OF ANALYTIC FUNCTIONS

1.6.1. Derivative and Analytic Functions

Definition 1.6.1.

A complex function $f(z)$ is *analytic* at a point z_0 provided there exists $\varepsilon > 0$ such that its *derivative* at z_0 exists, denoted by $f'(z_0)$, i.e.,

$$f'(z_0) = \lim_{\Delta z \to 0} \frac{f(z_0 + \Delta z) - f(z_0)}{\Delta z} = \lim_{z \to z_0} \frac{f(z) - f(z_0)}{z - z_0} \tag{1.6.1}$$

---

[8] *Brook Taylor*, August 18, 1685 – November 30, 1731, was an English mathematician and was born at Edmonton (or in then Middlesex). His name is attached to Taylor's Theorem and the Taylor series. He was born at Edmonton (at that time in Middlesex), entered St John's College, Cambridge, as a fellow-commoner in 1701, and took degrees of LL.B. and LL.D. respectively in 1709 and 1714. (http://en.wikipedia.org/wiki/Brook_Taylor)

[9] *Colin Maclaurin*, February, 1698 – June 14, 1746, was a Scottish mathematician, and was born in Kilmodan, Argyll. He entered the University of Glasgow at age eleven, not unusual at the time; but graduating MA by successfully defending a thesis on *the Power of Gravity* at age 14. After graduation he remained at Glasgow to study divinity for a period, and in 1717, aged nineteen, after a competition which lasted for ten days, he was elected professor of mathematics at Marischal College in the University of Aberdeen. He held the record as the world's youngest professor until March 2008. (http://en.wikipedia.org/wiki/Colin_Maclaurin) Side remark: Alia Sabur was three days short of her 19th birthday in February when she was hired to become a professor in the Department of Advanced Technology Fusion at Konkuk University, Seoul. She is the youngest college professor in history and broke Maclaurin's almost 300-year-old record. Alia Sabur was from New York and

exists for all z in a ε-neighborhood of z_0, i.e., $z \in D_\varepsilon(z_0) = \{z : |z - z_0| < \varepsilon\}, \varepsilon > 0$.

There are real analytic functions and complex analytic functions. For simplicity, it is common to use analytic functions exclusively for complex analytic functions. Simple examples of analytic functions include all complex-valued polynomial, exponential, and trigonometric functions. Two typical and classical examples of non-analytic functions are the absolute value function $g(z) = |z| = \sqrt{x^2 + y^2}$ and the complex conjugate function $h(z) = \overline{z} = x - iy$, where $z = x + iy$.

Definition 1.6.2.

A complex function $f(z)$ is *analytic* (or *holomorphic* or *regular*) on D if it is analytic at each point on D; it is *entire* if $f(z)$ is analytic at each point of the complex plane.

Example 1.6.1.

Show by definition that the function $f(z) = \overline{z} = x - iy$ for $z = x + iy$ is nowhere analytic.

Proof

By using the definition of $f'(z_0)$ in (1.6.1),

$$\lim_{\Delta z \to 0} \frac{f(z_0 + \Delta z) - f(z_0)}{\Delta z} = \lim_{\Delta z \to 0} \frac{\overline{(z_0 + \Delta z)} - \overline{z_0}}{\Delta z} = \lim_{\Delta z \to 0} \frac{\overline{\Delta z}}{\Delta z} = \lim_{\Delta z \to 0} \frac{\Delta x - i\Delta y}{\Delta x + i\Delta y}$$

which is -1 if $\Delta x = 0$ and 1 if $\Delta y = 0$, i.e., $f'(z_0)$ does not exist at any point z_0.

Example 1.6.2.

Show by definition that the function $f(z) = |z|$ for $z = x + iy$ is not analytic at any point z.

Proof

Again, by using (1.6.1),

$$\lim_{\Delta z \to 0} \frac{f(z_0 + \Delta z) - f(z_0)}{\Delta z} = \lim_{\Delta z \to 0} \frac{|z_0 + \Delta z| - |z_0|}{\Delta z}.$$

"Rationalize" the numerator of the expression $\dfrac{|z_0 + \Delta z| - |z_0|}{\Delta z}$ to get

$$\frac{|z_0 + \Delta z| - |z_0|}{\Delta z} = \frac{|z_0 + \Delta z| - |z_0|}{\Delta z} \cdot \frac{|z_0 + \Delta z| + |z_0|}{|z_0 + \Delta z| + |z_0|}$$

$$= \frac{|z_0 + \Delta z|^2 - |z_0|^2}{\Delta z(|z_0 + \Delta z| + |z_0|)} = \frac{z_0 \overline{\Delta z} + \overline{z_0} \Delta z + |\Delta z|^2}{\Delta z(|z_0 + \Delta z| + |z_0|)}$$

$$= \frac{z_0}{|z_0 + \Delta z| + |z_0|} \frac{\overline{\Delta z}}{\Delta z} + \frac{\overline{z_0} + \overline{\Delta z}}{|z_0 + \Delta z| + |z_0|}.$$

When $z_0 = 0$, it is $\dfrac{\overline{\Delta z}}{\Delta z}$, whose limit does not exist when $\Delta z \to 0$. When $z_0 \neq 0$, the second

term goes to $\dfrac{\overline{z_0}}{2|z_0|}$ when $\Delta z \to 0$, but the first term does not converge when $\Delta z \to 0$, also

due to $\lim\limits_{\Delta z \to 0} \dfrac{\overline{\Delta z}}{\Delta z}$ does not exist. Therefore, the function $|z|$ is nowhere analytic.

Example 1.6.3.

Find the derivative of $f(z) = \dfrac{z + i}{z - 2i}$ at $z = -i$.

Solution

Again, it follows from (1.6.1) that

$$f'(-i) = \lim_{\Delta z \to 0} \frac{f(-i + \Delta z) - f(-i)}{\Delta z} = \lim_{\Delta z \to 0} \frac{\dfrac{(-i + \Delta z) + i}{(-i + \Delta z) - 2i} - 0}{\Delta z}$$

$$= \lim_{\Delta z \to 0} \frac{1}{-3i + \Delta z} = \frac{1}{-3i} = \frac{i}{3}.$$

Another way of finding $f'(-i)$ is to use the quotient rule, i.e.,

$$f'(-i) = \left(\frac{z + i}{z - 2i} \right)' \bigg|_{z = -i} = \frac{-3i}{(z - 2i)^2} \bigg|_{z = -i} = \frac{i}{3}.$$

Back to definition (1.6.1), it can also be written as

$$f'(z_0) = \lim_{\Delta x + i\Delta y \to 0} \left(\frac{u(x_0 + \Delta x, y_0 + \Delta y) - u(x_0, y_0)}{\Delta x + i\Delta y} \right.$$
$$\left. + i \frac{v(x_0 + \Delta x, y_0 + \Delta y) - v(x_0, y_0)}{\Delta x + i\Delta y} \right), \qquad (1.6.2)$$

where $f(z) = u(x, y) + iv(x, y)$, $z_0 = x_0 + iy_0$, and $\Delta z = \Delta x + i\Delta y$. Then, when $\Delta x \to 0$, $\Delta y \to 0$ and $\Delta x \to 0$, $\Delta y \to 0$, (1.6.2) leads to

$$f'(z_0) = u_x(x_0, y_0) + iv_x(x_0, y_0), \quad f'(z_0) = v_y(x_0, y_0) - iu_y(x_0, y_0). \tag{1.6.3}$$

Theorem 1.6.1 (*Cauchy-Riemann[10] Equations and Analytic Functions*)

Let $f(z) = u(x, y) + iv(x, y)$ be a complex function, with u and v being continuously differentiable on an open set D. Then it is analytic on D if and only if u and v satisfy the *Cauchy-Riemann equations*

$$\frac{\partial u}{\partial x} = \frac{\partial v}{\partial y}, \tag{1.6.4}$$

$$\frac{\partial u}{\partial y} = -\frac{\partial v}{\partial x}. \tag{1.6.5}$$

The necessity follows from (1.6.3), The proof of sufficiency is omitted since it is beyond the scope of the book. We remark here that the Cauchy-Riemann equations in (1.6.3) has the following equivalent forms

$$\frac{\partial f}{\partial x} = -i\frac{\partial f}{\partial y}. \tag{1.6.6}$$

Moreover, if both u and v have continuous second order partial derivatives, we imply from (1.6.4)–(1.6.5) that both u and v are harmonic functions, meaning they satisfy the 2D Laplace's equation in (1.2.10). Two harmonic functions u and v are called a *harmonic conjugate pair* if they satisfy the Cauchy-Riemann equations (1.6.4)–(1.6.5).

It is now straightforward that both the functions $f(z) = \overline{z}$ and $f(z) = |z|$ are not analytic since their differentiable real and imaginary parts do not satisfy (1.6.4)–(1.6.5).

Example 1.6.4.

Let $u(x, y) = \ln(x^2 + y^2)$, $(x, y) \in \mathbb{R}^2 \setminus \{(0, 0)\}$, which is a harmonic function. Find its harmonic conjugate $v(x, y)$ such that the function $f(z) = u(x, y) + iv(x, y)$ is analytic.

Solution

It follows from (1.6.4)–(1.6.5) that

$$v_x = -\frac{2y}{x^2 + y^2}, \quad v_y = \frac{2x}{x^2 + y^2},$$

Hence, up to a constant, such a function v is given by

---

[10] *Georg Friedrich Bernard Riemann*, September 17, 1826 – July 20, 1866, was a German mathematician and was born in Breselenz, German. One of his most significant contributions is the Riemann hypothesis, a conjecture

$$v(x, y) = -2 \tan^{-1} \frac{x}{y}.$$

1.6.2. Line Integral in the Complex Plane

A definite integral of functions with one real variable in calculus can normally be calculated by applying the second fundamental theorem of calculus, if one of its antiderivatives can be explicitly given. Definite integrals for complex-valued functions, to be named complex line integrals, can also be similarly introduced.

Definition 1.6.3.

A *simple curve* is a curve that does not intersect or touch itself. A *simple closed curve* is a closed simple curve. It is also sometimes called a *contour*. A domain D is *simply connected* (also called 1-connected), if every simple closed curve within D can be shrunk continuously within D to a point in D (meaning without leaving D). A domain D is *multiple simply connected* with n holes (also called n-connected), if it is formed by removing n holes from a 1-connected domain.

Definition 1.6.4.

A *complex line integral* is a definite integral of a complex-valued *integrand* $f(z)$ with respect to a complex integral variable z over a given oriented curve C in the complex plane, denoted by

$$\int_C f(z)dz \quad \text{and} \quad \oint_C f(z)dz, \tag{1.6.7}$$

if C is closed. The curve C is also called the *path of integration*.

If $f(z)$ is continuous and the curve C is piecewise continuous, then the complex line integral in (1.6.7) exists. If C is in its parametric representation $z(t) = x(t) + iy(t)$, $t \in [a, b]$, the exact definition of a complex line integral is completely analogous to that of the Riemann integral in calculus, namely: partition $[a, b]$ into n subintervals according to the parameter t; form a Riemann sum; and take the limit when $n \to \infty$. We omit all details here. Furthermore, it is not hard to understand that the two integrals of $f(z)$ over the two opposite orientations of C only differ up to a sign. Most properties for definite integrals of one variable still carry over.

---

about thedistribution of zeros of the Rieman zeta-function $\zeta(s) = \sum_{n=1}^{\infty} (1/n^s)$: All non-trivial zeros of the Riemann zeta-function have real part $1/2$.

Definition 1.6.5.

The second line integral in (1.6.7) is also called a *contour integral*.

If $f(z)$ is analytic on a simply connected domain D, similar to the second theorem of calculus, then for any two points z_0 and z_1 in D,

$$\int_{z_0}^{z_1} f(z)dz = F(z_1) - F(z_0),$$

if $F'(z) = f(z)$, no matter how the curve C is formed from z_0 to z_1. This fact will be further proved by the Cauchy Integral Theorem in §1.6.3.

Theorem 1.6.2.

Let $u(x, y)$ be a harmonic function in a simply connected domain D. Then all its harmonic conjugates are given explicitly by

$$v(x, y) = \int_{z_0}^{z} (-u_y\, dx + u_x\, dy) + C, \tag{1.6.8}$$

where the integral is along any simple curve in D from a point z_0 to z, and $C \in \mathbb{R}$ is arbitrary.

Example 1.6.5.

Find the harmonic conjugates of $u(x, y) = e^{ax} \sin ay$, where $a \in \mathbb{R}$ is a constant.

Solution

By applying (1.6.8) directly,

$$v(x, y) = \int_{z_0}^{z} (-ae^{ax} \cos ay\, dx + ae^{ax} \sin ay\, dy) + C_1$$

$$= -\int_{z_0}^{z} d\left(e^{ax} \cos ay\right) + C_1$$

$$= -\left(e^{ax} \cos ay\right)\Big|_{z_0}^{z} + C_1$$

$$= -e^{ax} \cos ay + C.$$

1.6.3. Cauchy's Integral Theorem for Simply Connected Domains

The first look at the line integral (1.6.7) tells that it depends on not only the endpoints of C but also the choice of the path of C itself. The following theorem tells that a line integral is independent of the path if $f(z)$ is analytic in a simply connected domain D and C is a piecewise continuous curve in D.

Theorem 1.6.3 (*Cauchy's Integral Theorem*)

Let $f(z)$ be an analytic function in a simply connected domain D and C an arbitrary simple closed positively oriented contour that lies in D. Then

$$\oint_C f(z)dz = 0. \tag{1.6.9}$$

Example 1.6.6.

Evaluate $\oint_C e^z dz$ for any contour C.

Solution

Since e^z is entire, the line integral is 0 on any contour C. Indeed, if $f(z)$ is entire,

$$\oint_C f(z)dz = 0 \quad \text{on any contour} \quad C. \quad \text{Similarly,} \quad \oint_C \cos z\,dz = \oint_C \sin z\,dz = 0, \quad \text{and}$$

$$\oint_C z^n dz = 0 \text{ for } n \in Z_+^n.$$

Example 1.6.7.

Evaluate $\oint_C \frac{1}{z}dz$, where C is the counterclockwise unit circle.

Solution

It follows from the parametric form of the unit circle:
$$z(t) = \cos t + i \sin t = e^{it}, \quad t \in [0, 2\pi],$$

that

$$\oint_C \frac{1}{z}dz = \int_0^{2\pi} e^{-it} i e^{it} dt = 2\pi i.$$

Notice that the Cauchy's integral theorem cannot be applied on Example 1.6.7. The contour (unit circle) C can be included within the annulus $r < |z| < R$ for any $0 < r < 1$ and $R > 1$. The integrand $1/z$ is analytic on this annulus but it is not simply connected.

Example 1.6.8.

Evaluate $\oint_C z^a dz$, where C is the counterclockwise unit circle and a is a constant that is different from -1.

Solution

Similar to Example 1.6.7,

$$\oint_C z^a dz = \int_0^{2\pi} e^{iat} i e^{it} dt = i \int_0^{2\pi} e^{i(a+1)t} dt = i \frac{1}{i(a+1)} e^{i(a+1)t} \Big|_0^{2\pi} = \frac{e^{i2\pi a} - 1}{a+1},$$

which is 0 for any integer a different from -1. If we take the limit when $a \to -1$, its value is $2\pi i$ as it was shown in Example 1.6.7. It is not zero for any non-integer a.

Example 1.6.8 tells that the analytic property for $f(z)$ in Theorem 1.6.3 is sufficient but not necessary (for a contour integral to vanish).

Example 1.6.9.

Evaluate $\oint_C \bar{z} dz$, where C is the counterclockwise unit circle.

Solution

Again, similar to Example 1.6.7,

$$\oint_C \bar{z} dz = \int_0^{2\pi} e^{-it} i e^{it} dt = 2\pi i.$$

We knew in Section 1.6.1 that $\bar{z}$ was nowhere analytic. So result of Example 1.6.9 does not conflict with Theorem 1.6.3. However, $\oint_C |z| dz = 0$ when C is the counterclockwise unit circle, since $|z| \equiv 1$ on C and the constant 1 is entire.

1.6.4. Cauchy's Integral Theorem for Multiple Connected Domains

Cauchy's integral Theorem 1.6.3 for simply connected domains can be generalized to multiple connected domains.

Theorem 1.6.4. (Cauchy's Integral Theorem for Multiple Connected Domains)

Let $f(z)$ be an analytic function in a n-connected domain D, and let C an arbitrary simple closed positively oriented contour that lies in D. Then

$$\oint_C f(z)dz = \oint_{C_1} f(z)dz + \oint_{C_2} f(z)dz + \cdots + \oint_{C_n} f(z)dz, \tag{1.6.10}$$

where C_0 is the outer boundary (simple closed) curve with counterclockwise orientation and C_1, C_2, $\cdots$, C_n are the inner simple closed curves of the n holes inside D with clockwise orientation.

1.6.5. Cauchy's Integral Formula

A good starting point to understand analytic functions is the following, i.e., an integral can represent any analytic function.

Theorem 1.6.5. (Cauchy's Integral Formula)

Let $f(z)$ be an analytic function in the simply connected domain D and C a simple closed positively oriented contour that lies in D. If z_0 is a point that lies in interior to C, then

$$f(z_0) = \frac{1}{2\pi i} \oint_C \frac{f(z)}{z - z_0} dz. \tag{1.6.11}$$

Here, the *contour integral* is taken on C, which is positively oriented, meaning *counterclockwise*.

Example 1.6.10.

Evaluate $\oint_C \dfrac{e^{i\pi z}}{3z^2 - 7z + 2} dz$, where C is the unit circle with positive orientation.

Solution

It follows from $3z^2 - 7z + 2 = 3(z - 1/3)(z - 2)$ that the only zero of $3z^2 - 7z + 2$ that lies in the interior of $C : |z| = 1$ is $z_0 = 1/3$. So by letting $f(z) = e^{i\pi z}/(z - 2)$, the integral

$$\oint_C \frac{e^{i\pi z}}{3z^2 - 7z + 2} dz = \frac{1}{3} \oint_C \frac{e^{i\pi z}}{z - 2} \frac{1}{z - 1/3} dz = \frac{1}{3} (2\pi i) f\left(\frac{1}{3}\right)$$

$$= \frac{1}{3} (2\pi i) \frac{e^{i\pi/3}}{1/3 - 2} = \frac{\sqrt{3}}{5} \pi - i\frac{\pi}{5}.$$

It also follows from this example that

$$\oint_C \frac{\cos \pi z}{3z^2 - 7z + 2} dz = -\frac{\pi}{5} i \quad \text{and} \quad \oint_C \frac{\sin \pi z}{3z^2 - 7z + 2} dz = -\frac{\sqrt{3}}{5} \pi i.$$

Example 1.6.11.

Evaluate $\oint_C \frac{e^z}{z^3 - 1} dz$, where C is the circle $C : |z| = 2$ with positive orientation.

Solution

Introduce $f(z) = e^z$. Then, due to the fact that

$$\frac{1}{z^3 - 1} = \frac{1}{(z - 1)(z - e^{i2\pi/3})(z - e^{i4\pi/3})} = \frac{1}{3}\frac{1}{z - 1} + \frac{1}{6}\frac{-1 + \sqrt{3}i}{z - e^{i2\pi/3}} + \frac{1}{6}\frac{-1 - \sqrt{3}i}{z - e^{i4\pi/3}},$$

it follows from Theorem 1.6.5 that

$$\oint_C \frac{e^z}{z^3 - 1} dz$$

$$= \frac{1}{3} \oint_C \frac{e^z}{z - 1} dz + \frac{-1 + \sqrt{3}i}{6} \oint_C \frac{e^z}{z - e^{i2\pi/3}} dz + \frac{-1 - \sqrt{3}i}{6} \oint_C \frac{e^z}{z - e^{i4\pi/3}} dz$$

$$= 2\pi i \left[\frac{1}{3} f(1) + \frac{-1 + \sqrt{3}i}{6} f(e^{i2\pi/3}) + \frac{-1 - \sqrt{3}i}{6} f(e^{i4\pi/3}) \right]$$

$$= 2\pi i \left[\frac{1}{3} e + \frac{-1 + \sqrt{3}i}{6} e^{(-1+\sqrt{3}i)/2} + \frac{-1 - \sqrt{3}i}{6} e^{(-1-\sqrt{3}i)/2} \right]$$

$$= 2\pi i \left[\frac{1}{3} e - \frac{1}{3} e^{-1/2} \left(\cos \frac{\sqrt{3}}{2} + \sqrt{3} \sin \frac{\sqrt{3}}{2} \right) \right].$$

1.6.6. Cauchy's Integral Formula for Derivatives

For derivatives of analytic functions, see the following.

Theorem 1.6.6. (Cauchy's Integral Formula for Derivatives)

Let $f(z)$ be an analytic function in the simply connected domain D, and C a simple closed positively oriented contour that lies in D. If z_0 is a point that lies in interior to C, then for any nonnegative integer n,

$$f^{(n)}(z) = \frac{n!}{2\pi i} \oint_C \frac{f(\xi)}{(\xi - z)^{n+1}}\, dz. \tag{1.6.12}$$

Example 1.6.12.

Evaluate $\oint_C \frac{\cos 2\pi z}{(2z - 1)^3}\, dz$, where C is the unit circle with counterclockwise orientation.

Solution

Since $\cos 2\pi z$ is entire and the unit circle encloses $1/2$, by a direct application of (1.6.12) in Theorem 1.6.6,

$$\oint_C \frac{\cos 2\pi z}{(2z - 1)^3}\, dz = \frac{1}{8} \oint_C \frac{\cos 2\pi z}{(z - 1/2)^3}\, dz = \frac{1}{8} \frac{2\pi i}{2!} (\cos 2\pi z)'' \Big|_{z=1/2}$$

$$= \frac{1}{8}(\pi i)(-4\pi^2 \cos \pi) = \frac{1}{2}\pi^3 i.$$

Example 1.6.13.

Evaluate $\oint_C \frac{\sinh z}{z(z - 3i)^2}\, dz$, where C consists of the circle $|z - i| = 3$ (counterclockwise) and the circle $|z| = 1$ (clockwise).

Solution

Denote by D the annulus enclosed by C and let $f(z) = \dfrac{\sinh z}{z} = \dfrac{-i \sin iz}{z}$ (see Table 1.3.1). Then $f(z)$ is analytic in D and the point $3i$ is enclosed in the interior of C. It follows from Theorem 1.6.6 that

$$\oint_C \frac{\sinh z}{z(z - 3i)^2}\, dz = 2\pi i \left(\frac{\sinh z}{z}\right)' \Big|_{z=3i}$$

$$= 2\pi i(-i) \left(\frac{i \cos iz}{z} - \frac{\sin iz}{z^2}\right) \Big|_{z=3i} = \frac{2\pi}{9}(3 \cos 3 - \sin 3).$$

1.6.7. Taylor and Maclaurin Series of Complex-Valued Functions

The real-valued Taylor series can now be extended to the complex-valued.

Definition 1.6.6.

The *Taylor series* of a function $f(z)$ centered at $z = \alpha$ is defined as the power series

$$\sum_{n=0}^{\infty} \frac{f^{(n)}(\alpha)}{n!}(z - \alpha)^n = f(\alpha) + f'(\alpha)(z - \alpha) + \frac{f''(\alpha)}{2!}(z - \alpha)^2 + \cdots. \qquad (1.6.13)$$

Definition 1.6.7.

Similarly, the *Maclaurin series* is when $\alpha = 0$:

$$\sum_{n=0}^{\infty} \frac{f^{(n)}(0)}{n!} z^n = f(0) + f'(0)z + \frac{f''(0)}{2!} z^2 + \cdots. \qquad (1.6.14)$$

Table 1.6.1 includes a few common but important Maclaurin series.

Table 1.6.1. A Few Important Maclaurin Series

$f(z)$	Maclaurin series of $f(z)$	Radius of Convergent	Convergent Interval for real-valued z	Convergent region for complex-valued z
$\dfrac{1}{1-z}$	$\displaystyle\sum_{n=0}^{\infty} z^n = 1 + z + z^2 + \cdots$	1	$(-1, 1)$	$|z| < 1$
$\dfrac{1}{1+z}$	$\displaystyle\sum_{n=0}^{\infty} (-z)^n = 1 - z + z^2 - \cdots$	1	$(-1, 1)$	$|z| < 1$
e^z	$\displaystyle\sum_{n=0}^{\infty} \frac{z^n}{n!} = 1 + \frac{z}{1!} + \frac{z^2}{2!} + \cdots$	∞	$(-\infty, \infty)$	C
$\sin z$	$\displaystyle\sum_{n=0}^{\infty} (-1)^n \frac{z^{2n+1}}{(2n+1)!} = z - \frac{z^3}{3!} + \frac{z^5}{5!} - \cdots$	∞	$(-\infty, \infty)$	C
$\cos z$	$\displaystyle\sum_{n=0}^{\infty} (-1)^n \frac{z^{2n}}{(2n)!} = 1 - \frac{z^2}{2!} + \frac{z^4}{4!} - \cdots$	∞	$(-\infty, \infty)$	C
$\tan^{-1} z$	$\displaystyle\sum_{n=0}^{\infty} (-1)^n \frac{z^{2n+1}}{2n+1} = z - \frac{z^3}{3} + \frac{z^5}{5} - \cdots$	1	$[-1, 1]$	$|z| \le 1$

Theorem 1.6.7. (*Taylor Theorem*)

If $f(z)$ is analytic on D and $D_R(\alpha) = \{z : |z - \alpha| < R\} \subset D$, then the Taylor series of $f(z)$ converges to $f(z)$ for all $z \in D_R(\alpha) = \{z : |z - \alpha| < R\} \subset D$, i.e.,

$$f(z) = \sum_{n=0}^{\infty} \frac{f^{(n)}(\alpha)}{n!}(z - \alpha)^n, \quad z \in D_R(\alpha). \tag{1.6.15}$$

Furthermore, the convergence is uniform on $D_r(\alpha) = \{z : |z - \alpha| < r\}$ for any $0 < r < R$.

From Theorem 1.6.5, an analytic function can also be defined by a function that is locally given by a convergent power series.

Definition 1.6.8.

An *analytic function* $f(z)$ on an open set D is a complex-valued infinitely differentiable function such that its Taylor series $\sum_{n=0}^{\infty}(f^{(n)}(z_0)/n!)(z - z_0)^n$ at any point $z_0 \in D$ converges to $f(z)$ for all z in a neighborhood of z_0, i.e.,

$$f(z) = \sum_{n=0}^{\infty} \frac{f^{(n)}(z_0)}{n!}(z - z_0)^n, \quad |z - z_0| < r_{z_0}, \tag{1.6.16}$$

where r_{z_0} is a small positive number depending on z_0.

1.6.8. Taylor Polynomials and their Applications

Definition 1.6.9.

An nth degree *Taylor polynomial* of a real-valued function $f(x)$ is the nth degree polynomial truncated from or a partial sum of the Taylor series of $f(x)$, as the first $n + 1$ terms, namely,

$$P_n(x) = \sum_{n=0}^{n} \frac{f^{(m)}(a)}{m!}(x - a)^m$$

$$= f(a) + f'(a)(x - a) + \frac{f''(a)}{2!}(x - a)^2 + \cdots + \frac{f^{(n)}(a)}{n!}(x - a)^n. \tag{1.6.17}$$

Recall from Calculus I that if a function $f(x)$ is differentiable at a point $x = a$, the linear approximation $L(x) = f(a) + f'(a)(x - a)$ to $f(x)$ in a neighborhood of a was indeed

$P_1(x)$ in (1.6.17) when $n = 1$. In other words, the linear approximation $P_1(x)$ from Calculus I is now expended to any n^{th} degree polynomial approximation by using the nth degree Taylor polynomial $P_n(x)$.

EXERCISES

1.1. Functions of Several Variables

For Problems *1.1.1–10*, find and sketch the domain of the functions of two variables.

1.1.1. $f(x, y) = \sqrt{y - 3x}$.

1.1.2. $f(x, y) = \dfrac{\sqrt{16 - x^2 - y^2}}{x + 3y}$.

1.1.3. $f(x, y) = 3(x + y)\sqrt[4]{x^2 + y}$.

1.1.4. $f(x, y) = \dfrac{xy}{x^2 + y^2}$.

1.1.5. $f(x, y) = \dfrac{xy}{\sqrt{x^2 + y^2}}$.

1.1.6. $f(x, y) = \arctan(x - y)$.

1.1.7. $f(x, y) = \arccos(x + y)$.

1.1.8. $f(x, y) = \ln(1 - xy)$.

1.1.9. $f(x, y) = 2\ln x - 3\ln y$.

1.1.10. $f(x, y) = \ln xy$.

For Problems *1.1.11–21*, find the limits if they exist.

1.1.11. $\displaystyle\lim_{(x,y)\to(1,3)} (3x^2 - 4xy^3 + 7xy)$.

1.1.12. $\displaystyle\lim_{(x,y)\to(2,1)} \dfrac{x^2 + 4xy^3}{x^2 + y^2}$.

1.1.13. $\displaystyle\lim_{(x,y)\to(\pi,\pi)} y^2 \cos\frac{x+y}{2}$.

1.1.14. $\displaystyle\lim_{(x,y)\to(0,\pi)} (x+y^2)\sin\frac{x+y}{2}\cos\frac{x}{3}$.

1.1.15. $\displaystyle\lim_{(x,y)\to(0,0)} \frac{xy}{x^2+y^2}$.

1.1.16. $\displaystyle\lim_{(x,y)\to(0,0)} \frac{xy}{\sqrt{x^2+y^2}}$.

1.1.17. $\displaystyle\lim_{(x,y)\to(0,0)} \frac{x+y}{x^2+y^2}$.

1.1.18. $\displaystyle\lim_{(x,y)\to(0,0)} \frac{x+y+5}{\sqrt{x^2+y^2}-3}$.

1.1.19. $\displaystyle\lim_{(x,y)\to(0,0)} \frac{x+y}{\sqrt{x^2+y^2}}$.

1.1.20. $\displaystyle\lim_{(x,y)\to(0,0)} \frac{x+y}{\sqrt[4]{x^2+y^2}}$.

1.1.21. $\displaystyle\lim_{(x,y,z)\to(0,0,0)} \frac{2x^2+3y^2+4z^2}{x^2+y^2+z^2}$.

1.2. Partial Derivatives, Gradient, and Divergence

For *1.2.1–5*, find the indicated partial derivatives.

1.2.1. Let $f(x,y) = 20 - 5x^2 - y^2$. Find the values of $\dfrac{\partial f}{\partial x}$ and $\dfrac{\partial f}{\partial y}$ at the point $(1,3)$.

1.2.2. For $f(x,y) = xe^{-2y} + 4xy$, find $f_x(1,0)$ and $f_y(1,0)$.

1.2.3. Find $\dfrac{\partial z}{\partial x}(1,2)$ and $\dfrac{\partial z}{\partial y}(1,2)$ when $z = \dfrac{x^3+y^3}{x^2+y^2}$.

1.2.4. Find $\dfrac{\partial z}{\partial x}$ and $\dfrac{\partial z}{\partial y}$ if x, y, and z satisfy $xy + yz + zx = 0$.

1.2.5. Find $\dfrac{\partial R}{\partial R_i}$, $i = 1, 2, 3$, where R, R_1, R_2, and R_3 satisfy $\dfrac{1}{R} = \dfrac{1}{R_1} + \dfrac{1}{R_2} + \dfrac{1}{R_3}$.

For *1.2.6–10*, find the first order partial derivatives.

1.2.6. $z = \ln\left(x + \sqrt{x^2 + y^2}\right)$.

1.2.7. $f(x, y) = e^y \cot(x + y)$.

1.2.8. $u = z \cos \dfrac{x}{y + z}$.

1.2.9. $f(x, y, z) = x^{yz}$.

1.2.10. $u = x^3 y^2 z \ln(3x + 2y + z)$.

For *1.2.11–13*, find the second order partial derivatives.

1.2.11. $f(x, y) = x^2 y + x\sqrt{y}$.

1.2.12. $z = (x^2 + y^2)^{3/2}$.

1.2.13. $z = t \arctan \sqrt{x}$.

1.2.14. Find the gradient of the function $f(x, y) = \ln(x^2 + y^2)$ at the point $(1, 1)$.

1.2.15. Find the derivative of the function $f(x, y, z) = xy + yz + zx$ at the point $(1, -1, 2)$ in the direction $\mathbf{A} = 3\mathbf{i} + 6\mathbf{j} - 2\mathbf{k}$.

1.2.16. The volume of a cone is given by the equation $V = \dfrac{1}{3}\pi r^2 h$. Use the total differential to estimate the change in volume if the height h increases from 10 to 10.1 cm and the radius r decreases from 12 to 11.95 cm.

1.2.17. Find the linearization of $f(x, y) = e^{3y} \sin 3x$ at the point $\left(\dfrac{\pi}{6}, 0\right)$.

1.2.18. Three resistors are connected in parallel as shown:

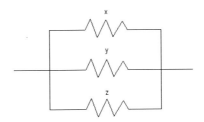

The resistance $R = \dfrac{xyz}{xy + yz + zx}$. Find the percent change in R if x, y, and z increase by $1\,\Omega$, $0\,\Omega$ and $-2\,\Omega$, and $(x, y, z) = (1, 2, 3)\,\Omega$.

For *1.2.19–22*, find the derivative of the function f at the given point P in the given direction **u**.

1.2.19. $f(x, y) = \sqrt{x + y}$, $P(3, 5)$, $\mathbf{u} = (10, 6)$.

1.2.20. $f(x, y) = xe^{xy}$, $P(1, 1)$, $\mathbf{u} = 3\mathbf{i} + 2\mathbf{j}$.

1.2.21. $f(x, y, z) = \sqrt{xyz}$, $P(1, 1, 2)$, and $\mathbf{u} = (3, 2, 3)$.

1.2.22. $f(x, y, z) = y \arctan \dfrac{x}{z}$, $P(2, 1, -2)$, and $\mathbf{u} = \mathbf{i} - \mathbf{j} + \mathbf{k}$.

1.3. Functions of a Complex Variable

For *1.3.1–4*, find $\operatorname{Re} f$ and $\operatorname{Im} f$.

1.3.1. $f(z) = 4z^2 + 3z + i - 1$.

1.3.2. $g(z) = \dfrac{2}{z}$.

1.3.3. $h(z) = \dfrac{z - i}{z^2 + 1}$.

1.3.4. $q(z) = \dfrac{3z^2 + 2}{|z - 1|}$.

For *1.3.5–8*, find the limits.

1.3.5. $\displaystyle\lim_{z \to 1 - 2i} (z + 4i)^2$.

1.3.6. $\lim\limits_{z \to 4i} \dfrac{z^2 + 16}{z - 4i}$.

1.3.7. $\lim\limits_{z \to i} \dfrac{z^2 + 1}{z^4 - 1}$.

1.3.8. $\lim\limits_{z \to 3} \dfrac{z^2 + 4}{iz}$.

For *1.3.9–17*, evaluate.

1.3.9. $e^{i\pi/2}$.

1.3.10. $\mathrm{Ln}(1 + i)$.

1.3.11. $\mathrm{Ln}\, i$.

1.3.12. $\cos(3 - i)$.

1.3.13. $\tan(2 - i)$.

1.3.14. $\sinh(-2 + i)$.

1.3.15. i^i.

1.3.16. $(-1)^{\sqrt{3}}$.

1.3.17. $(1 + i)^i$.

For *1.3.18–20*, solve the equations for z.

1.3.18. $\sin z = 2$.

1.3.19. $z^4 = 1$.

1.3.20. $z^3 = i$.

1.4. Power Series and their Convergent Behavior

1.4.1. Find the sums of both the finite series and infinite series in (1.4.1) and (1.4.2).

1.4.2. Derive by using (1.4.9) when the generic geometric series $\sum_{n=1}^{\infty} az^n$ converges and what it converges to. In addition, when does it diverge?

1.4.3. The sum of reciprocals of natural numbers is called the *harmonic series*, namely, $\sum_{n=1}^{\infty} \frac{1}{n}$. Show that the harmonic series is divergent.

1.4.4. By using the partial faction, show that the series $\sum_{n=1}^{\infty} \frac{1}{(n+K-1)(n+K)}$ converges for any fixed real number K. What does it converge to?

For *1.4.5–18*, determine whether the given series is convergent or divergent. If it is, find its sum.

1.4.5. $1 + .5 + .25 + .125 + .0625 + \cdots$.

1.4.6. $-\frac{1}{2} + \frac{1}{6} - \frac{1}{18} + \frac{1}{54} - \cdots$.

1.4.7. $\sum_{n=1}^{\infty} 4\left(-\frac{2}{3}\right)^{n-1}$.

1.4.8. $\sum_{n=1}^{\infty} \frac{1}{\pi^{3n}}$.

1.4.9. $\sum_{n=1}^{\infty} \frac{(-5)^n}{6^{n-1}}$.

1.4.10. $\sum_{n=1}^{\infty} \frac{1}{n(n+3)}$.

1.4.11. $\sum_{n=1}^{\infty} \frac{1}{\sqrt{n}}$.

1.4.12. $\sum_{n=0}^{\infty} \left(\frac{i}{7}\right)^n$.

1.4.13. $\sum_{n=0}^{\infty} \left(\frac{1+3i}{1+i}\right)^n$.

1.4.14. $\displaystyle\sum_{n=0}^{\infty} \frac{2}{(1+i)^n}$.

1.4.15. $\displaystyle\sum_{n=0}^{\infty} \frac{ni^n}{3n+2}$.

1.4.16. $\displaystyle\sum_{n=0}^{\infty} \left(\frac{2}{3i}\right)^n$.

1.4.17. $\displaystyle\sum_{n=0}^{\infty} \left(\frac{2}{3}\right)^{2n}$.

1.4.18. $\displaystyle\sum_{n=0}^{\infty} \frac{n!}{7^n}$.

For *1.4.19–24*, find the radius of convergence for the power series.

1.4.19. $\displaystyle\sum_{n=0}^{\infty} n^3 z^n$.

1.4.20. $\displaystyle\sum_{n=0}^{\infty} 3^n (z-2)^n$.

1.4.21. $\displaystyle\sum_{n=0}^{\infty} \frac{n}{2^n} (z+i)^n$.

1.4.22. $\displaystyle\sum_{n=0}^{\infty} \frac{(3-i)^n}{n^2} (z+1)^n$.

1.4.23. $\displaystyle\sum_{n=1}^{\infty} \frac{3^n}{4^n + 5^n} z^n$.

1.4.24. $f(z) = \dfrac{2z-1}{z+1} \displaystyle\sum_{n=0}^{\infty} \frac{(z+3)^{2n}}{4^n}$.

1.5. Real-Valued Taylor Series and Maclaurin Series

For *1.5.1–7*, find the Maclaurin series expansion of the functions.

1.5.1. $f(x) = \dfrac{2}{1+x}$.

1.5.2. $f(x) = \dfrac{2x - 1}{x + 1}.$

1.5.3. $f(x) = e^{-x^2}.$

1.5.4. $f(x) = x^4 \sin 4x.$

1.5.5. $f(x) = 3 \cos x + e^{-x}.$

1.5.6. $f(x) = \dfrac{x}{(2 - x)^2}.$

1.5.7. $f(x) = \sin x^2.$

1.5.8. Find the first five terms in the Taylor series about $x = 0$ for $f(x) = \dfrac{1}{1 - 2x}.$

1.5.9. Find the interval of convergence for the series in Problem 1.5.8.

1.5.10. Use partial fractions and the result from Problem 1.5.8 to find the first five terms in the Taylor series about $x = 0$ for $g(x) = \dfrac{1}{(1 - 2x)(1 - x)}.$

1.5.11. Let $f(x)$ be the function defined by $f(x) = \dfrac{1}{1 - x}.$ Write the first four terms and the general term of the Taylor series expansion of $f(x)$ about $x = 2.$

1.5.12. Use the result from Problem 1.5.11 to find the first four terms and the general term of the series expansion about $x = 2$ for $f(x) = \ln(x - 1),\ x > 1.$

1.5.13. Find the Taylor series expansion about $x = \dfrac{\pi}{4}$ for $f(x) = \cos x.$

1.6. Power Series Representation of Analytic Functions

For *1.6.1–4*, find the Derivatives of the functions.

1.6.1. $f(z) = 7z^3 + 5z^2 - iz + 11.$

1.6.2. $f(z) = \dfrac{z^2 - 16}{3iz^3 + z - \pi}.$

1.6.3. $f(z) = \dfrac{(z - 2)^2}{(z^2 - iz + 1)^3}.$

1.6.4. $f(z) = \cos(z^2 + iz + i).$

For *1.6.5–8*, verify the Cauchy-Riemann equations for the functions.

1.6.5. $f(z) = z^3$.

1.6.6. $f(z) = \sin z$.

1.6.7. $f(z) = \dfrac{2}{z}$.

1.6.8. $f(z) = \cos 2z$.

For *1.6.9–13*, evaluate the given integral, where C is the circle with positive orientation.

1.6.9. $\displaystyle\oint_C \frac{z^3}{z-i}\,dz, C : |z - 3i| = 3$.

1.6.10. $\displaystyle\oint_C \frac{\cos 2z}{z + 2i}\,dz, C : |z + i| = 2$.

1.6.11. $\displaystyle\oint_C \frac{e^{-2z}}{(z+1)^2}\,dz, C : |z| = 2$.

1.6.12. $\displaystyle\oint_C \frac{z}{z^4 - 1}\,dz, C : |z - 3| = 3$.

1.6.13. $\displaystyle\oint_C \frac{\sin 3z}{z^2(z - 5)}\,dz, C : |z| = 2$.

For *1.6.14–17*, find the Maclaurin series expansions of the functions.

1.6.14. $f(z) = \dfrac{2}{3 + z}$.

1.6.15. $f(z) = \dfrac{z - 1}{z + 1}$.

1.6.16. $f(z) = e^{-z^2}$.

1.6.17. $f(z) = 3\cos 2z + e^{-z}$.

1.6.18. Find the Taylor series expansion about $z = \dfrac{\pi}{4}$ for $f(z) = \cos z$.

1.6.19. Find the Taylor series expansion about $z = i$ for $f(z) = \dfrac{1 + z}{1 - z}$.

Chapter 2

FOURIER AND WAVELET ANALYSIS

Fourier analysis is a very well-developed classical subject area originated from expressing a generic function by infinite sums of simpler trigonometric functions such as sines and cosines, called Fourier series. Fourier series were introduced by Joseph Fourier[1] in around 1822 for solving the heat equation in a metal plate. Fourier analysis has been a vast area of research and found many applications in a variety of various branches of science, technology, engineering, and mathematics, with signal analysis and image processing in particular.

Wavelet analysis is a relatively new subject area and has also attracted researchers from diverse areas since late 1980's. Though it is safe to say that researchers from almost all different areas want to have a glimpse of wavelets, it still has, to some extent, great potential of developing and expanding, mainly due to its many powerful and successful practical applications. Two such examples are (1) Wavelet Scalar Quantization (WSQ) is chosen by FBI (U.S. Federal Bureau of Investigation) as the fingerprint image compression of fingerprint images, and (2) the newest image processing standard JPEG2000 is exclusively based upon wavelets. Wavelet analysis shed new lights for signal analysis by providing new powerful methods that complement the classical Fourier analysis.

Both Fourier analysis and wavelet analysis are used mainly for signal analysis. That is why they share many similarities and are included in the same context nowadays. We intend to give in this chapter a very brief introduction of both Fourier analysis and wavelet analysis.

2.1. VECTOR SPACES AND ORTHOGONALITY

We collect some fundamental notations of vector spaces and orthogonalility in this section. First, recall that the notion *inner product* or *dot product* on $\mathbb{C}^3 = \left\{ (x, y, z)^\top : x, y, z \in \mathbb{C} \right\}$ was simply defined by

$$\mathbf{u} \cdot \mathbf{v} = \mathbf{u}^* \mathbf{v}, \quad \mathbf{u}, \mathbf{v} \in \mathbb{C}^3, \tag{2.1.1}$$

---

[1] *Jean Baptiste Joseph Fourier*, March 21, 1768 – May 16, 1830, was a French mathematician and was born at Auxerre (now in the Yonne département of France. Both the Fourier series and the Fourier transform are named in his honor.

where the superscript "$*$" denotes the conjugate transpose (or Hermite transpose), meaning $\mathbf{u}^* = (\overline{\mathbf{u}})^\top = \overline{\mathbf{u}^\top}$. With the inner product in (2.1.1), the *amplitude* or *absolute value* (or *modulus*) of a vector $\mathbf{u}$ in $\mathbb{C}$ is

$$|\mathbf{u}| = \sqrt{\mathbf{u} \cdot \mathbf{u}}, \tag{2.1.2}$$

and the *orthogonality* of $\mathbf{u}$ and $\mathbf{v}$ is simply

$$\mathbf{u} \cdot \mathbf{v} = 0, \tag{2.1.3}$$

denoted by $\mathbf{u} \perp \mathbf{v}$. These notations can be generalized to vector spaces.

Definition 2.1.1.

A *vector space* X over a *field* K, which is either $\mathbb{R}$ or $\mathbb{C}$, is a nonempty set of elements, called *vectors*, accompanied by two algebraic operations $\oplus$ and $\odot$, called *vector addition* and *multiplication of vectors by scalars* or simply called *scalar multiplication*. The addition operation $\oplus$ satisfies

(1) $x \oplus y = y \oplus x$ (communitative property)
(2) $x \oplus (y \oplus z) = (x \oplus y) \oplus z$ (associative property)
(3) $x \oplus \theta = x$ (θ being the zero element)
(4) $x \oplus (-x) = \theta$ (negative element)   (2.1.4)

for any x, y in X; and the scalar multiplication operation $\odot$ satisfies

(1) $\alpha \odot (\beta \odot y) = (\alpha\beta) \odot x$ (associative property)
(2) $1 \odot x = x$ (scalr multiplication by 1)
(3) $\alpha \odot (x \oplus y) = (\alpha \odot x) \oplus (\alpha \odot y)$ (distributive property 1)
(4) $(\alpha + \beta) \odot x = (\alpha \odot x) \oplus (\beta \odot y)$ (distributive property 2)   (2.1.5)

for any x, y in X and α, β in K. X is called a *real vector space* if $K = \mathbb{R}$, and it is a *complex vector space* when $K = \mathbb{C}$. A nonempty subset Y of X is a *subspace* if

$$(\alpha \odot y_1) \oplus (\beta \odot y_2) \in Y, \quad \forall\, \alpha, \beta \in K \text{ and } \forall\, y_1, y_2 \in Y \subset X. \tag{2.1.6}$$

So Y itself is also a vector space under the same two algebraic operations $\oplus$ and $\odot$ of X. In many applications, the two algebraic operations $\oplus$ and $\odot$ are the usual addition and scalar multiplication. For convenience, we may not use the two notations $\oplus$ and $\odot$ too much to avoid any confusion they may cause. In other words, $x + y$ means $x \oplus y$, αx means $\alpha \odot x$, and the zero element θ is simply denoted by 0.

Example 2.1.1.

$\mathbb{R}^n$ and $\mathbb{C}^n$, the common n-dimensional real and complex *Euclidean spaces* with the *Euclidean metric* defined by

$$d(x,y) = \sqrt{|x_1 - y_1|^2 + \cdots + |x_n - y_n|^2},$$
$$x = (x_1, \cdots, x_n), \ y = (y_1, \cdots, y_n) \in \mathbb{R}^n \text{ or } \mathbb{C}^n. \tag{2.1.7}$$

They are vector spaces under the usual two algebraic operations defined in the usual fashion

$$x + y = (x_1 + y_1, \cdots, x_n + y_n); \quad \alpha x = (\alpha x_1, \cdots, \alpha x_n),$$
$$x = (x_1, \cdots, x_n), \ y = (y_1, \cdots, y_n) \in \mathbb{R}^n \text{ or } \mathbb{C}^n: \quad \alpha \in \mathbb{R} \text{ or } \mathbb{C}. \tag{2.1.8}$$

Example 2.1.2.

$C[a,b]$, the collection of all real-valued continuous functions on $[a,b]$. Due to the fact that the sum of two real-valued continuous functions on $[a,b]$ is also continuous on $[a,b]$, the space $C[a,b]$ is also a vector space under the normal addition and scalar multiplication of functions

$$(f+g)(t) = f(t) + g(t);$$
$$(\alpha f)(t) = \alpha f(t), \quad f(t), \ g(t) \in C[a,b]. \tag{2.1.9}$$

Example 2.1.3.

ℓ^2, the collection of all sequences that are square summable, i.e.,

$$\ell^2 = \left\{ (x_1, x_2, \cdots) : \sum_{j=1}^{\infty} |x_j|^2 < \infty \right\}. \tag{2.1.10}$$

It is real or complex depending on either $x_j \in \mathbb{R}$ or $x_j \in \mathbb{C}$, for all $j = 1, 2, \cdots$. With the usual addition and scalar multiplication, namely,

$$x + y = (x_1 + y_1, x_2 + y_2, \cdots); \quad \alpha x = (\alpha x_1, \alpha x_2, \cdots),$$
$$x = (x_1, x_2, \cdots), \ y = (y_1, y_2, \cdots) \in \ell^2; \quad \alpha \in \mathbb{R} \text{ or } \mathbb{C}. \tag{2.1.11}$$

the real or complex ℓ^2 is a vector space due to the Minkowski's[2] inequality

$$\left(\sum_{j=1}^{\infty}|x_j+y_j|^2\right)^{1/2} \le \left(\sum_{j=1}^{\infty}|x_j|^2\right)^{1/2} + \left(\sum_{j=1}^{\infty}|y_j|^2\right)^{1/2}.$$

$$x=(x_1,x_2,\cdots),\ y=(y_1,y_2,\cdots)\in\ell^2. \qquad (2.1.12)$$

We also use $x_1, x_2, \ldots$, to represent vectors in a vector space.

Definition 2.1.2.

A *linear combination* of vectors x_1, x_2, ..., x_m of a vector space X is an expression of the form

$$\alpha_1 x_1 + \alpha_2 x_2 + \cdots + \alpha_m x_m = \sum_{k=1}^{m}\alpha_k x_k, \quad \alpha_1, \alpha_2, \cdots, \alpha_m \in K. \qquad (2.1.13)$$

For an nonempty set M of X, the set of all possible linear combinations of vectors of M is called the *span* of M, denoted by $\operatorname{span} M$. If $Y = \operatorname{span} M$, we say that Y is *spanned* or *generated* by M.

Definition 2.1.3.

Let $M = \{x_1, x_2, \ldots, x_m\} \subset X$. M is *linearly independent* if

$$\alpha_1 x_1 + \alpha_2 x_2 + \cdots + \alpha_m x_m = \sum_{k=1}^{m}\alpha_k x_k = 0 \implies \alpha_1 = \alpha_2 = \cdots \alpha_m = 0,$$

$$(2.1.14)$$

where $\alpha_1, \alpha_2, \cdots, \alpha_m \in K$. Otherwise, M is *linearly dependent*.

Definition 2.1.4.

A vector space X is *finite dimensional* if X can be spanned by n linearly independent vectors x_1, x_2, $\cdots$, $x_n \in X$. In other words, any vector in X can be spanned by these n linearly independent vectors whereas any set of $n+1$ or more vectors of X is linearly dependent. The positive integer n is the *dimension* of X, denoted by $n = \dim X$, such a

---

[2] *Herman Minkowski*, June 22, 1864 – January 12, 1909, was a German mathematician. He was born in Aleksotas, a suburb of Kaunas, Lithuania, which was then part of the Russian Empire. Minkowski was one of Einstein's teachers. He was awarded the Mathematics Prize of the French Academy of Sciences for his manuscript on the theory of quadratic forms.

group of n vectors x_1, x_2, $\cdots$, x_n are called a *basis* of X. Otherwise, if X is not finite dimensional, it is said to be *infinite dimensional*.

Definition 2.1.5.

A *norm* on a vector space X over a *field* K is a real-valued function, denoted by

$$\|x\|, \quad x \in X, \tag{2.1.15}$$

which satisfies the properties

(1) $\|x\| \geq 0$; and $\|x\| = 0 \Leftrightarrow x = 0$;

(2) $\|\alpha x\| = |\alpha| \, \|x\|$; and

(3) $\|x + y\| \leq \|x\| + \|y\|$ (triangular inequality), $\tag{2.1.16}$

for any x, y in X and $\alpha \in K$. X is a *normed space* if it is a vector space with a norm defined on it. A *metric* on X *induced by the norm* is

$$d(x, y) = \|x - y\| = \sqrt{\langle x - y, x - y \rangle}. \tag{2.1.17}$$

Example 2.1.4.

$\mathbb{R}^n$ and $\mathbb{C}^n$, The n-dimensional Euclidean spaces are normed spaces with the norm

$$\|x\| = \sqrt{|x_1|^2 + \cdots |x_n|^2}, \quad x = (x_1, \cdots, x_n) \in \mathbb{R}^n \text{ or } \mathbb{C}^n, \tag{2.1.18}$$

and the metric induced by this norm is the Euclidean metric defined in (2.1.17). The *canonical bases* or *standard basis* or *natural basis* for $\mathbb{R}^n$ are

$$\{e_j : j = 1, \cdots, n\},$$

where e_j is the unit vector with all zero entries but entry 1 in the jth coordinate.

Example 2.1.5.

$C[a, b]$. It is a normed space with a norm defined by

$$\|x\| = \max_{t \in [a,b]} |x(t)|, \tag{2.1.19}$$

Since every continuous function on a closed interval $[a, b]$ attains its maximum value.

Example 2.1.6.

ℓ^2, a normed space with a norm defined by

$$\|x\| = \left(\sum_{j=1}^{\infty} |x_j|^2 \right)^{1/2}, \quad x = (x_1, x_2, \cdots) \in \ell^2, \tag{2.1.20}$$

since the triangular inequality in (2.1.16) is guaranteed, again, by the Minkowski's inequality in (2.1.12).

Example 2.1.7.

$L^2[a, b]$, the collection of all integrable real- or complex-valued functions on $[a, b]$, i.e.,

$$L^2[a, b] = \left\{ f(t) : \ t \in [a, b] \subset \mathbb{R} \to \mathbb{R} \ \text{or} \ \mathbb{C} \ \middle| \ \int_a^b |f(t)|^2 dt < \infty \right\}. \tag{2.1.21}$$

By defining

$$\|f\|_2 = \left(\int_a^b |f(t)|^2 dt \right)^{1/2}, \quad f(t) \in L^2[a, b], \tag{2.1.22}$$

it is a normed space due to, again, Minkowski's integral inequality

$$\left(\int_I |f(t) + g(t)|^2 \, dt \right)^{1/2} \leq \left(\int_I |f(t)|^2 \, dt \right)^{1/2} + \left(\int_I |g(t)|^2 \, dt \right)^{1/2}. \tag{2.1.23}$$

where I is either a finite interval or infinity interval (see the following Example 2.1.8).

Example 2.1.8.

$L^2(-\infty, \infty)$ or $L^2(\mathbb{R})$ or L^2, the collection of all integrable real- or complex-valued functions on $(-\infty, \infty)$, i.e.,

$$L^2(\mathbb{R}) = \left\{ f(t) : \ \mathbb{R} \to \mathbb{R} \ \text{or} \ \mathbb{C} \ \middle| \ \int_{-\infty}^{\infty} |f(t)|^2 dt < \infty \right\}. \tag{2.1.24}$$

By defining

$$\|f\|_2 = \left(\int_{-\infty}^{\infty} |f(t)|^2 dt \right)^{1/2}, \quad f(t) \in L^2(\mathbb{R}), \tag{2.1.25}$$

it is a normed space too by using (2.1.23).

Definition 2.1.6.

An *inner product* on a vector space X is a mapping of $X \times X \to K$, denoted by $\langle \cdot, \cdot \rangle$, or $\langle x, y \rangle$ when x, y in X, satisfying

(1) $\langle x + y, z \rangle = \langle x, z \rangle + \langle y, z \rangle$;

(2) $\langle \alpha x, y \rangle = \alpha \langle x, y \rangle$;

(3) $\langle x, y \rangle = \overline{\langle y, x \rangle}$; and

(4) $\langle x, x \rangle \geq 0$, and $\langle x, x \rangle = 0 \Leftrightarrow x = 0$. $\tag{2.1.26}$

An *inner product space* is a vector space X with an inner product defined on X.

Definition 2.1.7.

The *norm induced by an inner product* $\langle \cdot, \cdot \rangle$ is simply

$$\|x\| = \sqrt{\langle x, x \rangle}. \tag{2.1.27}$$

Example 2.1.9.

$\mathbb{R}^n$ and $\mathbb{C}^n$, inner product spaces with the usual dot product

$$\langle x, y \rangle = x_1 \overline{y_1} + x_2 \overline{y_2} + \cdots + x_n \overline{y_n},$$
$$x = (x_1, \cdots, x_n), \ y = (y_1, \cdots, y_n) \in \mathbb{R}^n \ \text{or} \ \mathbb{C}^n. \tag{2.1.28}$$

Example 2.1.10.

$L^2[a, b]$ and $L^2(\mathbb{R})$, inner product spaces with the (weighted) inner products

$$\langle f, g \rangle = \int_a^b f(t) \overline{g(t)} w(t) \, dt, \quad f(t), \ g(t) \in L^2[a, b], \tag{2.1.29}$$

or

$$\langle f, g \rangle = \int_{-\infty}^{\infty} f(t)\overline{g(t)}w(t)\,dt, \quad f(t),\, g(t) \in L^2(\mathbb{R}), \tag{2.1.30}$$

where $\overline{g}$ indicates complex conjugation, and $w(t)$ is a *weight function*, $w(t) \geq 0$, $t \in [a, b]$. By defining

$$\|f\|_2 = \left(\int_a^b |f(t)|^2 w(t)\,dt \right)^{1/2}, \quad f(t) \in L^2[a, b], \tag{2.1.31}$$

or

$$\|f\|_2 = \left(\int_{-\infty}^{\infty} |f(t)|^2 w(t)\,dt \right)^{1/2}, \quad f(t) \in L^2(\mathbb{R}), \tag{2.1.32}$$

they are also normed spaces. We remark here that *both* the interval $[a, b]$ and weight function $w(t)$ in (2.1.29)—(2.1.32) can be simply identified from the context. That is why we do not distinguish notations for the two norms in (2.1.31) and (2.1.32).

Definition 2.1.8.

Two vectors x and y in X are *orthogonal* with inner product in (2.1.26), if

$$\langle x, y \rangle = 0. \tag{2.1.33}$$

Definition 2.1.9.

A subset $Y \subset X$ is an *orthogonal set*, if all elements of Y are pairwise orthogonal. An *orthonormal* set $Y \subset X$ is an orthogonal set whose elements have norm 1, i.e., for $y_1, y_2 \in Y$,

$$\langle y_1, y_2 \rangle = \begin{cases} 0, & \text{if } y_1 \neq y_2, \\ 1, & \text{if } y_1 = y_2. \end{cases} \tag{2.1.34}$$

Definition 2.1.10.

For a subset M of X under metric (2.1.17), a point $x_0 \in X$ is a *limit point* or *accumulation point* of M, if every neighborhood of x_0 contains at least one point $y_0 \in M$ with $y_0 \neq x_0$. Here x_0 may not be a point of M. The *closure* of a subset M of X under metric (2.1.17), denoted by

$$\overline{M}, \quad \text{or} \quad \text{Clos}\,M, \tag{2.1.35}$$

is a set containing all the limit points of M and M itself, i.e.,

$$\overline{M} = M \cup \{x \in X : x \text{ is a limit point of } M\}. \tag{2.1.36}$$

A subset M of X under metric (2.1.17) is said to be *dense* in X, if

$$\overline{M} = M \cup \{x \in X : x \text{ is a limit point of } M\} = X. \tag{2.1.37}$$

Definition 2.1.11.

A subset M of X under metric (2.1.17) is *total* or *fundamental*, if span M (Definition 2.1.2) is dense in X, meaning

$$\overline{\text{span } M} = X.$$

If $M \subset X$ is both orthonormal and total, it is called an *orthonormal basis* of X.

2.2. FOURIER SERIES AND ITS CONVERGENT BEHAVIOR

A Fourier series is an expansion of a periodic function or periodic signal in terms of sine and cosine functions.

It is easy to check that the collection

$$\{1, \cos t, \sin t, \cos 2t, \sin 2t, \cdots\} \tag{2.2.1}$$

is an orthogonal set due to the fact that

$$\int_{-L}^{L} \cos \frac{m \pi t}{L} \cos \frac{n \pi t}{L} \, dt = L \delta_{mn},$$

$$\int_{-L}^{L} \cos \frac{m \pi t}{L} \sin \frac{n \pi t}{L} \, dt = 0,$$

$$\int_{-L}^{L} \sin \frac{m \pi t}{L} \sin \frac{n \pi t}{L} \, dt = L \delta_{mn}, \quad m = 0, 1, \cdots, \quad n = 1, 2, \cdots. \tag{2.2.2}$$

where δ_{mn} denotes the *Kronecker delta*, named after Leopold Kronecker[3], meaning

---

[3] *Leopold Kronecker*, December 7, 1823 – December 29, 1891, was a German mathematician, and was born in Liegnitz, Prussia (now Legnica, Poland).

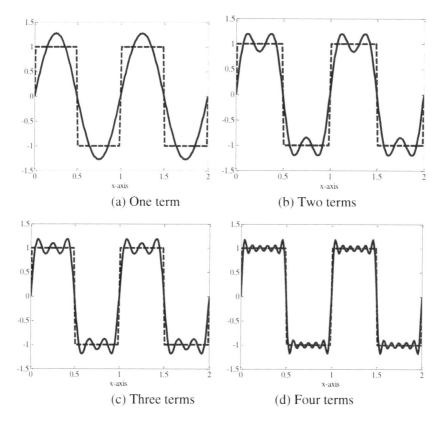

(a) One term   (b) Two terms

(c) Three terms   (d) Four terms

Figure 2.2.1. Graphs of the partial sums of the Fourier series of the 1-periodic Haar function ι, where $\iota(t) = 1$ for $t \in (0, 1/2)$, and $\iota(t) = -1$ for $t \in (1/2, 1)$. The partial sums are $S_n(t) = 4/\pi \sum_{k=0}^{n-1}(1/(2k+1))\sin(4k+2)\pi t$, $n \geq 1$. Graphs of ι is in dashed line and the graphs of S_n's are in solid line. (a) ι and S_1; (b) ι and S_2; (c) ι and S_3; (d) ι and S_6. All plots are illustrated on the length of two periods, namely, $t \in (0, 2)$.

$$\delta_{mn} = \begin{cases} 1, & \text{if } m = n. \\ 0, & \text{if } m \neq n. \end{cases} \qquad (2.2.3)$$

Definition 2.2.1.

For a 2π–periodic real-valued function f that is integrable on $[-\pi, \pi]$, its *Fourier series* is defined by

$$f \sim \frac{a_0}{2} + \sum_{n=1}^{\infty} [a_n \cos nt + b_n \sin nt], \qquad (2.2.4)$$

where the *Fourier coefficients* $\{a_n\}_{n=0}^{\infty}$ and $\{b_n\}_{n=1}^{\infty}$ of f are given explicitly by

$$a_n = \frac{1}{\pi} \int_{-\pi}^{\pi} f(t) \cos nt \, dt, \quad n = 0, 1, \cdots ; \quad \text{and}$$

$$b_n = \frac{1}{\pi} \int_{-\pi}^{\pi} f(t) \sin nt \, dt, \quad n = 1, 2, \cdots . \tag{2.2.5}$$

The *partial sum* of the Fourier series in (2.2.4) is defined by

$$S_n(t) = \frac{a_0}{2} + \sum_{k=1}^{n} [a_k \cos kt + b_k \sin kt]. \tag{2.2.6}$$

Definition 2.2.2.

For a $2L$–periodic real-valued function f that is integrable on $[-L, L]$, its *Fourier series* is defined by

$$f \sim \frac{a_0}{2} + \sum_{n=1}^{\infty} \left[a_n \cos \frac{n\pi t}{L} + b_n \sin \frac{n\pi t}{L} \right], \tag{2.2.7}$$

where

$$a_n = \frac{1}{L} \int_{-L}^{L} f(t) \cos \frac{n\pi t}{L} \, dt, \quad n = 0, 1, \cdots ; \quad \text{and}$$

$$b_n = \frac{1}{L} \int_{-L}^{L} f(t) \sin \frac{n\pi t}{L} \, dt, \quad n = 1, 2, \cdots , \tag{2.2.8}$$

are the *Fourier coefficients* of f. Similarly, the *partial sum* of the Fourier series in (2.2.7) is defined by

$$S_n(t) = \frac{a_0}{2} + \sum_{k=1}^{n} \left[a_k \cos \frac{k\pi t}{L} + b_k \sin \frac{k\pi t}{L} \right]. \tag{2.2.9}$$

Notice first that the only reason $a_0/2$ was being used in both (2.2.4) and (2.2.7) is to make the general formulas for a_n in both (2.2.5) and (2.2.8) applicable without change when $n = 0$. Secondly, most of the results presented for 2π-periodic functions can be extended easily to $2L$-periodic functions. So sometimes we may state just results for the case of 2π-periodic functions.

The next natural question is: When does the Fourier series in (2.2.4) or (2.2.7) converge, and if so, does it converge to f? In other words, when does the partial sum in (2.2.6) or (2.2.9) converge, and in what sense? There are a lot of results in the literature regarding the convergence of (2.2.4) or (2.2.7). We just state the most useful one here.

Definition 2.2.3.

A function $f(t)$ on $[a, b]$ is said to be *piecewise continuous*, if there is a partition of $[a, b]$:

$$a = t_0 < t_1 < \cdots < t_{n-1} < t_n = b, \qquad (2.2.10)$$

such that (1) $f(t)$ is continuous on $[a, b]$ except may be the points $t_0, \cdots, t_n$; and (2) all the left- and right-limits of $f(t)$ at knots $t_0, \cdots, t_n$, namely, $f(t_0+)$, $f(t_j-)$ and $f(t_j+)$, $j = 1, \cdots, n - 1$, and $f(t_n-)$, exist (meaning finite). The function $f(t)$ is said to be *piecewise smooth*, if $f'(t)$ is piecewise continuous.

Theorem 2.2.1. (*Uniform Convergence*)

Let $f(t)$ be a 2π–periodic continuous function with $f'(t)$ piecewise continuous. Then the Fourier series of $f(t)$ in (2.2.4) *converges uniformly* to $f(t)$ for all $t \in \mathbb{R}$. More precisely,

$$\frac{a_0}{2} + \sum_{n=1}^{\infty} [a_n \cos nt + b_n \sin nt] = f(t), \quad t \in \mathbb{R}. \qquad (2.2.11)$$

Theorem 2.2.2. (*Fourier Series of Piecewise Smooth Functions*)

Let $f(t)$ be a piecewise smooth real-valued function on $[-\pi, \pi]$. Then the Fourier series of $f(t)$ in (2.2.4) converges. More precisely,

$$\frac{a_0}{2} + \sum_{n=1}^{\infty} [a_n \cos nt + b_n \sin nt] = \frac{f(t-) + f(t+)}{2}, \quad t \in \mathbb{R}, \qquad (2.2.12)$$

where $f(t)$ on the right-hand side of (2.2.12) means $f(t)$'s 2π–periodic extension on $\mathbb{R}$.

2.3. FOURIER COSINE AND SINE SERIES AND HALF-RANGE EXPANSIONS

For an even $2L$–periodic real-valued function f, its Fourier coefficients $a_n = 0$, $n = 0, 1, \cdots$; and, for an odd $2L$–periodic real-valued function f, its Fourier coefficients $b_n = 0$, $n = 1, 2, \cdots$. Hence, the Fourier series of even functions contain only cosine terms (or cosine series) and the Fourier series of odd functions contain only sine terms (or sine series). This is exactly what the Fourier cosine and sine series are for.

Definition 2.3.1.

For an even real-valued function f that is defined and integrable on $[-L, L]$, its *Fourier cosine series* is

$$f \sim \frac{a_0}{2} + \sum_{n=1}^{\infty} a_n \cos \frac{n\pi t}{L}, \tag{2.3.1}$$

where

$$a_n = \frac{2}{L} \int_0^L f(t) \cos \frac{n\pi t}{L} \, dt, \quad n = 0, 1, \cdots. \tag{2.3.2}$$

For an odd real-valued function f that is defined and integrable on $[-L, L]$, its *Fourier sine series* is

$$\sum_{n=1}^{\infty} b_n \sin \frac{n\pi t}{L}, \tag{2.3.3}$$

where

$$b_n = \frac{2}{L} \int_0^L f(t) \sin \frac{n\pi t}{L} \, dt, \quad n = 1, 2, \cdots. \tag{2.3.4}$$

Definition 2.3.2.

A function f defined on $(0, L)$ can be extended to an even function on $(-L, L)$ and then $2L$–periodic. In other words, let $f_{\text{even}}(t)$ be such an extension. Then

$$f_{\text{even}}(t) = \begin{cases} f(-t), & \text{if } t \in (-L, 0), \\ f(t), & \text{if } t \in (0, L); \end{cases}$$
$$f_{\text{even}}(t) = f_{\text{even}}(t + 2L), \quad t \in \mathbb{R}. \tag{2.3.5}$$

The Fourier series of $f_{\text{even}}(t)$ is a Fourier cosine series, called the *cosine half-range expansion* of f. Similarly, f defined on $(0, L)$ can also be extended to an odd function on $(-L, L)$ and then $2L$–periodic, namely, if $f_{\text{odd}}(t)$ is such an extension, then

$$f_{\text{odd}}(t) = \begin{cases} -f(-t), & \text{if } t \in (-L, 0), \\ f(t), & \text{if } t \in (0, L); \end{cases}$$
$$f_{\text{odd}}(t) = f_{\text{odd}}(t + 2L), \quad t \in \mathbb{R}. \tag{2.3.6}$$

The Fourier series of $f_{\mathrm{odd}}(t)$ is a Fourier sine series of f, called the *sine half-range expansion* of f.

2.4. FOURIER SERIES AND PDEs

Definition 2.4.1.

The 3D *heat equation* is a (parabolic) partial differential equation (PDE) that describes the heat distribution $u(x, y, z, t)$ in a region $D \subset \mathbb{R}^3$ over time t:

$$\frac{\partial u}{\partial t} = k \left(\frac{\partial^2 u}{\partial x^2} + \frac{\partial^2 u}{\partial y^2} + \frac{\partial^2 u}{\partial z^2} \right) = c^2 \, \nabla^2 u, \quad (x, y, z) \in D \subset \mathbb{R}^3,$$

where $k = \dfrac{K}{\sigma \rho}$ is a positive constant, with K the thermal conductivity, σ the specific heat, and ρ the density of the material.

The heat equation is parabolic. It is useful for the study of Brownian motion in statistics. The 1D heat equation for $u(x, t)$ is simply

$$u_t = k \, u_{xx}, \quad x \in [a, b] \subset \mathbb{R}, \tag{2.4.1}$$

which describes the evolution of temperature $u(x, t)$ inside, e.g., a homogeneous metal rod. With appropriate boundary conditions, the PDE (2.4.1) can be solved using Fourier series. The 1D heat equation with initial condition (IC) and the Dirichlet[4]-type boundary conditions (BCs) is

$$
\begin{aligned}
u_t(x, t) &= k u_{xx}(x, t), \quad x \in (0, L), \ t > 0, \\
\mathrm{IC}: \quad u(x, 0) &= \varphi(x), \quad x \in (0, L), \\
\mathrm{BC}: \quad u(0, t) &= f_1(t), \quad u(L, t) = f_2(t), \quad t > 0.
\end{aligned}
\tag{2.4.2}
$$

The 1D heat equation with IC and the Neumann[5]-type BCs is

$$
\begin{aligned}
u_t(x, t) &= k u_{xx}(x, t), \quad x \in (0, L), \ t > 0, \\
\mathrm{IC}: \quad u(x, 0) &= \xi(x), \quad x \in (0, L), \\
\mathrm{BC}: \quad u_x(0, t) &= g_1(t), \quad u_x(L, t) = g_2(t), \quad t > 0.
\end{aligned}
\tag{2.4.3}
$$

The 1D heat equation with IC and the mixed BCs is

---

[4] *Johann Peter Gustav Lejeune Dirichlet*, February 13, 1805 – May 5, 1859, was a German mathematician and was born in Düren, French Empire (now Germany).

[5] *Carl Gottfried Neumann*, May 7, 1832 – March 27, 1925, was a German mathematician and was born in Königsberg, Prussia.

$$u_t(x,t) = ku_{xx}(x,t), \quad x \in (0,L), \ t > 0,$$

IC : $\quad u(x,0) = \eta(x), \quad x \in (0,L),$

BC : $\quad \alpha_1 u(0,t) + \beta_1 u_x(0,t) = h_1(t),$

$$\alpha_2 u(L,t) + \beta_2 u_x(L,t) = h_2(t), \quad t > 0, \tag{2.4.4}$$

for some constants α_1, α_2, β_1, and β_2. When the BCs in all three types PDEs in (2.4.2), (2.4.3), and (2.4.4) are reduced to $u(0,t)=u(L,t)=0$ (laterally insulated bars) or $u_x(0,t)=u_x(L,t)=0$ (end insulated bars), they all can be solved by simply using Fourier series.

In fact, (2.4.2) with $f_1(t) = f_2(t) = 0$ has solutions of the form

$$u(x,t) = \sum_{n=1}^{\infty} b_n e^{-k(n\pi/L)^2 t} \sin \frac{n\pi x}{L}, \tag{2.4.5}$$

where $\{b_n\}_{n=1}^{\infty}$ can be determined from the Fourier sine series of the initial function φ, i.e.,

$$\varphi(t) = \sum_{n=1}^{\infty} b_n \sin \frac{n\pi x}{L}, \tag{2.4.6}$$

$$b_n = \frac{2}{L} \int_0^L \varphi(x) \sin \frac{n\pi x}{L} \, dx, \quad n \geq 1. \tag{2.4.7}$$

Similarly, (2.4.3) with $g_1(t) = g_2(t) = 0$ has solutions of the form

$$u(x,t) = \frac{a_0}{2} + \sum_{n=1}^{\infty} a_n e^{-k(n\pi/L)^2 t} \cos \frac{n\pi x}{L}, \tag{2.4.8}$$

where $\{a_n\}_{n=0}^{\infty}$ can be determined from the Fourier cosine series of the initial function ξ, i.e.,

$$\xi(t) = \frac{a_0}{2} + \sum_{n=1}^{\infty} a_n \cos \frac{n\pi x}{L}. \tag{2.4.9}$$

$$a_n = \frac{2}{L} \int_0^L \xi(x) \cos \frac{n\pi x}{L} \, dx, \quad n \geq 0. \tag{2.4.10}$$

Example 2.4.1.

Solve the following 1D heat equation with Dirichlet-type BCs

$$u_t(x,t) = 2u_{xx}(x,t), \quad x \in (0,8), \ t > 0,$$

IC : $\quad u(x,0) = \varphi(x) = \begin{cases} 3x, & \text{if } 0 < x < 2, \\ 8 - x, & \text{if } 2 < x < 8, \end{cases}$

BC : $\quad u(0,t) = u(8,t) = 0, \quad t > 0.$

Solution

With $k = 2$, it follows from (2.4.6)-(2.4.7) and (2.4.5) that

$$b_n = \frac{2}{8} \left[\int_0^2 3x \sin \frac{n\pi x}{8} \, dx + \int_2^8 (8-x) \sin \frac{n\pi x}{8} \, dx \right]$$

$$= \frac{64 \sin \frac{n\pi}{4}}{n^2 \pi^2}, \quad n \geq 1,$$

and

$$u(x,t) = \sum_{n=1}^\infty b_n e^{-2(n\pi/8)^2 t} \sin \frac{n\pi x}{8}$$

$$= \frac{64}{\pi^2} \sum_{n=1}^\infty \frac{\sin \frac{n\pi}{4}}{n^2} e^{-n^2\pi^2 t/32} \sin \frac{n\pi x}{8}.$$

Example 2.4.2.

Solve the following heat equation with Neumann-type BCs:

$$u_t(x,t) = \frac{3}{2} u_{xx}(x,t), \quad x \in (0,12), \ t > 0,$$

IC : $\quad u(x,0) = \xi(x) = \begin{cases} 2x, & \text{if } 0 < x < 4, \\ 8, & \text{if } 4 < x < 12, \end{cases}$

BC : $\quad u_x(0,t) = u_x(12,t) = 0, \quad t > 0.$

Solution

By using (2.4.9)-(2.4.10) and (2.4.8),

$$a_0 = \frac{2}{12} \left[\int_0^4 2x \, dx + \int_4^{12} 8 \, dx \right] = \frac{40}{3},$$

$$a_n = \frac{2}{12} \left[\int_0^4 2x \cos \frac{n\pi x}{12} \, dx + \int_4^{12} 8 \cos \frac{n\pi x}{12} \, dx \right]$$

$$= -\frac{48 \left(1 - \cos \frac{n\pi}{3}\right)}{n^2 \pi^2}, \quad n \geq 1,$$

and

$$u(x,t) = \frac{a_0}{2} + \sum_{n=1}^{\infty} a_n e^{-3/2(n\pi/12)^2 t} \cos \frac{n\pi x}{12}$$

$$= \frac{20}{3} - \frac{48}{\pi^2} \sum_{n=1}^{\infty} \frac{1 - \cos \frac{n\pi}{3}}{n^2} e^{-n^2 \pi^2 t/96} \cos \frac{n\pi x}{12}.$$

2.5. FOURIER TRANSFORM AND INVERSE FOURIER TRANSFORM

Definition 2.5.1.

An *integral transform* is an operator (or transform) T of a function to a new function

$$Tf(\nu) = \int_a^b K(\mu, \nu) f(u) \, du, \quad \nu \in \mathbb{R},$$

where $K(\mu, \nu)$ is a *kernel function*, or simply called a *kernel*.

Definition 2.5.2.

For an integrable function $f : \mathbb{R} \to \mathbb{C}$, its *Fourier transform,* FT for short, is defined by

$$\widehat{f}(\omega) = \int_{-\infty}^{\infty} f(t) e^{-i\omega t} \, dt. \tag{2.5.1}$$

and the *inverse Fourier transform* is defined by

$$f(t) = \frac{1}{2\pi} \int_{-\infty}^{\infty} \widehat{f}(\omega) e^{it\omega} \, d\omega. \tag{2.5.2}$$

Observe that the kernels are $K(t, \omega) = e^{-i\omega t}$ in (2.5.1) and $K(\omega, t) = e^{i\omega t}$ in (2.5.2), respectively.

2.6. PROPERTIES OF FOURIER TRANSFORM AND CONVOLUTION THEOREM

Definition 2.6.1.

The *convolution* of two integrable functions f and g, denoted by $f * g$, is defined by

$$(f * g)(t) = \int_{-\infty}^{\infty} f(u) \, \overline{g(t - u)} \, du, \quad t \in \mathbb{R}. \tag{2.6.1}$$

Theorem 2.6.1. (*Convolution Theorem*)

The Fourier transform of the convolution of two functions f and g is the product of their Fourier transforms, i.e.,

$$(f * g)^{\wedge}(\omega) = \widehat{f}(\omega)\widehat{g}(\omega),$$
$$(fg)^{\wedge}(\omega) = \left(\widehat{f} * \widehat{g}\right)(\omega).$$

For convenience, almost all basic properties of Fourier transform are summarized in Table 2.6.1, including the *Plancherel identity*, *Parseval's identity*, and the uncertainty property.

2.7. DISCRETE FOURIER TRANSFORM
AND FAST FOURIER TRANSFORM

The Fourier transform in (2.5.1) is sometimes called *continuous Fourier transform*, CFT for short, or continuous-time Fourier transform (CTFT). We introduce in this section the discrete Fourier transform, DFT for short.

Table 2.6.1. Properties of the (continuous) Fourier transform

	Property name	Expression
(1)	Linearality	$(af(\cdot) + bg(\cdot))^{\wedge}(\omega) = a\widehat{f}(\omega) + b\widehat{g}(\omega), \ a, b \in \mathbb{C}$
(2)	Time reversal	$(f(-\cdot))^{\wedge}(\omega) = \overline{\widehat{f}(-\omega)}$
(3)	Conjugation	$\left(\overline{f(\cdot)}\right)^{\wedge}(\omega) = \overline{\widehat{f}(-\omega)}$
(4)	Time shift	$(f(\cdot - t_0))^{\wedge}(\omega) = e^{-i\omega t}\widehat{f}(\omega)$
(5)	Time scaling	$(f(\cdot - t_0))^{\wedge}(\omega) = e^{-i\omega t_0}\widehat{f}(\omega)$
(6)	Frequency shifting or modulation	$\left(f(\cdot)e^{i\omega_0 \cdot}\right)^{\wedge}(\omega) = \widehat{f}(\omega - \omega_0);$ $(f(t)\cos\omega_0 t)^{\wedge}(\omega) = (1/2)\left[\widehat{f}(\omega + \omega_0) + \widehat{f}(\omega - \omega_0)\right];$ $(f(t)\sin\omega_0 t)^{\wedge}(\omega) = (i/2)\left[\widehat{f}(\omega + \omega_0) - \widehat{f}(\omega - \omega_0)\right]$

	Property name	Expression
(7)	Time derivatives	$(f^{(n)}(\cdot))^\wedge(\omega) = (i\omega)^n \widehat{f}(\omega),\ n \in \mathbb{Z}_+$
(8)	Frequency derivatives	$i^n \widehat{f}^{(n)}(\omega) = (t^n f(t))^\wedge(\omega),\ n \in \mathbb{Z}_+$
(9)	Convolution	$(f * g)^\wedge(\omega) = \widehat{f}(\omega)\widehat{g}(\omega)$ $(fg)^\wedge(\omega) = \left(\widehat{f} * \widehat{g}\right)(\omega)$
(10)	Integration	$\left(\int_{-\infty}^{t} f(u)\,du\right)^\wedge(\omega) = \dfrac{1}{i\omega}\widehat{f}(\omega) + \pi\widehat{f}(0)\delta(\omega)$
(11)	Plancherel identity	$\langle f, g\rangle = \dfrac{1}{2\pi}\langle \widehat{f}, \widehat{g}\rangle$
(12)	Parseval's identity (Energy preservation)	$\|f\|^2 = \dfrac{1}{2\pi}\|\widehat{f}\|^2$
(13)	Uncertainty property	$\Delta\omega\,\Delta t = \Delta = \text{constant; and}$ $\Delta \geq 1/2,\ \text{equality only for Gaussian } f(t) = f_0 e^{-at^2}$

Definition 2.7.1.

A sequence of N complex numbers $f_0, \cdots, f_{N-1}$ can be transformed into a sequence of N new complex numbers $F_0, \cdots, F_{N-1}$ by the *discrete Fourier transform* (DFT), or discrete-time Fourier transform (DTFT), according to the rule

$$F_k = \sum_{n=0}^{N-1} f_n \exp\left(-\frac{2k\pi i}{N}n\right), \quad k = 0, \cdots, N-1, \tag{2.7.1}$$

where

$$w_N = \exp\left(-\frac{2\pi i}{N}\right) \tag{2.7.2}$$

is a primitive N^{th} root of unity, i.e., $\left(\exp\left(-\dfrac{2\pi i}{N}k\right)\right)^N = 1,\ k = 0, \cdots, N-1$. On the other hand, the N complex numbers $F_0, \cdots, F_{N-1}$ can be transformed into a sequence of N complex numbers $f_0, \cdots, f_{N-1}$ by the *inverse discrete Fourier transform* (IDFT) according to the rule

$$f_n = \sum_{k=0}^{N-1} F_k \exp\left(\frac{2n\pi i}{N}k\right), \quad n = 0, \cdots, N-1. \tag{2.7.3}$$

It is clear from (2.7.1) and (2.7.3) that

$$F_k = F_{k+N}, \quad \text{and} \quad f_k = f_{k+N},$$

i.e., the sequences $\{F_k\}$ and $\{f_n\}$ are N-periodic. Moreover, the Plancherel identity and the Parseval's identity become

$$\sum_{n=0}^{N-1} f_n g_n^* = \frac{1}{N} \sum_{k=0}^{N-1} F_k G_k^*; \quad \text{and}$$

$$\sum_{n=0}^{N-1} |f_n|^2 = \frac{1}{N} \sum_{n=0}^{N-1} |F_k|^2. \tag{2.7.4}$$

Definition 2.7.2.

For two N-periodic sequences $\{f_n\}$ and $\{g_n\}$, the *circular convolution* of $\{f_n\}$ and $\{g_n\}$, denoted by $\{(f * g)_n\}$, is defined by

$$(f * g)_n = \sum_{m=0}^{N-1} f_m g_{n-m}. \tag{2.7.5}$$

If $\{F_k\}$ and $\{G_k\}$ are the DFTs of $\{f_n\}$ and $\{g_n\}$, then the DFT of $\{(f * g)_n\}$ is $\{F_k G_k\}$.

Observe also that both (2.7.1) and (2.7.3) can be written into matrix forms. More precisely, introduce the Vandermonde matrix (named after Alexandre-Théophile Vandermonde[6])

$$W = \begin{bmatrix} w_N^{0 \cdot 0} & w_N^{0 \cdot 1} & \cdots & w_N^{0 \cdot (N-1)} \\ w_N^{1 \cdot 0} & w_N^{1 \cdot 1} & \cdots & w_N^{1 \cdot (N-1)} \\ \vdots & \vdots & \ddots & \vdots \\ w_N^{(N-1) \cdot 0} & w_N^{(N-1) \cdot 1} & \cdots & w_N^{(N-1) \cdot (N-1)} \end{bmatrix}, \tag{2.7.6}$$

where w_N is given by (2.7.2). In other words, the generic entry $w_{j,k}$ of the unitary matrix W in (2.7.6) is $w_{j,k} = w_N^{jk} = \exp\left(-\frac{2\pi i j k}{N}\right)$. It is also easy to prove that W in (2.7.6) satisfies

$$W^{-1} = \frac{1}{N} W^*. \tag{2.7.7}$$

By using W in (2.7.6), the DFT in (2.7.1) is an orthogonal transformation

---

[6] *Alexandre-Théophile Vandermonde,* February 28, 1735 – January 1, 1796, was a French musician and chemist, and was born in Paris. His name is now principally associated with determinant in mathematics, namely, Vandermonde determinant.

$$\begin{bmatrix} F_0 \\ F_1 \\ \vdots \\ F_{N-1} \end{bmatrix} = W \begin{bmatrix} f_0 \\ f_1 \\ \vdots \\ f_{N-1} \end{bmatrix} \qquad (2.7.8)$$

and the IDFT in (2.7.3) is

$$\begin{bmatrix} f_0 \\ f_1 \\ \vdots \\ f_{N-1} \end{bmatrix} = \frac{1}{N} W^* \begin{bmatrix} F_0 \\ F_1 \\ \vdots \\ F_{N-1} \end{bmatrix}. \qquad (2.7.9)$$

The DFT adapts the continuous Fourier transform in such a way that it applies to discrete data.

Definition 2.7.3.

For a discrete signal $\{x[n]\}_{n=0}^{L-1}$, its *z-transform* is defined by

$$X(n) = \mathcal{Z}\{x(n)\} = \sum_{n=0}^{L-1} x[n]z^{-n}. \qquad (2.7.10)$$

For a discrete signal $\{x[n]\}_{n=0}^{\infty}$, its *unilateral z-transform* is defined by

$$X(n) = \mathcal{Z}\{x(n)\} = \sum_{n=0}^{\infty} x[n]z^{-n}.$$

For a discrete signal $\{x[n]\}_{n=-\infty}^{\infty}$, its *bilateral z-transform* is defined by

$$X(n) = \mathcal{Z}\{x(n)\} = \sum_{n=-\infty}^{\infty} x[n]z^{-n}.$$

The DFT is the z-transform in (2.7.10) with z restricted to the unit circle.

The direct evaluation of $F_0, \cdots, F_{N-1}$ for the DFT in (2.7.1) requires N^2 operations: There are N outputs $F_0, \cdots, F_{N-1}$ in (2.7.1), with each output F_k requiring N complex multiplications. Similarly, the evaluation of $f_0, \cdots, f_{N-1}$ for the IDFT in (2.7.3) also requires N^2 operations.

Definition 2.7.4.

A *fast Fourier transform*, FFT for short, is an algorithm to evaluate both $F_0, \cdots, F_{N-1}$ in (2.7.1) and $f_0, \cdots, f_{N-1}$ in (2.7.3) with only $2N \log_2 N$ operations.

A DFT can be computed using an FFT by means of the Danielson-Lanczos Lemma if the number of points N is a power of two. If N is not a power of two, a transform can be performed on sets of points corresponding to the prime factors of N which is slightly degraded in speed. Base-4 and base-8 FFTs use optimized code, and can be 20-30% faster than base-2 FFTs. Prime factorization is slow when the factors are large, but DFT can be made fast for $N = 2, 3, 4, 5, 7, 8, 11, 13$, and 16, using the Winograd transform (Press *et al.* 1992).

FFT algorithms are usually classified into two categories: decimation in time and decimation in frequency. The Cooley-Tukey (1965) FFT algorithm first rearranges the input elements in bit-reversed order, then builds the output transform (decimation in time). The basic idea is to break up a transform of length N into two transforms of length $N/2$ using the identity

$$
\sum_{n=0}^{N-1} a_n \exp \left(-\frac{2k\pi i}{N} n \right)
$$

$$
= \sum_{n=0}^{N/2-1} a_{2n} \exp \left(-\frac{2k\pi i}{N} 2n \right) + \sum_{n=0}^{N/2-1} a_{2n+1} \exp \left(-\frac{2k\pi i}{N}(2n+1) \right)
$$

$$
= \sum_{n=0}^{N/2-1} a_{2n} \exp \left(-\frac{2k\pi i}{(N/2)} n \right)
$$

$$
+ \exp \left(-\frac{2k\pi i}{N} \right) \sum_{n=0}^{N/2-1} a_{2n+1} \exp \left(-\frac{2k\pi i}{(N/2)} n \right).
$$

2.8. CLASSICAL HAAR SCALING FUNCTION AND HAAR WAVELETS

Wavelet is a *small wave*. Mathematically it means if $f(t)$ is a wavelet then at least its average is zero, i.e., $\int_{-\infty}^{\infty} f(t) \, dt = 0$. Generally speaking, wavelets are generated in such a way that they have specific useful or desirable properties in order for them to be useful for signal processing. The classical Haar wavelets (Haar 1910) are considered as the first known wavelets in the literature. Define

$$
\phi_{\text{Haar}}(t) = \begin{cases} 1, & \text{if } t \in [0, 1), \\ 0, & \text{otherwise} \end{cases} \tag{2.8.1}
$$

and

$$\psi_{\mathrm{Haar}}(t) = \begin{cases} 1, & \text{if } t \in [0, 1/2). \\ -1, & \text{if } t \in [1/2, 1). \\ 0, & \text{otherwise} \end{cases}$$

(2.8.2)

See Figure 2.8.1 for graphs of $\phi_{\mathrm{Haar}}(t)$, $\psi_{\mathrm{Haar}}(t)$, $e^{i\omega/2}\widehat{\phi}_{\mathrm{Haar}}(\omega)$, and $-ie^{i\omega/2}\widehat{\psi}_{\mathrm{Haar}}(\omega)$.

Definition 2.8.1.

The family of functions $\left\{ (\psi_{\mathrm{Haar}})_{j,k} \right\}_{j,k=-\infty}^{\infty}$ are *Haar wavelets*, where

$$(\psi_{\mathrm{Haar}})_{j,k}(t) = 2^{j/2}\psi_{\mathrm{Haar}}(2^{j}t - k), \quad j, k \in \mathbb{Z}.$$

(2.8.3)

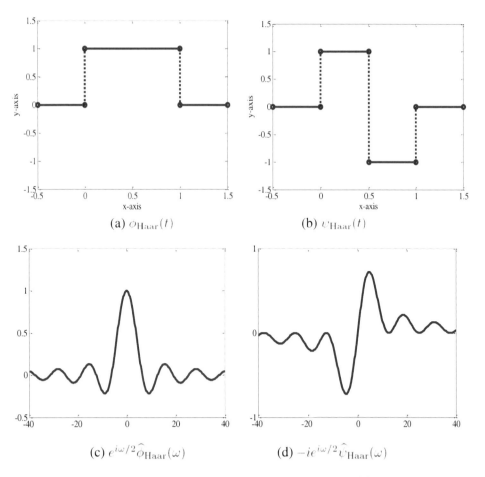

(a) $\phi_{\mathrm{Haar}}(t)$

(b) $\psi_{\mathrm{Haar}}(t)$

(c) $e^{i\omega/2}\widehat{\phi}_{\mathrm{Haar}}(\omega)$

(d) $-ie^{i\omega/2}\widehat{\psi}_{\mathrm{Haar}}(\omega)$

Figure 2.8.1. Graphs of (a) the Haar scaling function (father wavelet) $\phi_{\mathrm{Haar}}(t)$; (b) the Haar mother wavelet $\psi_{\mathrm{Haar}}(t)$; (c) the Fourier transform $e^{i\omega/2}\widehat{\phi}_{\mathrm{Haar}}(\omega)$; and (d) the Fourier transform $-ie^{i\omega/2}\widehat{\psi}_{\mathrm{Haar}}(\omega)$.

First of all, all the functions in $\left\{ (\psi_{\text{Haar}})_{j,k} \right\}_{j,k=-\infty}^{\infty}$ are *dilations* and *dyadic shifts* of a single function $\psi_{\text{Haar}}(t)$ in (2.8.2), which is called the *mother wavelet* of the Haar wavelets. Corresponding, $\phi_{\text{Haar}}(t)$ in (2.8.1) is called the *father wavelet* of the Haar wavelets, and all father wavelets are defined by $\left\{ (\phi_{\text{Haar}})_{j,k} \right\}_{j,k=-\infty}^{\infty}$, where

$$(\phi_{\text{Haar}})_{j,k}(t) = 2^{j/2}\phi_{\text{Haar}}(2^j t - k). \quad j,k \in \mathbb{Z}. \tag{2.8.4}$$

Secondly, both $\left\{ (\phi_{\text{Haar}})_{j,k} \right\}_{j,k=-\infty}^{\infty}$ and $\left\{ (\psi_{\text{Haar}})_{j,k} \right\}_{j,k=-\infty}^{\infty}$ are mutually orthonormal families (see Definition 2.1.9), meaning, $\phi_{j,k} = (\phi_{\text{Haar}})_{j,k}$ and $\psi_{j,k} = (\psi_{\text{Haar}})_{j,k}$ satisfy

$$\langle \phi_{j,k}, \phi_{m,n} \rangle = \delta_{jm}\,\delta_{kn};$$
$$\langle \phi_{j,k}, \psi_{m,n} \rangle = 0;$$
$$\langle \psi_{j,k}, \psi_{m,n} \rangle = \delta_{jm}\,\delta_{kn}. \quad j,k,m,n \in \mathbb{Z}. \tag{2.8.5}$$

Thirdly, both the father wavelet $\phi_{\text{Haar}}(t)$ and the mother wavelet $\psi_{\text{Haar}}(t)$ of the Haar wavelets satisfy the *two-scale equations*

$$\phi(t) = \phi(2t) + \phi(2t - 1); \quad \text{and}$$
$$\psi(t) = \phi(2t) - \phi(2t - 1). \quad t \in \mathbb{R}. \tag{2.8.6}$$

On the other hand, they also satisfy the *decomposition equations*

$$\phi(2t) = \frac{1}{2}\phi(t) + \frac{1}{2}\psi(t - 1); \quad \text{and}$$
$$\phi(2t - 1) = \frac{1}{2}\phi(t) - \frac{1}{2}\psi(t - 1). \quad t \in \mathbb{R}. \tag{2.8.7}$$

Both (2.8.6) and (2.8.7) are clear from their graphs in Figure 2.8.1(a) and Figure 2.8.1(b).

2.9. DAUBECHIES[7] ORTHONORMAL SCALING FUNCTIONS AND WAVELETS

The Haar wavelets in Section 2.8 are discontinuous. Is there any continuous compactly supported orthonormal wavelet? Daubechies answered this in her well-known paper (Daubechies 1992). It was due to this paper the topic wavelets has attracted many researchers

---

[7] *Ingrid Daubechies*, August 17, 1954—, was born in Houthalen, Belgium. Daubechies is best-known for her contribution to the construction of compactly supported continuous wavelets, named after name as the orthogonal Daubechies wavelets. She is also widely associated with the biorthogonal CDF wavelet (based upon a joint 1992 paper by A. Cohen, I. Daubechies, and J.C. Feauveau), which is now used for the JPEG2000 standard (http://www.jpeg.org/jpeg2000/). Daubechies has been since 1993 a full professor at Princeton University, the first female full professor of mathematics at Princeton.

in a wide range of diverse areas and have since then found many successful real world applications.

Definition 2.9.1.

For each $m \in \mathbb{Z}$, with $m \geq 2$, the *Daubechies orthonormal wavelets* are $\{(\psi_{2m}^D)_{j,k}(t)\}_{j,k=-\infty}^{\infty}$, which are generated from the dilations and integer shifts of a single function ψ_{2m}^D, called a *mother wavelet*, satisfying

$$\psi_{2m}^D(t) = \sum_{k=0}^{2m-1} (q_{2m}^D)_k\, \phi_{2m}^D(2t - k), \quad t \in \mathbb{R}, \tag{2.9.1}$$

where $\{(q_{2m}^D)_k\}$ is the *two-scale sequence* of ψ_{2m}^D; ϕ_{2m}^D is a *scaling function*, called a *father wavelet*, satisfying

$$\phi_{2m}^D(t) = \sum_{k=0}^{2m-1} (p_{2m}^D)_k\, \phi_{2m}^D(2t - k), \quad t \in \mathbb{R}, \tag{2.9.2}$$

with $\{(p_{2m}^D)_k\}$ being the *two-scale sequence* of ϕ_{2m}^D; and $\{(q_{2m}^D)_k\}$ is given explicitly in terms of $\{(p_{2m}^D)_k\}$ by

$$(q_{2m}^D)_k = (-1)^k (p_{2m}^D)_k, \quad k = 0, \cdots, 2m - 1. \tag{2.9.3}$$

Moreover, by taking the Fourier transforms both sides, (2.9.2) is equivalent to

$$\widehat{\phi_{2m}^D}(\omega) = P_{2m}^D\left(e^{-i\omega/2}\right) \widehat{\phi_{2m}^D}\left(\frac{\omega}{2}\right), \quad \omega \in \mathbb{R}, \tag{2.9.4}$$

with

$$P_{2m}^D\left(e^{-i\omega/2}\right) = \frac{1}{2}\sum_{k=0}^{2m-1} (p_{2m}^D)_k\, z^k = \left(\frac{1+z}{2}\right)^m S_{m-1}^D(z), \tag{2.9.5}$$

called the two-scale symbol of ϕ_{2m}^D, and $S_{m-1}^D(z)$ is a degree $m - 1$ polynomial extracted from

$$\left|S_{m-1}^D(z)\right|^2 = \sum_{\ell=0}^{m-1} \binom{m+\ell-1}{\ell} \left[\frac{1}{2}\left(1 - \frac{z + z^{-1}}{2}\right)\right]^\ell \tag{2.9.6}$$

by using the spectral factorization (or Riesz Lemma).

Definition 2.9.2.

The positive integer m represents the *vanishing moments* of ψ_{2m}^D, meaning

$$\int_{-\infty}^{\infty} t^\ell \psi_{2m}^D(t)\, dt = 0, \quad \ell = 0, \cdots, m-1.$$

This is equivalent to the fact that ϕ_{2m}^D has *polynomial preservation* of order m. The subscripts $2m$ of both ϕ_{2m}^D and ψ_{2m}^D indicate the lengths of both the lowpass filter $\{(p_{2m}^D)_k\}$ and the highpass filter $\{(q_{2m}^D)_k\}$, i.e., they are $2m$-tap filters. By using (2.9.4) repeatedly, say, N times,

$$\widehat{\phi_{2m}^D}(\omega) = \left[\prod_{j=1}^{N} P_{2m}^D\left(e^{-i\omega/2^j}\right) \right] \widehat{\phi_{2m}^D}\left(\frac{\omega}{2^N}\right),$$

$$\widehat{\psi_{2m}^D}(\omega) = \left[Q_{2m}^D\left(e^{-i\omega/2}\right) \prod_{j=1}^{N-1} P_{2m}^D\left(e^{-i\omega/2^{j+1}}\right) \right] \widehat{\phi_{2m}^D}\left(\frac{\omega}{2^N}\right),$$

for any positive integer N. In other words, with the normalization condition $\widehat{\phi_{2m}^D}(0) = 1$,

$$\widehat{\phi_{2m}^D}(\omega) = \prod_{j=1}^{\infty} P_{2m}^D\left(e^{-i\omega/2^j}\right),$$

$$\widehat{\psi_{2m}^D}(\omega) = Q_{2m}^D\left(e^{-i\omega/2}\right) \prod_{j=1}^{\infty} P_{2m}^D\left(e^{-i\omega/2^{j+1}}\right).$$

Hence, the Fourier transforms $\widehat{\phi_{2m}^D}$ and $\widehat{\psi_{2m}^D}$ of ϕ_{2m}^D and ψ_{2m}^D can be approximated by

$$\widehat{\phi_{2m}^D}(\omega) \approx \prod_{j=1}^{N} P_{2m}^D\left(e^{-i\omega/2^j}\right),$$

$$\widehat{\psi_{2m}^D}(\omega) \approx Q_{2m}^D\left(e^{-i\omega/2}\right) \prod_{j=1}^{N-1} P_{2m}^D\left(e^{-i\omega/2^{j+1}}\right),$$

for sufficiently large positive integer N.

Example 2.9.1.

Find ϕ_4^D and ψ_4^D of the Daubechies orthonormal scaling function and wavelet ($m = 2$).

Solution

It follows from (2.9.6) that

$$|S_1(z)|^2 = (1 + 2t)\Big|_{t=\frac{1}{2}\left(1-\frac{z+z^{-1}}{2}\right)} = -\frac{1}{2z}(z^2 - 4z + 1) = -\frac{1}{2z}(z - r_1)\left(z - \frac{1}{r_1}\right),$$

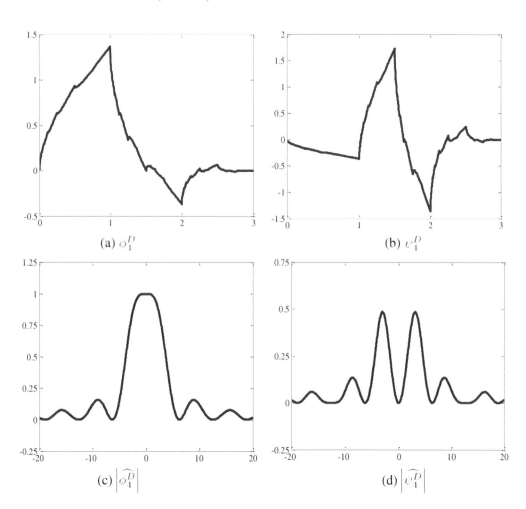

(a) ϕ_4^D

(b) ψ_4^D

(c) $\left|\widehat{\phi_4^D}\right|$

(d) $\left|\widehat{\psi_4^D}\right|$

Figure 2.9.1. Graphs of (a) the Daubechies orthonormal scaling function (father wavelet) ϕ_4^D; (b) the Daubechies orthonormal mother wavelet ψ_4^D; (c) the magnitude spectra $\left|\widehat{\phi_4^D}\right|$; and (d) the magnitude spectra $\left|\widehat{\psi_4^D}\right|$.

where $r_1 = 2 + \sqrt{3}$. Hence, by ensuring $S_1(1) = 1$ and choosing r_1 (due to the fact that $r_1 > 1$), we have

$$S_1(z) = -\frac{1}{\sqrt{2r_1}}(z - r_1) = -\frac{\sqrt{3}-1}{2}\left(z - 2 - \sqrt{3}\right),$$

so that

$$P_4^D(z) = \left(\frac{1+z}{2}\right)^2 S_1(z) = \frac{1}{2}\left(\frac{1+\sqrt{3}}{4} + \frac{3+\sqrt{3}}{4}z + \frac{3-\sqrt{3}}{4}z^2 + \frac{1-\sqrt{3}}{4}z^3\right),$$

which leads to

$$\left\{\left(p_4^D\right)_k\right\} = \left\{\frac{1+\sqrt{3}}{4}, \frac{3+\sqrt{3}}{4}, \frac{3-\sqrt{3}}{4}, \frac{1-\sqrt{3}}{4}\right\}.$$

$$\left\{\left(q_4^D\right)_k\right\} = \left\{\frac{1-\sqrt{3}}{4}, -\frac{3-\sqrt{3}}{4}, \frac{3+\sqrt{3}}{4}, -\frac{1+\sqrt{3}}{4}\right\}.$$

(2.9.7)

Hence, the two-scale equations of both ϕ_4^D and ψ_4^D are

$$\phi_4^D(t) = \frac{1+\sqrt{3}}{4}\phi_4^D(2t) + \frac{3+\sqrt{3}}{4}\phi_4^D(2t-1)$$

$$+ \frac{3-\sqrt{3}}{4}\phi_4^D(2t-2) + \frac{1-\sqrt{3}}{4}\phi_4^D(2t-3).$$

$$\psi_4^D(t) = \frac{1-\sqrt{3}}{4}\phi_4^D(2t) - \frac{3-\sqrt{3}}{4}\phi_4^D(2t-1)$$

$$+ \frac{3+\sqrt{3}}{4}\phi_4^D(2t-2) - \frac{1+\sqrt{3}}{4}\phi_4^D(2t-3).$$

With values of ϕ_4^D at 1 and 2, the only two integers inside the support of ϕ_4^D, namely,

$$\phi_4^D(1) = \frac{1+\sqrt{3}}{2}, \quad \phi_4^D(2) = \frac{1-\sqrt{3}}{2}.$$

we plot the graphs of both ϕ_4^D and ψ_4^D in Fig. 2.9.1, where the magnitude spectra are also included.

Example 2.9.2.

Find ϕ_6^D and ψ_6^D of the Daubechies orthonormal scaling function and wavelet ($m = 3$).

Solution

Again, it follows from (2.9.6) that

$$|S_2(z)|^2 = (1 + 3t + 6t^2)\Big|_{t=\frac{1}{2}\left(1 - \frac{z+z^{-1}}{2}\right)}$$

$$= 6\left[-\frac{1}{4z}\left(z^2 + \left(-3 + \frac{\sqrt{15}}{3}i\right)z + 1\right)\right]\left[-\frac{1}{4z}\left(z^2 + \left(-3 - \frac{\sqrt{15}}{3}i\right)z + 1\right)\right]$$

$$= 6\left[-\frac{1}{4z}(z - z_1)\left(z - \frac{1}{z_1}\right)\right]\left[-\frac{1}{4z}(z - \overline{z_1})\left(z - \frac{1}{\overline{z_1}}\right)\right].$$

where

$$z_1 = \frac{3}{2} + \frac{\sqrt{12\sqrt{10} + 15}}{6} - i\left(\frac{\sqrt{15}}{6} + \frac{\sqrt{12\sqrt{10} - 15}}{6}\right).$$

Hence, by requiring $S_2(1) = 1$ and choosing z_1 (due to the fact that $|z_1| > 1$) for $S_2(z)$, we have

$$S_2(z) = \frac{\sqrt{6}}{4|z_1|}(z - z_1)(z - \overline{z_1}),$$

so that

$$P_6^D(z) = \left(\frac{1 + z}{2}\right)^3 S_2(z) = \frac{1}{2}\sum_{k=0}^{5}\left(p_6^D\right)_k z^k,$$

with $\left\{\left(p_6^D\right)_k\right\}$ being explicitly given by

$$\left(p_6^D\right)_0 = \frac{1 + \sqrt{10}}{16} + \frac{2 + \sqrt{10}}{96}\sqrt{8\sqrt{10} - 10}.$$

$$\left(p_6^D\right)_1 = \frac{5 + \sqrt{10}}{16} + \frac{2 + \sqrt{10}}{32}\sqrt{8\sqrt{10} - 10}.$$

$$\left(p_6^D\right)_2 = \frac{5 - \sqrt{10}}{8} + \frac{2 + \sqrt{10}}{48}\sqrt{8\sqrt{10} - 10}.$$

$$\left(p_6^D\right)_3 = \frac{5 - \sqrt{10}}{8} - \frac{2 + \sqrt{10}}{48}\sqrt{8\sqrt{10} - 10}.$$

$$\left(p_6^D\right)_4 = \frac{5 + \sqrt{10}}{16} - \frac{2 + \sqrt{10}}{32}\sqrt{8\sqrt{10} - 10}.$$

$$\left(p_6^D\right)_5 = \frac{1 + \sqrt{10}}{16} - \frac{2 + \sqrt{10}}{96}\sqrt{8\sqrt{10} - 10}. \tag{2.9.8}$$

Values of ϕ_6^D at integers inside the support of ϕ_6^D are listed in the following,

$$\phi_6^D(1) = \frac{2 + \sqrt{10}}{8} + \frac{29 + 7\sqrt{10}}{312}\sqrt{8\sqrt{10} - 10}.$$

$$\phi_6^D(2) = \frac{2 - \sqrt{10}}{8} - \frac{35 - 5\sqrt{10}}{312}\sqrt{8\sqrt{10} - 10}.$$

$$\phi_6^D(3) = \frac{2 - \sqrt{10}}{8} + \frac{35 - 5\sqrt{10}}{312}\sqrt{8\sqrt{10} - 10}.$$

$$\phi_6^D(4) = \frac{2 + \sqrt{10}}{8} - \frac{29 + 7\sqrt{10}}{312}\sqrt{8\sqrt{10} - 10}.$$

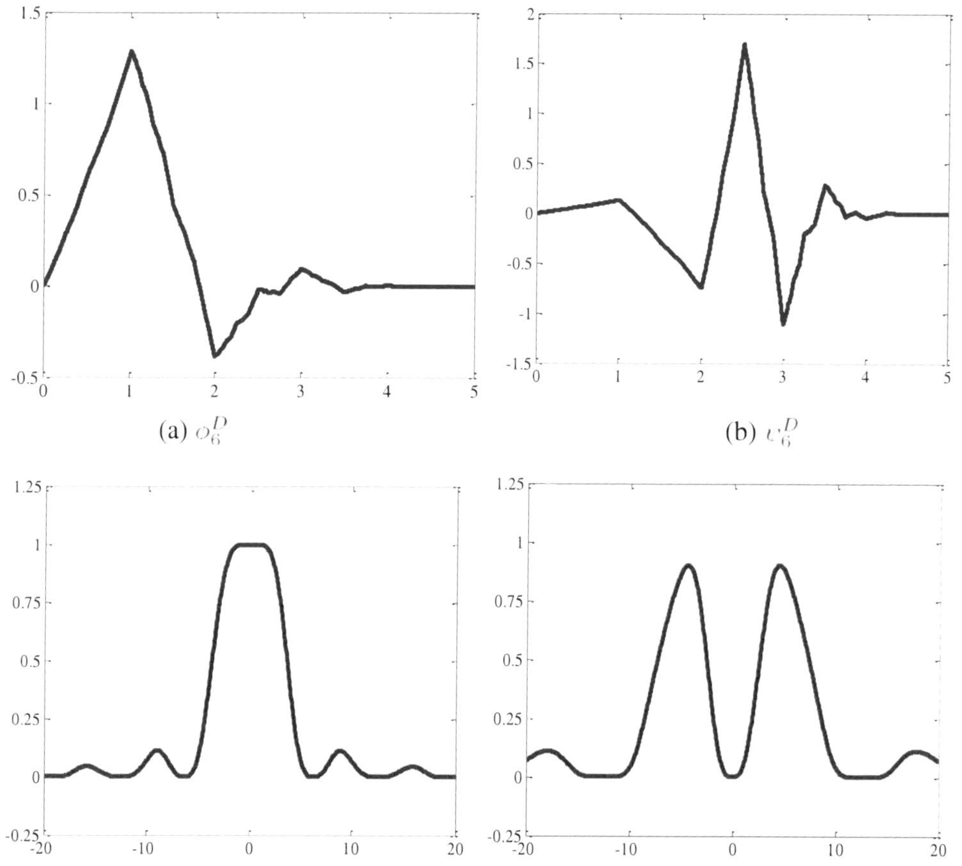

(a) ϕ_6^D

(b) ψ_6^D

(c) $\left|\widehat{\phi_6^D}\right|$

(d) $\left|\widehat{\psi_6^D}\right|$

Figure 2.9.2. Graphs of (a) the Daubechies orthonormal scaling function (father wavelet) ϕ_6^D; (b) the Daubechies orthonormal mother wavelet ψ_6^D; (c) the magnitude spectra $\left|\widehat{\phi_6^D}\right|$; and (d) the magnitude spectra $\left|\widehat{\psi_6^D}\right|$.

See Figure 2.9.2 for the graphs of both ϕ_6^D, ψ_6^D, and their magnitude spectra.

Table 2.9.1 includes coefficients or lowpass filters $\left\{\left(p_{2m}^D\right)_k\right\}_{k=0}^{2m-1}$ for $m = 1, \cdots, 10$.

Figure 2.9.3 shows the graphs of both ϕ_{2m}^D, ψ_{2m}^D, and their magnitude spectra $\widehat{\phi_{2m}^D}$, $\widehat{\psi_{2m}^D}$ for

$m = 4, 5, 6$ and Figure 2.9.4 includes the graphs of both ϕ_{2m}^D, ψ_{2m}^D, and their magnitude spectra $\widehat{\phi_{2m}^D}$, $\widehat{\psi_{2m}^D}$ for $m = 7, 8, 9$.

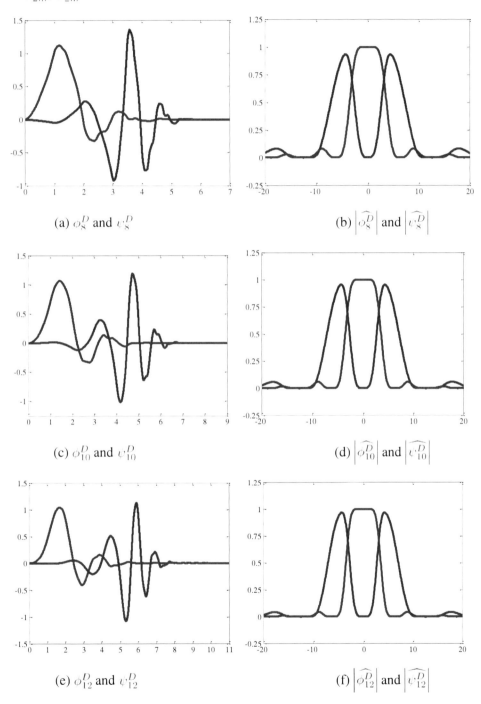

(a) ϕ_8^D and ψ_8^D (b) $\left|\widehat{\phi_8^D}\right|$ and $\left|\widehat{\psi_8^D}\right|$

(c) ϕ_{10}^D and ψ_{10}^D (d) $\left|\widehat{\phi_{10}^D}\right|$ and $\left|\widehat{\psi_{10}^D}\right|$

(e) ϕ_{12}^D and ψ_{12}^D (f) $\left|\widehat{\phi_{12}^D}\right|$ and $\left|\widehat{\psi_{12}^D}\right|$

Figure 2.9.3. Graphs of the Daubechies orthonormal scaling function (father wavelet) ϕ_{2m}^D and the Daubechies orthonormal mother wavelet ψ_{2m}^D, and their magnitude spectra $\left|\widehat{\phi_{2m}^D}\right|$ and $\left|\widehat{\psi_{2m}^D}\right|$ for $m = 4, 5, 6$.

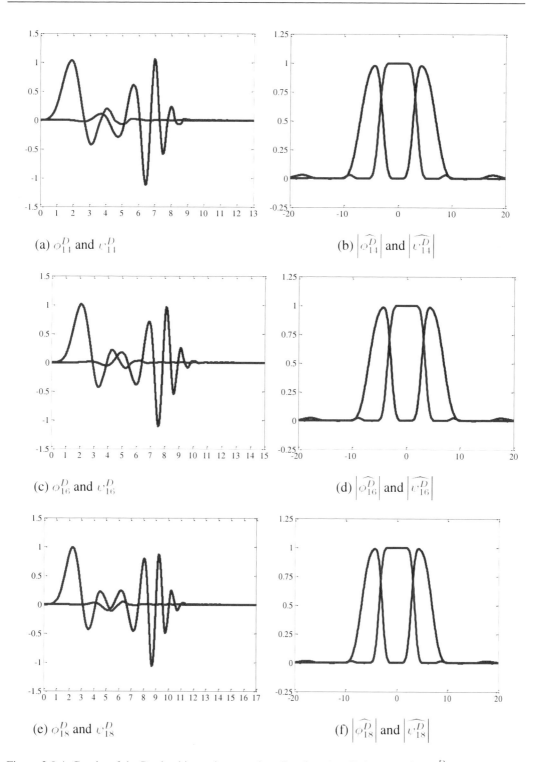

Figure 2.9.4. Graphs of the Daubechies orthonormal scaling function (father wavelet) ϕ_{2m}^D and the Daubechies orthonormal mother wavelet ψ_{2m}^D, and their magnitude spectra $\left|\widehat{\phi_{2m}^D}\right|$ and $\left|\widehat{\psi_{2m}^D}\right|$ for $m = 7, 8, 9$.

Table 2.9.1 Lowpass filters of Daubechies wavelets

m	k	$\left(p^D_{2m}\right)_k$	n	$o^D_{2m}(n)$	m	k	$\left(p^D_{2m}\right)_k$	n	$o^D_{2m}(n)$
1	0	1.		Haar	8	0	.0769556221081525		
	1	1.				1	.4424672471522498	1	.1368599843859938
2	0	.6830127018922193				2	.9554861504277474	2	.9915315055498689
	1	1.1830127018922193	1	1.3660254037844386		3	.8278165324223918	3	-.1686694280698240
	2	.3169872981077807	2	-.3660254037844386		4	-.0223857353337604	4	.0711140258408707
	3	-.1830127018922193				5	-.4016586327809781	5	-.0489069985868463
3	0	.4704672077841637				6	.0006681940924402	6	.0236025290831866
	1	1.1411169158314436	1	1.2863350694256967		7	.1820763568473155	7	-.0062059189812288
	2	.6503650005262325	2	-.3858369610458756		8	-.0245639010456968	8	.0007404149968362
	3	-.1909344155683274	3	.0952675460037808		9	-.0623502066502789	9	-.0000960519616915
	4	-.1208322083103962	4	.0042343456163981		10	.0197721592967015	10	.0000282786008600
	5	.0498174997368837				11	.0123688448196318	11	.0000016175870413
4	0	.3258034280512983				12	-.0068877192568836	12	.0000000417835973
	1	1.0109457150918289	1	1.0071699777256023		13	-.0005540045489588	13	-.0000000002287021
	2	.8922001382467596	2	-.0338369540528356		14	.0009552297112993	14	.0000000000000380
	3	-.0395750262356446	3	.03961046271590332		15	-.0001661372613732		
	4	-.2645071673690397	4	-.0117643582057267	9	0	.0538503495893256		
	5	.0436163004`41773	5	-.0011979575961770		1	.3448343038139559	1	.0696256148532700
	6	.0465036010709818	6	.0000188294132335		2	.8553490643594151	2	.8470941187123929
	7	-.0149869893303615				3	.9295457143662947	3	.1928879454874467
5	0	.2264189825835584				4	.1883695495063676	4	-.1615888385538713
	1	.8539435427050283	1	.6961360550943195		5	-.4147517618018770	5	.0621454565631310
	2	1.0243269442591971	2	.4490576048094863		6	-.1369535490247662	6	-.0058427119101699
	3	.1957669613478093	3	-.1822540053180363		7	.2100683422790124	7	-.0081011691539699
	4	-.3426567153829349	4	.0372318229038899		8	.0434526754612291	8	.0048407891813804
	5	-.0456011318835473	5	.0015175751497141		9	-.0956472641201941	9	-.0011332698708961
	6	.1097026586421336	6	-.0017267962430080		10	.0003548928132330	10	.0000518834218559
	7	-.0088268001083583	7	.0000375694705184		11	.0316241658525117	11	.0000227923963427
	8	-.0177918701019542	8	.0000001741331161		12	-.0066796202262772	12	-.0000025330020103
	9	.0047174279390679				13	-.0060549605750901	13	-.0000000802700433
6	0	.1577424320029014				14	.0026129672804945	14	.0000000021503760
	1	.6995038140752357	1	.4369121729450449		15	.0003258146713522	15	-.0000000000052346
	2	1.0622637598817380	2	.8323089728429706		16	-.0003563297590216	16	-.0000000000000003
	3	.4458313229300355	3	-.3847544237760172		17	.0000556455140343		
	4	-.3199865988921228	4	.1427984121126822	10	0	.0377171575922414		
	5	-.1835180640602952	5	-.0255073973408614		1	.2661221827938418	1	.0335440825684840
	6	.1378880929747446	6	-.0035305203040519		2	.7455750714864668	2	.6526806277842654
	7	.0389232097083293	7	.0017597663397331		3	.9736281107336399	3	.5552231947770544
	8	-.0446637483301891	8	.0000155911507261		4	.3976377417690174	4	-.3806874409343156
	9	.0007832511522972	9	-.0000025779244968		5	-.3533362017941126	5	.2022660795889995
	10	.0067560623629279	10	.0000000039542704		6	-.2771098787209663	6	-.0803945048003100
	11	-.0015235338056025				7	.1801274485333933	7	.0174035722935793
7	0	.1100994307456237				8	.1316029871010700	8	.0017888111543487
	1	.5607912836255251	1	.2531225927394706		9	-.1009665719967794	9	-.0022622919803254
	2	1.0311484916361973	2	1.0097568015527715		10	-.0416592480876016	10	.0003861859807100
	3	.6643724822110793	3	-.3922026570661041		11	.0469698140973971	11	.0000774649011710
	4	-.2035138224626925	4	.1845944600259790		12	.0051004369678145	12	-.0000259573513242
	5	-.3168350112806657	5	-.0667140607455830		13	-.0151790023358565	13	.0000000745576680
	6	.1008464650093882	6	.0102809473212267		14	.0019733253649632	14	.0000001064649412
	7	.1140034451597432	7	.0017705593880646		15	.0028178865901947	15	-.0000000050183483
	8	-.0537824525896909	8	-.0005441637518526		16	-.0009699478398564	16	.0000000000135063
	9	-.0234399415642066	9	-.0000667131499081		17	-.0001647090060908	17	.0000000000000959
	10	.0177497923793615	10	.0000022728924642		18	.0001323543668511	18	.0000000000000000
	11	.0006075149954021	11	-.0000000391869762		19	-.0000187584156275		
	12	-.0025479047181874	12	-.0000000000195526					
	13	.0005002268531225							

2.10. MULTIRESOLUTION ANALYSIS IN GENERAL

Definition 2.10.1.

A *multiresolution analysis*, MRA for short, is a sequence $\{V_j\}_{j\in\mathbb{Z}}$ of closed subspaces of $L^2(\mathbb{R})$, i.e., $V_j \subset L^2(\mathbb{R})$. $j \in \mathbb{Z}$, satisfying the following six properties

(1) (nested) $V_{j-1} \subset V_j$. $j \in \mathbb{Z}$;

(2) (separation) $\lim_{j\to-\infty} V_j = \bigcap_{j\in\mathbb{Z}} V_j = \{0\}$;

(3) (density) $\lim_{j\to+\infty} V_j = \overline{\bigcup_{j\in\mathbb{Z}} V_j} = L^2(\mathbb{R})$;

(4) (scaling) $f(t) \in V_j \Leftrightarrow f(2^{-j}t) \in V_0$. $j \in \mathbb{Z}$;

(5) (translation invariant) $f(t) \in V_j$. $j \in \mathbb{Z} \Leftrightarrow f(t - 2^{-j}k) \in V_j$. $(j,k) \in \mathbb{Z}^2$; and

(6) (Riesz basis) There exists a function $\phi \in V_0$ such that $\{\phi(\cdot - k) : k \in \mathbb{Z}\}$ is an **Riesz basis** for V_0, namely, there are constants $0 < A \le B < \infty$, such that

$$A \|\{a_n\}\|_2 \le \left\| \sum_{j\in\mathbb{Z}} a_n \phi(\cdot - n) \right\|_2 \le B \|\{a_n\}\|_2. \quad \forall \{a_n\} \in \ell^2, \tag{2.10.1}$$

where $\|\{a_n\}\|_2$ denotes the ℓ^2-norm, i.e., $\|\{a_n\}\|_2 = (\sum_n |a_n|^2)^{1/2}$, and $\left\| \sum_{j\in\mathbb{Z}} a_n \phi(\cdot - n) \right\|_2$ is the usual L^2-norm defined by $\|f\|_2 = \left(\int_{-\infty}^{\infty} |f(t)|^2 \, dt \right)^{1/2}$. The Riesz basis property in (2.10.1) is equivalent to

$$A \le \sum_{n\in\mathbb{Z}} \left| \hat{\phi}(\cdot + 2\pi n) \right|^2 \le B. \tag{2.10.2}$$

Common MRAs in wavelet theory are generated by a single or several scaling functions. The latter is referred as multiscaling functions (Chui & Lian 1996) in the wavelet literature. We focus on the single scaling functions here. In general,

$$V_j = \mathrm{clos}_{L^2} \mathrm{span} \left\{ 2^{j/2} \phi(2^j \cdot -k) : k \in \mathbb{Z} \right\}. \quad j \in \mathbb{Z}. \tag{2.10.3}$$

Observe that, due to the fact that $V_0 \subset V_1$, there must be a L^2-subspace $W_0 \subset V_1$ such that

$$V_1 = V_0 \oplus^{\perp} W_0, \tag{2.10.4}$$

where $\oplus^{\perp}$ means the direct orthogonal sum. In other words, W_0 is the orthogonal complement to V_0 in V_1. Hence, for each MRA $\{V_j\}_{j\in\mathbb{Z}}$ generated by a scaling function ϕ, there must be a function $\psi \in V_1$ that generates the space W_0. Both ϕ and ψ satisfy similar two-scale equations like (2.9.2) and (2.9.1). Moreover,

$$L^2(\mathbb{R}) = \text{clos} \bigoplus_{k \in \mathbb{Z}} W_k. \tag{2.10.5}$$

2.11. WAVELET TRANSFORM AND INVERSE WAVELET TRANSFORM

Definition 2.11.1.

An *(integral)* *wavelet* *transform* of a function $f(t) \in L^2(\mathbb{R})$ is a function $W_\psi f : \mathbb{R}^2 \to \mathbb{R}$ (or an integral transform) by using dilations and translates of a mother wavelet ψ for the kernel function, namely,

$$(W_\psi f)(a, b) = \frac{1}{\sqrt{|a|}} \int_{-\infty}^{\infty} \overline{\psi\left(\frac{t-b}{a}\right)} f(t) \, dt. \tag{2.11.1}$$

The wavelet coefficients $c_{j,k}$ are defined by

$$(W_\psi f)\left(\frac{1}{2^j}, \frac{k}{2^j}\right), \quad j, k \in \mathbb{Z}. \tag{2.11.2}$$

The mother wavelet ψ has to satisfy two criteria:

$$\int_{-\infty}^{\infty} \psi(t) \, dt = 0; \quad \text{and}$$

ψ is continuous and has *exponential decay*, i.e., there are constants $a, C > 0$ such that $|\psi(t)| < Ce^{-a|t|}, t \in \mathbb{R}$.

Definition 2.11.2.

The *inverse wavelet transform* of a function $W_\psi f : \mathbb{R}^2 \to \mathbb{R}$ is a function $f : \mathbb{R} \to \mathbb{R}$:

$$f(t) = \frac{1}{C_\psi} \int_{\mathbb{R}^2} (W_\psi f)(b, a) \left\{ \frac{1}{\sqrt{|a|}} \psi\left(\frac{t-b}{a}\right) \right\} \frac{da \, db}{a^2}, \tag{2.11.3}$$

where C_ψ is a positive finite constant defined by

$$C_\psi = \int_{-\infty}^{\infty} \frac{\left|\widehat{\psi}(\omega)\right|^2}{|\omega|} \, d\omega < \infty, \tag{2.11.4}$$

which is called *Grossmann-Morlet condition* [Meyer, 1993, p.7] or *admissibility condition* [Daubechies, 1992, p.24].

2.12. OTHER WAVELETS

There are various other wavelets in the wavelet literature. We include some of them in this section that have demonstrated many successful applications.

2.12.1. Compactly Supported Spline Wavelets

Definition 2.12.1.

The mth order *cardinal B-spline* N_m is defined by

$$N_m(t) = (N_{m-1} \star N_1)(t), \quad m \geq 2, \tag{2.12.1}$$

where $N_1(t) = \chi_{[0,1)}(t)$, the *characteristic function* of the interval $[0, 1)$, i.e., $N_1(t) = 1$ if $0 \leq t < 1$ and 0 otherwise. So N_m in (2.12.1) can also be defined by

$$N_m(t) = \int_0^1 N_{m-1}(t)\, dt, \quad m \geq 2. \tag{2.12.2}$$

The cardinal *B*-splines N_m are scaling functions:

$$N_m(t) = \sum_{j=0}^m p_j N_m(2t - j) = \frac{1}{2^{m-1}} \sum_{j=0}^m \binom{m}{j} N_m(2t - j), \tag{2.12.3}$$

and they also generate MRA's in (2.10.3) with $\phi = N_m$. For each $\phi = N_m$, a special generator, denoted by ψ_m and called a *semi-orthogonal spline wavelet*, if it generates the wavelet subspaces W_j's in (2.10.4), where all integer shifts of ψ_m do not form an orthogonal family. More precisely,

$$V_0 = \text{clos}_{L^2} \text{span} \{N_m(\cdot - k) : k \in \mathbb{Z}\},$$
$$W_0 = \text{clos}_{L^2} \text{span} \{\psi_m(\cdot - k) : k \in \mathbb{Z}\},$$

with

$$\langle N_m(\cdot - j), \psi_m(\cdot - k) \rangle = 0, \quad j, k \in \mathbb{Z}.$$

The explicit expressions of ψ_m are given by

$$\psi_m(t) = \sum_{n=0}^{3m-2} q_n N_m(2t - n),$$ (2.12.4)

$$q_n = \frac{(-1)^n}{2^{m-1}} \sum_{j=0}^{m} \binom{m}{j} N_{2m}(n+1-j), \quad n = 0, 1, \cdots, 3m - 2.$$ (2.12.5)

Indeed, the two-scale sequence $\{q_n\}_{n=0}^{3m-2}$ in (2.12.5) was from ψ_m'd two-scale symbol

$$Q_m(z) = \frac{1}{2} \sum_{n=0}^{3m-2} q_n z^n = \left(\frac{1-z}{2}\right)^m \sum_{\ell=0}^{2m-2} N_{2m}(\ell + 1)(-z)^\ell.$$

When $m = 1$, N_1 and ψ_1 are the Haar wavelet pair ϕ_{Haar} and ψ_{Haar}. When $m = 2$, N_2 is the linear B-spline, and the two-scale sequences of N_2 and ψ_2 are

$$(p_0, p_1, p_2) = \left(\frac{1}{2}, 1, \frac{1}{2}\right), \quad (q_0, \cdots, q_4) = \left(\frac{1}{12}, -\frac{1}{2}, \frac{5}{6}, -\frac{1}{2}, \frac{1}{12}\right);$$

when $m = 3$, N_3 is the quadratic B-spline, and the two-scale sequences of N_3 and ψ_3 are

$$(p_0, \cdots, p_3) = \left(\frac{1}{4}, \frac{3}{4}, \frac{3}{4}, \frac{1}{4}\right),$$

$$(q_0, \cdots, q_7) = \left(\frac{1}{480}, -\frac{29}{480}, \frac{49}{160}, -\frac{101}{160}, \frac{101}{160}, -\frac{49}{160}, \frac{29}{480}, \frac{1}{480}\right);$$

when $m = 4$, N_4 is the cubic B-spline, and the two-scale sequences of N_4 and ψ_4 are

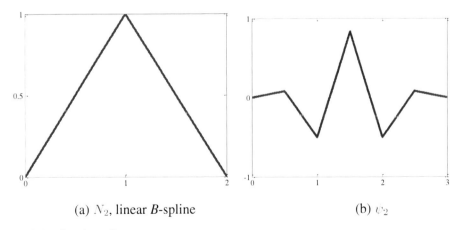

(a) N_2, linear B-spline   (b) ψ_2

Figure 2.12.1. (Continued)

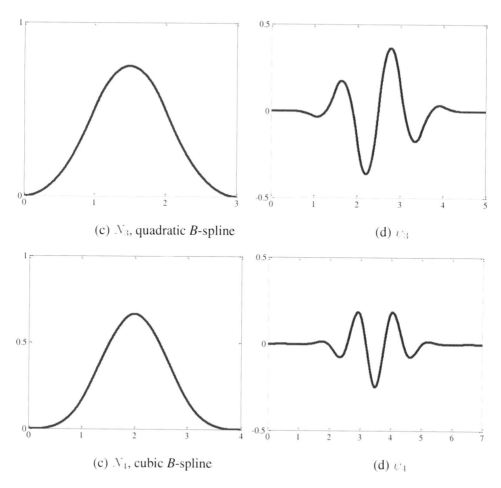

(c) N_3, quadratic B-spline   (d) ψ_3

(c) N_4, cubic B-spline   (d) ψ_4

Figure 2.12.1. Graphs of the cardinal B-splines N_m in (2.12.3) and their corresponding semi-orthogonal spline wavelets ψ_m in (2.12.4)-(2.12.5) for $m = 2, 3, 4$, called linear, quadratic, and cubic B-spline wavelets, respectively.

$$(p_0, \cdots, p_4) = \left(\frac{1}{8}, \frac{1}{2}, \frac{3}{4}, \frac{1}{2}, \frac{1}{8} \right),$$

$$(q_0, \cdots, q_{10}) = \left(\frac{1}{40320}, -\frac{31}{10080}, \frac{559}{13440}, -\frac{247}{1260}, \frac{9241}{20160}, \right.$$
$$\left. -\frac{337}{560}, \frac{9241}{20160}, -\frac{247}{1260}, \frac{559}{13440}, -\frac{31}{10080}, \frac{1}{40320} \right).$$

The graphs of N_m and ψ_m for $m = 2, 3, 4$ are plotted in Figure 2.12.1.

Definition 2.12.2.

The 1D *Gaussian function* or *Gaussian* is the *probability density function* of the *normal distribution*

$$G(x) = \frac{1}{\sigma\sqrt{2\pi}} e^{-(x-\mu)^2/(2\sigma^2)}.$$

2.12.2. Morlet Wavelets

Morlet[8] was one of the early and distinguished pioneers who were responsible for the introduction of the notion "wavelet." Although the idea of wavelets existed in the mathematical and engineering literature prior to Morlet's work, Morlet wavelets are still viewed as the "original"wavelets. Morlet wavelets have their own good characteristics. For instance, Morlet wavelets have a very intuitive nature and definition and are naturally robust against shifting a feature in time, which favors spectral accuracy. They were used extensively and successfully to analyze speech, music, and seismic signals.

Morlet wavelets [Goupilaud, Grossmann & Morlet, 1984], named in honor of Jean P. Morlet, was formulated by localizing a Gaussian.

(1) For *complex-valued* Morlet wavelets, write

$$\psi_\sigma(t) = c_1 e^{-t^2/2} \left(e^{i\sigma t} - c_2 \right), \tag{2.12.6}$$

whose Fourier transform is given by

$$\widehat{\psi_\sigma}(\omega) = c_1\sqrt{2\pi} \left(e^{-(\sigma-\omega)^2/2} - c_2 e^{-\omega^2/2} \right), \tag{2.12.7}$$

where the parameter σ is used for controlling the tradeoffs between ψ_σ and $\widehat{\psi_\sigma}$, and c_1 and c_2 are two to-be-determined constants. First of all, the condition $\int_{\mathbb{R}} \psi_\sigma(t)\,dt = 0$ implies $c_2 = e^{-\sigma^2/2}$. Secondly, the condition $\|\psi_\sigma\|_2 = 1$ implies

$$c_1 = \pi^{-1/4} \left(1 - 2e^{-3\sigma^2/4} + e^{-\sigma^2} \right)^{-1/2}.$$

Hence,

$$\psi_\sigma(t) = \pi^{-1/4} \left(1 - 2e^{-3\sigma^2/4} + e^{-\sigma^2} \right)^{-1/2} e^{-t^2/2} \left(e^{i\sigma t} - e^{-\sigma^2/2} \right), \tag{2.12.8}$$

and

---

[8] *Jean P. Morlet*, January 13, 1931 – April 27, 2007, was a French geophysicist. He invented the term wavelet. He was awarded in 1997 the Fessenden Award for his discovery of the wavelet transform.

$$\widehat{\psi_\sigma}(\omega) = \sqrt{2}\pi^{1/4} \left(1 - 2e^{-3\sigma^2/4} + e^{-\sigma^2}\right)^{-1/2} \left(e^{-(\sigma-\omega)^2/2} - e^{-(\sigma^2+\omega^2)/2}\right). \quad (2.12.9)$$

(2) For *real-valued* Morlet wavelets, we simply drop the imaginary part $c_1 e^{-t^2/2} \sin \sigma t$ in (2.12.6), and denote the Morlet wavelets by ψ_γ, i.e.,

$$\psi_\gamma(t) = d_1 e^{-t^2/2} (\cos \gamma t - d_2).$$

Similar calculation yields $d_2 = e^{-\gamma^2/2}$ and

$$d_1 = \sqrt{2}\,\pi^{-1/4} \left(1 - 4e^{-3\gamma^2/4} + 3e^{-\gamma^2}\right)^{-1/2}.$$

Therefore,

$$\psi_\gamma(t) = \sqrt{2}\,\pi^{-1/4} \left(1 - 4e^{-3\gamma^2/4} + 3e^{-\gamma^2}\right)^{-1/2} e^{-t^2/2} \left(\cos \gamma t - e^{-\gamma^2/2}\right),$$

$$(2.12.10)$$

and

$$\widehat{\psi_\gamma}(\omega) = 2\pi^{1/4} \left(1 - 4e^{-3\gamma^2/4} + 3e^{-\gamma^2}\right)^{-1/2} e^{-(\gamma^2+\omega^2)/2} (\cosh \gamma w - 1).$$

$$(2.12.11)$$

(3) If we require the ratio of the largest and second largest values of ψ_σ in (2.12.8) or ψ_γ in (2.12.10) to be approximately $1/2$, then σ in (2.12.8) or γ in (2.12.10) can be chosen as 5. When $\sigma = 5$ or $\gamma = 5$, the second terms in both (2.12.8) and (2.12.10) are already sufficient small and can be neglected in practical applications. That is why the real-valued Morlet wavelet in the *Wavelet Toolbox* in *MATLAB* is simply defined by, up to a normalization constant,

$$\psi_M(t) = Ce^{-t^2/2} \cos 5t. \quad (2.12.12)$$

Normalization of (2.12.12) implies

$$C = \frac{\sqrt{2}}{\pi^{1/4} \sqrt{1 + e^{-25}}}. \quad (2.12.13)$$

The Fourier transform of (2.12.12)-(2.12.13) is

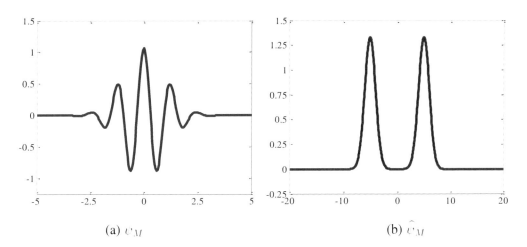

(a) ψ_M (b) $\widehat{\psi}_M$

Figure 2.12.2. Graphs of the Morlet wavelet ψ_M in (2.12.12)-(2.12.13) and its Fourier transform.

$$\widehat{\psi}_M(\omega) = \frac{\pi^{1/4}}{\sqrt{1 + e^{-25}}} \left(e^{-(\omega-5)^2/2} + e^{-(\omega+5)^2/2} \right). \tag{2.12.14}$$

By doing so, ψ_M in (2.12.12)-(2.12.13) does not satisfy the "small wave" condition, or it is not theoretically admissible. Indeed, mathematically,

$$\int_{\mathbb{R}} \psi_M(t)\,dt = \widehat{\psi}_M(0) = \frac{2\pi^{1/4} e^{-25/2}}{\sqrt{1 + e^{-25}}} \approx 9.922850 \times 10^{-6}.$$

See Figure 2.12.2 for graphs of the Morlet wavelet ψ_M in (2.12.12)-(2.12.13) and its Fourier transform.

2.12.3. Gaussian Wavelets

The family of *Gaussian wavelets* of order m is defined by

$$\psi_{G,m}(t) = C_m \frac{d^m}{d^m t} e^{-t^2/2}, \quad m \in \mathbb{Z}_+, \tag{2.12.15}$$

where C_m is the normalization constant. In other words, Gaussian wavelets of order m is from the mth order derivative of Gaussian (DOG). Here the order is for the vanishing moments of $\psi_{G,m}$. When m is even it is symmetric, and when m is odd, it is antisymmetric. (2.12.16) is when $m = 2$. Gaussian wavelets are not compactly supported in both time and frequency domains, and result in good time localization but poor for frequency.

By taking the second order derivative of the Gaussian $e^{-t^2/2}$ and normalization, we arrive at the *Mexican hat* function

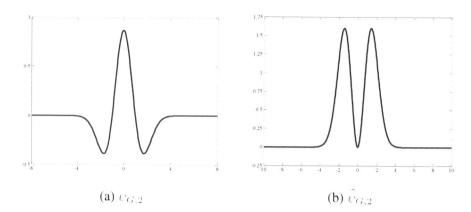

(a) $\psi_{G,2}$ (b) $\widehat{\psi}_{G,2}$

Figure 2.12.3. Graphs of the Mexican hat or Gaussian wavelet $\psi_{G,2}$ of order 2 in (2.12.16) and its Fourier transform.

$$\psi(t) = \frac{2\pi^{-1/4}}{\sqrt{3}}(1 - t^2)e^{-t^2/2}$$
(2.12.16)

with Fourier transform

$$\widehat{\psi}(\omega) = \frac{2\sqrt{6}\pi^{1/4}}{3}\omega^2 e^{-\omega^2/2}.$$

See Figure 2.12.3 for Graphs of the Mexican hat or Gaussian wavelet $\psi_{G,2}$ of order 2 in (2.12.16) and its Fourier transform.

2.12.4. Biorthogonal Wavelets

Cohen-Daubechies-Feauveau wavelets [CDF, 1992] are the first family of *biorthogonal* wavelets. They are not the same as the orthogonal Daubechies wavelets in Section 2.11. However their construction is quite similar.

Let $\{\phi, \psi\}$ and $\{\widetilde{\phi}, \widetilde{\psi}\}$ be two pairs of compactly supported scaling functions and wavelets that form a *biorthogonal system*, meaning:

$$\langle \phi(\cdot), \widetilde{\psi}(\cdot - k)\rangle = 0; \quad \langle \psi(\cdot), \widetilde{\phi}(\cdot - \ell)\rangle = 0.$$
(2.12.17)

$$\langle \phi(\cdot), \widetilde{\phi}(\cdot - k)\rangle = \delta_{k,0}; \quad \langle \psi(\cdot), \widetilde{\psi}(\cdot - \ell)\rangle = \delta_{\ell,0}.$$
(2.12.18)

for $k, \ell \in \mathbb{Z}$. First, by using the two-scale equations of ϕ, ψ, $\widetilde{\phi}$, and $\widetilde{\psi}$:

$$\widehat{\phi}(\omega) = P\left(e^{-i\omega/2}\right)\widehat{\phi}\left(\frac{\omega}{2}\right), \quad \widehat{\psi}(\omega) = Q\left(e^{-i\omega/2}\right)\widehat{\phi}\left(\frac{\omega}{2}\right);$$
(2.12.19)

$$\widehat{\widetilde{\phi}}(\omega) = \widetilde{P}\left(e^{-i\omega/2}\right)\widehat{\widetilde{\phi}}\left(\frac{\omega}{2}\right), \quad \widehat{\widetilde{\psi}}(\omega) = \widetilde{Q}\left(e^{-i\omega/2}\right)\widehat{\widetilde{\phi}}\left(\frac{\omega}{2}\right),$$
(2.12.20)

and requiring Q and $\widetilde{Q}$ to be determined from $\widetilde{P}$ and P, namely,

$$Q(z) = \varepsilon z^{2K-1}\overline{\widetilde{P}(-z)}, \quad \widetilde{Q}(z) = \varepsilon z^{2K-1}\overline{P(-z)}, \tag{2.12.21}$$

for some $K \in \mathbb{Z}$ and $\varepsilon = \pm 1$, the identities in (2.12.17) and (2.12.18) are equivalent to the only requirement

$$P(z)\overline{\widetilde{P}(z)} + P(-z)\overline{\widetilde{P}(-z)} = 1, \quad |z| = 1. \tag{2.12.22}$$

Secondly, by requiring ϕ and $\widetilde{\phi}$ to have polynomial preservation of orders n and $\tilde{n}$, respectively, we can write P and $\widetilde{P}$ as

$$P(z) = \left(\frac{1+z}{2}\right)^n S_{n_1}(z), \tag{2.12.23}$$

$$\widetilde{P}(z) = \left(\frac{1+z}{2}\right)^{\tilde{n}} \widetilde{S}_{\tilde{n}_1}(z), \tag{2.12.24}$$

for some polynomials S_{n_1} and $\widetilde{S}_{\tilde{n}_1}$ of degrees n_1 and $\tilde{n}_1$, satisfying $S_{n_1}(1) = \widetilde{S}_{\tilde{n}_1}(1) = 1$. For both ϕ and $\widetilde{\phi}$ to be symmetric, we can choose S_{n_1} and $\widetilde{S}_{\tilde{n}_1}$ to be reciprocal, meaning $S_{n_1}(z) = z^{n_1} S_{n_1}(z^{-1})$ and $\widetilde{S}_{\tilde{n}_1}(z) = z^{\tilde{n}_1} \widetilde{S}_{\tilde{n}_1}(z^{-1})$. Thirdly, by substituting P in (2.12.23) and $\widetilde{P}$ in (2.12.24) into (2.12.22), and for n and $\tilde{n}$ having the same parities, i.e., either $n = 2\ell$ and $\tilde{n} = 2\tilde{\ell}$ or $n = 2\ell+1$ and $\tilde{n} = 2\tilde{\ell}+1$ for some $\ell, \tilde{\ell} \in \mathbb{Z}_+$, the identity (2.12.22) yields the classical Bezout-type polynomial identity

$$(1-t)^m f(t) + t^m f(1-t) = 1, \quad t \in \mathbb{R}, \tag{2.12.25}$$

where either $m = \ell + \tilde{\ell}$ or $m = \ell + \tilde{\ell} + 1$. Fourthly, (2.12.25) has solutions of the form

$$f(t)\Big|_{t = \frac{1-(z+1/z)/2}{2}} = z^L S_{n_1}(z)\overline{\widetilde{S}_{\tilde{n}_1}(z)} \tag{2.12.26}$$

for some $L \in \mathbb{Z}_+$, so that the right hand side of (2.12.26) is a function of $z + z^{-1}$. Finally, there is another construction procedure that is a little different from the aforementioned procedure. Write P the same as in (2.12.23), include $(1-z)^{\tilde{n}}$ as a factor for Q, and require

$$\det \begin{bmatrix} P(z) & P(-z) \\ Q(z) & Q(-z) \end{bmatrix} = \varepsilon z^{2K-1}, \quad |z| = 1, \tag{2.12.27}$$

for some $K \in \mathbb{Z}$ and $\varepsilon = \pm 1$.

2.12.5. CDF 5/3 Wavelets

Due to the fact that the JPEG2000 compression standard exclusively uses the biorthogonal CDF 5/3 wavelet (also called the LeGall 5/3 wavelet) for lossless compression and CDF 9/7 wavelet for lossy compression, we only include the biorthogonal CDF 5/3 and 9/7 wavelets here and the following subsection.

By setting $n = \tilde{n} = 2$ and writing

$$P(z) = \left(\frac{1+z}{2}\right)^2 (a_0 + (1 - a_0)z + a_0 z^2),$$

$$\tilde{Q}(z) = -\left(\frac{1-z}{2}\right)^2,$$

the identity (2.12.27) with $K = 2$ and $\varepsilon = -1$ yields $a_0 = -1/2$. Therefore,

$$P(z) = \frac{1}{2}\sum_{k=0}^{4} p_k z^k, \quad Q(z) = \frac{1}{2}\sum_{k=0}^{2} q_k z^k,$$

$$\widetilde{P}(z) = \frac{1}{2}\sum_{k=0}^{3} \tilde{p}_k z^k, \quad \widetilde{Q}(z) = \frac{1}{2}\sum_{k=-1}^{3} \tilde{q}_k z^k,$$

with

$$(p_0, p_1, \cdots, p_4) = \left(-\frac{1}{4}, \frac{1}{2}, \frac{3}{2}, \frac{1}{2}, -\frac{1}{4}\right); \tag{2.12.28}$$

$$(q_0, q_1, q_2) = \left(-\frac{1}{2}, 1, -\frac{1}{2}\right); \tag{2.12.29}$$

$$(\tilde{p}_0, \tilde{p}_1, \tilde{p}_2, \tilde{p}_3) = \left(0, \frac{1}{2}, 1, \frac{1}{2}\right);$$

$$(\tilde{q}_0, \tilde{q}_1, \cdots, \tilde{q}_4) = \left(-\frac{1}{4}, -\frac{1}{2}, \frac{3}{2}, -\frac{1}{2}, -\frac{1}{4}\right).$$

i.e., the lowpass and highpass filters induced from the CDF 5/3 or LeGall 5/3 wavelets, up to a constant multiple, are in (2.12.28) and (2.12.29), which have lengths 5 and 3. The corresponding $\tilde{\phi}$ is the second order cardinal B-spline N_3, so $\tilde{\psi}$ is piecewise linear too. However, ϕ and ψ are discontinuous.

2.12.6. CDF 9/7 Wavelets

The CDF 9/7 is derived by asking $n = \tilde{n} = 4$. By writing

$$P(z) = \left(\frac{1+z}{2}\right)^4 (s_0 + s_1 z + (1 - 2(s_0 + s_1))z^2 + s_1 z^3 + s_0 z^4), \tag{2.12.30}$$

$$\widetilde{P}(z) = \left(\frac{1+z}{2}\right)^4 z(\tilde{s}_0 + \tilde{s}_1 z + \tilde{s}_0 z^2), \tag{2.12.31}$$

the identity (2.12.22) leads to the following exact and real-valued solution

$$
\begin{aligned}
s_0 &= \frac{5}{24} + \frac{\sqrt{15} - 5}{48}\alpha + \frac{2\sqrt{15} - 5}{336}\alpha^2, \\
s_1 &= -\frac{3}{2} + \frac{6 - \sqrt{15}}{12}\alpha - \frac{\sqrt{15} - 3}{24}\alpha^2, \\
\tilde{s}_0 &= -\frac{1}{3} - \frac{1}{12}\alpha + \frac{3\sqrt{15} - 11}{168}\alpha^2, \\
\tilde{s}_1 &= \frac{5}{3} + \frac{1}{6}\alpha - \frac{3\sqrt{15} - 11}{84}\alpha^2,
\end{aligned} \tag{2.12.32}
$$

where α is defined by

$$\alpha = \sqrt[3]{154 + 42\sqrt{15}}. \tag{2.12.33}$$

To be more precise and explicit, if we write

$$P(z) = \frac{1}{2}\sum_{k=0}^{8} p_k z^k, \quad Q(z) = \frac{1}{2}\sum_{k=0}^{6} q_k z^k,$$

$$\widetilde{P}(z) = \frac{1}{2}\sum_{k=0}^{7} \tilde{p}_k z^k, \quad \widetilde{Q}(z) = \frac{1}{2}\sum_{k=-1}^{7} \tilde{q}_k z^k,$$

it follows from (2.12.30)-(2.12.33) and (2.12.21) that $\{p_k\}_{k=0}^{8}$ is given by

$$
\begin{aligned}
p_0 &= \frac{5}{192} - \frac{5 - \sqrt{15}}{384}\alpha + \frac{2\sqrt{15} - 5}{2688}\alpha^2, \\
p_1 &= -\frac{1}{12} + \frac{1}{96}\alpha - \frac{3\sqrt{15} - 11}{1344}\alpha^2, \\
p_2 &= -\frac{7}{48} + \frac{7 - \sqrt{15}}{96}\alpha + \frac{16 - 5\sqrt{15}}{672}\alpha^2, \\
p_3 &= \frac{7}{12} - \frac{1}{96}\alpha - \frac{11 - 3\sqrt{15}}{1344}\alpha^2, \\
p_4 &= \frac{119}{96} + \frac{3\sqrt{15} - 23}{192}\alpha + \frac{18\sqrt{15} - 59}{1344}\alpha^2, \\
p_5 &= p_3, \quad p_6 = p_2, \quad p_7 = p_1, \quad p_8 = p_0;
\end{aligned} \tag{2.12.34}
$$

$\{q_k\}_{k=0}^{6}$ is given by

$$q_0 = \frac{1}{24} + \frac{1}{96}\alpha - \frac{3\sqrt{15} - 11}{1344}\alpha^2,$$

$$q_1 = \frac{1}{24} - \frac{1}{48}\alpha + \frac{3\sqrt{15} - 11}{672}\alpha^2,$$

$$q_2 = -\frac{13}{24} - \frac{1}{96}\alpha + \frac{3\sqrt{15} - 11}{1344}\alpha^2,$$

$$q_3 = \frac{11}{12} + \frac{1}{24}\alpha - \frac{3\sqrt{15} - 11}{336}\alpha^2,$$

$$q_4 = q_2, \quad q_5 = q_1, \quad q_6 = q_0; \tag{2.12.35}$$

$\{\widetilde{p}_k\}_{k=0}^7$ is given by

$$\widetilde{p}_0 = 0,$$

$$\widetilde{p}_1 = -q_0 = -\frac{1}{24} - \frac{1}{96}\alpha + \frac{3\sqrt{15} - 11}{1344}\alpha^2,$$

$$\widetilde{p}_2 = q_1 = \frac{1}{24} - \frac{1}{48}\alpha + \frac{3\sqrt{15} - 11}{672}\alpha^2,$$

$$\widetilde{p}_3 = -q_2 = \frac{13}{24} + \frac{1}{96}\alpha - \frac{3\sqrt{15} - 11}{1344}\alpha^2,$$

$$\widetilde{p}_4 = q_3 = \frac{11}{12} + \frac{1}{24}\alpha - \frac{3\sqrt{15} - 11}{336}\alpha^2,$$

$$\widetilde{p}_5 = -q_4 = \widetilde{p}_3, \quad \widetilde{p}_6 = q_5 = \widetilde{p}_2, \quad \widetilde{p}_7 = -q_6 = \widetilde{p}_1;$$

and $\{\widetilde{q}_k\}_{k=-1}^7$ is given by

$$\widetilde{q}_{-1} = p_0 = \frac{5}{19} - \frac{5 - \sqrt{15}}{384}\alpha + \frac{2\sqrt{15} - 5}{2688}\alpha^2,$$

$$\widetilde{q}_0 = -p_1 = \frac{1}{12} - \frac{1}{96}\alpha + \frac{3\sqrt{15} - 11}{1344}\alpha^2,$$

$$\widetilde{q}_1 = p_2 = -\frac{7}{48} + \frac{7 - \sqrt{15}}{96}\alpha + \frac{16 - 5\sqrt{15}}{672}\alpha^2,$$

$$\widetilde{q}_2 = -p_1 = -\frac{7}{12} + \frac{1}{96}\alpha - \frac{3\sqrt{15} - 11}{1344}\alpha^2,$$

$$\widetilde{q}_3 = p_2 = \frac{119}{96} - \frac{23 - 3\sqrt{15}}{192}\alpha + \frac{18\sqrt{15} - 59}{1344}\alpha^2,$$

$$\widetilde{q}_4 = -p_5 = \widetilde{q}_2, \quad \widetilde{q}_5 = p_6 = \widetilde{q}_1, \quad \widetilde{q}_6 = -p_7 = \widetilde{q}_0, \quad \widetilde{q}_7 = p_8 = \widetilde{q}_{-1}.$$

The length-9 lowpass filter in (2.12.34) and the length-7 highpass filter in (2.12.35) are induced from the CDF 9/7 wavelets.

To plot ϕ, ψ, $\widetilde{\phi}$, and $\widetilde{\psi}$, the values of ϕ and $\widetilde{\phi}$ at integers inside their supports are useful. To this end, we list them in the following.

$$\phi(1) = -\frac{1}{288} + \frac{9\sqrt{15}-5}{40320}\alpha + \frac{25-6\sqrt{15}}{40320}\alpha^2.$$

$$\phi(2) = -\frac{23}{720} + \frac{141\sqrt{15}-587}{20160}\alpha + \frac{15\sqrt{15}-8}{20160}\alpha^2.$$

$$\phi(3) = \frac{149}{1440} + \frac{5861-1209\sqrt{15}}{40320}\alpha + \frac{719-306\sqrt{15}}{40320}\alpha^2.$$

$$\phi(4) = \frac{311}{360} - \frac{2341-459\sqrt{15}}{10080}\alpha + \frac{141\sqrt{15}-364}{10080}\alpha^2.$$

$$\phi(5) = \phi(3), \quad \phi(6) = \phi(2), \quad \phi(7) = \phi(1):$$

$$\tilde{\phi}(2) = -\frac{97}{9360} + \frac{21\sqrt{15}-71}{18720}\alpha + \frac{82-9\sqrt{15}}{131040}\alpha^2.$$

$$\tilde{\phi}(3) = \frac{227}{2340} - \frac{59+21\sqrt{15}}{4680}\alpha + \frac{204\sqrt{15}-797}{32760}\alpha^2.$$

$$\tilde{\phi}(4) = \frac{3869}{4680} + \frac{307+63\sqrt{15}}{9360}\alpha + \frac{3106-807\sqrt{15}}{65520}\alpha^2.$$

$$\tilde{\phi}(5) = \tilde{\phi}(3), \quad \tilde{\phi}(6) = \tilde{\phi}(2).$$

Here the normalization conditions $\sum_{k\in\mathbb{Z}} \phi(k) = \sum_{k\in\mathbb{Z}} \tilde{\phi}(k) = 1$ have been used. The decimal values of the four filters are listed in Table 2.12.1, while those values of ϕ and $\tilde{\phi}$ at integers (inside their supports) are included in Table 2.12.2.

Table 2.12.1. Numerical values of $\{p_k\}_{k=0}^{8}$, $\{q_k\}_{k=0}^{6}$, $\{\tilde{p}_k\}_{k=0}^{7}$,
and $\{\tilde{q}_k\}_{k=-1}^{7}$ for the CDF 9/7 .

p_0	.0534975148216202	q_0	.0912717631143501
p_1	−.0337282368857499	q_1	−.0575435262285002
p_2	−.1564465330579805	q_2	−.5912717631142501
p_3	.5337282368857499	q_3	1.1150870524570004
p_4	1.2058980364727207	q_4	−.5912717631142501
p_5	.5337282368857499	q_5	−.0575435262285002
p_6	−.1564465330579805	q_6	.0912717631142501
p_7	−.0337282368857499		
p_8	.0534975148216202		
		$\tilde{q}_{-1}$	.0534975148216202
$\tilde{p}_0$	.0	$\tilde{q}_0$	.0337282368857499
$\tilde{p}_1$	−.0912717631142501	$\tilde{q}_1$	−.1564465330579805
$\tilde{p}_2$	−.0575435262285002	$\tilde{q}_2$	−.5337282368857499
$\tilde{p}_3$	.5912717631142501	$\tilde{q}_3$	1.2058980364727207
$\tilde{p}_4$	1.1150870524570004	$\tilde{q}_4$	−.5337282368857499
$\tilde{p}_5$	.5912717631142501	$\tilde{q}_5$	−.1564465330579805
$\tilde{p}_6$	−.0575435262285002	$\tilde{q}_6$	.0337282368857499
$\tilde{p}_7$	−.0912717631142501	$\tilde{q}_7$	.0534975148216202

Table 2.12.2. Numerical values of ϕ and $\tilde\phi$ at integers inside their supports

$\phi(1)$	.0036054310908294		
$\phi(2)$	.0696674590803585	$\tilde\phi(2)$	.0101129804661569
$\phi(3)$	$-.2343940145008981$	$\tilde\phi(3)$	$-.1171766235026279$
$\phi(4)$	1.3222422486594204	$\tilde\phi(4)$	1.2141272860729420
$\phi(5)$	$-.2343940145008981$	$\tilde\phi(5)$	$-.1171766235026279$
$\phi(6)$	.0696674590803585	$\tilde\phi(6)$	.0101129804661569
$\phi(7)$	.0036054310908294		

Finally, we plot ϕ, ψ, $\tilde\phi$, and $\tilde\psi$ in Figure 2.12.4.

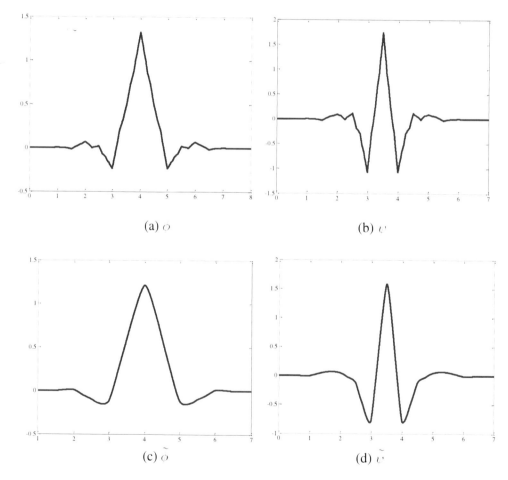

(a) ϕ (b) ψ

(c) $\tilde\phi$ (d) $\tilde\psi$

Figure 2.12.4. Graphs of the CDF 9/7 wavelet pairs $\{\phi, \psi\}$ and $\{\tilde\phi, \tilde\psi\}$. (a) ϕ; (b) ψ; (c) $\tilde\phi$; and (d) $\tilde\psi$.

EXERCISES

2.1. Vector Spaces and Orthogonality

2.1.1. Show that the set of functions

$$1, \sin x, \cos x, \sin 2x, \cos 2x, \ldots,$$

form an orthogonal set in the interval $(-\pi, \pi)$.

2.1.2. Determine the normalizing constant for the set in Problem 2.1.1 so that it is orthonormal in $(-\pi, \pi)$.

2.1.3. Show that the set of functions

$$1, \sin \frac{\pi x}{L}, \cos \frac{\pi x}{L}, \sin \frac{2\pi x}{L}, \cos \frac{2\pi x}{L}, \ldots$$

form an orthogonal set in the interval $(-L, L)$.

2.1.4. Determine the normalizing constant for the set in Problem 2.1.3 so that it is orthonormal in $(-L, L)$.

2.1.5. Show that the "taxicab norm"

$$\|x\| \equiv |x_1| + |x_2| + \cdots + |x_n| = \sum_{k=1}^{n} |x_k|$$

satisfies the requirements of a norm.

2.2. Fourier Series and its Convergent Behavior

2.2.1. Prove

$$\int_{-L}^{L} \sin \frac{k\pi x}{L} \, dx = \int_{-L}^{L} \cos \frac{k\pi x}{L} \, dx = 0, \quad k = 1, 2, \cdots$$

2.2.2. Prove

$$\int_{-L}^{L} \sin \frac{m\pi x}{L} \cos \frac{n\pi x}{L} \, dx = 0, \quad m, n = 1, 2, \cdots$$

2.2.3. Prove

$$\int_{-L}^{L} \sin \frac{m\pi x}{L} \sin \frac{n\pi x}{L} \, dx = \int_{-L}^{L} \cos \frac{m\pi x}{L} \cos \frac{n\pi x}{L} \, dx = L\delta_{mn},$$

$$m, n = 1, 2, \cdots .$$

2.2.4. Expand $f(x) = x$, $-\pi < x < \pi$, in a Fourier series.

2.2.5. (a) Graph $g(x)$ as a 2π- periodic function:

$$g(x) = \begin{cases} -1 & \text{if } -\pi < x < 0, \\ 1 & \text{if } 0 < x < \pi. \end{cases}$$

(b) Find the Fourier coefficients corresponding to the function $g(x)$.

(c) Write the corresponding Fourier series.

(d) How should $g(x)$ be defined at $x = -\pi$, $x = 0$ and $x = \pi$ in order that the Fourier series will converge to $g(x)$ for $-\pi \le x \le \pi$?

2.2.6.

(a) Graph $f(x)$ as a $2L$-periodic function:

$$f(x) = \begin{cases} 0 & \text{if } -3 < x < 0, \\ 5 & \text{if } 0 < x < 3, \end{cases} \quad L = 3.$$

(b) Find the Fourier coefficients corresponding to the function $f(x)$.

(c) Write the corresponding Fourier series.

(d) How should $f(x)$ be defined at $x = -3$, $x = 0$ and $x = 3$ in order that the Fourier series will converge to $f(x)$ for $-3 \le x \le 3$?

2.2.7. Expand

$$f(x) = \begin{cases} -x & \text{if } -\pi < x < 0, \\ x & \text{if } 0 < x < \pi, \end{cases}$$

in a Fourier series.

2.2.8. Expand $f(x) = x^2$, $-\pi < x < \pi$, in a Fourier series.

2.2.9. Expand

$$f(x) = \begin{cases} 1 - x & \text{if } -\pi < x < 0, \\ x - 2 & \text{if } 0 < x < \pi, \end{cases}$$

in a Fourier series.

2.2.10. Expand $f(x) = |\sin x|$, $-\pi < x < \pi$ in a Fourier series.

2.3. Fourier Cosine and Sine Series and Half-Range Expansions

2.3.1. Show that an even function has no sine terms in its Fourier expansion.

2.3.2. Show that an odd function has no cosine terms in its Fourier expansion.

2.3.3. Expand $f(x) = \sin x$, $0 < x < \pi$, in a Fourier cosine series.

2.3.4. (a) Expand $f(x) = \cos x$, $0 < x < \pi$, in a Fourier sine series.
        (b) How should $f(x)$ be defined at $x = 0$ and $x = \pi$ so that the series will converge to $f(x)$ for $0 \le x \le \pi$?

2.3.5. Expand
$$f(x) = \begin{cases} 0 & \text{if } 0 < x < \dfrac{\pi}{2}, \\ 1 & \text{if } \dfrac{\pi}{2} < x < \pi, \end{cases}$$
        in a half-range
        (a) Sine series. (b) Cosine series.

2.3.6. Expand $g(x) = x$, $0 < x < 8$, in a half-range
        (a) Sine series. (b) Cosine series.

2.3.7. Expand
$$f(x) = \begin{cases} x & \text{if } 0 < x < 2, \\ 4 - x & \text{if } 2 < x < 4, \end{cases}$$
        in a half-range
        (a) Sine series. (b) Cosine series.

2.3.8. Expand $f(x) = x(\pi - x)$, $0 < x < \pi$, in a half-range
        (a) Sine series. (b) Cosine series.

2.4. Fourier Series and PDEs

2.4.1. Solve the heat equation with Dirichlet-type boundary conditions:
$$u_t = 2u_{xx}, \quad 0 < x < 3; \ t > 0,$$
$$u(0, t) = u(3, t) = 0,$$
$$u(x, 0) = 3\sin 6\pi x - 7\sin 8\pi x + 4\sin 12\pi x.$$

2.4.2. Solve the heat equation with Dirichlet-type boundary conditions:
$$u_t = u_{xx}, \quad 0 < x < 4; \ t > 0,$$
$$u(0, t) = u(4, t) = 0,$$
$$u(x, 0) = 3\sin 5x - 7\sin 7x.$$

2.4.3. Solve the heat equation with Neumann-type boundary conditions:

$$u_t = u_{xx}, \quad 0 < x < 4; \ t > 0,$$
$$u_x(0, t) = u_x(4, t) = 0,$$
$$u(x, 0) = 3 \sin 5x - 7 \sin 7x.$$

2.4.4. Solve the heat equation with Neumann-type boundary conditions:

$$u_t = u_{xx}, \quad 0 < x < 6; \ t > 0,$$
$$u_x(0, t) = u_x(6, t) = 0,$$
$$u(x, 0) = x + 1.$$

2.4.5. Solve the heat equation with Dirichlet-type boundary conditions:

$$u_t = u_{xx}, \quad 0 < x < 4, \ t > 0,$$
$$u(0, t) = u(4, t) = 0,$$
$$u(x, 0) = \begin{cases} 1 & \text{if } 0 < x < 2 \\ 2 & \text{if } 2 < x < 4 \end{cases}$$

2.5. Fourier Transform and Inverse Fourier Transform

2.5.1. (a) Find the Fourier transform $\widehat{f}(\omega)$ of the function

$$f(x) = \begin{cases} 1 & \text{if } |x| < 1, \\ 0 & \text{if } |x| > 1. \end{cases}$$

(b) Graph $f(x)$ and $\widehat{f}(\omega)$.

2.5.2. (a) Find the Fourier transform $\widehat{f}(\omega)$ of the function

$$f(x) = \begin{cases} 1 & \text{if } |x| < 3, \\ 0 & \text{if } |x| > 3. \end{cases}$$

(b) Graph $f(x)$ and $\widehat{f}(\omega)$.

2.5.3. Find the Fourier transform $\widehat{f}(\omega)$ of the function $f(x) = e^{-x}, x > 0$.

2.5.4. Find the Fourier transform $\widehat{f}(\omega)$ of the function $f(x) = e^{-|x|}$.

2.6. Properties of Fourier Transform and Convolution Theorem

2.6.1. Verify the convolution theorem for the functions:

$$f(x) = g(x) = \begin{cases} 1 & \text{if } |x| < 1, \\ 0 & \text{if } |x| > 1. \end{cases}$$

2.6.2. Verify the convolution theorem for the functions:

$$f(x) = g(x) = e^{-4x^2}.$$

2.8. Classical Haar Scaling Function and Haar Wavelets

2.8.1. Plot the functions $(\phi_{\text{Haar}})_{2,4}(t)$ and $(\psi_{\text{Haar}})_{2,4}(t)$.

2.8.2. Show (2.8.6) and (2.8.7) by graphs.

2.9. Daubechies Orthonormal Scaling Functions and Wavelets

2.9.1. For the two-scale sequences $\{(p_4^D)_k\}_{k=0}^3$ and $\{(q_4^D)_k\}_{k=0}^3$ of ϕ_4^D and ψ_4^D, given by (2.9.7), show that

$$\sum_{k\in\mathbb{Z}} p_k\, p_{k+2\ell} = 2\delta_{\ell,0},$$

$$\sum_{k\in\mathbb{Z}} p_k\, q_{k+2\ell} = 0,$$

$$\sum_{k\in\mathbb{Z}} q_k\, q_{k+2\ell} = 2\delta_{\ell,0}, \quad \ell \in \mathbb{Z},$$

provide that $p_k = (p_4^D)_k$ and $q_k = (q_4^D)_k$ with $p_k = q_k = 0$ for either $k < 0$ or $k \geq 4$.

2.9.2. For the two-scale sequences $\{(p_6^D)_k\}_{k=0}^3$ of ϕ_6^D, given by (2.9.8), and $\{(q_6^D)_k\}_{k=0}^3$ of ψ_6^D, determined by (2.9.3) with $m = 3$, show that

$$\sum_{k\in\mathbb{Z}} p_k\, p_{k+2\ell} = 2\delta_{\ell,0},$$

$$\sum_{k\in\mathbb{Z}} p_k\, q_{k+2\ell} = 0,$$

$$\sum_{k\in\mathbb{Z}} q_k\, q_{k+2\ell} = 2\delta_{\ell,0}, \quad \ell \in \mathbb{Z},$$

provide that $p_k = (p_6^D)_k$ and $q_k = (q_6^D)_k$ with $p_k = q_k = 0$ for either $k < 0$ or $k \geq 6$.

2.12. Other Wavelets

2.12.1. Find the piecewise expressions of N_m for $m = 2, 3$ and 4 by using (2.12.2).

2.12.2. Show by graph that N_m for $m = 2, 3$ and 4 satisfy (2.12.3).

2.12.3. Plot the Morlet wavelets ψ_γ in (2.12.10) for $\gamma = 4, 5$, and 6.

2.12.4. Plot the Morlet wavelets ψ_M in (2.12.12)-(2.12.13).

2.12.5. Plot the Gaussian wavelets $\psi_{G,m}$ in (2.12.15) for $m = 3, 4$, and 5. Show that $\psi_{G,3}$, $\psi_{G,4}$, and $\psi_{G,5}$ have 3, 4, and 5 vanishing moments, respectively.

2.12.6. Find the biorthogonal wavelet pairs $\{\phi, \psi\}$ and $\{\widetilde{\phi}, \widetilde{\psi}\}$, with two-scale symbols P, $\widetilde{P}$, Q, and $\widetilde{Q}$ in (2.12.23), (2.12.24), and (2.12.21), with $n = \tilde{n} = 3$ in (2.12.23)-(2.12.24).

Chapter 3

LAPLACE TRANSFORM

Laplace transform has been a powerful method for solving linear ODEs, their corresponding initial-value problems, and systems of linear ODEs. To obtain functions from their Laplace transforms, the inverse Laplace transform has to be introduced at the same time.

3.1. DEFINITIONS OF LAPLACE TRANSFORM AND INVERSE LAPLACE TRANSFORM

Definition 3.1.1.

The *Laplace transform* is an integral operator to a function $f(t)$ on $[0, \infty)$, denoted by $F(s)$ or $\mathcal{L}(f)(s)$, and is defined by

$$F(s) = \mathcal{L}(f)(s) = \mathcal{L}\{f(t)\}(s) = \int_0^\infty e^{-st} f(t) dt. \tag{3.1.1}$$

The *inverse Laplace transform* of $F(s)$, denoted by $f(t)$ or $\mathcal{L}^{-1}(F)(t)$, is defined by

$$f(t) = \mathcal{L}^{-1}(F)(t) = \mathcal{L}^{-1}\{F(s)\}(t), \tag{3.1.2}$$

if the function $F(s)$ is determined from the function $f(t)$ in (3.1.1). The Laplace and inverse Laplace transforms are named after P. Laplace[1].

---

[1] *Pierre-Simon, marquis de Laplace*, March 23, 1749 – March 5, 1827, was a French mathematician and astronomer, and was born in Beaumont-en-Auge, Normandy. He formulated Laplace's equation, and pioneered the Laplace transform which appears in many branches of mathematical physics, a field that he took a leading role in forming. The Laplacian differential operator, widely used in applied mathematics, is also named after him. He is remembered as one of the greatest scientists of all time, sometimes referred to as a French Newton or Newton of France, with a phenomenal natural mathematical faculty superior to any of his contemporaries (http://en.wikipedia.org/wiki/Pierre-Simon_Laplace).

First of all, for convenience, $\mathcal{L}\{f(t)\}$ will be used for $\mathcal{L}\{f(t)\}(s)$ in (3.1.1), and $\mathcal{L}^{-1}\{F(s)\}$ will also be used for $\mathcal{L}^{-1}\{F(s)\}(t)$ in (3.1.2). Secondly, Laplace transform is a linear operation, meaning

$$\mathcal{L}\{af(t) + bg(t)\} = a\mathcal{L}\{f(t)\} + b\mathcal{L}\{g(t)\}$$

for any constants a and b, provided that both $\mathcal{L}\{f(t)\}$ and $\mathcal{L}\{g(t)\}$ exist. Thirdly, as a simple example, for $f(t) = e^{at}, t \geq 0$, with a being a constant,

$$\mathcal{L}(e^{at}) = \int_0^\infty e^{-st} e^{at} dt = \frac{1}{s-a}, \quad s > a. \tag{3.1.3}$$

For monomials t^n, it is easy to see that

$$\mathcal{L}(t^n) = \int_0^\infty e^{-st} t^n dt = -\frac{1}{s} t^n e^{-st} \Big|_{t=0}^{t=\infty} + \frac{n}{s} \int_0^\infty e^{-st} t^{n-1} dt = \frac{n}{s}\mathcal{L}(t^{n-1}). \tag{3.1.4}$$

It then follows immediately from (3.1.3) and (3.1.4) that $\mathcal{L}(1) = 1/s$ and

$$\mathcal{L}^{-1}\left(\frac{n!}{s^{n+1}}\right) = t^n, \quad n = 0, 1, \cdots. \tag{3.1.5}$$

Finally, in practical applications, we do not often actually evaluate integrals to get either the Laplace transform or inverse Laplace transform. Instead, we rely on their properties as well as a table that includes Laplace transforms of commonly used elementary functions. Table 3.1.1 is the first of such a table. A more comprehensive table will be given in Section 3.10.

Table 3.1.1. Laplace Transforms of 8 Elementary Functions

	$f(t)$	$\mathcal{L}(f)$		$f(t)$	$\mathcal{L}(f)$
1	$\dfrac{t^a}{\Gamma(a+1)}, \ a \geq 0$	$\dfrac{1}{s^{a+1}}$	2	$e^{at}, \ a \geq 0$	$\dfrac{1}{s-a}, \ s > a$
3	$\cos\omega t$	$\dfrac{s}{s^2 + \omega^2}$	4	$\sin\omega t$	$\dfrac{\omega}{s^2 + \omega^2}$
5	$\cosh at$	$\dfrac{s}{s^2 - a^2}$	6	$\sinh at$	$\dfrac{a}{s^2 - a^2}$
7	$e^{at}\cos\omega t$	$\dfrac{s-a}{(s-a)^2 + \omega^2}$	8	$e^{at}\sin\omega t$	$\dfrac{\omega}{(s-a)^2 + \omega^2}$

Regarding the generic existence of the Laplace transform for a given function $f(t)$, a sufficient condition is that the function $f(t)$ does not grow too fast.

Theorem 3.1.1. *(Existence of Laplace Transform)*

Let $f(t)$ be a function that (1) it is defined on $[0, \infty)$; (2) it is piecewise continuous on $[0, \infty)$; and (3) it satisfies

$$|f(t)| \le Me^{kt}, \quad t \in [0, \infty), \tag{3.1.6}$$

for some constants M and k, with $M > 0$. Then $f(t)$'s Laplace transform $\mathcal{L}(f)$ exists for $s > k$.

Proof

The piecewise continuity of $f(t)$ implies that $e^{-st}f(t)$ is integrable on $[0, \infty)$. Therefore,

$$|\mathcal{L}(f)| = \left| \int_0^\infty e^{-st} f(t) dt \right| \le \int_0^\infty e^{-st}|f(t)|dt \le \int_0^\infty e^{-st} \cdot Me^{kt} dt = \frac{M}{s-k}, \quad s > k.$$

Example 3.1.1.

Find the Laplace transform of both $\cos \omega t$ and $\sin \omega t$.

Solution

One quick way of getting the Laplace transforms for both $\cos \omega t$ and $\sin \omega t$ is to use the Euler formula $e^{i\omega t} = \cos \omega t + i \sin \omega t$:

$$\mathcal{L}(e^{i\omega t}) = \frac{1}{s - i\omega} = \frac{s + i\omega}{(s - i\omega)(s + i\omega)} = \frac{s + i\omega}{s^2 + \omega^2} = \frac{s}{s^2 + \omega^2} + i\frac{\omega}{s^2 + \omega^2},$$

so that

$$\mathcal{L}(\cos \omega t) = \frac{s}{s^2 + \omega^2}, \quad \mathcal{L}(\sin \omega t) = \frac{\omega}{s^2 + \omega^2}.$$

Example 3.1.2.

Find the inverse Laplace transform of $\dfrac{s - 21}{s^2 - 2s - 15}$.

Solution

To get the inverse Laplace transform of the given rational function, we can use its partial fraction form. Due to the fact that

$$\frac{s-21}{s^2-2s-15}=\frac{3}{s+3}-\frac{2}{s-5},$$

it is easy to see that

$$\mathcal{L}^{-1}\left\{\frac{s-21}{s^2-2s-15}\right\}=3\mathcal{L}^{-1}\left\{\frac{1}{s+3}\right\}-2\mathcal{L}^{-1}\left\{\frac{1}{s-5}\right\}=2e^{-3t}-2e^{5t}.$$

Example 3.1.3.

Find the inverse Laplace transform of $\dfrac{s^2+6+2s^3+8s}{s^4+5s^2+4}$.

Solution

Again, it follows from its partial fraction

$$\frac{s^2+6+2s^3+8s}{s^4+5s^2+4}=\frac{5}{3(s^2+1)}+\frac{2s}{s^2+1}-\frac{2}{3(s^2+4)}$$

that

$$\mathcal{L}^{-1}\left\{\frac{s^2+6+2s^3+8s}{s^4+5s^2+4}\right\}=\frac{5}{3}\sin t+2\cos t-\frac{1}{3}\sin 2t.$$

3.2. FIRST SHIFTING THEOREM

From the definition of Laplace transform in (3.1.1), it is straightforward that a shift of s corresponds multiplying $f(t)$ by e^{at}. This is exactly the first shifting or s-shifting theorem of the Laplace transform.

Theorem 3.2.1. (*First Shifting or s-Shifting Theorem*)

Let $F(s)$, $s>k$, be the Laplace transform of $f(t)$. Then

$$\mathcal{L}\left\{e^{at}f(t)\right\}=F(s-a),\quad s-a>k;\tag{3.2.1}$$

or equivalently,

$$\mathcal{L}^{-1}\left\{F(s-a)\right\}=e^{at}f(t).\tag{3.2.2}$$

Proof

By using definition (3.1.1),

$$\mathcal{L}\left\{e^{at}f(t)\right\} = \int_0^\infty e^{-st}e^{at}f(t)\,dt = \int_0^\infty e^{-(s-a)t}f(t)\,dt = F(s-a).$$

The Laplace transforms of $e^{at}\cos\omega t$ and $e^{at}\sin\omega t$ in Table 3.1.1 were indeed obtained by using the first shifting theorem and the Laplace transforms of $\cos\omega t$ and $\sin\omega t$, respectively.

3.3. LAPLACE TRANSFORM OF DERIVATIVES

Direct calculation leads us to

$$\mathcal{L}(f') = \int_0^\infty e^{-st}f'(t)dt = e^{-st}f(t)\Big|_{t=0}^{t=\infty} + s\int_0^\infty e^{-st}f(t)dt = s\mathcal{L}(f) - f(0).$$

Hence, we have the following.

Theorem 3.3.1. *(Laplace Transform of First Order Derivative) .*

If $f(t)$ satisfies all three conditions in Theorem 3.1.1, and $f'(t)$ is piecewise continuous on $[0, \infty)$, then

$$\mathcal{L}(f') = s\mathcal{L}(f) - f(0). \tag{3.3.1}$$

Proof

It follows from (3.1.1) that

$$\mathcal{L}\left\{f'(\cdot)\right\} = \int_0^\infty e^{-st}f'(t)\,dt$$
$$= e^{-st}f(t)\Big|_0^{t=-\infty} + s\int_0^\infty e^{-st}f(t)\,dt$$
$$= sF(s) - f(0) = s\mathcal{L}(f) - f(0),$$

under the assumption (3.1.6).

Laplace transforms of higher order derivatives are natural consequences of Theorem 3.3.1.

Theorem 3.3.2. (Laplace Transform of High Order Derivatives)

If $f(t)$ and its all up to $(n-1)^{\text{st}}$ order derivatives $f'(t), \cdots, f^{(n-1)}(t)$ satisfy all three conditions in Theorem 3.1.1, and $f^{(n)}(t)$ is piecewise continuous on $[0, \infty)$, then

$$\mathcal{L}(f^{(n)}) = s^n \mathcal{L}(f) - s^{n-1} f(0) - s^{n-2} f'(0) - \cdots - s f^{(n-2)}(0) - f^{(n-1)}(0).$$

$$(3.3.2)$$

Proof:

Analogous to the proof of Theorem 3.3.1, (3.3.2) can be proved by mathematical induction.

3.4. SOLVING INITIAL-VALUE PROBLEMS BY LAPLACE TRANSFORM

Theorem 3.3.2 can be used to solving initial-value problems for ODE, as shown in the following example.

Example 3.4.1.

Solve the initial-value problem $y'' - y' - 2y = 4e^{3t+6}$, $y(-2) = 2$, $y'(-2) = -1$.

Solution

To use Theorem 3.3.2, we introduce

$$\tilde{t} = t + 2, \quad \tilde{y}(t) = y(t - 2).$$

Then $y(t) = \tilde{y}(\tilde{t})$, $y'(t) = \tilde{y}'(\tilde{t})$, $y''(t) = \tilde{y}''(\tilde{t})$, and $y(-2) = \tilde{y}(0)$, $y'(-2) = \tilde{y}'(0)$. Hence, the ODE becomes $\tilde{y}'' - \tilde{y}' - 2\tilde{y} = 4e^{3\tilde{t}}$, $\tilde{y}(0) = 2$, $\tilde{y}'(0) = -1$. By using

$$\mathcal{L}(\tilde{y}'') = s^2 \mathcal{L}(\tilde{y}) - s\tilde{y}(0) - \tilde{y}'(0), \quad \mathcal{L}(\tilde{y}') = s\mathcal{L}(\tilde{y}) - \tilde{y}(0),$$

$$(3.4.1)$$

and taking the Laplace transforms both sides, we arrive at

$$\mathcal{L}(\tilde{y}'' - \tilde{y}' - 2\tilde{y}) = (s^2 \mathcal{L}(\tilde{y}) - 2s + 1) - (s\mathcal{L}(\tilde{y}) - 2) - 2\mathcal{L}(\tilde{y}) = \frac{4}{s-3},$$

i.e.,

$$\mathcal{L}(\tilde{y}) = \frac{1}{s^2 - s - 2}\left(\frac{4}{s-3} + 2s - 3\right) = \frac{2}{s+1} - \frac{1}{s-2} + \frac{1}{s-3},$$

where $\mathcal{L}(\tilde{y})$ was given in its partial fraction form in the last equality. Finally, by taking the inverse Laplace transform, we have $\tilde{y} = 2e^{-\tilde{t}} - e^{2\tilde{t}} + e^{3\tilde{t}}$ so that $y = 2e^{-t-2} - e^{2t+4} + e^{3t+6}$.

3.5. HEAVISIDE FUNCTION AND SECOND SHIFTING THEOREM

One of the important functions in engineering is the Heaviside function. It is also sometimes called the unit step function. It was named after O. Heaviside[2].

Definition 3.5.1.

The *Heaviside function* or *unit step function*, denoted by $u(t - a)$, is defined by

$$u(t - a) = \begin{cases} 0, & \text{if } t < a; \\ 1, & \text{if } t > a. \end{cases} \tag{3.5.1}$$

The value of $u(0)$ is not important though. One application of the Heaviside function is for the on and off switch in electrical and mechanical engineering. Another application is to express piecewise functions. For example, the hat function can be simply written as

$$f(t) = \begin{cases} t, & \text{if } 0 \le t < 1 \\ 2 - t, & \text{if } 1 \le t < 2 = tu(t) - 2(t-1)u(t-1) + (t-2)u(t-2). \\ 0, & \text{otherwise} \end{cases}$$

In other words, $f(t)[u(t - a) - u(t - b)]$ truncates the function $f(t)$ to its restriction on the interval (a, b). Hence, functions on $[0, \infty)$ with only jump discontinuous can always be written in terms of the shifts of the Heaviside function. It is also easy to evaluate its Laplace transform:

$$\mathcal{L}\{u(t - a)\} = \frac{e^{-as}}{s}, \quad s > 0. \tag{3.5.2}$$

We then have the following.

---

[2] *Oliver Heaviside*, May 18, 1850 – February 3, 1925, was a English mathematician and engineer, and was born in Camden Town, London. It was Heaviside who changed the face of mathematics and science for years to come. http://en.wikipedia.org/wiki/Oliver_Heaviside.

Theorem 3.5.1. *(The Second Shifting or t-Shifting Theorem)*

Let $F(s)$ be the Laplace transform of a function $f(t)$. Then

$$\mathcal{L}\{f(t-a)u(t-a)\} = e^{-as}F(s), \tag{3.5.3}$$

or equivalently,

$$\mathcal{L}^{-1}\{e^{-as}F(s)\} = f(t-a)u(t-a). \tag{3.5.4}$$

Example 3.5.1.

Find the Laplace transform of $1 - e^{-3t}$ $(0 < t < 2)$.

Solution
Write the function as

$$1 - e^{-3t}(0 < t < 2) = (1 - e^{-3t})(1 - u(t-2)) = 1 - e^{-3t} - u(t-2) + e^{-6}e^{-3(t-2)}u(t-2).$$

Then,

$$\mathcal{L}\{1 - e^{-3t}(0 < t < 2)\} = \frac{1}{s} - \frac{1}{s+3} - \frac{e^{-2s}}{s} + \frac{e^{-2(s+3)}}{s+3}.$$

Example 3.5.2.

Find the inverse Laplace transform of $\dfrac{2(1 - e^{-\pi s})}{s^2 + 25}$.

Solution
Write

$$\frac{2(1 - e^{-\pi s})}{s^2 + 25} = 2 \cdot \frac{1}{5} \cdot \frac{5}{s^2 + 25} - 2 \cdot \frac{1}{5}e^{-\pi s} \cdot \frac{5}{s^2 + 25}.$$

Then

$$\mathcal{L}^{-1}\left(\frac{2(1 - e^{-\pi s})}{s^2 + 25}\right) = \frac{2}{5}\sin 5t - \frac{2}{5}u(t-\pi)\sin 5(t-\pi) = \frac{2}{5}\sin 5t + \frac{2}{5}u(t-\pi)\sin 5t.$$

3.6. SOLVING INITIAL-VALUE PROBLEMS
WITH DISCONTINUOUS INPUTS

Due to the special property of the Heaviside function, we can use the Laplace transform to solve initial-value problems with discontinuous inputs.

Example 3.6.1.

Solve the initial-value problem $y'' + 3y' + 2y = 1$, if $t \in (0, 1)$; and 0 if $t \in (1, \infty)$, $y(0) = \alpha$, $y'(0) = \beta$.

Solution

The initial function is the (discontinuous) unit box function, which can be written as $1 - u(t - 1)$. Hence, by taking the Laplace transforms both sides of the ODE yields

$$\mathcal{L}(y) = \frac{1}{(s+1)(s+2)} \left(3\alpha + \beta + \alpha s + \frac{1}{s} - \frac{e^{-s}}{s} \right),$$

i.e.,

$$\mathcal{L}(y) = \frac{3}{2s} + \frac{2\alpha + \beta - 1}{s+1} + \frac{-2(\alpha + \beta) + 1}{2(s+2)} - e^{-s} \left(\frac{1}{2s} - \frac{1}{s+1} + \frac{1}{2(s+2)} \right).$$

Therefore,

$$y = \frac{3}{2} + (2\alpha + \beta - 1)e^{-t} + \frac{-2(\alpha + \beta) + 1}{2}e^{-2t} - u(t - 1) \left(\frac{1}{2} - e^{-(t-1)} + \frac{1}{2}e^{-2(t-1)} \right).$$

Example 3.6.2.

Solve the initial-value problem $y'' - 3y' + 2y = 6e^{-t}$, if $t \in (0, 2)$; and 0 if $t \in (2, \infty)$, $y(0) = 0$, $y'(0) = 1$.

Solution

For the left-hand side,
$$\mathcal{L}(y'' - 3y' + 2y) = (s^2\mathcal{L}(y) - sy(0) - y'(0)) - 3(s\mathcal{L}(y) - y(0)) + 2\mathcal{L}(y)$$
$$= (s-1)(s-2)\mathcal{L}(y) - s + 5;$$

and for the right-hand side,

$$\mathcal{L}(6e^{-t} - 6e^{-t}u(t-2)) = \frac{6}{s+1} - \frac{6e^{-2(s+1)}}{s+1}.$$

Therefore,

$$\mathcal{L}(y) = \frac{s^2 - 4s + 1}{(s+1)(s-1)(s-2)} - e^{-2(s+1)}\frac{6}{(s+1)(s-1)(s-2)}$$

$$= \frac{1}{s+1} + \frac{1}{s-1} - \frac{1}{s-2} - e^{-2(s+1)}\left(\frac{1}{s+1} - \frac{3}{s-1} + \frac{2}{s-2}\right).$$

so that

$$y = e^{-t} + e^t - e^{2t} - e^{-t}u(t-2) + 3e^{t-4}u(t-2) - 2e^{2t-6}u(t-2)$$

$$= \begin{cases} e^{-t} + e^t - e^{2t}, & \text{if } 0 < t < 2, \\ e^t - e^{2t} + 3e^{t-4} - 2e^{2t-6}, & \text{if } t > 2. \end{cases}$$

3.7. SHORT IMPULSE AND DIRAC'S DELTA FUNCTIONS

To describe the action of voltage over a very short period of time, or the phenomena of an *impulsive* nature, we need the Dirac's delta function.

Definition 3.7.1.

The *Dirac's delta function*, denoted by $\delta(t)$, is defined by both

$$\delta(t-a) = \begin{cases} \infty, & \text{if } t = a \\ 0, & \text{otherwise} \end{cases} \tag{3.7.1}$$

and the requirement of

$$\int_0^\infty \delta(t-a)dt = 1. \tag{3.7.2}$$

The impulse value of a function $f(t)$ at a can then be evaluated by

$$f(a) = \int_0^\infty f(t)\delta(t-a)dt. \tag{3.7.3}$$

The Dirac's delta function can be approximated by the sequence of functions $\{f_k(t)\}_{k=1}^\infty$, where

$$f_k(t) = \frac{1}{k}\left[u(t) - u(t-k)\right], \quad k = 1, 2, \cdots$$

It follows from

$$\mathcal{L}\{f_k(t)\} = \frac{1}{ks}\left[1 - e^{-ks}\right] = \frac{1 - e^{-ks}}{ks}, \quad k = 1, 2, \cdots$$

that

$$\mathcal{L}\{\delta(t-a)\} = e^{-as}. \tag{3.7.4}$$

3.8. SOLVING INITIAL-VALUE PROBLEMS WITH IMPULSE INPUTS

Initial-value problems with the input functions being impulsive Dirac's delta functions can now be solved by using the Laplace transform too.

Example 3.8.1.

Solve the initial-value problem $y'' + 5y' + 6y = \delta(t-a) + \cos t$, $y(0) = \alpha$, $y'(0) = \beta$. Here a, α, and β are constants with $a > 0$.

Solution

By using (3.4.1), taking the Laplace transforms both sides, and applying (3.7.4), we have

$$(s^2 + 5s + 6)\mathcal{L}(y) - sy(0) - (y'(0) + 5y(0)) = e^{-as} + \frac{s}{s^2 + 1},$$

which yields

$$\mathcal{L}(y) = \frac{3\alpha + \beta}{s+2} - \frac{2\alpha + \beta}{s+3} + e^{-as}\left(\frac{1}{s+2} - \frac{1}{s+3}\right) - \frac{2}{5}\frac{1}{s+2} + \frac{3}{10}\frac{1}{s+3} + \frac{1}{10}\frac{s+1}{s^2+1}.$$

Therefore,

$$y = \left(-\frac{2}{5} + 3\alpha + \beta\right)e^{-2t} + \left(\frac{3}{10} - 2\alpha - \beta\right)e^{-3t}$$
$$+ \frac{\cos t + \sin t}{10} + \left(e^{-2(t-a)} - e^{-3(t-a)}\right)u(t-a).$$

3.9. APPLICATION OF LAPLACE TRANSFORM
TO ELECTRIC CIRCUITS

Laplace transform can also be applied to find the capacity and/or current of various electric circuits such as capacitor (C), capacity discharge, resistor-capacitor (RC), resistor-inductor (RL), and resistor-inductor-capacitor (RLC) circuits.

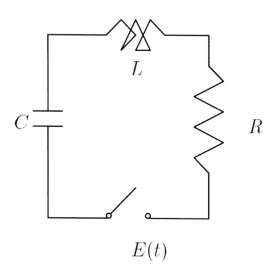

Let Q be the charge on the capacitor at time t, denoted by $Q(t)$. Then the current $I(t)$ of a typical RLC circuit at time t is $I(t) = Q'(t)$ satisfying the ODE

$$L I' + R I + \frac{1}{C} \int I \, dt = E(t), \tag{3.9.1}$$

where L is the inductance, R represents the resistance, C denotes the capacitance, and $E(t)$ is the electromotive force which is normally given by $E_0 \sin \omega t$ with E_0 a constant. Take the derivative of (3.9.1) to get

$$L Q'' + R Q' + \frac{1}{C} Q = E'(t). \tag{3.9.2}$$

With appropriate initial conditions, e.g., the charge on the capacitor and current in the circuit are Q_0 and I_0, i.e., $Q(0) = Q_0$ and $I(0) = Q'(0) = I_0$, the ODE (3.9.2) can be solved by using the Laplace transform, as shown in Section 3.4.

3.10. TABLE OF LAPLACE TRANSFORMS

We summarize in this section the Laplace transform by simply including the Laplace transforms of some commonly used functions in Table 3.10.1.

Table 3.10.1. Laplace Transforms of Some Elementary Functions

	$f(t)$	$\mathcal{L}(f)$		$f(t)$	$\mathcal{L}(f)$
1	$\dfrac{t^{n-1}}{(n-1)!}$, $n=1,2,\cdots$	$\dfrac{1}{s^n}$, $n=1,2,\cdots$	2	$\dfrac{t^{a-1}}{\Gamma(a)}$, $a \geq 0$	$\dfrac{1}{s^a}$, $a>0$
3	$u(t-a)$	$\dfrac{1}{s}e^{-as}$	4	$\delta(t-a)$	e^{-as}
5	e^{at}, $a\geq 0$	$\dfrac{1}{s-a}$, $s>a$	6	$\dfrac{1}{\Gamma(b)}t^{b-1}e^{at}$, $b>0$	$\dfrac{1}{(s-a)^b}$, $b>0$
7	$\dfrac{(ac+d)e^{at}-(bc+d)e^{bt}}{a-b}$	$\dfrac{cs+d}{(s-a)(s-b)}$, $a\neq b$	8	$\dfrac{1}{a-b}(ae^{at}-be^{bt})$	$\dfrac{s}{(s-a)(s-b)}$, $a\neq b$
9	$\dfrac{1}{t}(e^{at}-e^{bt})$	$\ln\dfrac{s-b}{s-a}$	10	$t^a \ln t$, $a>-1$	$\dfrac{\Gamma'(a+1)-\Gamma(a+1)\ln s}{s^{a+1}}$
11	$\cos\omega t$	$\dfrac{s}{s^2+\omega^2}$	12	$\sin\omega t$	$\dfrac{\omega}{s^2+\omega^2}$
13	$t\cos\omega t$	$\dfrac{s^2-\omega^2}{(s^2+\omega^2)^2}$	14	$t\sin\omega t$	$\dfrac{2\omega s}{(s^2+\omega^2)^2}$
15	$t^2\cos\omega t$	$\dfrac{2(s^3-3\omega^2 s)}{(s^2+\omega^2)^3}$	16	$t^2\sin\omega t$	$\dfrac{2\omega(3s^2-\omega^2)}{(s^2+\omega^2)^3}$
17	$t^3\cos\omega t$	$\dfrac{6(s^4-6\omega^2 s^2+\omega^4)}{(s^2+\omega^2)^4}$	18	$t^3\sin\omega t$	$\dfrac{24\omega(s^3-\omega^2 s)}{(s^2+\omega^2)^4}$
19	$e^{at}\cos\omega t$	$\dfrac{s-a}{(s-a)^2+\omega^2}$	20	$e^{at}\sin\omega t$	$\dfrac{\omega}{(s-a)^2+\omega^2}$
21	$\dfrac{1}{\omega^2}(1-\cos\omega t)$	$\dfrac{1}{s(s^2+\omega^2)}$	22	$\dfrac{1}{\omega^3}(\omega t-\sin\omega t)$	$\dfrac{1}{s^2(s^2+\omega^2)}$
23	$\dfrac{1}{2\omega^3}(\sin\omega t-\omega t\cos\omega t)$	$\dfrac{1}{(s^2+\omega^2)^2}$	24	$\dfrac{1}{2\omega}(\sin\omega t+\omega t\cos\omega t)$	$\dfrac{s^2}{(s^2+\omega^2)^2}$
25	$\dfrac{1}{t}\sin\omega t$	$\arctan\dfrac{\omega}{s}$	26	$\dfrac{1}{b^2-a^2}(\cos at-\cos bt)$	$\dfrac{s}{(s^2+a^2)(s^2+b^2)}$
27	$\cosh at$	$\dfrac{s}{s^2-a^2}$	28	$\sinh at$	$\dfrac{a}{s^2-a^2}$
29	$t\cosh at$	$\dfrac{s^2+a^2}{(s^2-a^2)^2}$	30	$t\sinh at$	$\dfrac{2as}{(s^2-a^2)^2}$
31	$t^2\cosh at$	$\dfrac{2(s^3+3a^2 s)}{(s^2-a^2)^3}$	32	$t^2\sinh at$	$\dfrac{2a(3s^2+a^2)}{(s^2-a^2)^3}$
33	$t^3\cosh at$	$\dfrac{6(s^4+6a^2 s^2+a^4)}{(s^2-a^2)^4}$	34	$t^3\sinh at$	$\dfrac{24a(s^3+a^2 s)}{(s^2-a^2)^4}$

EXERCISES

3.1. Definitions of Laplace Transform and Inverse Laplace Transform

For *3.1.1-10*, find the Laplace transform of each function $f(t)$.

3.1.1. $f(t) = \begin{cases} 2, & \text{if } t \in [0, 2] \\ 0, & \text{otherwise.} \end{cases}$

3.1.2. $f(t) = 1 - 2e^{-2t}$.

3.1.3. $f(t) = t^2 e^t + 2te^{-t} + 3\cos 3t$.

3.1.4. $f(t) = \cos^2 3t$.

3.1.5. $f(t) = \left(e^{-2t} - 1\right)^2$.

3.1.6. $f(t) = \sin 2t \cos 2t$.

3.1.7. $f(t) = t^2$.

3.1.8. $f(t) = \cos^3 3t$.

3.1.9. $f(t) = \sin^2 3t$.

3.1.10. $f(t) = \sin(3t - 4)$.

For *3.1.11-20*, find the inverse Laplace transform of each function $F(s)$.

3.1.11. $F(s) = \dfrac{2}{s+1}$.

3.1.12. $F(s) = \dfrac{s+1}{s^2 + 2s}$.

3.1.13. $F(s) = \dfrac{2}{(s-3)^4}$.

3.1.14. $F(s) = \dfrac{2}{s^2 + 3s + 2}$.

3.1.15. $F(s) = \dfrac{1}{(s+2)^3}$.

3.1.16. $F(s) = \dfrac{3s + 5}{s^2 + 7}.$

3.1.17. $F(s) = \dfrac{5}{s^2 - 9}.$

— 3.1.18. $F(s) = \dfrac{s + 1}{s^2(s + 2)^3}.$

3.1.19. $F(s) = \dfrac{3(s + 3)}{s^2 + 6s + 8}.$

3.1.20. $F(s) = \dfrac{s + 8}{s^2 + 4s + 5}.$

3.2. First Shifting Theorem

For *3.2.1-5*, find the Laplace transform of each function $f(t)$ by using the First Shifting Theorem.

3.2.1. $f(t) = t^3 e^{7t}.$

3.2.2. $f(t) = e^{5t} \cos 3t.$

3.2.3. $f(t) = e^{5t} \sin 3t.$

3.2.4. $f(t) = e^{4t} \sinh 3t.$

3.2.5. $f(t) = e^t \cosh 9t.$

For *3.2.6-10*, find the inverse Laplace transform of each function $F(s)$ by using the First Shifting Theorem.

3.2.6. $F(s) = \dfrac{2s + 1}{s^2 + 2s + 4}.$

3.2.7. $F(s) = \dfrac{3s + 7}{s^2 + 9s + 13}.$

3.2.8. $F(s) = \dfrac{1}{s^2 - 2s + 5}.$

3.2.9. $F(s) = \dfrac{s + 3}{s^2 + 6s + 11}.$

3.2.10. $F(s) = \dfrac{1}{s^2 + 2s + 4}$.

3.3. Laplace Transform of Derivatives

For *3.3.1-3*, find the Laplace transform of $f'(t)$, i.e., $\mathcal{L}\{f'(t)\}$, for each of the following function $f(t)$ by using Theorem 3.3.1.

3.3.1. e^{5t}.

3.3.2. $e^{-7t} + 5$.

3.3.3. $t^2 + 3t - 1$.

For 3.3.4-6, find the Laplace transform of $f''(t)$, i.e., $\mathcal{L}\{f''(t)\}$, for each of the following function $f(t)$ by using Theorem 3.3.2.

3.3.4. $\sinh 5t$.

3.3.5. $\cosh 3t + 4t^5$.

3.3.6. $4t^3 - \sin 2t$.

3.4. Solving Initial-Value Problems by Laplace Transform

For *3.4.1-8*, solve the IVP's.

3.4.1. $y'' + 2y' + 5y = 0$, $y(0) = 2$, $y'(0) = -1$.

— 3.4.2. $y'' - 3y' + 2y = 0$, $y(0) = 3$, $y'(0) = 1$.

— 3.4.3. $y'' + 4y' + 4y = 1$, $y(0) = 5$, $y'(0) = -2$.

3.4.4. $y'' + y' = t$, $y(0) = y'(0) = 0$.

— 3.4.5. $y'' + 4y = \cos 3t$, $y(0) = y'(0) = 2$.

— 3.4.6. $y'' + 4y' + 3y = 7$, $y(-3) = 1$, $y'(-3) = 2$.

3.4.7. $y^{(4)} - y = 0$, $y(0) = 1$, $y'(0) = y''(0) = y'''(0) = 0$.

3.4.8. $y'' - 4y' - 5y = 2 + e^{-t}, y(0) = y'(0) = 0.$

For *3.4.9-10*, solve the systems of differential equations by Laplace transform.

3.4.9. $\begin{cases} x' + y = 0, \\ y' - 2x - 2y = 0, \end{cases} x(0) = y(0) = 2.$

3.4.10. $\begin{cases} x' = y - z, \\ y' = x + y, \quad x(0) = 2, y(0) = 3, z(0) = 4. \\ z' = x + z, \end{cases}$

3.5. Heaviside Function and Second Shifting Theorem

Find the Laplace transform or the inverse Laplace transforms of the following functions:

3.5.1. $f(t) = \begin{cases} t, & \text{if } 0 \le t < 3, \\ 6 - t, & \text{if } 3 \le t < 6, \\ 0, & \text{if } 6 \le t. \end{cases}$

3.5.2. $f(t) = \begin{cases} e^{-2t}, & \text{if } 0 \le t < 2, \\ 0, & \text{if } 2 \le t. \end{cases}$

3.5.3. $f(t) = \begin{cases} 2, & \text{if } 0 \le t < 3, \\ 6, & \text{if } 3 \le t < 6, \\ 1, & \text{if } 6 \le t. \end{cases}$

3.5.4. $f(t) = \begin{cases} 2 - t, & \text{if } 0 \le t < 4, \\ 6t - 1, & \text{if } 4 \le t < 7, \\ 2t, & \text{if } 7 \le t. \end{cases}$

3.5.5. $f(t) = \begin{cases} \sin t, & \text{if } 0 \le t < 4\pi, \\ 0, & \text{if } 4\pi \le t. \end{cases}$

3.5.6. $f(t) = \begin{cases} \cos t, & \text{if } 0 \le t < 3\pi, \\ -1, & \text{if } 3\pi \le t. \end{cases}$

3.5.7. $f(t) = \begin{cases} 2 - e^{-5t}, & \text{if } 0 \le t < 2, \\ 0, & \text{if } 2 \le t. \end{cases}$

3.5.8. $F(s) = \dfrac{e^{-s}}{s^2 + 9}.$

3.5.9. $F(s) = \dfrac{e^{-2s}}{s^2 + 6s + 9}.$

3.5.10. $F(s) = \dfrac{2e^{-5s}}{s^2 + 5s + 6}.$

3.6. Solving Initial-Value Problems with Discontinuous Inputs

3.6.1. $y'' + y = f(t) = \begin{cases} 2, & \text{if } 0 \le t < \pi, \\ 0, & \text{if } t \ge \pi. \end{cases}$ $y(0) = y'(0) = 0.$

3.6.2. $y'' - 3y' = f(t) = \begin{cases} 2, & \text{if } 0 \le t < 2 \\ 4, & \text{if } t \ge 2, \end{cases}$ $y(0) = 1, \ y'(0) = -1.$

3.6.3. $y'' + 3y' + 2y = f(t) = \begin{cases} 1, & \text{if } 0 \le t < 1 \\ -1, & \text{if } t \ge 1, \end{cases}$ $y(0) = y'(0) = 0..$

3.6.4. $y'' + 4y = f(t) = \begin{cases} \sin t, & \text{if } 0 \le t < \pi \\ \cos t, & \text{if } t \ge \pi, \end{cases}$ $y(0) = y'(0) = 0.$

3.6.5. $y'' + 4y' + 4y = f(t) = \begin{cases} e^t, & \text{if } 0 \le t < 1 \\ e^t - 1, & \text{if } t \ge 1, \end{cases}$ $y(0) = 1, \ y'(0) = 0.$

3.8. Solving Initial-Value Problems with Impulse Inputs

3.8.1. $y'' + y' = \delta(t - 1) + \delta(t - 2), \quad y(0) = 1, \ y'(0) = -2.$

3.8.2. $y'' + 9y = e^{-t} + \delta(t - 1), \quad y(0) = 0, \ y'(0) = 3.$

3.8.3. $y'' + 16y = \cos t + \delta(t - \pi), \quad y(0) = 0, \ y'(0) = 1.$

3.8.4. $y'' + 3y' + 2y = 1 + \delta(t - 1), \quad y(0) = -1, \ y'(0) = 1.$

3.9. Application of Laplace Transform to Electric Circuits

For *3.9.1-4*, by using

$$L Q'' + R Q' + \frac{1}{C} Q = E'(t),$$

find the charge Q and current $I = \dfrac{dQ}{dt}$ in the given *RLC* circuit if at $t = 0$ the charge on the capacitor and current in the circuit are zero.

3.9.1. $\dfrac{1}{10} Q'' + 4Q' + 200Q = 0.$

3.9.2. $\dfrac{1}{10} Q'' + 4Q' + 200Q = \cos 4t.$

3.9.3. $\dfrac{1}{20} Q'' + 6Q' + 100Q = 2\sin 3t.$

3.9.4. $\dfrac{1}{5} Q'' + 6Q' + 45Q = \sin 3t + \cos 3t.$

Chapter 4

PROBABILITY

After three chapters of being in a deterministic world, we now change the gear and walk into a world with everything stochastic (probabilistic). We will start this chapter by defining some terms, which will make up our basic vocabularies.

4.1. INTRODUCTION

Definition 4.1.1.

By an *outcome* of an experiment we mean any result of performing that experiment. An experiment for which occurrences of outcomes are not certain or completely predictable is called a *chance or random experiment*. In a chance experiment, certain sets of outcomes are considered primitive (or simple), that is, they will not be broken down into smaller sets. Such sets are called *elementary* (or *simple*) *events*. In many (but not all) cases, the elementary events are just singletons (a set with a single element, such as $\{a\}$). Any combination of elementary events is just called an *event*. In all cases, an *event* is a set of outcomes. We say an *event occurs* when any of its outcomes takes place. The union of all events for a chance experiment (i.e., the set of all possible outcomes) is called the *sample space*. In this terminology an *event* is a subset of the sample space. Sometimes a particular outcome, i.e., an element of the sample space, is also called a *sample point*. We, generally, denote the *sample space* by Ω, events by capital letters such as A, B, C, and outcomes by lower case letters such as a, b, c, $\cdots$.

Example 4.1.1.

Suppose you get up in the morning and it is cloudy. Then, you could anticipate that it may or may not rain that day. You will indeed perform a chance experiment for which the sample space is the set containing "falling of rain" and "not falling of rain", i.e., $\Omega = \{$falling of rain, not falling of rain$\}$. There are two simple events in this case and they are $\{$falling of rain$\}$ and $\{$not falling of rain$\}$.

Example 4.1.2.

Consider an automatic switch, which has only two states "on" and "off", Figure 4.1.1.

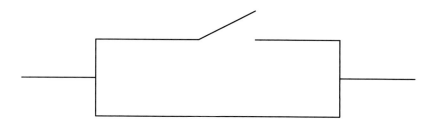

Figure 4.1.1. An automatic switch.

Let the states of the switch be numbered as 0 and 1; 0 being the "off" state and 1 being the "on" state of the switch. Then, the sample space consists of two outcomes 0 and 1, i.e., Ω = {0, 1}. There are four events {0}, {1}, {0, 1}, $\varnothing$ (the empty set), three of which are simple. They are {0}, {1}, and $\varnothing$. The "empty event" is a simple event because it cannot be broken into further events.

Example 4.1.3.

Suppose there are two automatic off-on switches S_1 and S_2 that work together. Then, using the notation as in Example 4.1.2, the possible outcomes are:

(0, 0), i.e.,   *"both off"*,

(0, 1), i.e.,   *"S_1 off and S_2 on"*,

(1, 0), i.e.,   *"S_2 off and S_1 on"*,

(1, 1), i.e.,   *"both on"*.

The sample space, in this case, is the set of these four ordered pairs, i.e., Ω = {(0, 0), (0, 1), (1, 0), (1, 1)}. See Figure 4.1.2.

A sample space being a set, it may be finite or infinite. Examples 4.1.1 through 4.1.3 are examples of finite sample spaces. An infinite sample space may be discrete or continuous. Examples 4.1.4 through 4.1.18 are examples of infinite sample spaces.

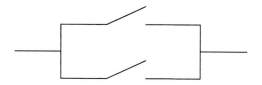

Figure 4.1.2. Two automatic off-on switches.

Note that even though there are only a finite number of people or objects in the world, when we are experimenting with a large number of people or objects, the sample space is considered infinite.

Examples 4.1.4, 4.1.5, and 4.1.6 are examples of discrete infinite sample spaces.

Example 4.1.4.

The number of customers waiting in a line at a cash register in a store.

Example 4.1.5.

The number of airplanes trying to land in very crowded international airports such as JFK in New York or Houston Intercontinental in Texas. Even though, there are only a finite number of airplanes coming to the United States from all over the world and try to land, yet the sample space is considered the set of nonnegative integers.

Example 4.1.6.

The working state of a machine and observing the number of days it is working. The sample space will be $\Omega = \{1, \ 2, \ \cdots\}$, which, again, is infinite and discrete.

Examples 4.1.7 and 4.1.18 are examples of continuous sample spaces.

Example 4.1.7.

Suppose a machine is working and we are waiting for it to break down for the first time. The sample space in this case is the entire half-open interval $[0,\infty)$ of possible real-valued breakdown times. Here the symbol infinity, ∞, means that the machine never breaks down. If we want to include that possibility, then the sample space would have to be $[0,\infty]$.

Example 4.1.8.

An instrument that can uniformly (with the same chance) generate numbers on a set is called a *random generator*. Suppose we have a random generator that generates numbers on the interval [0, 1]. The selected number is either less than $1/\sqrt{3}$ or greater than or equal to $1/\sqrt{3}$. In this chance experiment the outcomes are the occurrence of the subinterval

$\left[0, 1/\sqrt{3}\right)$ or $\left[1/\sqrt{3}, 1\right]$. These are the elementary events. Thus, the sample space is $\Omega =$ [0, 1], which is infinite, but the collection of events has just four members. They are $\varnothing$, Ω, $\left[0, 1/\sqrt{3}\right)$, and $\left[1/\sqrt{3}, 1\right]$. So, the sample space can be infinite while the collection of events of interest may be finite.

Definition 4.1.2.

Consider a sample space, say Ω. Two events, say A_1 and A_2, are called *mutually exclusive* if their intersection is the empty set. In other words, mutually exclusive events are disjoint subsets of the sample space. Let $A_1, A_2, \cdots, A_n$ be mutually exclusive events such that $A_1 \cup A_2 \cup \cdots \cup A_n = \Omega$. Then, the set of $\{A_1, A_2, \cdots, A_n\}$ is called a *partition* of Ω.

Example 4.1.9.

The two events $A_1 = \left[0, \dfrac{1}{\sqrt{3}}\right)$ and $A_2 = \left[\dfrac{1}{\sqrt{3}}, 1\right]$ (mentioned in Example 4.1.8) are

mutually exclusive events. In this case, the set $\{A_1, A_2\}$ is a partition of $\Omega = [0, 1]$.

Example 4.1.10. (Non-mutually Exclusive Events)

Suppose a machine is to select two digits from 1 to 9 at random. Then, the sample space is $\Omega = \{(1, 1), (1, 2), \cdots, (1, 9), (2, 1), (2, 2), \cdots, (2, 9), (3, 1), (3, 2), \cdots, (3, 9), \cdots, (9, 1), (9, 2), \cdots, (9, 9)\}$. If the event, A, of interest is the sum of two digits selected to be eight, then this event, i.e., "sum is eight" will be $A = \{(1, 7), (2, 6), (3, 5), (4, 4), (5, 3), (6, 2), (7, 1)\}$. Now, if the event, B, of interest is the two digits to be even numbers, then this event, i.e., "both are even" will be $B = \{(2, 2), (4, 4), (6, 6), (8, 8)\}$. We see that for the outcome (4, 4), the sum is eight and both are even, i.e., this outcome belongs to both events. Thus, A and B are not mutually exclusive.

Definition 4.1.3.

The *probability of an event E,* denoted by $P(E)$, is a number between 0 and 1 (inclusive) describing the likelihood of the event E to occur.

The association of probability to an event is to *measure* how likely it is for the event to occur. An event with probability 1 is called an *almost sure event* or *certain event*. An event with probability 0 is called a *null* or an *impossible event*. In other words, measure of the almost sure event is 1 and that of the impossible event is 0.

Definition 4.1.4.

If all outcomes of a finite sample space with n elements have the same chance to occur, then each member is assigned probability $1/n$ and the sample space is called *equiprobable*. The idea of *equiprobable* measure may be extended to a countably infinite sample space, which is referred to as *uniform measure*.

The equiprobable measure is an example of a discrete probability measure. Sometimes the expression "*at random*" is used for equiprobable sample space. Hence, by saying: we choose a digit at random from 1 to 9, we mean that every digit of $\{1, 2, \cdots, 9\}$ has the same chance to be picked, i.e., all elementary events in $\{1\}, \{2\}, \cdots,$ and $\{9\}$ are equiprobable. In that case, we may associate probability 1/9 to each digit singleton.

Example 4.1.11.

Consider Example 4.1.2. If the statistical history of the machine shows that the machine is "on" four times as much as it is "off", then we may associate probability 1/5 to the "off" state, and 4/5 to the "on" state.

We can define a "*probability*" in different ways, as we will see below.

Definition 4.1.5.

If an experiment is performed repeatedly, then the chance of occurrence of an outcome, intuitively, will be approximated by the proportion of the number of times that the outcome occurs, i.e., the ratio of occurrences of the outcome to the total number or repetitions of the experiment. This ratio that is called *relative frequency* is defined as *empirical probability*, for large sample sizes. (We will revisit this term in the next chapter.)

Definition 4.1.6.

For any sample space that has n *equiprobable points*, each point corresponds to a simple event with probability $1/n$. In this case, the *probability of an event* with k points will be k/n, $k \leq n$.

Example 4.1.12.

In controlling the quality of a product, suppose we choose a sample of 20 items. Each item is checked to meet a certain criteria. Suppose 5 items were reported defective. Then, we conclude that the relative frequency of defective items is 5/20 or 1/4. Therefore, we can say that probability of a defective item in the sample is 1/4.

Definition 4.1.7.

Calling the sets as the equiprobable sample points in a combinatorial problem, the "classical" *probability of an event E* is given by:

$$P(E) = \frac{number\ of\ ways\ the\ event\ E\ can\ occur}{total\ number\ of\ ways\ any\ outcome\ can\ occur}. \tag{4.1.1}$$

Definition 4.1.8.

By "*size*" of a finite set A we mean the number of elements in A. Then, *probability of an event A* in an equiprobable space Ω can be written as:

$$P(A) = \frac{size\ of\ A}{size\ of\ \Omega}. \tag{4.1.2}$$

Example 4.1.13.

Consider interval, area, and volume as entities with countably infinite points. Let $B = [a, c)$ be a sub-interval of the interval $I = [a, b]$, A be a subarea of Ω, v be a sub-volume of V. Then, based on Definition 4.1.8, for the probability of a subinterval, a subarea, or a subvolume, assuming that probabilities are distributed uniformly over the entire interval, area, or volume, respectively, i.e., equiprobability for all points, we have

$$P(A) = \frac{length\ of\ A}{length\ of\ I}, \ P(A) = \frac{area\ of\ A}{area\ of\ \Omega}, \ and\ P(A) = \frac{volume\ of\ A}{volume\ of\ V}. \tag{4.1.3}$$

(Note that probabilities defined by (4.1.3) are examples of continuous probability measures that will be discussed in details later.)

Example 4.1.14.

Suppose a point (a, b) is chosen at random in the rectangle $0 \le a \le 3$, $0 \le b \le 1$. What is the probability that the equation $ax^2 + 3bx + 1 = 0$ has two real roots?

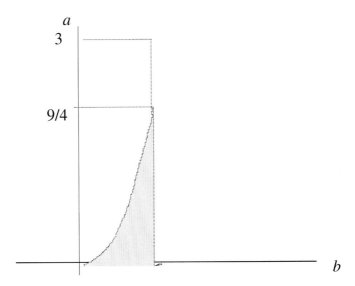

Figure 4.1.4. Example 4.1.14.

Solution

To answer the question note that the equation will have two real roots if and only if the discriminant of the equation is nonnegative, i.e., $9b^2 - 4a \geq 0$. Hence, we must have $a \leq 9b^2/4$ with $0 \leq a \leq 3$ and $0 \leq b \leq 1$. Thus, the pair (a, b) must be placed below or on the parabola $a = 9b^2/4$, i.e., located in the shaded area of Figure 4.1.4. But, the area of the region is $\dfrac{9}{4}\displaystyle\int_0^1 u^2 du = \dfrac{3}{4}$. Thus, by (4.1.3), the probability in question will be the ratio of this area to the given rectangle, i.e., $(3/4)/3 = 1/4$.

We now summarize properties of an event, E, as *axiom of probability*. The axioms involve three components of the *probability space* defined below. The set of these axioms is a fundamental formulation originated by the Russian mathematician Kolmogorov (1933a).

Note that an axiom is a statement that cannot be proved or disproved. A mathematician may or may not accept such a statement. Although all probabilists accept the three axioms of probability, there are axioms in mathematics such as the Axiom of Choice that is yet controversial and not accepted by some prominent mathematicians.

Definition 4.1.9.

The triple $(\Omega, \mathcal{B}, P)$ is called the *probability space*, where Ω is the sample space, $\mathcal{B}$ is the set function containing all possible events drawn from Ω, and P is the probability (measure) of an event.

Axioms of probability:

Axiom 1. $0 \le P(E) \le 1$ for each event E is $\mathfrak{B}$.

Axiom 2. $P(\Omega) = 1$.

Axiom 3.   If E_1 and E_2 are *mutually exclusive* events in $\mathfrak{B}$, then

$$P(E_1 \cup E_2) = P(E_1) + P(E_2).$$

The first axiom states that (no matter how the probability of an event, E, is defined) it is between 0 and 1 (inclusive). The second axiom states probability of the sample space is 1 (and due to Axiom 2, the probability of the null event is 0). The third axiom states that probability may be viewed as an additive operator.

4.2. COUNTING TECHNIQUES

We can sometimes define probability in terms of the number of ways an outcome can occur. See Example 4.1.6, formula (4.1.1). However, we need to know how to calculate the number of ways. The main computational tool for the equiprobable measure is *counting*. In other words, in assigning probabilities to outcomes, we sometime need to count the number of outcomes of a chance experiment and/or the number of ways an outcome may occur.

Definition 4.2.1.

If n is a non-negative integer, then by *n factorial*, denoted by *n!*, we mean:

$$n! = n(n\text{-}1)(n\text{-}2)...(2)(1), \tag{4.2.1}$$

with $0! \equiv 1$, and $1! \equiv 1$.

Example 4.2.1.

$5! = (5)(4)(3)(2)(1) = 120.$

Rule 4.2.1. The Fundamental Principle of Counting

Suppose an experiment can be performed in n_1 different ways, a second experiment can be performed in n_2 different ways independently of the first experiment, and so on. Then, all of these experiments can be performed together in $n_1 \times n_2 \cdots$ different ways as long as each experiment can be performed in a number of ways which is independent of all preceding experiments in the order given, see Figure 4.2.1.

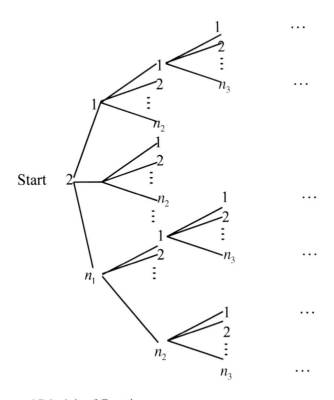

Figure 4.2.1. Fundamental Principle of Counting.

Example 4.2.2.

Suppose we are to code products of a manufacturer by a letter, another letter different from the first choice, another letter different from the first two choices, a non-zero digit, and finally two digits. How many different codes could we have?

Solution

Here is the answer to the question. The first letter can be chosen in 26 different ways, the second letter in 25 different ways, and the third in 24 different ways. The first digit in 9 different ways, the second and the third in 10 different ways each. Thus, there will be (26)(25)(24)(9)(10)(10) = 14,040,000 different ways coding the products of that manufacturer.

It can be shown that for large n, $n!$ can be approximated by *Stirling Formula*:

$$n! \approx \sqrt{2\pi}\, n^{n+\frac{1}{2}} e^{-n}, \qquad\qquad (4.2.2)$$

where the sign $"\approx"$ means approximately equal to, i.e.,

$$\lim_{n\to\infty}\frac{\sqrt{2\pi}\, n^{n+\frac{1}{2}}e^{-n}}{n!}=1.$$ (4.2.3)

The accuracy of (4.2.2) is good for many practical purposes.

Example 4.2.3.

$2! = 2$, while its Stirling approximation (4.2.2) is 1.919. The relative error in this case is 4%. Also, $10! = 3,628,800$, while its approximation (4.2.2) is $3,598,600$ with a relative error of 0.8%.

Sometimes outcomes of an experiment will be a selection of things which are to be arranged in a certain order. This selection may be done *with or without replacement*. For instance, when we were selecting digits in Example 4.2.2, we first selected a digit and "put it back" and selected again. This is a selection *with replacement*. On the other hand, choosing letters in that example were *selections without replacement* since after we chose a letter the next letter had to be different.

If the order is important and selection is without replacement, then this arrangement is called a *permutation*.

Definition 4.2.2.

If r of n distinct objects, $r \le n$, are to be arranged in a specific order, then the arrangement is called *r-permutation* or a *permutation of n objects taken r at a time*. Generally speaking, an arrangement of n objects is a *permutation*. The number of *r-permutation of n* objects is denoted by $(n)_r$ or $P_{r,n}$ or $_nP_r$. We use the first of these options. In case that $r = n$, then the ordered arrangement of n objects taken n at a time without replacement is called a *permutation or sampling without replacement*.

Theorem 4.2.1.

$(n)_r$ is given by:

$$(n)_r = \frac{n!}{(n-r)!}, r \le n.$$ (4.2.4)

Note that $(n)_n = n!$, $(n)_1 = n$, and by definition, $(n)_0 = 1$.

Proof

To choose the first object we have n ways (choices), for the second, n-1 choices, etc. Hence, by the Fundamental Principle of Counting (Rule 4.2.1) we have:

$$n(n-1)\cdots(n-r+1) = \frac{n(n-1)\cdots(n-r+1)(n-r)!}{(n-r)!} = \frac{n!}{(n-r)!}.$$

Example 4.2.4.

Consider the set $\{1, 2, 3\}$. The permutation of the three digits taken two at a time is $(3)_2$ = 3!/(3-2)! = 3! = 6. These are (1, 2), (1, 3), (2, 1), (2, 3), (3, 1), (3, 2). In this example, a permutation without repetition of the 3 numbers taken all 3 at a time, i.e., $(3)_3$ is also 6, i.e., 3!. They are (1,2,3), (1,3,2), (2,1,3), (2,3,1), (3,1,2), (3,2,1).

A permutation is defined when an arrangement is ordered. Now, we would like to see what happens if the order of arrangement is not counted, i.e., when we are enumerating subsets. We use the name *combination* or *unordered sampling without replacement*. In this case and define it as follows:

Definition 4.2.3.

The number of *combinations* (or *unordered sampling without replacement*) of n objects taken r at a time, without repetition, (i.e., a subset of r things from a collection n), is denoted by $\binom{n}{r}$ or C_r^n or $_nC_r$. Each of the $\binom{n}{r}$ permutations of n objects, r of one type and n-r of another type, is called a *distinguishable permutation*. See Definition 4.2.4.

Note that $\binom{n}{r}$ cannot be larger than $(n)_r$, by definition, since any given combination can be put in order $r!$ different ways.

It can easily be seen that $(n)_r = \binom{n}{r}r!$, where $(n)_r$ is given by (4.2.4). This can be done by constructing an ordered set in two stages: First select the subset of n (combinations), $\binom{n}{r}$ ways. Then, order it, $r!$ ways. Thus, we have:

Theorem 4.2.2.

$$\binom{n}{r} = \frac{(n)_r}{r!} = \frac{n(n-1)\cdots(n-r+1)}{1\cdot 2\cdots\cdots(r-1)\cdot r} = \frac{n!}{r!(n-r)!}.$$ \hfill (4.2.5)

Example 4.2.5.

Suppose a box of 24 light bulbs contains 5 defective bulbs. Two bulbs are taken from the box at random without replacement. What is the probability that neither one of the two are defective?

Solution

The total number of ways we can take the bulbs in order is $(24)_2$. For the first draw, the number of ways of having no defective is $\binom{19}{1}$ and for the second will be $\binom{18}{1}$. Thus, the probability in question is $\dfrac{\binom{19}{1}\binom{18}{1}}{(24)_2} = \dfrac{\left(\frac{19!}{18!1!}\right)\left(\frac{18!}{17!1!}\right)}{\left(\frac{24!}{22!}\right)} = \dfrac{57}{92}$.

Note that we could solve this problem by simply taking simply probability of choosing 19 non-defectives out of 24 total in the first round and then 18 out of 23 at the second round, then using the multiplication law to obtain $\dfrac{19}{24} \times \dfrac{18}{23} = \dfrac{57}{92}$.

Definition 4.2.4.

Let n, k_1, k_2, $\cdots$, k_r be non-negative integers such that $n = k_1 + k_2 + \cdots + k_r$. Then, the *multinomial coefficient* is define as:

$$\binom{n}{k_1,\ k_2,\ \cdots,\ k_r} = \frac{n!}{k_1!k_2!\cdots k_r!}.$$ \hfill (4.2.6)

In other words, if there are n objects with k_1 duplicates of one kind, k_2 duplicates of a second kind, ..., k_r duplicates of a rth kind, then the number of *distinguishable permutations* of these n objects is given by (4.2.6).

If all k's but one, say k, are zero, then we will have

$$\binom{n}{k} = \frac{n(n-1)\cdots(n-k+1)}{k!}, \tag{4.2.7}$$

which is called the *binomial coefficients*.

Now if the order of arrangement is important and selection is with replacement, then Theorem 4.2.1 takes a different form which is given by the following theorem:

Theorem 4.2.3.

The number of permutations of n objects with n_1 alike, n_2 alike, $\cdots$, n_k alike (called *permutation of distinguishable permutations of n objects or permutation with repetition* or *sampling repetition*) is given by the multinomial coefficient, formula (4.2.6).

Proof

Suppose we have a set of n elements such that k_1 of one kind and indistinguishable, k_2 of another kind and indistinguishable, $\cdots$, and yet k_n of another kind and alike, $n_1 + n_2 + \cdots + n_k = n$. We like to determine the number of distinct permutations of all n elements. Now to select the first k_1 elements we have $\binom{n}{k_1}$ ways. Having k_1 selected, it remains $n - k_1$ elements to select the second group of k_2. Thus, we have $\binom{n-k_1}{k_2}$ ways to select the second group. For the next selection we have $n - k_1 - k_2$ elements left of which k_3 has to be taken, i.e., $\binom{n-k_1-k_2}{k_3}$ ways. This process continues until $n - k_1 - k_2 - \cdots - k_{n-2}$ is left for the last group, i.e., k_{n-1}, to be selected from. Hence, for this last selection we have $\binom{n-k_1-k_2-\cdots-k_{n-2}}{k_{n-1}}$ choices.

Now by the Principle of Counting (Rule 4.2.1) the number of distinct permutations of all n elements will be the product of these, i.e.,

$$\binom{n}{k_1}\binom{n-k_1}{k_2}\binom{n-k_1-k_2}{k_3}\cdots\binom{n-k_1-k_2-\cdots-k_{n-2}}{k_{n-1}}.$$

From the *Binomial Theorem*, i.e.,

$$(a+b)^n = \sum_{k=0}^{n} \binom{n}{k} a^k b^{n-k} \,, \tag{4.2.8}$$

we have:

$$\frac{n!}{k_1(n-k_1)!} \frac{(n-k_1)!}{k_2(n-k_1-k_2)!} \cdots \frac{(n-k_1-k_2-\cdots-k_{n-2})}{k_{n-1}(n-k_1-k_2-\cdots-k_{n-2}-k_{n-1})!},$$

which reduces to $\dfrac{n!}{k_1! k_2! \cdots k_n!}$, i.e., the multinomial coefficient, formula (4.2.6).

Example 4.2.6.

Suppose there are 3 officers in a committee of 10 people. We choose two people at random from the committee.

(a) What is the probability that both of these two are officers?
(b) What is the probability that neither of these two is an officer?
(c) What is the probability that at least one of the two is an officer?

Solution

To answer these questions, let:

$A = \{$both persons are officers$\}$,
$B = \{$neither persons is an officer$\}$, and
$C = \{$at least one person is an officer$\}$.

Then, using (4.2.7) we obtain:

$$\text{a) } P(A) = \frac{\binom{3}{2}}{\binom{10}{2}} = \frac{\dfrac{3!}{2!1!}}{\dfrac{10!}{2!8!}} = \frac{3}{45} = \frac{1}{15},$$

$$\text{b) } P(B) = \frac{\binom{7}{2}}{\binom{10}{2}} = \frac{\dfrac{7!}{2!5!}}{\dfrac{10!}{2!8!}} = \frac{21}{45} = \frac{7}{15}, \text{ and}$$

c) $P(C) = P(B^c) = 1 - P(B) = 1 - \dfrac{7}{15} = \dfrac{8}{15}$.

Alternative Solution

We could also use permutations to answer these questions as follows:

a) $P(A) = \dfrac{P_2^3}{P_2^{10}} = \dfrac{\frac{3!}{1!}}{\frac{10!}{8!}} = \dfrac{1}{15}$.

b) $P(B) = \dfrac{P_2^7}{P_2^{10}} = \dfrac{\frac{7!}{5!}}{\frac{10!}{8!}} = \dfrac{7}{15}$.

c) The same as above.

4.3. TREE DIAGRAMS

The calculation of probability of an event can sometimes be visualized with a *tree diagram*. We illustrate this method with several examples.

Example 4.3.1.

To control quality of a product, we choose an item. Either this item passes the specific standards of the test, in which case we call it a *success* and denote it by s, or *fails* to do so and we denote it by f. We repeat the selection three times in sequence. Now after the first experiment we either have "s" or "f", depending upon success or failure of the trial. After the second trial, we have "s" or "f" for each possible outcome of the first experiment , that is, if on the first trial we had "s", we now may have "s" or "f", and for "f" on the first trial. Summarizing in a diagram we will have Figure 4.3.1:

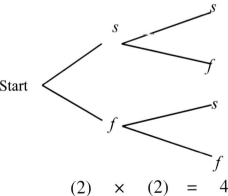

$$(2) \quad \times \quad (2) \quad = \quad 4$$

Figure 4.3.1. Tree diagram for sampling of size 2.

We now repeat the test for the third time. Then, the diagram will look like Figure 4.3.2.

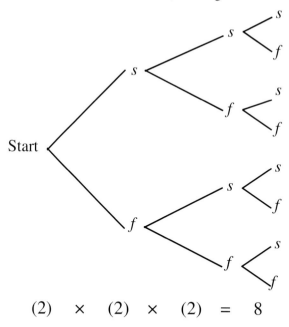

$$(2) \quad \times \quad (2) \quad \times \quad (2) \quad = \quad 8$$

Figure 4.3.2. Tree diagram for sampling of size 3.

Thus, the outcomes of this repeated experiment will be (s, s, s), (s, f, f), (s, f, s), (s, s, f), (f, s, s), (f, s, f), (f, f, s), (f, f, f). Hence, the sample space consists of $2^3 = (2)(2)(2) = 8$ ordered triples, where components of triples are possible successes or failures of the tests. Now if the test is so bad that the sample points are equiprobable, then the probability of a complete success is $P((s, s, s)) = 1/8$, and probability of two successes is $P\{(s ,s, f),\ (s, f, s),\ (f, s, s)\}$ = 3/8.

4.4. CONDITIONAL PROBABILITY AND INDEPENDENCE

We might ask how the probability of occurrence of an event would be affected if we knew another event had already occurred. Let us start with an example.

Example 4.4.1.

Suppose a machine with three components A, B, and C, is malfunctioning and only one component can be blamed for the problem.

a. If components A and B have the same chance to cause the problem and the chance that the problem is with C is 2/3 of the chance of A or B, what is the probability of each part to be responsible for malfunctioning of the machine?

b. Suppose that the component B has been ruled out and the chance that the problem is with C is 2/3 of the chance of A. What is the probability that A is the cause? What is the probability that C is the cause?

Solution

a. Let E_A, E_B, and E_C be the events that A, B, or C caused the problem with the machine, respectively. Then, $P(E_A) = P(E_B)$, $P(E_C) = (2/3) P(E_A)$. Since $P(E_A) + P(E_B) + P(E_C) = 1$, we will have $P(E_A) = 3/8$, $P(E_B) = 3/8$ and $P(E_C) = 1/4$.

b. If there is no other information available, then, we will have $P(E_A) + P(E_C) = 1$. Since $P(E_C) = (2/3) P(E_A)$, we should have $P(E_A) = 3/5$ and $P(E_C) = 2/5$.

Notice the change of probabilities of E_A and E_C due to information about E_B. Also notice that at the beginning, the space was $E_A \cup E_B \cup E_C$ (partition is $\{E_A, E_B, E_C\}$), while after B was ruled out it was reduced to $E_A \cup E_C$ (partition is $\{E_A, E_C\}$). Hence, if the occurrence of an event definitely does have influence in the occurrence of the other events under consideration, then we should expect a change in probabilities of those events. In other words, the probability of an event may be restricted or conditioned on the occurrence of another event.

We will now define the probability of the event A given that the occurrence of the event B is known.

Definition 4.4.1.

Let $(\Omega, \mathcal{B}, P)$ be a probability space and B an event (i.e., $B \in \mathcal{B}$) with positive probability, $P(B) > 0$. Then, the *conditional probability*, denoted by $P(\cdot|B)$, defined on $\mathcal{B}$ is given by:

$$P(A|B) = \frac{P(AB)}{P(B)} \text{ for any event } A \text{ in } \mathcal{B}. \tag{4.4.1}$$

The conditional probability $P(A|B)$ is read as: *the probability of A given B*. If $P(B) = 0$, then $P(A|B)$ is not defined. Figure 4.4.1 shows the Venn diagram for the conditional sample space.

Note that although $P(\cdot|B)$ is a probability measure on all of Ω, under the condition given, the effective sample space is smaller, i.e., since $P(B|B) = 1$, the sample space Ω may be replaced by B. Thus, we now have a new triple, i.e., a new probability space

$\left(\Omega, \mathcal{B}, P\left(\cdot|B\right)\right)$. This space is called the *conditional probability space induced on* $\left(\Omega, \mathcal{B}, P\right)$, *given B*. Note also that conditional probability has the same properties as ordinary probability, but restricted to a smaller space.

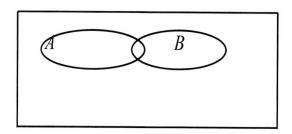

Figure 4.4.1. Conditional sample space.

Example 4.4.2.

Consider Example 4.4.1. In part (a), for events E_A, E_B, and E_C, we found $P(E_A) = 3/8$, $P(E_B) = 3/8$ and $P(E_C) = 1/4$. In part (b) it was assumed that B is ruled out. Hence, either A or C, and only one of them, could be the problem. Let us denote this event by $E_D = E_A \cup E_C$. Thus, $P\left(E_B|E_D\right) = 0$. Let us find the following probabilities of

(a) the component A is the problem, and
(b) the component C is the problem.

Solution

To answer these questions, we have to find $P\left(E_A|E_D\right)$ and $P\left(E_C|E_D\right)$. But, $P\left(E_D\right)$

$$= P\left(E_A\right) + P\left(E_C\right) = \frac{3}{8} + \frac{1}{4} = \frac{5}{8}. \text{ Therefore,}$$

$$P\left(E_A|E_D\right) = \frac{P\left(E_A E_D\right)}{P\left(E_D\right)} = \frac{P\left(E_A\right)}{P\left(E_D\right)} = \frac{\frac{3}{8}}{\frac{5}{8}} = \frac{3}{5},$$

$$P\left(E_C|E_D\right) = \frac{P\left(E_C E_D\right)}{P\left(E_D\right)} = \frac{P\left(E_C\right)}{P\left(E_D\right)} = \frac{\frac{1}{4}}{\frac{5}{8}} = \frac{2}{5}.$$

Note that these values agree with those in Example 4.4.1.b.

There are many cases that occurrences of events do not influence occurrences of other events, which lead to the idea of *independence* of two events. Let us consider the following example:

Example 4.4.3.

If two machines are working independently of each other and one stops, does this affect the probabilities for the working condition of the other? In other words, if being in working condition for the first machine is denoted by A and its failure by $\tilde{A}$ and being in working condition for the second machine is denoted by B and its failure by $\tilde{B}$, then what is $P\left(\tilde{B}\middle|\tilde{A}\right)$?

Solution

The answer is $P\left(\tilde{B}\middle|\tilde{A}\right) = P\left(\tilde{B}\right)$, as we will see below.

Definition 4.4.2.

Two events A and B are *independent* if and only if

$$P(AB) = P(A)P(B). \tag{4.4.2}$$

Note that Definition 4.4.2 can be extended to more than two events.

As an immediate consequence of this definition we have:

Theorem 4.4.1.

If events A and B are independent and $P(B) > 0$, then:

$$P\left(A\middle|B\right) = P\left(A\right), \tag{4.4.3}$$

and conversely, if $P(B) > 0$ and (4.4.3) is true, then A and B are independent.

Proof

From (4.4.1) and (4.4.2) we have $P\left(AB\right) = P\left(A\middle|B\right)P\left(B\right) = P\left(A\right)P\left(B\right)$, from which the first part of the theorem follows. Conversely if $P\left(A\middle|B\right) = P\left(A\right)$, then by multiplying both sides by $P(B)$ and using (4.4.1) we have:

$$P\left(A\middle|B\right)P\left(B\right) = P\left(A\right)P\left(B\right) = P\left(AB\right),$$

from which the second part of the theorem follows.

Note that pairwise independence is not a transitive relation, that is, if $\{A, B\}$ and $\{B, C\}$ are two pairs of independent events, it does not imply that $\{A, C\}$ is independent. Note also that pairwise independence does not imply total independence either.

Example 4.4.4.

Consider the set of all families of two children. Suppose that boys and girls have the same chance of being chosen. Taking the order of birth as the order of the sequence, the sample space will be $\Omega = \{bb, bg, gb, gg\}$, where b and g stands boy and girl, respectively.

Now let us choose a family at random. Let the events A, B, and C be defined as follows:

A = The first child is a girl.
B = One child is a boy and one is a girl.
C = The first child is a boy.

From S we see that: $AB = \{g\}$, $BC = \{bg\}$, $AC = \varnothing$. Now $P(AB) = 1/4$, $P(AC) = 0$, $P(A) = 1/2$, $P(A) = 1/2$, and $P(C) = 1/2$. Hence, $P(AB) = 1/4 = P(A)P(B)$, which shows A and B are independent; but $P(AC) = 0 \neq P(A)P(C) = 1/4$ shows that A and C are not independent.

In general, we have the following:

Definition 4.4.3.

Events A_1, A_2, $\cdots$, A_n are *independent* if and only if the probability of the intersection of any subset of them is equal to the product of corresponding probabilities. In symbols, A_1, A_2, $\cdots$, A_n are *independent* if and only if for every subset $\{i_1, i_2, \cdots, i_k\}$ of $\{1, 2, \cdots, n\}$ we have

$$P\left(\left\{A_{i_1}, A_{i_2}, \cdots A_{i_k}\right\}\right) = P\left(A_{i_1}\right)P\left(A_{i_2}\right)\cdots P\left(A_{i_k}\right). \tag{4.4.4}$$

Example 4.4.5.

A machine selects digits 1 through 9. Suppose digits have the same chance to appear at every selection. When the machine selects digits at random, the probability of having 2 and 8 is $(2)(1/9)(1/9)$. This is because 2 may appear first or second and 8 in the same way. Probability of getting 2, 6, and 8 is $(6)(1/9)(1/9)(1/9) = 6/729$. This is because there are six ways of getting these three numbers together.

4.5. The Law of Total Probability

When we defined conditional probability by (4.4.1) we noted that $P(B)$ must be positive. However, as long as the conditional probability $P(A|B)$ or $P(B|A)$ exists, we will have the following theorem known as the Multiplicative Law.

Theorem 4.5.1. (The Multiplicative Law)

For any two events A and B with conditional probability $P(B|A)$ or $P(A|B)$ we have

$$P(AB) = P(B|A)P(A) = P(A|B)P(B). \qquad (4.5.1)$$

Proof

The theorem follows from (4.4.1).

Note that (4.5.1) is always true even when $P(A)$ or $P(B)$ is zero, because in any of these cases $P(AB)$ becomes zero.

Example 4.5.1.

Suppose a box of light bulbs contains 48 bulbs 3 of which are defective. If we draw two bulbs at random, one after the other, and notice that the first is defective, what is the probability that the second one is also defective?

Solution
Assume all bulbs are equiprobable. Let A_1 be the event that the first bulb picked is a defective and A_2 the event that the second bulb is a defective. Then, from (4.5.1.) we have:

$$P(A_1 A_2) = P(A_1)P(A_2|A_1) = \left(\frac{3}{48}\right)\left(\frac{2}{47}\right) = \frac{1}{376}.$$

The Multiplicative Law can be extended as follows:

Theorem 4.5.2. (The Multiplicative Law)

Let A_1, A_2, $\cdots$, A_n, be n events, then:

$$P(A_1 A_2 \cdots A_n) = P(A_1) P(A_2 | A_1) P(A_3 | A_1 A_2) \cdots P(A_n | A_1 A_2 \cdots A_{n-1}). \qquad (4.5.2)$$

Proof

We apply (4.4.1) to each conditional probability on the right side of (4.5.2). Thus, we will have:

$$P(A_1) \frac{P(A_1 A_2)}{P(A_1)} \frac{P(A_1 A_2 A_3)}{P(A_1 A_2)} \cdots \frac{P(A_1 A_2 \cdots A_n)}{P(A_1 A_2 \cdots A_{n-1})}.$$

Successive terms of numerators and denominators of this expression cancel, except the last numerator term which is the left side of (4.5.2) and that proves the theorem.

Example 4.5.2.

Consider Example 4.5.1. Now instead of two, suppose we pick three bulbs at random, one at a time without replacement. What is the probability that all three are defective?

Solution

If A_1, A_2, and A_3 are the events that the first, the second, and the third bulbs picked are defective, respectively, then from (4.5.2) we will have (3/48)(2/47)(1/46) = 1/17296.

This result can be obtained by tree diagram. Let D and $\tilde{D}$ stand for defective and non-defective light bulb, respectively. Then, we will have Figure 4.5.1.

We have seen some properties of conditional probability of which the definition of independence is one of the important ones. Another very useful property of the conditional probability is the *Law of Total Probability* which will discussed below.

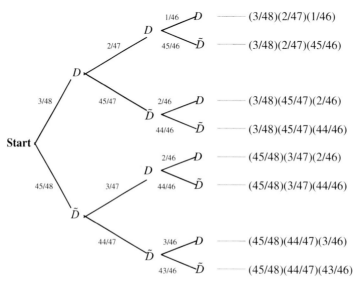

Figure 4.5.1.

Theorem 4.5.3. (The Law of Total Probability)

Let A_1, A_2, $\cdots, A_n$ be a partition of the sample space Ω. Then, for any given event B we have:

$$P(B) = \sum_{i=1}^{n} P(A_i) P(B|A_i) .$$ (4.5.3)

Proof

By definition, since A_1, A_2, $\cdots, A_n$ is a partition of Ω, we have $A_1 \cup A_2 \cup \cdots \cup A_n = \Omega$. Thus, from properties of sets an event B can be represented as follows:

$$B = B\Omega = B \cap (A_1 \cup A_2 \cup \cdots \cup A_n)$$
$$= (B \cap A_1) \cup (B \cap A_2) \cup \cdots \cup (B \cap A_n)$$

Therefore, using Axiom 3 of probability we have

$$P(B) = P(B \cap A_1) + P(B \cap A_2) + \cdots + P(B \cap A_n)$$

or

$$P(B) = \left[P(B|A_1) P(A_1) \right] + \left[P(B|A_2) P(A_2) \right] + \cdots + \left[P(B|A_n) P(A_n) \right],$$

and this will complete the proof.

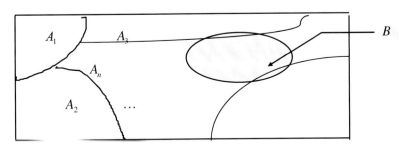

Figure 4.5.2. The Law of Total Probability. (The shaded area in Figure 4.5.2 shows the set B.)

Example 4.5.3.

Suppose we have three boxes A, B, and C of mixed colored (yellow and white) bulbs such that:

Box B_1 has 12 bulbs of which 3 are yellow.
Box B_2 has 24 bulbs of which 5 are yellow.
Box B_3 has 36 bulbs of which 7 are yellow.

A box is selected at random and from it a bulb is chosen at random. What is the probability that the chosen bulb is yellow?

Solution

The tree diagram and probability of each branch is given by Figure 4.5.3, assuming equiprobable property for boxes and the bulbs. In this figure W and Y stand for white and yellow, respectively.

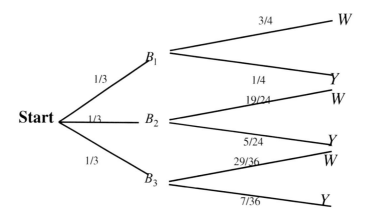

Figure 4.5.3. Tree diagram for the Multiplicative Law and the Law of Total Probability.

Hence, by Multiplicative Law (Theorem 4.5.1) we can find probability that any particular path occurs. This can be done by, simply, multiplying the corresponding probabilities on that branch. For instance, if we know that box B_2 were selected, then the probability of choosing a yellow bulb will be $(1/3)(5/24) = 5/72$. However, without knowing which box was selected we have to sum the probabilities of the possible outcomes. Thus, by the Law of Total Probability (Theorem 4.5.3) the probability in question will be $(1/3)(1/4) + (1/3)(5/24) + (1/2)(7/36) = 47/216$.

Now let us answer this last question by the Theorem of Total Probability. Let A_1 be the event that the first child is a girl, A_2 be the event that the second child is a girl, and E the event that both the children are girls. As we see { A_1, A_2 } is a partition for S_2. Hence by the Law of Total Probability we will have:

$$P(E) = P(A_1)P(E|A_1) + P(A_2)P(E|A_2) = \left(\frac{1}{2}\right)\left(\frac{1}{3}\right) + \left(\frac{1}{2}\right)\left(\frac{1}{3}\right) = \frac{1}{3}.$$

As an immediate consequence of the Law Total Probability we have the following:

Theorem 4.5.4. (Bayes' Formula)

Let A_1, A_2, $\cdots$, A_n be a partition of the sample space Ω. If an event B occurs, the probability of any event A_j is:

$$P\left(A_j|B\right) = \frac{P\left(A_j\right)P\left(B|A_j\right)}{\sum\limits_{i=1}^{n} P\left(A_i\right)P\left(B|A_i\right)}, \; j = 1, \; 2, \; \cdots, \; n. \tag{4.5.4}$$

Proof

For two events A_j and B, from (4.4.1) we have:

$$P\left(A_j|B\right) = \frac{P\left(B \cap A_j\right)}{P(B)}, \; j = 1, \; 2, \; \cdots, \; n$$

$$= \frac{P\left(B|A_j\right)P\left(A_j\right)}{P(B)}, \; j = 1, \; 2, \; \cdots, \; n.$$

Substituting $P(B)$ from the Law of Total Probability (4.5.3) we will have:

$$P\left(A_j|B\right) = \frac{P\left(B|A_j\right)P\left(A_j\right)}{P\left(B|A_1\right)P\left(A_1\right) + P\left(B|A_2\right)P\left(A_2\right) + \cdots + P\left(B|A_n\right)P\left(A_n\right)},$$

$$j = 1, \; 2, \; \cdots, \; n,$$

and that completes the proof.

Equation (4.5.4) is called *Bayes' Formula*. This theorem is designed to give the probability of the "cause" A_j, on the basis of the observed "effect" B. $P\left(A_j\right)$ is called the *a priori (prior)* and $P\left(A_j|B\right)$ is called the *a posteriori (posterior)* probability of the cause.

Example 4.5.4.

Consider Example 4.5.3. Now suppose the bulb chosen is a yellow. We return this bulb to the box and chose another bulb at random from it. What is the probability that the second bulb is also yellow?

Solution

Assume that all three boxes B_1, B_2 and B_3 have equal chances to be selected, the probability of each one being selected, i.e., the *a priori* probability, will be 1/3. Let Y_1 and Y_2 be the following events:

Y_1 = the first bulb chosen is yellow.

Y_2 = the second bulb chosen is yellow.

Thus, from (4.5.4) we will have:

$$P(B_1|Y_1) = \frac{(1/3)(1/4)}{(1/3)(1/4) + (1/3)(5/24) + (1/3)(7/36)} = \frac{1/4}{47/72} = \frac{18}{47}$$

$$P(B_2|Y_1) = \frac{5/24}{47/72} = \frac{15}{47}$$

$$P(B_3|Y_1) = \frac{7/36}{47/72} = \frac{14}{47}.$$

Of course, sum of these *a posteriori* probabilities must be 1, as is. To compute the probability that the second bulb is also yellow we use the Law of Total probability (Theorem 4.5.2). In using (4.5.2) let:

$A_1 = Y_1$ *is from* B_1,

$A_2 = Y_1$ *is from* B_2, and
$A_3 = Y_1$ *is from* B_3.

Hence,

$$P(A_1) = P(B_1|Y_1) = \frac{1}{4},$$

$$P(A_2) = P(B_2|Y_1) = \frac{5}{24}, \text{ and}$$

$$P(A_3) = P(B_3|Y_1) = \frac{7}{36}.$$

On the other hand,

$$P(Y_2|A_1) = \frac{1}{4},$$

$$P(Y_2|A_2) = \frac{5}{24}, \text{ and}$$

$$P(Y_2|A_3) = \frac{7}{36}.$$

Thus,

$$P(Y_2|Y_1) = P(A_1)P(Y_2|A_1) + P(A_2)P(Y_2|A_2) + P(A_3)P(Y_2|A_3)$$

$$= (1/4)(1/4) + (5/24)(5/24) + (7/36)(7/36) = 1717/20736.$$

Note that this value would be different if we had asked to find the probability that the second bulb be yellow. In that case we would have had:

$$P(Y_2) = P(_1)P(Y_2|B_1) + P(B_2)P(Y_2|B_2) + P(B_3)P(Y_2|B_3)$$

$$= (1/3)(1/4) + (1/3)(5/24) + (1/3)(7/36) = 47/216.$$

As we have seen before, the tree diagram is a good pictorial method for calculating probabilities. This is especially useful when Bayes' formula is to be employed such as in this example. Let us take a look at it. The tree diagram we had with probabilities of events written on the right column is shown below. From it , for instance, the probability that the first bulb chosen is yellow and it comes from box number 1 is $P(B_1|Y_1)$ = (1/3)(1/4). Thus using the Bayes' formula [formula (4.5.4)] we obtain what we found as 18/47. See Figure 4.5.4.

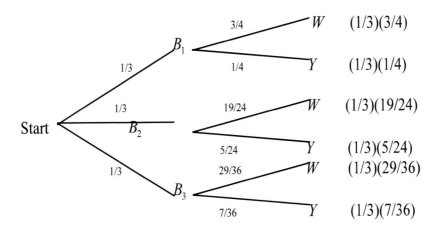

Figure 4.5.4. Example 4.5.4.

Now if we know the bulb is from which box, then we can see the probability that it is yellow or white. For instance, if we know it is from the first box, then the probability that it is yellow will be 1/4. Then given the color of the first bulb drawn we can calculate the probability of the color of the second bulb drawn from the following tree.

If we know that the first bulb is yellow, then for the probability of the second be yellow we just add the corresponding probabilities. Thus, we will have (See Figure 4.5.5)

$$P(Y_2|Y_1) = (1/4)(1/4) + (5/24)(5/24) + (7/36)(7/36) = 1717/20736.$$

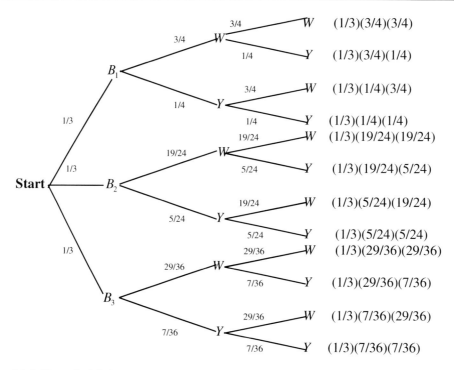

Figure 4.5.5. Example 4.5.4.

Example 4.5.5.

Suppose that one of three prisoners A, B, and C of a jail is to be executed at random and the other two to be freed. Prisoner A claims that he knows at least one of the prisoners B or C will be freed. Thus he asks the jailer to tell him which one of B or C is going to be freed. The jailer refuses to give him the information that if he does so, A will increase the chance of his execution from 1/3 to 1/2 since he will then be one of the two, one of which will be executed. Is the jailer's reasoning valid?

Solution

Suppose the jailer would have responded to the question of A by pointing out to the prisoner B. Then, probability that A be executed knowing that B is going to be freed is

$$\frac{P\{A \text{ be executed} \mid B \text{ is to be freed}\}}{P\{B \text{ is to be freed}\}}$$

$$= \frac{P\{A \text{ be executed}\}\, P\{B \mid A \text{ be executed}\}}{P\{B \text{ is to be freed}\}}$$

$$= \frac{(1/3)\, P\{B \mid A \text{ be executed}\}}{1/2}.$$

But,

$P\{B \mid A$ be executed$\} = 1/2$.

Thus,

$$P\{A \text{ be executed} \mid B \text{ is to be freed}\}$$
$$= P\{A \text{ be executed} \mid C \text{ is to be freed}\} = 1/3.$$

Therefore, the jailer's reasoning is valid.

Example 4.5.6.

Suppose a person had an HIV test and was told that he is infected. The good news is that the test is incorrect 5 percent of the time. What is the chance that the person is HIV free?

Solution

An immediate wrong answer is 5 percent based on the incorrectness of the test. To answer correctly, however, we have to note that the information included in the statement of the example is not complete. We may use the Bayes' Formula to answer the question correctly. To do that, the person tested must be chosen randomly, that the statement of the problem does not mention it. So let us assume this as a fact. Secondly, we need to have the *a priori* probability. So, let us assume that, based on what people believe, 99 percent of the person's demographic group is uninfected.

Now let us symbolize these. Let A_1 refer to the event "infected" and A_2 refer to the event "not infected". Thus, if we select a person at random from that demographic group, we know that the probability that he is infected is 0.01 or $P(A_1) = 0.01$ (the *a priori* probability). The prior probability that an individual is not infected, $P(A_2) = P$(un-infected), is 0.99 .

Let B denote the event "test shows infected". Thus, the probability that the individual was diagnosed infected and test confirm it is .95; i.e., $P(B \mid A_1) = 0.95$. Based on the given information, the probability that an individual actually is infection free but test indicates the infection is 0.05, i.e., $P(B \mid A_2) = 0.05$. Therefore, the probability that the person is HIV free is $P(A_1 \mid B)$ (the *a posteriori* probability). Then, according to the Bayes' Formula we have

$$P(A_1 \mid B) = \frac{P(A_1) P(B \mid A_1)}{P(A_1) P(B \mid A_1) + P(A_2) P(B \mid A_2)}$$

$$= \frac{(0.01)(0.95)}{(0.01)(0.95) + (.99)(.05)} = .161,$$

i.e., the chance that the individual is okay is 16 percent rather than 5 percent as a quick answer.

Example 4.5.7.

Suppose a manufactory three machines A, B, and C produce the same type of products and store in the warehouse. Of the total production, percent share of each of these machines are 20, 50 and 30. Of their production machines A, B, and C produce 1%, 4% and 2% defective items, respectively. For the purpose of quality control, a produce item is selected at random from the total items produced in a day. We want to answer the following two questions:

1. What is the probability of the item being defective?

2. Given that the item selected was defected, what is the probability that was produced by machine B?

Solution

To answer the first question, we denote the event of defectiveness of the item selected by E. Then by the Law of Total Probability, we will have:

$$P(E) = P(A)P(E \mid A) + P(B)P(E \mid B) + P(C)P(E \mid C)$$
$$= 0.20 \times 0.01 + 0.50 \times 0.04 + 0.30 \times 0.02$$
$$= 0.002 + 0.020 + .006 = 0.028.$$

That is, the probability of the produced item selected at random being defective is 2.8%.

To answer the second question, let the conditional probability in question be denoted by $P(B \mid E)$. Then, by Bayes' formula and answer to the first question, we have:

$$P(B \mid E) = \frac{P(B)P(E \mid B)}{P(E)} = \frac{0.50 \times 0.04}{0.028} = 0.714.$$

Thus, the probability that the defective item selected be produced by machine C is 71.4%.

We should note that *a priori* probability of 50% has increased to the *posteriori* 71.4%. This is because the second machine produces more defective products that the first and the third.

4.6. DISCRETE RANDOM VARIABLES

Definition 4.6.1.

A *random variable* quantifies a sample space (as is usually desirable). That is, a random variable assigns numerical (or set) labels to the sample points. In other words, a random variable is a function (or a mapping) on a sample space, whose domain is the sample space and its range is the set of assigned values to the sample points.

There are two main types of random variables, namely, *discrete* and *continuous*. We will discuss each in detail.

Definition 4.6.2.

A *discrete random variable* is a function, say X, from a countable sample space, Ω (that could very well be a numerical set), into the set of real numbers.

Example 4.6.1.

Suppose the color of hair of students in a class is black, brown and blond, i.e., the sample space Ω consists of the colors black, brown and blond. Let ω be an element of Ω. Then, ω is black, brown or blonde. Now let X map the color coders as $X(\text{black}) = 1$, $X(\text{brown}) = 2$, $X(\text{blond}) = 3$. Then, X is a finite discrete random variable.

We will use the following notations:

1. By $\left[X = x \right]$ we mean an event $\left\{ \omega \in \Omega; \ X(\omega) = x \right\}$.

2. By $P\left([X = x]\right)$ we mean the probability of an event $\left[X = x \right]$, which means $P\left(\omega \in \Omega : X(\omega) = x \right)$.

3. By $P\left([a \leq X \leq b]\right)$ we mean $P\left(\left\{ \omega \in \Omega : a \leq X(\omega) \leq b \right\} \right)$.

4. Similarly, for $P([X = a, Y = b])$, $P([X \leq a])$, $P([a \leq X \leq b, c \leq Y \leq d])$, etc., we mean

$$P\left([X = a, Y = b]\right) = P\left(\left\{ \omega_1, \omega_2 \in \Omega : X(\omega_1) = a, Y(\omega_2) = b \right\} \right),$$

$$P\left([X \leq a]\right) = P\left(\left\{ \omega \in \Omega : X(\omega) \leq a \right\} \right), \text{ and}$$

$$P\big([a \le X \le b, c \le Y \le d\big) = P\big(\{\omega_1, \omega_2 \in \Omega: a \le X(\omega_1) \le b, c \le Y(\omega_2) \le d\}\big),$$

etc., respectively.

Example 4.6.2.

Suppose we are to choose two digits from 1 to 6 such that the sum of the two numbers selected equals 7. Assuming that repetition is not allowed, the event under consideration will be {(1,6), (2,5), (3,4), (4,3), (5,2), (6,1)}. This event can also be described as {(i, j): $i + j = 7$, i, j = 1, 2, ..., 6}. Now, if X is the random variable defined by $X(i, j) = i + j$, then we are looking at all (i, j), for which $X((i, j)) = 7$, i.e., $[X = 7]$ contains six points since $X((1,6)) = 7$, $X((2,5)) = 7$, $X((3,4)) = 7$, $X((4,3)) = 7$, $X((5,2)) = 7$, and $X((6,1)) = 7$.

Note that we may extend our sample space to the entire real line with probability of all sample points not in Ω to be zero.

Example 4.6.3.

Consider the experiment of tossing a fair coin four times. Let H and T denote the occurrence of heads and tails, respectively. Then, the sample space for this experiment will be: Ω = {$HHHH$, $HHHT$, $HHTH$, $HTHH$, $THHH$, $HHTT$, $HTHT$, $HTTH$, $TTHH$, $THHT$, $THTH$, $HTTT$, $THTT$, $TTHT$, $TTTH$, $TTTT$}. If the number of tails occurring in the experiment is our concern, then a random variable X can be defined so that its values are 0, 1, 2, 3, and 4, depending upon the number of tails in each possible outcome, i.e., $[X=0]$, $[X=1]$, $[X=2]$, $[X=3]$, $[X=4]$. The following diagram shows this correspondence:

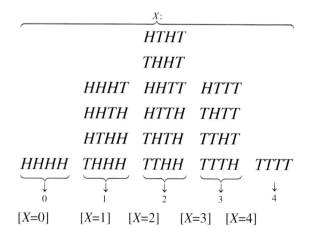

Example 4.6.4.

Consider an automobile manufacturing company. The cost of a certain model of a car is $9,000 per unit up to 10,000 units, $8,000 per unit for between 10,000 and 20,000 units and $5,000 per unit afterwards. Suppose the manufacturing initially made a batch of 10,000 cars and continuously make cars thereafter. Suppose also that the cars are sold at $19,000 each. The profit (or loss) in selling these cars is the subject of interest to this company.

Solution

A discrete random variable, denoted by X can describe the number of cars sold, since the number of cars sold will determine the amount of profit (or loss), Y. Now Y is a function of X. However, depending upon the value of a car as a non-integer real number or an integer, Y could be discrete or continuous. Thus, this is an example of a random variable that may not be discrete.

As it can be seen from the assumptions, the cost of making the initial 10,000 units is $(9000)(10000) = $90,0000,000 and the amount of sales is $19,000X$. Hence, the profit (or loss) of selling X units, $X \le 10,000$, is $(19,000X - 90,000,000)$. Then, after selling 10,000 units, the profit is $(19000 - 9000)(10000) = $100,000,000 plus the profit on the difference up to 20,000 units, i.e., $(19000-8000)(X - 10000) = $11,000(X - 10000)$. Finally, the profit from selling more than 20,000 units will be $(19000-9000)(10000) = $100,000,000 for the first 10,000 units, plus $(20000 - 10000)(19000 - 8000) = $110,000,000 on the selling of 20,000 units plus $(19000 - 5000)(X - 20000) = $14,000(X -20000)$ on the number of units over 20,000 units.

Note that, in this example, the profit random variable, Y, may be negative, representing loss for the company.

Summarizing this, we will have:

$$Y = \begin{cases} \$19,000X - 90,000,000, & \text{if} & X \le 10,000 \\ \$100,000,000 + 11,000(X - 10,000), & \text{if} & 10,000 < X \le 20,000 \\ \$210,000,000 + 14000(X - 20,000), & \text{if} & X > 20,000. \end{cases}$$

There is a special random variable that due to its importance we would like to single out and define it here.

Definition 4.6.3.

Let A be an event from a sample space Ω. The random variable $I_A(\omega)$ for $\omega \in A$ defined as:

$$I_A(\omega) = \begin{cases} 1, & \text{if } \omega \in A \\ 0, & \text{if } \omega \in A^c \end{cases} \tag{4.6.1}$$

is called the *indicator function* (or *indicator random variable*) .

Note that for every $\omega \in \Omega$, $I_\Omega(\omega) = 1$ and $I_\varnothing(\omega) = 0$.

Example 4.6.4.

Suppose we are to select an urn from among urns full of marbles. Our interest is in whether or not there is a defective (chipped off or out of normal shape) marble in the urn.

Let A be the event that there is a defective marble in the urn. Then, we can use the indicator random variable to describe the event of interest. We may assign value 1 to $I_A(\omega)$ if there is a defective one in the urn and 0 if there is none.

We now list some properties of random variables and leave the proof as exercises.

Theorem 4.6.1.

If X and Y are discrete random variables, then so are $X \pm Y$ and XY. Further if $[Y = 0]$ is empty, X/Y is also a random variable.

Proof

Left as exercise.

4.7. DISCRETE PROBABILITY DISTRIBUTIONS

Definition 4.7.1.

The way probabilities of a random variable are distributed across the possible values of that random variable is generally referred to as the *probability distribution* of that random variable.

Example 4.7.1.

Suppose a machine is in either "good repair" or "not good repair". Then, if we denote "good repair " by "1" and " not good repair " by "0", the sample space of states of this machine will be $\Omega = \{0, 1\}$. Let us describe the status of the machine by the random variable X. Then, $P([X = 1])$ means probability that the machine is in "good repair", and $P([X = 0])$ means the probability that the machine is not in "good repair".

Now if $P([X=1]) = 1/4$ and $P([X = 0]) = 3/4$, then we have a distribution for X. We can tabulate these probabilities as follows:

X	0	1
P(X = x)	3/4	1/4

Example 4.7.2.

Consider the number of customers waiting in a line for a cash register in a store. If we let Q be the random variable describing these events, then $P([Q = n])$ for all n will be the distribution of Q. For instance, we may have $P([Q < 3]) = 1/4$, $P([3 < Q < 5]) = 1/2$, and $P([Q > 5]) = 1/4$. Notice that the set $[Q < 3]$, for instance, in this case, has only three elements, namely, 0, 1, and 2.

As we have discussed before and it can be seen from the Example 4.7.2, the range of a random variable may be a finite or a countably infinite set of real numbers (it is possible to have a random variable with an uncountable range, although, this is a case we have not yet discussed.) For instance, the range of the random variable Q in example 4.7.2 is a set of nonnegative integers (if we assume infinite capacity for the store, otherwise a finite subset of that set). However, when we associate probabilities to the values of this random variable by writing, for instance, $P([Q < 3])$, we actually mean to associate probabilities to the numbers 0, 1, and 2 which belong to the range of Q for that example.

Now what if we include other values that do not belong to the range of Q (so that we do not have to specify Q is a set of integers)? It will do no harm (and can make our job easier) if we assign zero probabilities to such values. The sum of probabilities remains equal to 1. Therefore, we may define our *probability distribution* in two separate ways: *discrete* and *continuous*. In this section, we deal with the first case.

Definition 4.7.2.

Let X be a discrete random variable defined on a sample space Ω. Suppose x is a typical element of the range of X. Let p_x denote the probability that the random variable X takes the value x, i.e.,

$$p_x = P([X = x]) \text{ or } p_x = P(X = x). \tag{4.7.1}$$

Then, p_x is called the *probability mass function (pmf) of* X and also referred to as the (*discrete*) *probability density function (pdf) of* X.

Note that $\sum_x p_x = 1$, where x varies over all possible values for X.

Example 4.7.3.

Suppose an office has a two-line telephone system. Then, at any time either both lines are idle, both lines are busy, or one line is busy and the other idle. We are concerned with the number of lines that are busy.

Solution

We consider the number of busy lines as a random variable, denoted by X. Then, the range of X is the set $\{0, 1, 2\}$. Suppose our experience shows that both lines are idle half of the times, both lines are busy one-sixth of the times, and one line is busy one-third of the times. Then, using the notation defined by (4.7.1), $p_0 = 1/2, p_1 = 1/3$, and $p_2 = 1/6$. The probabilities p_0, p_1 and p_2 determine the distribution of X. We can tabulate this distribution in the following way.

x	0	1	2
p_x	1/2	1/3	1/6

Notice that for p_X we have

$$\sum_x p_x = p_0 + p_1 + p_2 = \frac{1}{2} + \frac{1}{3} + \frac{1}{6} = 1.$$

Definition 4.7.3.

Consider a finite population of size N. Let n represent a sample from this population that consists of two types of items n_1 of "type 1" and n_2 of "type 2", $n_1 + n_2 = n$. Suppose we are interested in probability of obtaining x items of type 1 in the sample. n_1 must be at least as large as x. Hence, x must be less than or equal the smallest of n and n_1. Then,

$$p_x \equiv P(X = x) = \frac{\binom{n_1}{x}\binom{N-n_1}{n-x}}{\binom{N}{n}}, \quad x = 0,1,2,\cdots,\min(n,n_1).. \tag{4.7.2}$$

defines the general form of *hypergeometric pmf* of the random variable X.

Note that p_x is the probability of waiting time for the occurrence of exactly x "type 1" outcomes. We could think of this scenario as an urn containing N white and green balls. From the urn, we select a random sample (a sample selected such that each element has the same chance to be selected) of size n, one ball at a time without replacement. The sample consists of n_1 white and n_2 green balls, $n_1 + n_2 = n$. What is the probability of having x white balls drawn in a row? This model is called an *urn model*.

To see (4.7.2), let $X(\omega)$ denote the number of "type 1" outcomes. Then, we want to find $p_x = P\{[X(\omega) = x]\}$. Notice that since we are looking for x number of "type 1" outcomes, the number of "type 2" outcomes will be $n - x$, since we are repeating the drawing n times. The number of ways that the "type 1" can occur is $\binom{n_1}{x}$ and for the "type 2" is $\binom{N - n_1}{n - x}$. Since any possible occurrence of x "type 1" may be combined with any occurrence of the "type 2", the number of ways both can occur is the product of these two, i.e., $\binom{n_1}{x}\binom{N - n_1}{n - x}$. Assuming that any one of the $\binom{N}{n}$ ways of selecting n objects from the total of N available objects has the same probability, the relation (4.7.2) follows. $\sum_x p_x = 1$ (why?) implies that (4.7.2), indeed, defines a *pmf*.

Example 4.7.4.

Suppose that there is a box containing of 120 decorative light bulbs, 5 of which are defective. We choose 6 bulbs at random without replacement. What is the probability that 2 of the bulbs chosen are defective?

Solution

In this example, we consider the defective bulbs as the "type 1" outcomes and non-defectives as the "type 2". Thus, in this case, $N = 120$, $n = 6$, $n_1 = 5$, $n_2 = 115$ and $x = 2$. Hence, from (4.7.2) we will have:

$$p_2 = \frac{\binom{5}{2}\binom{115}{4}}{\binom{120}{6}} = 0.0189 \approx 2\%.$$

Definition 4.7.4.

Let X be a discrete random variable, and x a real number from the infinite interval $(-\infty,\ x]$. We define $F_X(x)$ as:

$$F_X(x) = P([X \le x]) = \sum_{n=-\infty}^{x} p_n, \qquad (4.7.3)$$

where p_n is defined as $P([X = n])$ or $P(X = n)$. $F_X(x)$ is called the *cumulative distribution function (cdf) for X.*

Note that from the set of axioms of probability, we see that for all x:

$$p_x \ge 0, \text{ and } \sum_x p_x = 1. \qquad (4.7.4)$$

Example 4.7.5.

Consider Example 4.7.3. If we are interested in finding the probability of having at most one busy line, then we are talking about a value of the cumulative distribution function. Hence, $F_X(1) = \sum_{x=0}^{1} p_x = p_0 + p_1 = \dfrac{1}{2} + \dfrac{1}{3} = \dfrac{5}{6}$, since for all other points of $(-\infty, 1]$, probabilities are zero. We can tabulate the commutative probability functions and graph F_X as follows:

x	0	1	2
p_x	1/2	1/3	1/6
F_X	1/2	5/6	1

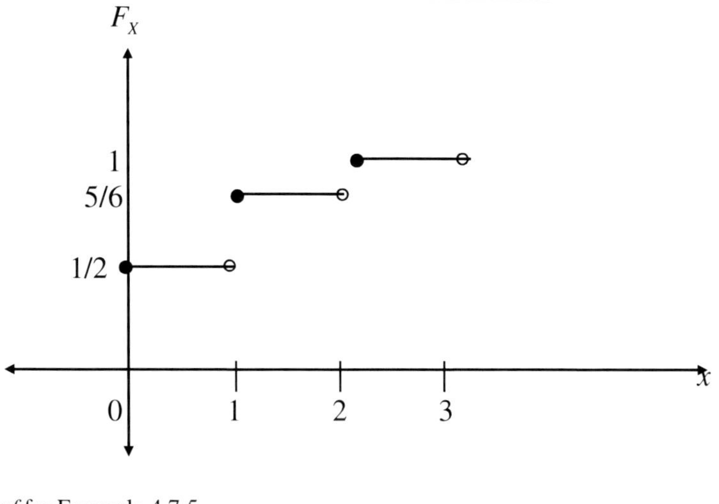

Figure 4.7.1. *cmf* for Example 4.7.5.

In Figure 4.7.1, we assume that p_x is 0 for values of X less than 0 or greater than 2.

As we mentioned, the probability mass function (*pmf*) is sometimes tabulated as we did in Examples 4.7.3 and 4.7.5. In general, the tabulation looks like the following:

x	x_1	x_2	$\cdots$	x_n	$\cdots$
p_x	p_{x_1}	p_{x_2}	$\cdots$	p_{x_n}	$\cdots$

As we did in Example 4.7.5, the cumulative probability function (*cpf*) is some times tabulated and graphed. It is possible to obtain the *pmf* from the *cdf*. The graph of *cdf* will be a step function. This is because, $F_X = 0$ for up to some values of X, then $F_X > 0$, and finally, when it reaches 1, it remains there, because the total probability is 1. The table and the graph in general look as follows:

x	x_1	x_2	$\cdots$	x_n	$\cdots$
p_x	p_{x_1}	p_{x_2}	$\cdots$	p_{x_n}	$\cdots$
F_X	p_{x_1}	$p_{x_1} + p_{x_2}$	$\cdots$	$p_{x_1} + \cdots + p_{x_n}$	$\cdots$

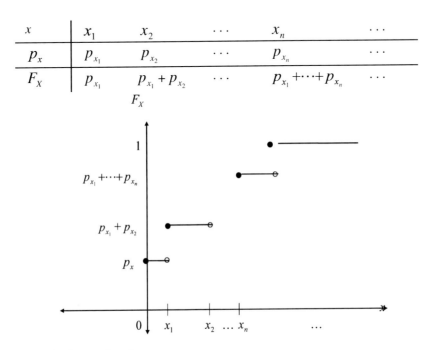

Figure 4.7.2. Graph of general *cmf*.

Example 4.7.6.

Suppose X is a random variable with *pmf*

$$p_x = \begin{cases} (1/2)^{x+1}, & x = 0, 1, 2, \cdots \\ 0, & otherwise. \end{cases}$$

The commutative distribution function of X is

$$F_X = \begin{cases} 0, & x < 0, \\ \displaystyle\sum_{k=0}^{x} (1/2)^{k+1} = 1 - (1/2)^x, & x = 0, 1, \cdots. \end{cases}$$

To see that, let us write the first x terms of the finite sum, i.e.,

$$S_x = \frac{1}{2} + \left(\frac{1}{2}\right)^2 + \left(\frac{1}{2}\right)^3 + \cdots + \left(\frac{1}{2}\right)^x.$$

Now multiply S_x by $(1/2)$ and subtract from S_x, i.e.,

$$\left(\frac{1}{2}\right) S_x = \left(\frac{1}{2}\right)^2 + \left(\frac{1}{2}\right)^3 + \cdots + \left(\frac{1}{2}\right)^{x+1}$$

and

$$S_x - \left(\frac{1}{2}\right) S_x = \left(1 - \frac{1}{2}\right) S_x = \left(\frac{1}{2}\right) - \left(\frac{1}{2}\right)^{x+1} = \left(\frac{1}{2}\right)\left[1 - \left(\frac{1}{2}\right)^x\right].$$

Thus, $S_x = 1 - \left(\frac{1}{2}\right)^x.$

We can display the *pdf* and *cdf* in a table and be graph as follows:

x	$x < 0$	0	1	2	$\ldots$	$x-1$	$\ldots$	$x \to \infty$
p_x	0	$1/2$	$1/4$	$1/8$	$\ldots$	$(1/2)^x$	$\ldots$	$\ldots$
F_X	0	$1/2$	$\frac{1}{2} + \frac{1}{4} = 3/4$	$\frac{1}{2} + \frac{1}{4} + \frac{1}{8} = 7/8$	$\ldots$	$1 - (1/2)^{x+1}$	$\ldots$	1

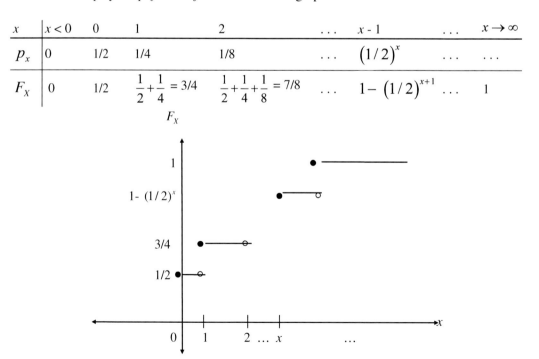

Figure 4.7.3. *cmf* for Example 4.7.6.

Example 4.7.7.

Suppose we are monitoring the working condition of a machine once each day. When we check, it is either working or not working. Let X represent the number of times we check the machine before it is observed to be nonworking. Also let the probability of the nonworking condition be q and of working condition be p, $q + p = 1$. Then, at the first time of check, if it is working, we will have $P([X = 1]) = p$. If we find it not working at the second check, then we will have $P([X = 2]) = qp$. However, if it were working we would have $P([X = 2]) = p^2$. This is because our experiment results in a 2-tuple $(0, 1)$, where the first component indicates the nonworking condition at the first check and the second component indicates the working condition at the second check. Thus, if we find the machine in a nonworking condition at the kth check for the first time, we would have a k-tuple with the first k-1 component as zeros and the kth as 1. Therefore, $P([X = k]) = p^{k-1}q$. Thus,

$$p_X = P([X = k])= \begin{cases} p^{k-1}q, & k = 1,\ 2,\ \cdots \\ 0, & otherwise. \end{cases}$$

This is a *pmf* since $p > 0$, $q > 0$, and, using the geometric series' property, we have $\sum_{k=1}^{\infty} p^{k-1}q = \dfrac{q}{1-p} = \dfrac{q}{q} = 1$. We can display this *pmf* and its *cdf* as follows:

k	$k \le 0$	1	2	3	...	k	...
p_k	0	p	p^2	p^3	...	$p^{k-1}q$	...
F_X	0	p	$(1+p)p$	$\left(1+p+p^2\right)p$	...	$\left(1+p+p^2+\cdots+p^{k-2}\right)pq$	...

We now discuss some of the common and one not so common distribution functions.

Definition 4.7.5.

A chance experiment is sometimes called a *trial*. If a trial has exactly two possible outcomes, it is called a *Bernoulli Trial*. The two possible outcomes of a Bernoulli trial are often referred to as *"success"* and *"failure"* denoted by *"s"* and *"f"*, respectively. Hence, if the probability of *"s"* is p, $0 \le p \le 1$, then the probability of *"f"* will be $q = 1 - p$.

If a Bernoulli trial is repeated independently n times with the same probabilities of success and failure for each trial, then, the scheme is called *Bernoulli Trials*. The sample space for each trial has two sample points. When there are n independent trials, the number of sample points will be 2^n.

Now let X be a random variable taking values 0 and 1, corresponding to failure and success, respectively, of the possible outcome of a Bernoulli trial. Thus, we will have:

$$P(X = k) = p^k q^{1-k}, \qquad k = 0, 1. \tag{4.7.5}$$

Formula (4.7.5), indeed, gives the probability distribution function of the random variable X. This is because first of all $\sum_{k=0}^{1} p^k q^{1-k} = p + q = 1$ and secondly that $p^k q^{1-k} > 0$.

Definition 4.7.6.

The distribution just discussed, defined by a *pmf* (4.7.5), is called a *Bernoulli distribution* and X is called a *Bernoulli random variable*.

Example 4.7.8.

If a repairman checks a machine for its working or nonworking condition three times a day, then the sample space for these repeated independent trials will have $2^3 = 8$ possible points. If we call the working condition a "success" and nonworking condition a "failure", the sample space will be $\Omega = \{sss, ssf, sfs, fss, ffs, fsf, sff, fff\}$, where the succession of letters indicates the success or failure in each trial. See Figure 4.7.4 for the tree diagram of three repeated Bernoulli trials.

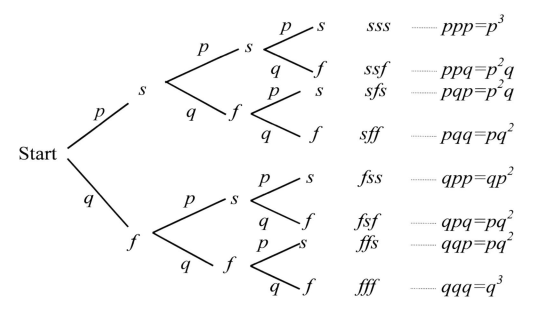

Figure 4.7.4. Tree diagram for three independent Bernoulli trials.

Now when we have n independent Bernoulli trials, the probability of any event will be the product of probabilities of simple events involved. For instance, in the example above, the probability of [*sff*], will be *pqq*. In case that working and nonworking status of the machine are equiprobable, then $P([sff])$ will be 1/8.

Note that in a Bernoulli trials sequence, if we let X_i be the random variable associated with the ith trial, then the sequence will consist of a sequence of ones and zeros.

Example 4.7.9.

Suppose that we test five different materials for strength/stress. If the probability of stress for each test is 0.2, then what is the probability that the 3rd material test is successful, i.e., does not break under pressure?

Solution

In this case, we have a sequence of five Bernoulli trials. Assuming 1 for a success and 0 for a failure (breakage), then we would have a 5-tuple (00100). Thus, the probability would be $(0.8)(0.8)(0.2)(0.8)(0.8) = 0.08192$.

Usually, out of repeated trials, we are interested in the probability of the total number of times that one of the two possible outcomes occurs regardless of the order of their occurrences. The following theorem gives us such a probability.

Theorem 4.7.1.

Let X_n be the random variable representing the number of successes in n independent Bernoulli trials. Then, *pmf* of X_n, denoted by $B_k = b(k; n, p)$, is:

$$B_k = \binom{n}{k} p^k q^{n-k}, \qquad k = 0, 1, 2, \cdots, n, \qquad (4.7.6)$$

where, and $q = 1 - p$.

Definition 4.7.7.

We say that $B_k = b(k; n, p)$ is the *binomial distribution function* with *parameters* n and p of the *random variable X*, where the parameters n, p and the number k refer to the number of independent trials, probability of *success* in each trial and the number of successes in n trials, respectively. In this case, X is called the *binomial random variable*. The notation $X \sim b(k; n, p)$ is used to indicate that X is a binomial random variable with parameters n and p.

Proof of Theorem 4.7.1.

As we mentioned before, the sample space consists of 2^n sample points. Each sample point consists of a succession of *s*'s and *f*'s. Since the trials are independent, the probability of each event (a succession of s's and f's) is a product of p's and q's. Since the interest is in the number of times s's occur in n trials regardless of their order, the number of ways we can have k s's in n trials will be $\binom{n}{k}$. For the k s's, the probability of occurrence is p^k and for the n-k f's the probability of occurrence will be q^{n-k}. Thus, the proof is completed.

Note that (4.7.6), indeed, defines a distribution function. This is because p and q being non-negative implies that $B_k > 0$, and

$$\sum_{k=0}^{n} B_k = \sum_{k=0}^{n} \binom{n}{k} p^k q^{n-k} = (p+q)^n = 1..$$

Example 4.7.10.

Let two identical machines be run together, each to select a digit from 1 to 9 randomly five times. What is the probability that a sum of 6 or 9 appears k times (k = 0, 1, 2, 3, 4, 5)?

Solution
We have five independent trials. The sample space for this example for one trial has 81 sample points and can be written in a matrix form as follows:

There are 13 sample points, where the sum of the components are 6 or 9 and is located in the matrix below. Hence, the probability of getting a sum as 6 or 9 on one selection of both machines together is $p = 13/81$. This will give the probability of success, $p = 13/81$.

$$\begin{bmatrix} (1,1) & (1,2) & \cdots & (1,8) & (1,9) \\ (2,1) & (2,2) & \cdots & (2,8) & (2,9) \\ \vdots & \ddots & \ddots & \vdots & \vdots \\ (8,1) & (8,2) & \ddots & (8,8) & (8,9) \\ (9,1) & (9,2) & \cdots & (9,8) & (9,9) \end{bmatrix}$$

Now let X be the random variable representing the total number of times a sum as 6 or 9 is obtained in 5 trials. Thus, from (4.7.6) we have

$$P([X = k]) = \binom{5}{k}(13/81)^k (68/81)^{5-k}, \qquad k = 0, 1, 2, 3, 4, 5.$$

For instance, the probability that the sum as 6 or 9 does not appear at all will be $(68/81)^5 = 0.42$, i.e., there is a 58 percent chance that we do get at least a sum as 6 or 9 during the five trials.

$$\begin{bmatrix} 0 & \cdots & & & (1,5) & \cdots & & (1,8) & 0 \\ 0 & & & (2,4) & \cdots & & (2,7) & & \\ \vdots & & \cdots & (3,3) & \cdots & & (3,6) & \cdots & \vdots \\ 0 & (4,2) & \cdots & & (4,5) & \cdots & & \vdots & 0 \\ (5,1) & \cdots & & (5,4) & \cdots & & \ddots & & \\ \vdots & & \cdots & (6,3) & \cdots & & \vdots & \ddots & \vdots \\ 0 & (7,2) & \cdots & & \ddots & & \ddots & & \\ (8,1) & \cdots & & \ddots & & \ddots & & \ddots & \vdots \\ 0 & \cdots & & & \cdots & & & \cdots & 0 \end{bmatrix}$$

Based on a sequence of independent Bernoulli trials we define two other important discrete random variables. So, consider a sequence of independent Bernoulli trials with p as probability of success in each trial, $0 \le p \le 1$. Suppose we are interested in the total number of trials required to have the rth success, r being a fixed positive integer.

Example 4.7.11.

Suppose in a lottery game the chance of winning is 15%. If, a person purchases 10 tickets, what is the probability of 3 winning tickets?

Solution

Let X be the number of winning tickets. Then, X has a binomial distribution with $n = 10$, $p = 0.15$, and $k = 3$. Thus,

$$P([X = 3]) = \binom{10}{3}(0.15)^3(0.85)^7 = 0.1298.$$

Definition 4.7.8.

Let X be a random variable with *pmf* as:

$$f(r;\ k,\ p) = P([X = r+k]) = \binom{r+k-1}{r}(1-p)^k p^r, \quad r = 0,\ 1,\ \cdots, \quad (4.7.7)$$

and k is a real, positive number. Alternatively, (4.7.7) can be written as:

$$f\left(r;\ k,\ p\right)=P([X=r])=\binom{r-1}{r-k}(1-p)^{k}p^{r-k},\quad r=k,\ k+1,\ \cdots.$$

Then, (4.7.7) is called a *negative binomial* (or *Pascal distribution*) *function* (or *binomial waiting time*). In particular, if $r=1$ in (4.7.7), then we will have

$$f\left(1;k,\ p\right)=P\left(X=k+1\right)=p(1-p)^{k},\ k=0,1,\cdots. \tag{4.7.8}$$

The *pmf* given by (4.7.8) is called a *geometric distribution*.

What the definition says is that X is a random variable representing the waiting time (failures) for the rth success to occur. The requirement of the rth success to occur implies that the number of trials must at least be r, i.e., $x=r,\ r+1,r+2,\ \cdots$. So, $X=r+k$ represents the occurrence of the rth success on the $(r+k)^{th}$ trial, $k=0,\ 1,\ \cdots$. This means we must have r-1 successes in the first $r+k$ - 1 trials and a success on the $(r+k)^{th}$ trial. Another way of saying this last statement is that we must have exactly k failures in $r+k$-1 trials and a success on the $(r+k)^{th}$ trial, i.e., k is the number of failures before the $(r+k)^{th}$ success.

Let Y be the random variable representing the number of failures before the rth success, i.e., $Y+r=X$. Also, let the *pmf* of Y be denoted by $f(r;\ k,\ p)$. Then,

$$P([Y=r])=f\left(r;\ k,\ p\right)=\binom{r+k-1}{r}p^{r-1}(1-p)^{k}p$$

$$=\binom{r+k-1}{r}p^{r}(1-p)^{k},\quad r=0,\ 1,\ 2,\ \cdots$$

Note that $\binom{r+k-1}{r}$ can be written as:

$$\binom{r+k-1}{r}=\frac{(r+k-1)(r+k-2)\cdots(r+k-1-r+1)}{r!}$$

$$=\frac{(r+k-1)(r+k-2)\cdots(k)}{r!}$$

$$=\frac{(-k)(-k-1)\cdots(-k-r+1)}{r!}(-1)^{r}$$

$$=\binom{-k}{r}(-1)^{r}.$$

Thus,

$$f(r;\ k,\ p)= P\big(X= r+ k\big)= \binom{-k}{r}\big(- 1\big)^r p^r (1-p)^k, \qquad r= 0,\ 1,\ \cdots.$$

This probability is indeed a distribution function since, first of all, $f(r;\ k,\ p) > 0$, and secondly using the Binomial Theorem we have:

$$f(k;\ r,\ p) = \sum_{k=0}^{\infty}\binom{-r}{k}\big(- 1\big)^k p^r q^k$$

$$= p^r \sum_{k=0}^{\infty}\binom{-r}{k}\big(- 1\big)^k q^k$$

$$= p^r \big(1- q\big)^{-r} = 1.$$

Example 4.7.12.

Suppose an automobile service company finds that 40% of cars attend for service need advanced-technology service. Suppose that on a particular day, all tags written are put in a pool and cars are drawn randomly for service. Finally, suppose that on that particular day, there are four positions available for advanced-technology services. What is the probability that the fourth car in need of advanced-technology service is found on the sixth tag picked?

Solution

Here we have independent trials with $p = 0.4$ as probability of success, i.e., in need of advanced-technology service, on any trial. Let X represent the number of the tags on which the fourth car in question is found. Then,

$$P\big(X= 6\big)= \binom{5}{3}\big(0.4\big)^4\big(0.6\big)^2 = 0.09216.$$

Another very important discrete random variable:

Definition 4.7.9.

A *Poisson random variable* is a nonnegative random variable X such that:

$$p_k = P\big([X= k]\big)= \frac{e^{-\lambda}\lambda^k}{k!}, \qquad k= 0,\ 1,\ \cdots, \tag{4.7.9}$$

where λ is a constant. Relation (4.7.9) is called a *Poisson distribution function with parameter* λ.

Later we will see that the parameter λ is the mean of this distribution function.

Note that probabilities given by (4.7.9) are nonnegative and

$$\sum_{k=0}^{\infty} \frac{e^{-\lambda} \lambda^k}{k!} = e^{-\lambda} \sum_{k=0}^{\infty} \frac{\lambda^k}{k!} = e^{-\lambda} e^{\lambda} = 1.$$

Thus, probabilities given by (4.7.9), indeed, define a probability mass function for the discrete random variable X.

Example 4.7.13.

Suppose that the number of telephone calls arriving to a switchboard of an institution every working day has a Poisson distribution with parameter 20. We want to respond to the following questions: What is the probability that there will be

(a) 30 calls in one day?
(b) at least 30 calls in one day?
(c) at most 30 calls in one day?

Solution

Using $\lambda = 20$ in (4.7.9) we will have

(a) $P_{30} = P([X = 30]) = \dfrac{\left(e^{-20}\right)\left(20^{30}\right)}{30!} = 0.0083$.

(b) $P([X \geq 30]) = \displaystyle\sum_{k=30}^{\infty} \frac{e^{-20} 20^k}{k!} = 0.0218$.

(c) $P([X \leq 30]) = 1 - P([X \geq 30]) + P([X = 30]) = 1 - .0218 + .0083 = .9865$.

Example 4.7.14.

Consider Example 4.7.10. Now suppose that the probability that an operator answers a call is 0.8. Let us find the distribution of the total number of calls answered by a particular operator.

Solution

Let X and Y be random variables representing the total number of calls arriving and the total number of calls answered by a particular operator, respectively. We want to find

$P([Y = j])$, $j = 0, 1, 2, \cdots$. If we know that n calls have arrived, then Y will have a binomial distribution $b(j; n, 0.8)$, i.e.,

$$P([Y = i|X = n]) = \binom{n}{j}(.8)^{j}(.2)^{n-j}, \qquad j = 0, 1, \cdots, n.$$

From the law of total probability (Theorem 4.5.3) we have:

$$
\begin{aligned}
P([Y = j]) &= \sum_{n=j}^{\infty} P([Y = j|X = n])P([X = n]) \\
&= \sum_{n=j}^{\infty} \binom{n}{j}(.8)^{j}(.2)^{n-j}\frac{e^{-20}20^{n}}{n!} \\
&= \frac{e^{-20}20^{j}(.8)^{j}}{j!}\sum_{n=j}^{\infty}\binom{n}{j}\frac{(.2)^{n-j}20^{n-j}}{(n-j)!} \\
&= \frac{e^{-20}20^{j}(.8)^{j}}{j!}\sum_{m=0}^{\infty}\binom{n}{j}\frac{(.2)^{m}20^{m}}{m!} \\
&= \frac{e^{-20}20^{j}(.8)^{j}}{j!}e^{(.2)(20)} = \frac{e^{-(.8)(20)}\left((.8)(20)\right)^{j}}{j!}.
\end{aligned}
$$

Hence, the distribution of Y is a Poisson with parameter $(.80)(20)$. For instance, the probability that 30 calls be answered by the end of the day will be:

$$P([Y = 30]) = \frac{e^{-(.8)(20)}\left((.8)(20)\right)^{30}}{30!} = \frac{e^{-16}16^{30}}{30!} = 0.0006.$$

One important property of the Poisson distribution is that it can be obtained as an approximation from a binomial distribution. In fact, the Poisson distribution approximates binomial distribution when the number of independent Bernoulli trials is large and the probability of success in each trial is small. Thus, we prove the following theorem:

Theorem 4.7.2.

Let X be a binomial random variable with distribution function B_{k}. Let $\lambda = np$ be fixed. Then,

$$\lim_{\substack{n\to\infty \\ p\to 0}}B_{k} = \frac{\lambda^{k}e^{-k}}{k!}, \qquad k = 0, 1, 2, \cdots. \tag{4.7.10}$$

Proof

From (4.7.6) and the hypothesis of the theorem, $\lambda = np$, we have

$$P(X = k) = \binom{n}{k} p^k q^{n-k} = \frac{n!}{k!(n-k)!} p^k (1-p)^{n-k}$$

$$= \frac{n!}{k!(n-k)!} \frac{n^k p^k}{n^k} \left(1 - \frac{np}{n}\right)^{n-k} = \frac{n! n^k}{k!(n-k)!} \left(\frac{\lambda}{n}\right)^k \left(1 - \frac{\lambda}{n}\right)^{n-k}$$

$$= \frac{n(n-1)\cdots(n-k+1)}{n^k} \frac{\lambda^k}{k!} \frac{\left(1 - \frac{\lambda}{n}\right)^n}{\left(1 - \frac{\lambda}{n}\right)^k}. \qquad (4.7.11)$$

Now for n large, p small, and knowledge of calculus, we will have

$$\lim_{n\to\infty} \left(1 - \frac{\lambda}{n}\right)^n = e^{-\lambda}, \ \lim_{n\to\infty} \frac{n(n-1)\cdots(n-k+1)}{n^k} = 1 \text{ and } \lim_{n\to\infty} \left(1 - \frac{\lambda}{n}\right)^k = 1.$$

Substituting these values in (4.7.11) will complete the proof.

Example 4.7.15.

Suppose a machine is making tools each of which needs a random number of parts of a particular type. We place 1000 of this particular part in a container designed for this purpose. Suppose that 100 tools have depleted the stock of these parts. Now suppose a tool is chosen at random. What is the probability that it has k number of these parts ($k = 0, 1, 2, \cdots$)?

Solution

Assume that all parts have the same chance to be used in a tool. Hence, the probability of a part to be used is .01. So the problem can be described as having $n = 1000$ independent Bernoulli trials. Each trial consists of trying to use a part in a tool. The probability of a success, p, is .01. Of course, $p = .01$ is small. Thus, we can use Poisson approximation, (4.7.11), with $\lambda = (1000)(.01) = 10$.

For 26 different values of k, the values of binomial, $b(k; 1000, .01)$, and Poisson, p_k, distributions with $\lambda = 10$ are given in the following table.

k	$b(k; 1000, .01)$	p_k
0	.0000	.0000
1	.0000	.0005
2	.0022	.0023
3	.0074	.0076
4	.0186	.0189
5	.0374	.0378
6	.0627	.0631
7	.0900	.0901
8	.1128	.1126
9	.1256	.1251
k	$b(k; 1000, .01)$	p_k
10	.1257	.1251
11	.1143	.1137
12	.0952	.0948
13	.0731	.0729
14	.0520	.0521
15	.0345	.0347
16	.0215	.0217
17	.0126	.0128
18	.0069	.0071
19	.0036	.0037
20	.0018	.0019
21	.0009	.0009
22	.0004	.0004
23	.0002	.0002
24	.0001	.0001
25	.0000	.0000

4.8. RANDOM VECTORS

Two distinct chance experiments with sample spaces Ω_1 and Ω_2, can conceptually be combined into a single one with only one sample space, Ω. The new sample space, Ω, is the Cartesian product of Ω_1 and Ω_2, i.e., $\Omega_1 \times \Omega_2$, which is the set of all ordered pairs (Ω_1, Ω_2) with $\omega_1 \in \Omega_1$ and , i.e., ω_1 and ω_2 are outcomes of the first and the second experiment, respectively. This idea may be extended to a finite number of sample spaces. In such cases we are talking of a random vector. In other words by a *random vector* we mean an n-tuple of random variables. More precisely, we have the following definition:

Definition 4.8.1.

A *discrete random vector* $\boldsymbol{X}(\omega) = [X_1(\omega), \cdots, X_r(\omega)]$, where $X_1, \cdots, X_r$ are r discrete random variables and $\omega \in \Omega$, is a function from the sample space Ω into the r-tuple

(r-dimensional) real line, $\mathbf{R}^r$, such that for any r real numbers $x_1, x_2, \cdots, x_r$, the set $\{\omega \in \Omega : X_i(\omega) = x_i, i = 1, 2, \cdots, r\}$ is an event.

Example 4.8.1.

Suppose two identical machines are to randomly select digits from 1 to 9 at the same time. Let X_1 be the random variable describing the sum of digits selected by the machines, and let X_2 be the absolute value of the difference of the digits. Thus, the sample space Ω is $\Omega = \{(i, j) : i, j = 1, 2, \cdots, 9\}$. If ω is a typical element of Ω, then $X_1(\omega) = i + j$ and $X_2(\omega) = |i - j|$. Hence, the range of X_1 and X_2 are $\{2, 3, \cdots, 18\}$ and $\{0, 1, 2, \cdots, 8\}$, respectively. Therefore, (X_1, X_2) is a random vector with values as $\{(x_1, x_2) : x_1 = 2, 3, \cdots, 18; x_2 = 0, 1, \cdots, 8\}$.

Example 4.8.2.

Suppose a manufacturing company wants to control the quality of its product by considering numerical factors $x_1, x_2, \cdots, x_n$. These n factors could be weight, height, volume, color, etc. This control may very well be done by the numerical value of the probability $P(X_1 \leq x_1, \cdots, X_n \leq x_n)$, where the random vector $(X_1, X_2, \cdots, X_n)$ will describe the joint factors concerned by the company.

When we have a random vector, the probability distribution of such a vector must give the probability for all the components at the same time. This is the joint probability distribution for the n component random variables which make up the random vector. So, the necessary probabilistic information can be transferred to the range value from the original probability space.

Definition 4.8.2.

Let X be a bivariate random vector, i.e., $X = (x_1, x_2)$. Suppose that x_1 and x_2 are two real numbers. Let $p_{x_1 x_2} = P(X_1 = x_1, X_2 = x_2)$. Then, p_{x_1, x_2} is called the *(discrete) joint probability mass function of* x_1 *and* x_2. The bivariate X can be extended to a random vector with discrete components $X_1, \cdots, X_r$. Then, the joint probability mass function of $x_1, x_2, \cdots, x_r$ is, similarly, defined as $p_{x_1 x_2 \cdots x_r} = P(X_1 = x_1, X_2 = x_2, \cdots, X_r = x_r)$.

Note that from the Axioms of probability, a joint probability mass function has the following properties:

(a) $p_{x_1 x_2 \cdots x_r} \geq 0,$

(b) $\displaystyle\sum_{x_1}\sum_{x_2}\cdots\sum_{x_r} p_{x_1 x_2 \cdots x_r} = 1.$

Example 4.8.3.

Suppose a repairman checks two identical machines at random for their working conditions. Let the working and not working conditions of the machines be denoted by 1 and 0, respectively. Thus, the sample space Ω is $\Omega = \{(0,0), (0,1), (1,0), (1,1)\}$. Now let X_1 and X_2 be two random variables defined as follows:

   X_1 = the number of machines working on the first check.
   X_2 = the number of machines working on the second check.

We want to find the joint distribution function of X_1 and X_2.

Solution
   As we mentioned, the values of the random vector are (0,0), (0,1), and (1,1). Assuming equiprobable property, each one of these values has probability 1/4. Now using Definition 4.8.1, we observe that:

$$\{\omega \in \Omega : X_1(\omega) = x_1,\ X_2(\omega) = x_2\} = \begin{cases} \varnothing, & \text{if} \quad x_1 < 0 \ \text{ or } \quad x_2 < 0, \\ \{(0,0)\}, & \text{if} \quad 0 \leq x_1 < 1 \text{ and } 0 \leq x_2 \leq 1, \\ \{(0,0),(0,1)\}, & \text{if} \quad 0 \leq x_1 < 1 \text{ and } \quad x_2 \geq 1, \\ \Omega, & \text{if} \quad x_1 \geq 1 \text{ and } \quad x_2 \geq 1. \end{cases}$$

Therefore, the joint distribution function $F_{X_1 X_2}(x_1, x_2)$ will be

$$\begin{aligned} F_{X_1 X_2}(x_1, x_2) &= 0, && \text{if} \quad x_1 < 0 \text{ or } \quad x_2 < 0 \\ &= 1/4, && \text{if} \quad 0 \leq x_1 < 1 \text{ and } 0 \leq x_2 \leq 1 \\ &= 1/2, && \text{if} \quad 0 \leq x_1 < 1 \text{ and } \quad x_2 \geq 1 \\ &= 1, && \text{if} \quad x_1 \geq 1 \text{ and } \quad x_2 \geq 1. \end{aligned}$$

In case that the random vector is bivariate, say (X, Y), then the joint probability mass function may be displayed by a rectangular array as shown in Chart 4.8.1.

X \ Y	y_1	y_2	$\cdots$	y_n	$\cdots$
x_1	p_{x_1,y_1}	p_{x_1,y_2}	$\cdots$	p_{x_1,y_n}	$\cdots$
x_2	p_{x_2,y_1}	p_{x_2,y_2}	$\cdots$	p_{x_2,y_n}	$\cdots$
$\cdots$	$\cdots$	$\cdots$	$\cdots$	$\cdots$	$\cdots$
x_m	p_{x_m,y_1}	p_{x_m,y_2}	$\cdots$	p_{x_m,y_n}	$\cdots$
$\cdots$	$\cdots$	$\cdots$	$\cdots$	$\cdots$	$\cdots$

Chart 4.8.1. Discrete joint probability distribution, bivariate vector.

Example 4.8.4.

Suppose the joint probability mass function of X and Y is given by: $p_{1,1} = 1/11$, $p_{1,2} = 2/11$, $p_{2,1} = 3/11$, and $p_{2,2} = 5/11$. Then, we can arrange these probabilities in a rectangular array as follows:

X \ Y	1	2	Total
1	1/11	2/11	3/11
2	3/11	5/11	8/11
Total	4/11	7/11	1

Again, in the discrete bivariate, p_{x_i,y_j} means the probability that $X = x_i$, $Y = y_j$, and P is defined on the set of ordered pairs $\{(x_i, y_j), \; j \leq i \leq m, \; i \leq j \leq n\}$ by $p_{x_i,y_j} = P\big([X = x_i] \text{ and } [Y = y_j]\big)$. Now if $A_i = X = x_i$ and $B_j = Y = y_j$ are events, then $A_i B_j$, $i = 1, 2, \cdots, m$, are mutually exclusive events and $A_i = \bigcup A_i B_j$. Thus, for each $i = 1, 2, \cdots, m$, $i = 1, 2, \cdots, n$, we have:

$$p_{x_i} = P(A_i) \sum_{j=1}^{n} P(A_i B_j) = \sum_{j=1}^{n} p_{x_i,y_j} \tag{4.8.1}$$

Similarly, we will have:

$$p_{y_j} = P(B_j) \sum_{i=1}^{m} P(A_i B_j) = \sum_{i=1}^{m} p_{x_i,y_j} \tag{4.8.2}$$

Definition 4.8.3.

Each probability mass function p_X and p_Y, defined by (4.8.1) and (4.8.2), respectively, is called the *marginal probability mass function.*

In other words, a marginal probability mass function, p_X or p_Y, can be obtained from the joint distribution function p_{XY} of X and Y by summing up the joint distribution over all values of the Y or X, respectively.

As in Chart 4.8.1, we can display the marginal distribution of a bivariate random vector in a rectangular array as shown in Chart 4.8.2.

Example 4.8.5.

Suppose a box contains 9 balls: 3 blacks, 2 greens, and 4 blues. We draw 6 balls at random, one at a time and replacing after registering the color. We want to find the probability of getting 3 blacks, 1 green, and 2 blues.

X \ Y	y_1	y_2	$\cdots$	y_n	p_X
x_1	p_{x_1,y_1}	p_{x_1,y_2}	$\cdots$	p_{x_1,y_n}	p_{x_1}
x_2	p_{x_2,y_1}	p_{x_2,y_2}	$\cdots$	p_{x_2,y_n}	p_{x_2}
$\cdots$	$\cdots$	$\cdots$	$\cdots$	$\cdots$	$\cdots$
x_m	p_{x_m,y_1}	p_{x_m,y_2}	$\cdots$	p_{x_m,y_n}	p_{x_m}
p_Y	p_{y_1}	p_{y_2}	$\cdots$	p_{y_n}	1 (total)

Chart 4.8.2. Discrete marginal distribution function of a bivariate random vector.

Solution

Let the random variables X_1, X_2, and X_3 represent the number of black, green, and blue color balls, respectively, drawn in the six independent trials. The probability of drawing these colors on each draw (trial) will be: 1/3, 2/9, and 4/9, respectively. Now the number of ways to obtain a black, a green and a blue is 3!/1!2!. Hence, the joint probability mass function in question will be

$$P\left(X_1 = 3,\ X_2 = 1,\ X_3 = 2\right) = \frac{6!}{3!1!2!}\left(1/3\right)^3\left(2/9\right)^1\left(4/9\right)^2 = 0.98.$$

The above example is an illustration of independent trials, where in each trial there is a number of possible outcomes instead of just two, "success" and "failure", that we had in a Bernoulli trial. Thus, we can extend the binomial distribution to *multinomial distribution.*

But, first recall the Binomial Theorem [formula (4.2.8)]. We now extend this theorem by statement and without proof as follows:

Theorem 4.8.1. Multinomial Theorem

$$\left(x_1 + x_2 + \cdots + x_n\right)^n = \sum_{r=1}^{n} \frac{n!}{n_1! \, n_2! \cdots n_r!} \left(x_1\right)^{n_1} \left(x_2\right)^{n_2} \cdots \left(x_n\right)^{n_r}, \tag{4.8.3}$$

where the sum ranges over all ordered r-tuples $n_1, n_2, \cdots n_r$, such that $n_1 \geq 0$, $n_2 \geq 0$, $n_r \geq 0$, and $n_1 + n_2 + \cdots + n_r = n$.

Note that when $r = 2$, this theorem reduces to the Binomial Theorem [formula (4.2.8)] (why?).

Now suppose $X_1, X_2, \cdots, X_r$ are r random variables representing the occurrence of r outcomes among $X_1, X_2, \cdots, X_r$ possible outcomes of a chance experiment which is being repeated independently n times. Let the corresponding probabilities of these outcomes be $p_1, p_2, \cdots, p_r$, respectively. Then, the joint pmf of these random variables will be:

$$P\left(X_1 = x_1, \cdots X_r = x_r\right) = \frac{n!}{n_1! \cdots n_{r!}} p_1^{n_1} \cdots p_r^{n_r}, \tag{4.8.4}$$

where n_j ranges over all possible integral values subject to (4.8.4).

Definition 4.8.4.

The relation (4.8.4) does, indeed, represent a probability mass function (why?) and it is called the *multinomial probability mass function* for the random vector $(X_1, X_2, \cdots, X_r)$. We denote this distribution similar to the binomial distribution as: $m\left(n; \; r; \; n_1, \; \cdots, \; n_r\right)$ subject to $p_1 + \cdots + p_r = 1$.

Marginal mass function for each one of the random variables $X_1, X_2, \cdots, X_r$ alone will be a binomial probability mass function and can be obtained from (4.8.4). For instance, for X_1 we have

$$P\left(X_1 = x_1\right) = b\left(n; \; p, \; n\right) = \frac{n!}{n_1!\left(n - n_1\right)!} p_1^{n_1} \left(1 - p_1\right)^{n - n_1}. \tag{4.8.5}$$

4.9. CONDITIONAL DISTRIBUTION AND INDEPENDENCE

Earlier, we defined the conditional probability of the event B, given the event A (Definition 4.4.1) as: $P(B|A) = P(AB)/P(A)$. We also discussed the Law of Total Probability in Theorem 4.5.1. This law can be restated in terms of a random variable as follows:

Theorem 4.9.1. (The Law of Total Probability)

$$P(A) = \sum_x P(A|X = x) P(X = x).$$

Proof

Left as an exercise.
We now discuss the conditional mass (density) functions.

Definition 4.9.1.

Let (S, Ω, P) be a probability space and X and Y be two random variables defined on S. Then, the conditional *pmf* of X given Y is defined as: If X and Y are elementary with marginal probability mass functions p_X and p_Y, respectively, and joint mass (density) function $p_{X,Y}$, then the *conditional mass (density) function of X, given Y*, denoted by $p_{X,Y}$ is:

$$p_{X|Y} = \frac{p_{X,Y}}{p_Y}, \tag{4.9.1}$$

provided that $p_Y > 0$. $p_{X|Y}$ in this case means $P(X = x|Y = y)$.

Example 4.9.1.

Let X and Y be random variables with joint probability mass function as:

$$p_{X,Y} = \begin{cases} \dfrac{1}{16}(x + y), & x = -2, -1, 0, 1; \ y = 2, 3, \\ 0, & \textit{otherwise.} \end{cases}$$

We want to find the conditional distribution of X, given Y.

Solution

As we did before, the joint *pmf* of X and Y as well as the marginal distributions of X and Y are shown in Table 4.9.1.

Table 4.9.1. The joint distribution of X and Y, and the marginal distribution of X and Y

Y \ X	- 2	-1	0	1	p_Y
2	0	1/16	1/8	3/16	3/8
3	1/16	1/8	3/16	1/4	5/8
p_X	1/16	3/16	5/16	7/16	1 (total)

Now the marginal distribution function of Y is:

$$p_Y = \sum_X p_{X,Y}$$
$$= p_{-2,y} + p_{-1,y} + p_{0,y} + p_{1,y}$$
$$= (1/16)(-2 + y - 1 + y + 0 + y + 1 + y)$$
$$= (1/16)(-2 + 4y), \, y = 2, 3.$$

Thus, from (4.9.1) we have:

$$p_{X|Y} = (x + y)/(-2 + 4y), \, x = -2, -1, 0, 1; \, y = 2, 3.$$

The conditional *pmf* for each given Y is shown in Table 4.9.2.

Table 4.9.2. Conditional *pmf* , $p_{X|Y}$, with $Y = 2, 3$

Y \ X	- 2	-1	0	1	Total
			$p_{X|Y}$		
2	0	1/6	1/3	1/2	1
3	1/10	1/5	3/10	2/5	1

As we did with events, we can define independent random variables. Recall that in section 8 we mentioned that a random variable generates a partition $\mathcal{P}_X$ (refer to Definition 4.1.2). Now two partitions are independent as defined below:

Definition 4.9.2.

Two partitions P_1 and P_2 of a sample space S are *independent* if and only if any event from P_1 is independent of any event of P_2. Hence, if $P_1 = (A_1, \cdots, A_m)$ and $P_2 = (B_1, \cdots, B_n)$, then P_1 and P_2 are independent if and only if:

$$P(A_i B_j) = P(A_i) P(B_j), \; 1 \le i \le m, \; 1 \le j \le n. \tag{4.9.2}$$

Thus, we can define independence of two discrete random variables as follows:

Definition 4.9.3.

Let (S, Ω, P) be an elementary probability space and let the random vector $X = (X_1, X_2)$ be defined on the sample space S. Then, the random variables X_1 and X_2 are *independent* if the partitions generated by X_1 and X_2 are independent.

Theorem 4.9.2.

If X and Y are two independent discrete random variables, then the joint cumulative probability distribution function of X and Y is the product of the marginal distributions of X and Y. The same is true for probability mass functions, i.e., two random variables X and Y are independent if and only if

$$p_{X,Y} = p_X p_Y, \tag{4.9.3}$$

where p_X and p_Y are *pmf* of X and Y, respectively.

Proof

Let $P_1 = (A_1, \cdots, A_m)$ and $P_2 = (B_1, \cdots, B_n)$ be partitions generated by the random variables X and Y, respectively. Thus, if X and Y are independent, then $P(A_i B_j) = P(A_i) P(B_j), \; 1 \le i \le m, \; 1 \le j \le n$. Now let the probability mass function of X and Y be denoted by p_X and p_Y, respectively, with the joint distribution function $p_{X,Y}$. Then, indeed, $P(A_i) = p_X(x_i)$, $P(B_j) = p_Y(y_j)$, and $P(A_i B_j) = p_{X,Y}(x_i y_j)$. Thus, if X and Y are independent, then $p_{X,Y}(x_i, y_j) = p_X(x_i) p_Y(y_j), \; 1 \le i \le m, \; 1 \le j \le n$.

From (4.9.3), if two random variables X and Y are independent, then the probability mass function of sum of these random variables can be written as follows:

$$P(X + Y = k) = \sum_{i=0}^{k} P(X = i, Y = k - i) = \sum_{i=0}^{k} P(X = i) P(Y = k - i). \quad (4.9.4)$$

That is, the probability mass function of sum of two independent variables is the convolution of their *pmf*'s.

Theorem 4.9.3.

Let X and Y be two independent random variables with marginal mass (density) functions $p_X(x)$ and $p_Y(y)$, respectively. Then, if $E(X)$ and $E(Y)$ exist, we will have:

$$E(XY) = E(X) E(Y). \quad (4.9.5)$$

Proof

From the independence of X and Y we have $p_{X,Y} = p_X(x) p_Y(y)$. Thus, from the definition of expectation we obtain:

$$E(XY) = \sum_x \sum_y xy p_X(x) p_Y(y)$$

$$= \left[\sum_x x p_X(x) \right] \left[\sum_y y p_Y(y) \right]$$

$$= E(X) E(Y).$$

Note that this theorem can be extended for a finite number of random variables, i.e., if $X_1, X_2, \cdots, X_n$ are n independent random variables, then

$$E(X_1 X_2 \cdots X_n) = E(X_1) E(X_2) \cdots E(X_n)$$

Theorem 4.9.4.

If X and Y are independent, then $Cov(X,Y) = 0$.

Proof

Left as exercise.

Note that the converse of Theorem 4.9.4 is not true, i.e., zero covariance does not imply independence of the random variables. The dependency of two random variables is measured by the coefficient of correlation as defined below.

Corollary 4.9.1.

If X and Y are independent random variables, then

$$Var(X + Y) = Var(X) + Var(Y). \tag{4.9.6}$$

Proof

Left as exercise.

4.10. DISCRETE MOMENTS

Definition 4.10.1.

The *arithmetic average* of numbers n_1, n_2, $\cdots$, n_k denoted, by $\overline{n}$, is defined as

$$\overline{n} = \frac{n_1 + n_2 + \cdots + n_k}{k}. \tag{4.10.1}$$

In other words, we add all n values with the same weight, namely, $1/k$.

Example 4.10.1.

If scores on three tests and the final exam are 70, 58, 88, and 100, respectively, then the arithmetic average of these scores is

$$\overline{n} = \frac{70 + 58 + 88 + 100}{4} = 79.$$

Example 4.10.2.

Now suppose the first three tests were counted equally and the final exam counted twice as much as of a test. Then, the arithmetic average of scores will be

$$\bar{n} = \frac{(1)(70) + (1)(58) + (1)(88) + (2)(100)}{5} = 83.2.$$

In this case we still added the scores, but with different weights. That is, the first three tests with weight 1/5 and the final exam with weight 2/5.

When weights are different, then the name "*weighted average*" is used, otherwise it is said "*the average*".

Now, since a random variable was defined as a real-valued function, then we can speak of its average. However, to each value of a random variable X, we associate a probability (often different for each value), hence these probabilities form weights for the values of X. Talking about a chance variable we can talk about "*expected value*" of the random variable X rather then the "average" or interchangeably so. Thus, we define "*expected value*" or "*expectation*" or "*mathematical expectation*" or "*weighted average*" or "*average*" or "*mean*" of a random variable X as follows:

Definition 4.10.2.

Let X be a discrete random variable defined on a sample space Ω with *pmf* p_X. Then, the *mathematical expectation* or simply *expectation* of X, denoted by $E(X)$, is defined as follows: if Ω is finite and the range of X is $\{x_1, x_2, \cdots, x_n\}$, then

$$E(X) = \sum_{i=1}^{n} x_i p_{X_i} \tag{4.10.2}$$

If $\{x_1, x_2, \cdots, x_n, \cdots\}$ is infinite, then

$$E(X) = \sum_{i=1}^{\infty} x_i p_{X_i}, \tag{4.10.3}$$

provided that the series converges.

Example 4.10.3.

Suppose 3 defective light bulbs are mixed with other 21 light bulbs in a box totaling 24 bulbs. Let a sample of 4 bulbs be selected from this box. What is the expected number of defective light bulbs in this sample?

Solution

Let the number of defective light bulbs in the sample be denoted by X. Then, the sample space Ω in this case is the set of 24 light bulbs. There are $\binom{24}{4} = (24!)/[(4!)(20!)] = 10626$ distinct ways of choosing the sample of size 4 from this sample space. We assume that all sample points are equiprobable. Now there are:

$$\binom{4}{0}\binom{21}{4} = 5985 \text{ samples with no defective bulbs,}$$

$$\binom{4}{1}\binom{21}{3} = 5320 \text{ samples with one defective bulbs,}$$

$$\binom{4}{2}\binom{21}{2} = 1260 \text{ samples with two defective bulbs,}$$

$$\binom{4}{3}\binom{21}{1} = 84 \text{ samples with three defective bulbs,}$$

$$\binom{4}{4}\binom{21}{0} = 1 \text{ samples with four defective bulbs.}$$

Hence, the probability of having no defective, one defective, two defective, three defective, and four defective bulbs is 5985/10626, 5320/10626, 1260/10626, 5320/10626, and 4854/10626, respectively. In other words, X has the following distribution and the expected number of defective light bulbs is:

$$x$$

X	0	1	2	3	4
p_X	5985/10626	5320/10626	1260/10626	84/10626	1/10626

$$E(X) = (0)(5985/10626) + (1)(5320/10626) + (2)(1260/10626) + (3)(84/10626) + (4)(1/10626) = 0.77$$

Of particular interest is the expectation of the indicator function. Recall that we defined the indicator random variable by (4.6.1) as: for each $A \in \Omega$:

$$I_{A(s)} = \begin{cases} 1, & \text{if } s \in A \\ 0, & \text{if } s \in A^c. \end{cases}$$

The expectation of such a random variable is given by the following theorem:

Theorem 4.10.1.

If I_A is an indicator function of an event A, then $E(I_A) = P(A)$.

Proof

Left as an exercise.

We now summarize some of the basic properties of expectation in the following theorems:

Theorem 4.10.2.

If c is a constant, then $E(c) = c$.

Proof

Left as an exercise.

Theorem 4.10.3.

Expectation is a linear operation, i.e., if c, c_1, and c_2 are constants and X and Y are two random variables, then

$E(cX) = cE(X)$, and $E(c_1X + c_2Y) = c_1 E(X) + c_2E(Y)$.

Proof

$$E(cX) = \sum_x cxp_X = c\sum_x xp_X = cE(X).$$

$$E(c_1X + c_2Y) = \sum_x \sum_y p_{X,Y}(c_1x + c_2y) = \sum_x \sum_y (c_1xp_{X,Y} + c_2yp_{X,Y})$$

$$= \sum_x c_1x \sum_y p_{X,Y} + \sum_x c_2y \sum_y p_{X,Y} = c_1 \sum_x xp_X + c_2 \sum_y yp_Y$$

$$= c_1 E(X) + c_2 E(Y).$$

Hence, the proof is completed.

Corollary 4.10.1.

If X_1, X_2, $\cdots$, X_n are n random variables, then

$$E(X_1 + X_2 + \cdots + X_n) = E(X_1) + E(X_2) + \cdots + E(X_n).$$

Proof

Left as an exercise.

Example 4.10.4.

Suppose products of a manufacture are boxed by 24 units in a box. These boxes are controlled by an inspector. The inspector chooses an item from a box for possible defectiveness. If a defective item is found in a box, the box packer must pay \$120. Suppose a packer found out that he left one defective item in a box by mistake and that box is under inspection. What is the expected pay of the packer?

Solution

Let the random variable X represents the amount of money that the packer has to pay. Then, X is 120 for exactly one item and 0 for all other items. Suppose all items in the box have the same chance to be picked up by the inspector. Hence, X is defined as:

$$X = \begin{cases} 120, & \textit{with probability } 1/24, \\ 0, & \textit{with probability } 23/24, \end{cases}$$

and the expected pay will be: $E(X) = (120)(1/24) + (0)(23/24) = \$5.$

Example 4.10.5.

Consider Example 4.10.4. Now suppose the inspector chooses two items from a box at random to inspect. What is the expected pay of the packer?

Solution

In this case we have two random variables, say, X_1, and X_2 each defined in the same way as in the Example 4.10.4. Then: $E(X_1 + X_2) = E(X_1) + (X_2) = 2E(X) = (2)(5) = \$10.$

Alternative Solution

We could do this problem in another way. Since X will be 120 for two exactly cases and 0 for all other cases, we have:

$$X = \begin{cases} 120, & \text{with probability} \quad 2/24, \\ 0, & \text{with probability} \quad 22/24, \end{cases}$$

and $E(X) = (120)(2/24) + (0)(22/24) = \$10$.

Note that if the inspector in Examples 4.10.4 and 4.10.5 inspects all 24 items, then the expected pay of the packer will be $(24)(5) = \$120$.

Now we define expectation of a particular random variable, namely, X^r, when r is a positive integer.

Definition 4.10.3.

Let X be a discrete random variable and r a positive integer, then

$E(X^r)$ is called the *rth moment* of X or *moment of order r* of X.

In symbols:

$$E(X^r) = \sum_{k=1}^{\infty} x_k^r P(X = x_k). \tag{4.10.3}$$

Note that if $r = 1$, $E(X^r) = E(X)$, i.e., the moment of first order or the first moment of X is just the expected value of X. The second moment, i.e., $E(X^2)$ is also important, as we will see below.

Let us denote $E(X)$ by μ. It is clear that if X is a random variable $X - \mu$, where μ is a constant, is also a random variable. However, $E(X - \mu) = E(X) - E(\mu) = \mu - \mu = 0$. In other words, we can *center* X by choosing the new random variable $X - \mu$. This leads to the following definition.

Definition 4.10.4.

The rth moment of the random variable $X - \mu$, i.e., $E\left[(X - \mu)^r\right]$ is called the *central moment* of X. Note that the new random variable $X - \mu$ measures the *deviation* of X from its expectation (or the mean). However, this deviation can be positive or negative according to the values of $X - \mu$. Hence, the absolute value of this will give an *absolute measure of deviation* of the random variable X from its mean μ. Since working with absolute value is not quite easy and desirable, the *mean square deviation*, i.e., $E\left[(X-\mu)^2\right]$ (which is the *second central moment of X*) is usually considered.

The important second moment has a particular name as follows:

Definition 4.10.5.

Let X be a random variable with a finite expectation $E(X)$. Then, the *variance* of X, denoted by $Var(X)$ or equivalently by $\sigma^2(X)$, or if there is no fear of confusion, just σ^2, is defined as the second central moment of X, i.e., $\sigma^2(X)=E\left[(X-\mu)^2\right]$. In other words, the variance measures the *average deviation* or *dispersion* of the random variable from its mean.

Theorem 4.10.4.

If X is a random variable and μ is finite, then

$$Var(X)=E(X^2)-\mu^2.$$ (4.10.4)

Proof

From Definition 4.10.5 we see that

$$\sigma^2(X)=E\left[(X-\mu)^2\right]=E(X^2-2\mu X+\mu^2)=E(X^2)-2E(\mu X)+E(\mu^2)$$
$$=E(X^2)-2\mu^2+\mu^2=E(X^2)-\mu^2.$$

As we see, the variance does measure the deviation from the mean, but it rather measures it by squares. To adjust for this squaring process we extract square root.

Definition 4.10.6.

The positive square root of the variance of a random variable X is called the *standard deviation* and is denoted by $\sigma(X)$.

Example 4.10.6.

Consider Example 4.10.3. Let us find the variance and the standard deviation for the random variable X. In that example we found that $\mu = 2.2403$. Now from Definition 4.10.2 we have:

$$E\left(X^2\right) = \sum_{x_k} x_k^2 p_{x_k}$$

$$= 0^2(5985/10626) + 1^2(5320/10626) + 2^2(1260/10626)$$
$$+ 3^2(5320/10626) + 4^2(1/10626)$$
$$= 0 + 0.5007 + 0.4743 + 4.5059 + 0.0015$$
$$= 5.4824.$$

Hence, from (4.10.4), the variance of X is $\sigma^2(X) = E\left(X^2\right) - \left(E(X)\right)^2 = 5.4824 -$ $5.0189 = 0.4635$, and, from Definition 4.10.6, the standard deviation of X is $\sigma(X) = 0.6808$.

Note that if the variance is small, it shows that the values of the random variable are clustered about its mean.

We have seen one property of the variance in the Theorem 4.10.4. Let us consider some more properties.

Theorem 4.10.5.

If X is a random variable and c is a real number, then

(a) $Var(X + c) = Var(X)$,
(b) $Var(cX) = c^2\, Var(X)$.

Proof

(a) $Var\left(X + c\right) = E\left[(X + c)^2\right] - \left[E(X + c)\right]^2 = E\left(X^2 + 2cX + c^2\right) - \left(\mu + c\right)^2$
$$= E\left(X^2\right) + 2c\mu + c^2 - \mu^2 - 2c\mu - c^2 = E\left(X^2\right) - \mu^2 = Var\left(X\right)$$

(b) $Var\left(cX\right) = E\left[(cX)^2\right] - \left[E(cX)\right]^2 = E\left(c^2 X^2\right) - \left(c\mu\right)^2$
$$= c^2 E\left(X^2\right) - c^2\mu^2 = c^2\left[E\left(X^2\right) - \mu^2\right] = c^2 Var\left(X\right).$$

A notion of importance that measures the dependency of random variables is the notion of covariance, which will be defined as follows:

Definition 4.10.7.

Let $E(X) = \mu_X$ and $E(Y) = \mu_Y$. Then, the *covariance* of two random variables X and Y, denoted by $Cov(X,Y)$, is defined by:

$$Cov(X,Y) \equiv E\left[(X - \mu_X)(Y - \mu_Y)\right]. \tag{4.10.5}$$

Theorem 4.10.6.

$$Cov(X,X) = Var(X). \tag{4.10.6}$$

Proof

Replaced Y with X in (4.10.5).

From (4.10.5) it can be easily shown that:

Theorem 4.10.7.

$$Cov(X,Y) = E\left[(X)(Y)\right] - \mu_X \mu_Y. \tag{4.10.7}$$

Proof

Left as exercise.

As other properties of the covariance, we have:

Theorem 4.10.8.

$$Cov(X + Y, Z) = Cov(X,Y) + Cov(X,Z). \tag{4.10.8}$$

Proof

Left as exercise.

Theorem 4.10.9.

$$Cov(X,Y) = Cov(Y,X)$$

Proof

Left as exercise.

Theorem 4.10.10.

If c is a real number, then $Cov(c\,X,Y) = c\,Cov(X,Y)$.

Proof

Left as exercise.

Theorem 4.10.11.

For two random variables X and Y we have:

$Var(X + Y) = Var(X) + Var(Y) + 2\,Cov(X,Y)$.

Proof

Left as exercise.

Definition 4.10.8.

The *coefficient of correlation* of two random variables X and Y denoted by $\rho(X,Y)$ is given by:

$$\rho(X,Y) \equiv \frac{Cov(X,Y)}{\sigma(X)\sigma(Y)}, \tag{4.10.9}$$

where $\sigma(X)$ and $\sigma(Y)$ are standard deviations of X and Y, respectively, provided that the denominator is not zero.

It can be shown that:

$$-1 \le \rho(X,Y) \le 1. \tag{4.10.10}$$

When ρ is negative, it means that the random variables are dependent oppositely, i.e., if the value of one increases, the value of the other one will decrease. When ρ is positive, it means that both random variables increase and decrease together. If $\rho = 0$, the random variables are called *uncorrelated*.

As other properties of the correlation coefficient, we list the following:

$$\rho(X,Y) = \rho(Y,X)$$

$$\rho(X,X) = 1$$

$$\rho(X,-X) = -1,$$

$$\rho(aX + b, cY + d) = \rho(X,Y), a, b, c, \text{ and d are real numbers and } a, c \neq 0$$

Example 4.10.7.

Let X and Y be two random variables describing success or failure of two trouble-shooting instruments that work together with the following joint distribution.

X \ Y	0	1	sum
0	1/4	1/3	7/12
1	1/3	1/12	5/12
sum	7/12	5/12	

The distributions of X and Y are as follows:

X	0	1
p_X	7/12	5/12

Y	0	1
p_Y	7/12	5/12

Hence,

$E(XY) = (0)(0)(1/4) + (0)(1)(1/3) + (1)(0)(1/3) + (1)(1)(1/2) = 1/12 = 0.083$
$E(X) = (0)(7/12) + (1)(5/12) = 5/12 = 0.416$
$E(Y) = (0)(7/12) + (1)(5/12) = 5/12 = 0.416$
$E(X^2) = (0)(7/12) + (1)(5/12) = 5/12 = 0.416$
$E(Y^2) = (0)(7/12) + (1)(5/12) = 5/12 = 0.416$
$Var(X) = 0.416 - 0.173 = 0.243$
$Var(Y) = 0.416 - 0.173 = 0.243$

$$\sigma(X) = 0.49$$

$$\sigma(Y) = 0.49$$

$Cov(XY) = E(XY) - \mu_X \mu_Y = 0.083 - (0.416)(0.416) = 0.083 - 0.172 = -0.089$

$$\rho(X,Y) = (-0.089)/(0.49)(0.49) = -0.37.$$

Definition 4.10.8.

Let X be a discrete random variable with *pmf* p_X. Suppose that for a real nonnegative t, $E\left(e^{tX}\right)$ exists. Then, the *moment generating function* of X, denoted by $M(t)$, is defined as:

$$M(t) = E\left(e^{tX}\right) = \sum_{x} e^{tX} p_x .$$ (4.10.11)

Note that t could be a complex number with real part nonnegative. Note also that from (4.10.11) we have:

(1) If $t = 0$, then $M\left(0\right) = \sum_{k=1}^{\infty} p_k = 1$.

(2) $\displaystyle\lim_{t \to 0^-} \frac{d^k M(t)}{dt^k} = \sum_{x} x^k p_x = E\left(X^k\right).$

Thus, the moment generating function generates all the moments of X.

Example 4.10.8.

Suppose that X is a discrete random variable over $\{1, 2, \cdots, n\}$ and $1 \le k \le n$. Then, $p_k = \dfrac{1}{n}, k = 1, 2, \cdots, n.$ Hence, the moment generating function of X is:

$$
\begin{aligned}
M\left(t\right) &= \sum_{k=1}^{n} e^{tk} \frac{1}{n} \\
&= \frac{e^t + e^{2t} + \cdots + e^{nt}}{n} = \frac{e^t}{n}\left(1 + e^t + e^{2t} + \cdots + e^{(n-1)t}\right).
\end{aligned}
$$ (4.10.12)

The right hand side is a geometric progression with ratio e^t. Hence,

$$M(t) = \frac{e^t}{n} \frac{e^{nt} - 1}{e^t - 1}.$$

Thus, the expectation and variance for this random variable, using (4.10.12), for example, can be obtained as follows:

$$E(X) = M'\left(0\right) = \frac{1}{n}\left(e^t + 2e^{2t} + \cdots + ne^{nt}\right)_{t=0} = \frac{1}{n}\left(1 + 2 + \cdots + n\right) = \frac{1}{n}\frac{n(n+1)}{2} = \frac{n+1}{2}.$$

$$M''(0) = \frac{1}{n}\left(e^t + 2^2 e^{2t} + \cdots + n^2 e^{nt}\right)_{t=0} = \frac{1}{n}\left(1 + 2^2 + \cdots + n^2\right) = \frac{1}{n}\frac{n(n+1)(2n+1)}{6}$$

$$= \frac{(n+1)(2n+1)}{6}.$$

$$Var(X) = M''(0) - \left[M'(0)\right]^2 = \frac{(n+1)(2n+1)}{6} - \left(\frac{n+1}{2}\right)^2 = \frac{(n^2-1)}{12}.$$

The random variable in this example is called a *discrete type uniform.*

It should also be noted that for computational purposes, it is more convenient to use the logarithm of $M(t)$. In that case, we will have:

$$E(X) = \frac{d\left[\log M(t)\right]}{dt}\Bigg)_{t=0}$$

and

$$Var(X) = \frac{d^2\left[\log M(t)\right]}{dt^2}\Bigg)_{t=0}.$$

4.11. Continuous Random Variables and Distributions

We start this section by defining a continuous sample space and, then discuss some continuous probability distribution functions.

Definition 4.11.1.

When the outcomes of a chance experiment can assume real values (not necessarily integers or rational), the sample space, Ω, is a *continuous sample space*, i.e., Ω is the entire real number set R or a subset of it (an interval).

Example 4.11.1.

Suppose we are to choose a number between 2 and 5, inclusive. In such an experiment, the sample space would be the closed interval [2, 5], a subset of R.

Definition 4.11.2.

We note that since the set consisting of all subsets of R is extremely large, it will be impossible to assign probabilities to all of them. It has been shown in the theory of probability that a smaller set, say $\mathscr{B}$, may be chosen that contains all events of our interest. In this case, $\mathscr{B}$ is referred to as the *Borel field*. The triplet $(\Omega, \mathscr{B}, P)$ is called the *probability space*, where P is the probability (measure) of evens.

Now, if $E_1, E_2, \cdots$ is a sequence of mutually exclusive events represented as intervals of R and $P(E_i)$ is the probability of the event E_i, $i = 1, 2, \cdots$, then, by the third axiom of probability, we will have:

$$P\left(\bigcup_{i=1}^{\infty} E_i \right) = \sum_{i=1}^{\infty} P(E_i).$$ (4.11.1)

For a random variable, X, defined on a continuous sample space, Ω, the probability associated with the points of Ω for which the values of X falls on the interval $[a, b]$ is denoted by $P(a \le X \le b)$. We define the *continuous random variable* and its distribution in the following definition:

Definition 4.11.3.

Let the function $f(x)$ be defined on the set of real numbers, R, such that $f(x) \ge 0$, for all real x, and $\int_{-\infty}^{\infty} f(x) \, dx = 1$. Then, $f(x)$ is called a *continuous probability density function* (*pdf*) *on R*.

Definition 4.11.4.

If X is a random variable that its probability is described by a continuous *pdf* denoted by $f_X(x)$ as

$$P(a \le X \le b) = \int_a^b f_X(t) \, dt \text{, for any interval } [a, b],$$ (4.11.2)

then X is called a *continuous random variable*.

Note that there is a significant difference between discrete *pdf* and continuous *pdf*. For a discrete *pdf*, $f_X = P(X = x)$ is a probability, while with a continuous *pdf*, $f_X(x)$ is not interpretable   as   a   probability.   The   best   we   can   do   is   to   say that $f_X(x)dx \approx P(x \le X \le x + dx)$ for all infinitesimally small dx. Note also that since in this text density functions considered are probability density functions, we may sometimes suppress the adjective "probability" and use the word "density" alone.

Example 4.11.2.

Random variables that measure the cholesterol level of a person, the time it takes for a medicine to affect on a patient, speed of a car, the tensile strength of alloy, or the amount of time a customer has to wait in a waiting line before it receives a service are examples of continuous random variables.

Definition 4.11.5.

Let X be a random variable defined on a sample space Ω with a probability density function $f(x)$. The *probability distribution function* of X, denoted by $F_X(x)$ is defined as

$$F_X(x) = P(X \le x) = \int_{-\infty}^{x} f(t)\,dt .$$ (4.11.3)

Note that if there is no fear of confusion, we will suppress the subscript "X" from $f_X(x)$ and $F_X(x)$ and write $f(x)$ and $F(x)$, respectively.

As it can be seen from (4.11.3), distribution function can be described as the area under the graph of the density function. Note from (4.11.2) and (4.11.3) that if $a = b = x$, then

$$P(x \le X \le x) = P(X = x) = \int_{x}^{x} f(t)dt = 0 .$$ (4.11.4)

In words, if X is a continuous random variable, then probability of any given point is zero. That is, for a continuous random variable to have a positive probability we have to choose an interval. Note also that from (4.11.4) we will have:

$$P(a \le X \le b) = P(a < X \le b) = P(a \le X < b) = P(a < X < b).$$ (4.11.5)

By the Fundamental Theorem of Integral Calculus, we have $f_X(x) = \dfrac{dF_X(x)}{dx}$. Hence, the density function of a continuous random variable can be obtained as the derivative of the distribution function, i.e.,

$$F_X'(x) = f_X(x).$$ (4.11.6)

Moreover, the cumulative distribution function can be recovered from the probability density function with (4.11.3). Note that $F_X(x)$ collects all the probabilities of values of X up to and including x. Thus, it is the *cdf* (*cumulative distribution function*) of X.

For $x_1 < x$, the intervals $(-\infty, x_1]$ and $(x_1, x]$ are disjoint and their union is $(-\infty, x]$. Hence, it is clear from (4.11.2) and (4.11.3) that if $a \le b$, then

$$P(a \le X \le b) = P(X \le b) - P(X \le a) = F_X(b) - F_X(a). \tag{4.11.7}$$

It can easily be verified that $F_X(-\infty) = 0$ and $F_X(+\infty) = 1$ (why?).

Note that, as it can be seen from (4.11.3), the *cdf* is a general expression that applies to both discrete and continuous random variables. If the random variable is discrete, then $F_X(x)$ is sum of f_X's. However, if the random variable is continuous, then the sum becomes a limit and eventually an integral of the density function. The most obvious difference between the *cdf* for a continuous and for a discrete random variable is that F_X is a continuous function if X is continuous, while it is a step function if X is discrete.

Example 4.11.3.

Consider the random variable T representing the length of conversation, t, of a call in a series of telephone conversations. Then, the probability, $F_T(t)$, that the length of a conversation is less than 2 minutes is $F_T(2) = P(T < 2) = \int_0^2 f_T(t)dt$, where $f_T(t)$ is the *pdf* of T.

We now discuss some continuous distributions of interest. Let us start with is *uniform* that is a very well known continuous distribution.

Definition 4.11.6.

A continuous random variable X has a *uniform distribution* over an interval $[a, b]$ if it has the following probability density function:

$$f_X(x) = \begin{cases} 1/(b-a), & \text{if } a \le x \le b, \\ 0, & \text{elsewhere.} \end{cases} \tag{4.11.8}$$

Obviously, $b > a$ implies that $f(x) \geq 0$. Also $\int_{-\infty}^{\infty} f(x)dx = \int_{a}^{b} [1/(b-a)]dx = 1$. Hence, relation (4.11.8) defines a probability density function. From (4.11.4) we can observe that the *uniform distribution function of X* is give by:

$$
F_X(x) = \begin{cases} 0, & \text{if} \quad x \leq a, \\ (x-a)/(b-a), & \text{if} \quad a \leq x \leq b, \\ 1, & \text{if} \quad x \geq b. \end{cases} \tag{4.11.9}
$$

Note that since the graphs of the uniform density function (4.11.8) and distribution function (4.11.9) have rectangular shapes, they are sometimes referred to as the *rectangular density functions* and *rectangular distribution functions*, respectively.

Example 4.11.4.

Suppose X is distributed uniformly over [0, 10]. We want to find the following:

(a) $P(X < 4)$.

(b) $P(4 < X < 6)$.

(c) $P(X > 6)$.

(d) $P(3 \leq X < 7)$.

(e) $P(2 < X \leq 8)$.

From (4.11.3) and (4.11.8) we have:

(a) $P(X < 4) = \dfrac{1}{10} \int_{0}^{4} dx = \dfrac{1}{10}(4-0) = \dfrac{2}{5}$.

(b) $P(4 < X < 6) = \dfrac{1}{10} \int_{4}^{6} dx = \dfrac{1}{10}(6-4) = \dfrac{1}{5}$

(c) $P(X > 6) = \dfrac{1}{10} \int_{6}^{10} dx = \dfrac{1}{10}(10-6) = \dfrac{2}{5}$.

(d) $P(3 \leq X < 7) = \dfrac{1}{10} \int_{3}^{7} dx = \dfrac{1}{10}(7-3) = \dfrac{2}{5}$.

(e) $P(2 < X \leq 8) = \dfrac{1}{10} \int_{2}^{8} dx = \dfrac{1}{10}(8-2) = \dfrac{3}{5}$.

The *cdf* and *pdf* for this example are shown in Figure 4.11.1 and Figure 4.11.2, respectively.

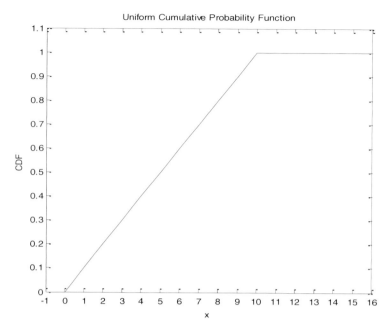

Figure 4.11.1. Uniform Cumulative Distribution, Example 4.11.4.

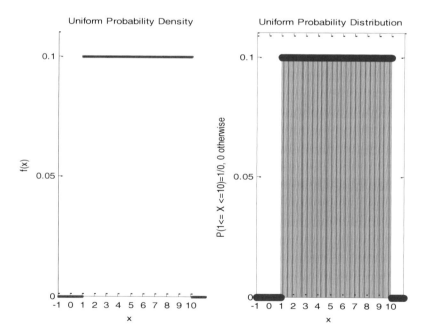

Figure 4.11.2. Graphs of uniform density and distribution functions, Example 4.11.4.

Example 4.11.5.

Suppose a Poisson counter with rate 4 starts counting at 8:00 am and registers 1 count in 30 minutes. What is the probability that the count occurred by 8:20 am?

Solution

To answer the question, we rephrase the problem in general terms. Suppose a Poisson counter with rate λ has registered 1 count in τ minutes. When did the event occur, i.e., what is the *pdf* of the occurrence at time t?

Now, we let $N(t)$ be the number of counts registered until time t. We also let T_1 be the time of occurrence of the first count. Then, using properties of the conditional probability and the Poisson distribution, the *cdf* is:

$$
\begin{aligned}
F(t) &= P\{T_1 \le t \,|\, N(\tau)=1, \quad 0 \le t \le \tau\} \\
&= P\{N(t)=1 \,|\, N(\tau)=1\} \\
&= P\{N(t)=1 \text{ and } N(\tau)=1\} / P\{N(\tau)=1\} \\
&= P\{N(\tau)=1 \,|\, N(t)=1\} \, P\{N(t)=1\} / P\{N(\tau)=1\} \\
&= P\{N(\tau-t)=1 \,|\, N(0)=1\} \, P\{N(t)=1\} / P\{N(\tau)=1\} \\
&= P\{N(\tau-t)=0 \,|\, N(0)=0\} \, P\{N(t)=1\} / P\{N(\tau)=1\} \\
&= \frac{e^{-(\tau-t)\lambda}\, \lambda t e^{-\lambda t}}{\lambda \tau e^{-\lambda \tau}} \\
&= \frac{t}{\tau}, \quad 0 < t < \tau.
\end{aligned}
$$

Therefore, T_1 is a uniform random variable in $(0, \tau)$. Hence, the first count happened before 8:20 am with probability $(20/30) = 66.67\%$.

Another well-known and especially useful continuous random variable is *exponential random variable*.

Definition 4.11.7

A continuous random variable X with *pdf*

$$
f(t) = \begin{cases} \mu e^{-\mu t}, & t \ge 0, \\ 0, & \text{elsewhere,} \end{cases} \tag{4.11.10}
$$

where $\mu > 0$ and *cdf*

$$
F(t) = \begin{cases} 1 - e^{-\mu t}, & t \ge 0 \\ 0 & \text{elsewhere,} \end{cases} \tag{4.11.11}
$$

is called *negative exponential* (or *exponential*) *random variable*. Relations (4.11.10) and (4.11.11) are called *exponential density function* and *exponential distribution function*, respectively. μ is the parameter for the *pdf* and *cdf*. Because of the parameter μ the *pdf* $f(t)$ defined by (4.11.10) may be written as $f\left(t\,|\,\mu\right)$.

Figure 4.11.3 shows graph of the exponential density function for different values of μ.

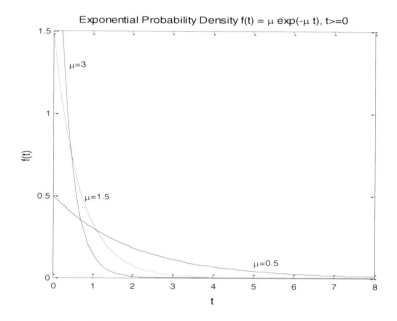

Figure 4.11.3. Exponential density function.

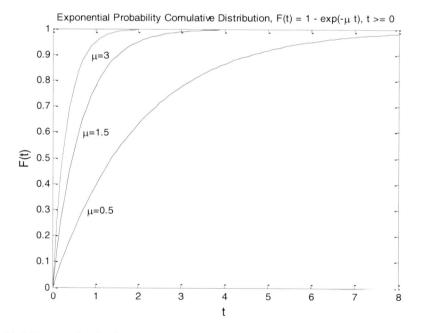

Figure 4.11.4. Exponential distribution function.

Example 4.11.6.

Let T represent the lifetime (the working duration) of a 100-watts light bulb, where the random variable, T, has an exponential density function as:

$$f(t) = \begin{cases} \dfrac{3}{2}e^{-\left(\frac{3}{2}\right)t}, & t > 0, \\ 0, & \text{elsewhere.} \end{cases}$$

See Figure 4.11.3 for the graph of this function. Then, the exponential distribution function for this random variable is:

$$F(t) = \int_{-\infty}^{t} f(x)dx = \begin{cases} \int_{0}^{t}\left(\dfrac{3}{2}\right)e^{-\left(\frac{3}{2}\right)x}dx, & t > 0, \\ 0, & \text{elsewhere} \end{cases}$$

$$= \begin{cases} 1 - e^{-\left(\frac{3}{2}\right)t}, & t > 0, \\ 0, & \text{elsewhere.} \end{cases}$$

See Figure 4.11.4 for the graph of this function.

Note that $f(t)$ defined above is a density function because, for a finite value of t,

$$f(x) \geq 0 \text{ and } \int_{-\infty}^{\infty}\left(\frac{3}{2}\right)e^{-\left(\frac{3}{2}\right)t}dt = 1.$$

The general form of the exponential random variable is called *gamma random variable* and will be defined below. But first we define the *gamma function* denoted by $\Gamma(\tau)$ as:

$$\Gamma(t) = \int_{0}^{\infty} x^{t-1}e^{-x}dx, \qquad\qquad (4.11.12)$$

where t is a positive real number. Note that $\Gamma(t)$ is a positive function of t. If t is a natural number, say $t = n$, then

$$\Gamma(n) = (n-1)!, \qquad n \geq 1, \qquad\qquad (4.11.13)$$

where $n!$ is defined by (4.2.1). $\Gamma(1)$, by convention, is equal to 1. It is not difficult to see that (4.11.13) is true (why?). It is because of (4.11.13) that the gamma function (4.11.12) is called the *generalized factorial*.

Using double integration and polar coordinates yield the integration formula

$$\int_0^\infty e^{-x^2}\, dx = \sqrt{2\pi} \ .$$
(4.11.14)

Hence, it can be seen that

$$\Gamma\left(\frac{1}{2}\right) = \sqrt{\pi}\ ,$$
(4.11.15)

(why?).

Example 4.11.7.

If it is known that the lifetime of a light bulb has an exponential distribution with parameter 1/250, what is the probability that the bulb works for:

(a) Between 100 and 250 hours?

(b) Between 250 and 400 hours?

(c) More then 300 hours, if it has already worked 200 hours?

(d) More than 100 hours?

Solution

Let X be the random variable representing the lifetime of a bulb. Then, from (4.11.10) and (4.11.11) we have:

$$f(x) = \begin{cases} \dfrac{1}{250}\, e^{-\frac{1}{250}x}, & x \geq 0, \\ 0, & x < 0, \end{cases}$$

and

$$F(x) = \begin{cases} 1 - e^{-\frac{x}{250}}, & x \geq 0, \\ 0, & x < 0. \end{cases}$$

Then,

(a) $P(100 < X < 250) = F(250) - F(100) = e^{-.4} - e^{-1} = .3024.$

(b) $P(250 < X < 400) = e^{-1} - e^{-1.6} = .1660$

(c) $P(X > 300 | X > 200) = \dfrac{P(X > 300 \, and \, X > 200)}{P(X > 200)}$

$$= \dfrac{P(X > 300)}{P(X > 200)} = \dfrac{e^{-1.2}}{e^{-.8}} = .6703 .$$

(d) $P(X > 100) = e^{-.4} = .6703$.

Note that in part (c) we have calculated the probability that the bulb works for more then 300 hours after the first 200 hours. In part (d), on the other hand, we have found the probability that the bulb works for more than 100 hours. Both values came out to be the same (except perhaps for rounding error). This suggests that the system forgets where it started from and forgets the history. This indeed is an important property of the exponential distribution function, i.e., the *memorylessness*.

Definition 4.11.8.

A continuous random variable, X, with probability density function $f(x)$ defined as:

$$f_X(x; \mu, t) = \begin{cases} \dfrac{\mu^t x^{t-1}}{\Gamma(t)} e^{-\mu x}, & x \ge 0, \\ 0, & x < 0, \end{cases} \qquad (4.11.16)$$

where $\Gamma(t)$ is defined by (4.11.12), $\mu > 0$, and $t \ge 1$, is called a *gamma random variable with parameters μ and t*. The corresponding distribution called *gamma distribution function* will, therefore, be:

$$F_X(x; \mu, t) = \begin{cases} \dfrac{1}{\Gamma(t)} \int_0^x \mu^t u^{t-1} e^{-\mu u} \, du, & \text{if } x \ge 0, \\ 0, & \text{if } x < 0. \end{cases} \qquad (4.11.17)$$

Note that $f_X(x; \mu, t) \ge 0$ and $\displaystyle\int_{-\infty}^{\infty} \dfrac{\mu^t x^{t-1}}{\Gamma(t)} e^{-\mu x} dx = \dfrac{1}{\Gamma(t)} \int_{-\infty}^{\infty} \mu^t x^{t-1} e^{-\mu x} dx =$

$\Gamma(t) / \Gamma(t) = 1$. Thus, $f_X(x; \mu, t)$ given by (4.11.16), indeed, defines a probability density function. Note also that in (4.11.14) if $t = 1$, we will obtain the exponential density function with parameter μ defined by (4.11.7). Note further that the parameter μ in (4.11.17) is called the *scale parameter*, since values other than 1 either stretch or compress the *pdf* in the x direction.

Definition 4.11.9.

The integral $\int_0^x u^{t-1}e^{-u}du$, $x>0$, from (4.11.12), denoted by $\Gamma(t,x)$, which is not an elementary integral, is called the *incomplete gamma function*. That is,

$$\Gamma(t,x) = \int_0^x u^{t-1}e^{-u}du, \ x>0.$$ (4.11.18)

The parameter t in (4.11.18) could be a complex number whose real part must be positive for the integral to converge. There are tables available for values of $\Gamma(t,x)$. As x approaches infinity, the integral in (4.11.18) becomes the gamma function defined by (4.11.12).

We saw that with $t = 1$, the exponential distribution function was obtained as a special case of the gamma distribution. Another special case is when t is a nonnegative integer.

Definition 4.11.10

In (4.11.17) if t is a nonnegative integer, say k, then the distribution obtained is called the *Erlang distribution of order k*, denoted by $E_k(t; x)$, i.e.,

$$E_k(x;\mu) = \begin{cases} \dfrac{\mu^k}{\Gamma(k)} \int_0^x t^{k-1}e^{-\mu t}dt, & \text{if } x \geq 0 \\ 0, & \text{if } x < 0. \end{cases}$$ (4.11.19)

The *pdf* in this case, denoted by $f(x;k,\mu)$ will be:

$$f(x;k,\mu) = \frac{\mu^k x^{k-1}e^{-\mu x}}{(k-1)!}, \qquad x, \mu \geq 0.$$ (4.11.20)

Note that (4.11.19) reduces to the exponential distribution (4.11.11) if $k = 1$.

Example 4.11.8

A firm starts its business at 10 in the morning. Suppose customers arrive to the firm according to a Poisson process of two each half-of an hour. What is the probability that the second customer arrives at 11.00?

Solution

Note that taking each half-of an hour as a unit of time, there are 2 units of time between 10 and 11 am. Thus, we are asked to have 2 customers within 2 units of time, the second of which is to arrive at $t = 2$.

Let T_k denote the arrival time of the kth customer. Let also $F_k(t)$ be the distribution function of T_k, i.e., $F_k(t) = P(T_k \leq t)$. Then,

$$1 - F_2(t) = P(T_2 > t) = P(N(t) \leq 2) = \sum_{j=0}^{2} P(N(t) = j) = \sum_{j=0}^{2} \frac{e^{-\lambda t}(\lambda t)^j}{j!}$$

$$= e^{-\lambda t}\left(1 + \lambda t + \frac{(\lambda t)^2}{2}\right).$$

Therefore, the *pdf* is

$$f_2(t) = F_2'(t) = \frac{\lambda^3 t^2 e^{-\lambda t}}{2},$$

which is an Erlang *pdf*, $f(x; 3, \lambda)$. Hence, to answer the question, we can use the Erlang distribution with $\lambda = 2$ and $t = 2$. Thus, from (4.11.19) we have

$$E_3(2; 2) = \frac{2^3}{\Gamma(3)} \int_0^2 t^{3-1} e^{-2t} dt = 0.227$$

.

Yet, another special case of the gamma function, *Chi-square*, which is defined as follows:

Definition 4.11.11.

In (4.11.17) if $t = r/2$, where r is a positive integer, and if $\mu = 1/2$, then the random variable X is called the *Chi-square* random variable with r degrees of freedom, denoted by $X^2(r)$. The *pdf* and *cdf*, in this case, with shape parameter r is

$$f(x) = \frac{1}{\Gamma\left(\frac{r}{2}\right) 2^{\frac{r}{2}}} x^{\frac{r}{2}-1} e^{\frac{-x}{2}}, \quad 0 \leq x < \infty, \tag{4.11.21}$$

$$F(x) = \int_0^x \frac{1}{\Gamma\left(\dfrac{r}{2}\right)2^{\frac{r}{2}}} v^{\frac{r}{2}-1} e^{\frac{-v}{2}} dv,$$ (4.11.22)

where $\Gamma\left(\dfrac{r}{2}\right)$ is the gamma function with parameter $r/2$.

Due to importance of X^2 distribution, tables are available for values of the distribution function (4.11.22) for selected values of r and x.

We leave as exercises to show the following properties of X^2 random variable: $\mu = r$, $\sigma^2 = 2r$, moment generating function, $M(t)$, $= (1-2t)^{-r/2}$, $t < 1/2$, and the Laplace transform of the $pdf = (1+2s)^{-r/2}$. Note that mean and variance of the chi-square random variable is its degrees of freedom and twice of it, respectively.

Example 4.11.9.

Suppose customers arrive at a chain store at the rate on average of 45 per hour. Assume that arrival of customers is according to Poisson process. We want to find the probability that it takes longer than 9.591 minutes before the first 10 customer arrives to the store.

Solution
Assuming λ as the parameter for the Poisson distribution of arrivals, we will have $\lambda = 3/4$. From what is given we have $r/2 = 10$. Let X represent the waiting time until the 10th arrive. Then, X has a chi-square distribution with 20 degrees of freedom. Thus, we use the chi-square table and find $P(X > 9.591) = 1 - 0.25 = 0.975$. To see the value 0.975, go to the Table 3 in the Appendix, look on the first column on the left for $r = 20$, which is the number of degrees of freedom. Then go across to look for 9.591 and pick the number above it on the first row, which are the probability. In case the chi-square value is not in the table, then one needs to interpolate.

Weibull Probability Distribution is another important continuous distribution with two parameters, as such; it is some time referred to as the family of Weibull distribution. The name refers to Swedish physicist Waloddi Weibull who published a paper in this regard in 1951. It is mostly used to fit observed data for particular values of α and β, $\alpha, \beta > 0$

The Weibull distribution is commonly used to model a life length and in the study of breaking strengths of materials because of the properties of its failure rate function. Resistors used in the construction of an aircraft guidance system have lifetime lengths that follow a Weibull distribution with $\alpha = 2$, and $\beta = 10$, (with measurements in thousands of hours).

Definition 4.11.12.

A continuous random variable X with *pdf* denoted by $f(x; \alpha, \beta)$ with two real parameters α and β, $\alpha, \beta > 0$ as

$$f(x; \alpha, \beta) = \begin{cases} \dfrac{\alpha}{\beta^{\alpha}} x^{\alpha-1} e^{-\left(\frac{x}{\beta}\right)^{\alpha}}, & x \geq 0, \\[2mm] 0, & x < 0, \end{cases} \qquad (4.11.23)$$

is called a *Weibull* random variable. If X is a random variable with pdf as in (4.11.23), then we write $X \sim Weibull(\alpha, \beta)$.

Note:

1. When $\alpha = 1$, (4.11.23) reduces to the exponential density function with $\mu = 1/\beta$. Changing α and β will lead to obtain variety of distribution shapes. See Figures 4.11.5 and 4.11.6.

2. Changing values of β stretches or compresses the curves in the x direction and so it is called a *scale parameter*.

3. Integrating (4.11.23) leads to the Weibull cumulative distribution function (*cdf*) as

$$F(x) = \begin{cases} 1 - e^{-\left(\frac{x}{\beta}\right)^{\alpha}}, & x \geq 0, \\[2mm] 0, & x < 0. \end{cases} \qquad (4.11.24)$$

The mean and variance of the Weibull random variable are, respectively,

$$\mu_X = \frac{1}{\beta} \Gamma\left(1 + \frac{1}{\alpha}\right) \text{ and } \sigma_X^2 = \frac{1}{\beta^2} \left\{ \Gamma\left(1 + \frac{2}{\alpha}\right) - \left[\Gamma\left(1 + \frac{1}{\alpha}\right)\right]^2 \right\}.$$

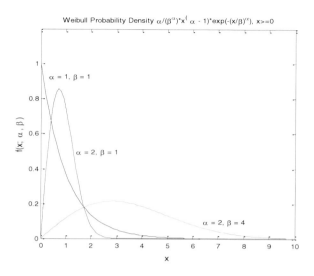

Figure 4.11.5. Weibull Density Function.

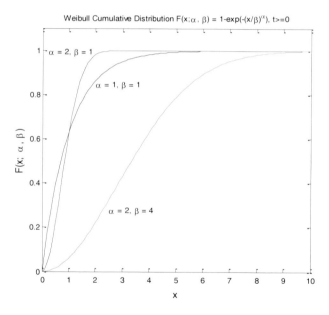

Figure 4.11.6. Weibull Density Function.

Example 4.11.10

Suppose $X \sim$ Weibull(0.4, 0.2). We want to find the probability that X is between 1 and 5.

Solution

Using (4.11.24), we get

$$P(1 < X < 5) = P(X \le 5) - P(X \le 1)$$

$$= \left(1 - e^{-[(0.2)(5)]^{0.4}}\right) - \left(1 - e^{-[(0.2)(1)]^{0.4}}\right)$$

$$= e^{-[(0.2)(1)]^{0.4}} - e^{-[(0.2)(5)]^{0.4}} = e^{-(0.2)^{0.4}} - e^{-1^{0.4}}$$

$$= 0.591 - 0.368 = 0.223.$$

The following two distributions are widely used in engineering applications.

Definition 4.11.13.

A continuous random variable X with *pdf*

$$f(x) = \frac{e^{-x}}{\left(1 + e^{-x}\right)^2}, \qquad -\infty < x < \infty. \tag{4.11.25}$$

is called a *logistic* random variable. Generally, *cdf* and *pdf* of a two-parameter logistic distribution are given, respectively, as:

$$F_X(x) = \frac{1}{1 + e^{-\frac{x-\alpha}{\beta}}}, \quad f_X(x) = \frac{e^{-\frac{x-\alpha}{\beta}}}{\beta \left(1 + e^{-\frac{x-\alpha}{\beta}}\right)^2}, \quad -\infty < x < \infty, \ -\infty < \alpha < \infty, \ \beta > 0,$$

$$\tag{4.11.26}$$

where α is the *location parameter*, β is the scale parameter. Letting α equal to 0 in (4.11.26), we obtain a *one-parameter Logistic cdf* and *pdf*. Mean and variance of logistic random variable, respectively, are as follows:

$$E(X) = M_o = M_e = \alpha.$$

$$Var(X) = \frac{\pi^2 \beta^2}{3} = 3.289868\beta^2.$$

Graph of logistic *pdf* in (4.11.26) for $\alpha = 0$ and $\beta = 1, 2$, and 3 are shown in Figure 4.11.7.

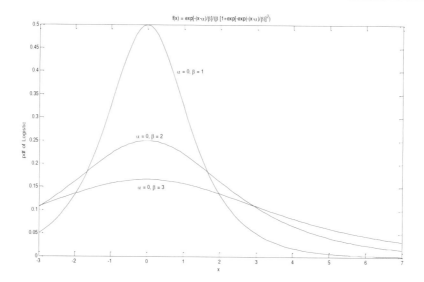

Figure 4.11.7. Graph of logistic *pdf* for $\alpha = 0$ and $\beta = 1, 2,$ and 3.

Now the *extreme-value* distribution. Gumbel was first to bring engineers and statisticians' attentions to possible application of "*extreme-value*" theory to some empirical distributions they were using such as radioactive emission, strength of materials, flood analysis, and rainfall analysis. The first application in the United States happened in 1941, treating annual flood flows.

There are three types of extreme-value distribution as we define below:

Definition 4.11.14.

A random variable X has an *extreme-value distribution*, if the distribution is of one the following three forms:

1. Type 1 (or *double exponential* or *Gumbel-type* distribution) with υ as the location parameter and θ as the scale parameter:

$$F_X(x;\upsilon,\theta) \equiv P(X \leq x) = e^{-e^{-(x-\upsilon)/\theta}}. \qquad (4.11.27)$$

2. Type 2 (or *Fréchet-type distribution*) with three parameters:

$$F_X(x;\upsilon,\theta,\alpha) \equiv P(X \leq x) = e^{-\left(e^{-(x-\upsilon)/\theta}\right)^{\alpha}}, \qquad x \geq \upsilon, \quad \theta,\alpha > 0. \qquad (4.11.28)$$

3. Type 3 (or *Weibull-type distribution*) with two parameters:

$$F_X(x;\upsilon,\theta,\alpha) \equiv P\big(X \le x\big) = \begin{cases} e^{-\left(e^{-(x-\upsilon)/\theta}\right)^\alpha}, & x \le \upsilon, \quad \theta, \alpha > 0, \\ 1, & x > \upsilon. \end{cases} \qquad (4.11.29)$$

Note:

1. If X is an extreme-value random variable, so is $(-X)$.

2. Type 2 and Type 3 can be obtained from each other by changing the sign of the random variable.

3. Type 2 can be transformed to type 1 by the following transformation
$$Z = \log\big(X - \upsilon\big).$$

4. Type 3 can be transformed to type 1 by the following transformation
$$Z = -\log\big(\upsilon - X\big).$$

5. Of the three types, type 1 is the most commonly used as *the extreme-value distribution*. The *pdf* for type 1 is:

$$f_X(x;\upsilon,\theta) = \frac{1}{\theta} e^{-e^{-(x-\upsilon)/\theta}} e^{-e^{-(x-\upsilon)/\theta}}. \qquad (4.11.30)$$

Graph of (4.11.30) for $\upsilon = 0$ and $\theta = 1, 2,$ and 3 are shown in Figure 4.11.8.

6. From (4.11.27), we have

$$-\log\big[-\log P\big(X < x\big)\big] = \frac{x-\upsilon}{\theta}. \qquad (4.11.31)$$

7. The moment generating function $[M(t)]$, mean, variance, and standard deviation (*STD*) of type 1 are, respectively, as follows:

$$M(t) = E\big(e^{tX}\big) = e^{t\upsilon}\Gamma(1-\theta t), \quad \theta|t| < 1. \qquad (4.11.32)$$

$$E(X) = \upsilon - \theta\varphi(1) = \upsilon + \gamma\theta = \upsilon + 0.57722\theta, \qquad (4.11.33)$$

where $\varphi(\cdot)$ is the *digamma function* defined by

$$\varphi(t) = \upsilon t - \log\big[\Gamma(1-\theta t)\big] \qquad (4.11.34)$$

and γ is *Euler's constant* equal to 0.57722.

$$Var(X) = \frac{1}{6}\pi^2\theta^2 = 1.64493\theta^2. \tag{4.11.35}$$

$$STD(X) = 1.28255\theta. \tag{4.11.36}$$

8. All three types of extreme-value distributions can be obtained from the following *cdf*:

$$F_X(x;\upsilon,\theta,\alpha) \equiv P\big(X \le x\big) = \left[1 + \alpha\left(\frac{x-\upsilon}{\theta}\right)\right]^{-1/\alpha},$$

$$1 + \alpha\left(\frac{x-\upsilon}{\theta}\right) > 0, \ -\infty < \alpha < \infty, \ \theta > 0. \tag{4.11.37}$$

Relation (4.11.37) is called *generalized extreme-value* or *von Mises type* or *von Mises-Jenkinson type* distribution.

9. Type 1 can be obtained from (4.11.37), if $\alpha \to \pm\infty$. This is the reason for naming (4.11.37) as above.

10. Type 2 can be obtained from (4.11.37), if $\alpha > 0$.

11. Type 3 can be obtained from (4.11.37), if $\alpha < 0$.

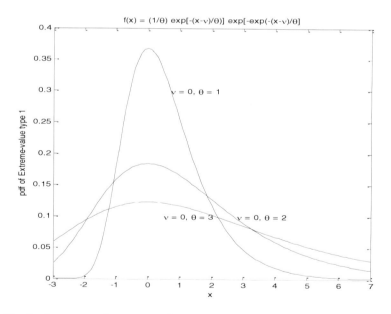

Figure 4.11.8. Graph of extreme-value type 1 pdf for $\upsilon = 0$ and $\theta = 1, 2,$ and 3.

A very well-known and widely used continuous random variable is *normal*.

Definition 4.11.15.

A continuous random variable X with *pdf* of $f(x)$ with two real parameters μ, $-\infty < \mu < \infty$ and $\sigma > 0$, where

$$f(x) = \frac{1}{\sigma\sqrt{2\pi}} e^{-\frac{(x-\mu)^2}{2\sigma^2}}, \qquad -\infty < x < \infty \tag{4.11.38}$$

is called a *Gaussian* or *normal* random variable. This is the most important distribution in statistics. The notation $N(\mu,\sigma^2)$ or $\Phi(\mu,\sigma^2)$ is usually used to show that a random variable has a normal distribution with mean μ and σ^2 variance. The normal *pdf* has a "bell-shaped" curve and it is *symmetric* about the line $f(x) = \mu$. it is also asymptotic, i.e., the tails of the curve from both sides get very close to the horizontal axis, but never touch it. We will see later that μ is the *mean* and σ^2 is the *variance* of this density function. The smaller the value of σ^2, the higher the peek of the "bell".

To see that $f(x)$ is indeed a *pdf*, we have to show the two principles, simply, the positiveness of $f(x)$ and that its sum equal to 1. We will do that after we define the *standard normal* density function.

Definition 4.11.16

A continuous random variable Z with $\mu = 0$ and $\sigma^2 = 1$ is called a *standard normal* random variable. The *cdf* of Z, $P(Z \leq z)$, is denoted by $\Phi(z)$. The notation $N(0,1)$ or $\Phi(0,1)$ is used to show that a random variable has a standard normal distribution function. That means, with parameters 0 and 1. The *pdf* of Z, denoted by $\varphi(z)$, therefore, will be:

$$\varphi(z) = \frac{1}{\sqrt{2\pi}} e^{-\frac{z^2}{2}}, \qquad -\infty < z < \infty. \tag{4.11.39}$$

Note that any normally distributed random variable X with parameters μ and $\sigma >$ can be standardized using a substitution

$$Z = \frac{(X - \mu)}{\sigma} \tag{4.11.40}$$

The *cdf* of Z is:

$$\Phi(z) = P(Z \le z) = \frac{1}{\sqrt{2\pi}} \int_{-\infty}^{z} e^{-\frac{u^2}{2}} du. \tag{4.11.41}$$

Now we will show that $\varphi(x)$ is a *pdf*. Obviously, $\varphi(x)$ is positive. To see that its integral over the entire real line is 1, we let

$$I = \frac{1}{\sqrt{2\pi}} \int_{-\infty}^{\infty} e^{-z^2} dz.$$

Then,

$$I^2 = \left[\frac{1}{\sqrt{2\pi}} \int_{-\infty}^{\infty} e^{-\frac{z^2}{2}} dz \right]^2 = \left[\frac{1}{\sqrt{2\pi}} \int_{-\infty}^{\infty} e^{-\frac{z^2}{2}} dz \right]\left[\frac{1}{\sqrt{2\pi}} \int_{-\infty}^{\infty} e^{-\frac{y^2}{2}} dy \right]$$

$$= \frac{1}{2\pi} \int_{-\infty}^{\infty} e^{-\frac{(z^2+y^2)}{2}} dz\, dy.$$

Using polar coordinates: $z = r\cos(\theta)$, $y = r\sin(\theta)$, $r^2 = z^2 + y^2$, $\theta = \tan^{-1}\left(\dfrac{y}{z}\right)$, $dz\, dy = r\, dr\, d\theta$, $0 < r < \infty$, and $0 \le \theta \le 2\pi$, it will yield:

$$I^2 = \frac{1}{2\pi} \int_0^{2\pi} \int_0^{\infty} e^{-\frac{r^2}{2}} r\, dr\, d\theta = \frac{1}{2\pi} \int_0^{2\pi} \left(\frac{e^{-r^2}}{-2} \right)_0^{\infty} d\theta$$

$$= \frac{1}{2\pi} \int_0^{2\pi} d\theta = \frac{2\pi}{2\pi} = 1.$$

Thus, $I = 1$ and $\varphi(x)$ is a *pdf*.

A practical way of finding normal probabilities is to find the value of z from (4.11.40), then use available tables for values of the area under the $\varphi(x)$. A table is given for values of the standard normal random variable Z in the Appendix.

The following figures show the normal density function with different parameters. We note that since the *pdf* of a normal random variable is symmetric, for negative values of z, we can use the following method:

$$\Phi(-z) = P(Z \le -z) = P(Z > z) = 1 - P(Z \le z) = 1 - \Phi(z),\, z > 0. \tag{4.11.42}$$

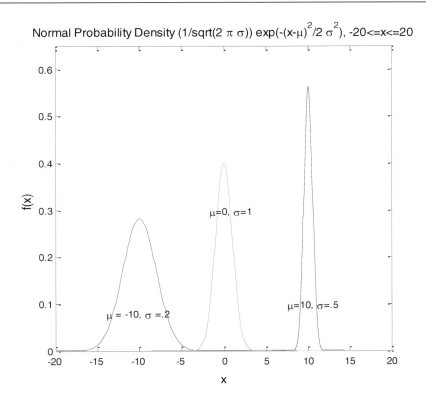

Figure 4.11.9. Normal Density Function.

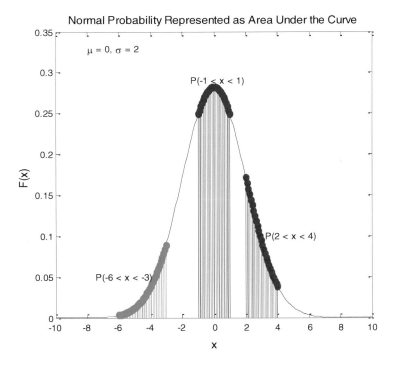

Figure 4.11.10. Normal Distribution Function.

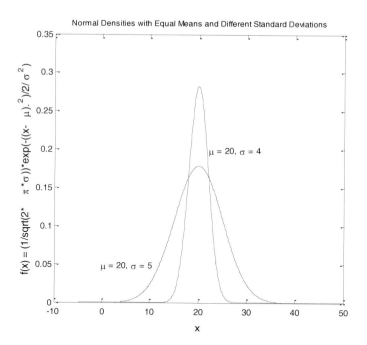

Figure 4.11.11. Normal Density Function.

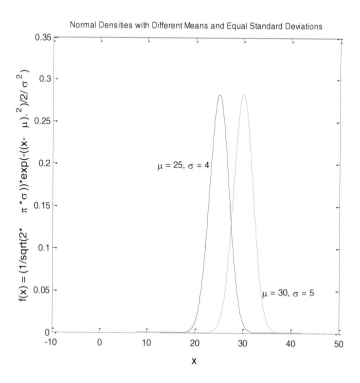

Figure 4.11.12. Normal Density Function.

We note that about 68 percent (68.26%) of the area under the normal density curve falls within plus and minus one standard deviation, about 95 percent (95.44%) within 2 and almost all (99.74%) within plus and minus 3 standard deviations.

Example 4.11.9.

Suppose that a periodical maintenance cost of a firm is known to have a normal distribution with parameters μ = \$400 and $\square$ = \$20. What is the probability that the actual cost exceeds \$450 in a particular period?

Solution

In this case the probability in question, using the Z-table (see Appendix), is:

$$P(X > 450) = P\left(\left[\frac{(X - \mu)}{\sigma}\right] > \left[\frac{(450 - \mu)}{\sigma}\right]\right) = P\left(Z > \left[\frac{(450 - 400)}{20}\right]\right).$$

Since the area under the pdf is 1, to find values of z we could either add or subtract probability of z between 0 and z to or from 0.5 or go directly for the entire area. Hence, we can have

$$P(X > 450) = P(Z > 2.5) = 0.5 - P(0 \leq Z \leq 2.5) = 0.5 - 0.4938 = 0.0062,$$

or

$$P(X > 450) = P(Z > 2.5) = 1 - P(Z \leq 2.5) = 1 - 0.9938 = 0.0062.$$

Example 4.11.10.

Suppose Z is a standard normal random variable. We want to find the probabilities of Z for the following values:

(a) Between 0.87 and 1.28;
(b) Between - 0 .34 and 0.62;
(c) Grater than 0.85;
(d) Grater than -0.65.

Solution

From the standard normal table, we will have the following:

(a) $F(1.28) - F(0.87) = 0.8997 - 0.8078 = 0.0919,$

(b) $F(0.62) - F(-0.34) = 0.7324 - 0.3669 = 0.3655,$

where,

$$\Phi(-0.34) = 1 - \Phi(0.34) = 1 - 0.6331 = 0.3669.$$

(c) $1 - F(0.85) = 1 - 0.8023 = 0.1977,$

(d) $P(Z > -0.65) = 1 - P(Z \leq -.65) = 1 - P(Z > 0.65)$
$$= 1 - P(Z \leq 0.65) = \Phi(0.65) = 0.7422.$$

Or, it could be obtained alternatively as follows:

$$P(Z > -0.65) = 1 - \Phi(-0.65) = 1 - [1 - \Phi(0.65)] = \Phi(0.65) = 0.7422.$$

Remark 4.11.1.

We saw in earlier sections, the probability of having k successes in n independent Bernoulli trials with probability of success in each trial as p followed a binomial distribution with parameters n and p. We also saw that if the product np remains fixed, then the limiting binomial probability will be a Poisson. In other words, Poisson distribution is an approximation to binomial. Another approximation of the binomial distribution is normal distribution when n is very large. This can be done by taking $\mu = np$ and $\sigma = \sqrt{np(1-p)}$.

Example 4.11.11.

Suppose a fair coin is tossed 20 times. What is the probability of having 9 heads?

Solution
This is actually a binomial problem with $n = 20$, $k = 9$, and $p = 0.5$. Thus, letting X to represent the number of heads and using the binomial distribution, we will have:

$$P(X = 9) = b(k; n, p) = \frac{20!}{9!\,11!}(0.5)^9(0.5)^{11} = 0.1602.$$

Alternatively, we could use normal distribution as follows:

$$\mu = np = (20)\,(0.5) = 10$$

$$\sigma = \sqrt{(np(1-p))} = \sqrt{(20)(0.5)(0.5)} = \sqrt{5} \approx 2.24.$$

Now, recall that for a continuous random variable, the probability of a point is zero. To avoid this problem, since the binomial *pmf* is true for integer values of X, we will choose an interval in the neighborhood of the point. Thus, we will choose

$P(X = 9) \approx P(8.5 < X < 9.5)$.

Now

$$z_1 = \frac{x - \mu}{\sigma} = \frac{8.5 - 10}{2.24} = -0.67, \text{ and } z_2 = \frac{x - \mu}{\sigma} = \frac{9.5 - 10}{2.24} = -0.22.$$

The looking up the probabilities for $z = -0.67$ (which is 0.2486) and for $z = -0.22$ (which is 0.087), we will have:

$P(X = 9) \approx P(8.5 < X < 9.5)$
$\approx P(-0.67 < Z < -0.22)$
$\approx 0.2486 - 0.0871 = 0.1615$

Thus, from the binomial *pmf* we find the probability in question to be 0.1602 and from the normal *pdf* we found it to be 0.1615, that are very close.

4.12. CONTINUOUS RANDOM VECTOR

Let X and Y be two continuous random variables with $f_X(x)$ and $f_Y(y)$. Then, the *joint bivariate pdf of X and Y* is denoted by $f_{X,Y}(x, y)$. From *pmf* we will have

$$P(X = x \text{ and } Y = y) \approx f_{X,Y}(X, Y) dx dy. \tag{4.12.1}$$

In other words,

Definition 4.12.1.

The *joint pdf of two continuous random variables X and Y* is an integrable function, say $f_{X,Y}(x, y)$ or just $f(x, y)$, with the following properties:

(a) $f_{X,Y}(x, y) \geq 0$.

(b) $\int_{-\infty}^{\infty} \int_{-\infty}^{\infty} f_{X,Y}(x, y) \, dx \, dy = 1$.

(c) $P\{(X,Y) \in A\} = \iint\limits_{A} f_{X,Y}(x,y)\,dx\,dy$, where $\{(X,Y) \in A\}$ is an event defined in

the xy-plane. Hence, properties of discrete joint *pmf* can be extended to continuous case using the approximation (4.12.1). For instance, since

$$f_{X|Y}(x \mid y)\,dx = \frac{f_{X|Y}(x \mid y)\,dx\,dy}{f_Y(y)\,dy},$$

we can define the *conditional probability density function of X given Y* as:

$$f_{X|Y}(x, y) \equiv \frac{f_{X,Y}(x, y)}{f_Y(y)}.\qquad(4.12.2)$$

The *marginal pdf* of X (or Y) can be obtained from the joint *pdf* $f_{X,Y}(x,y)$ as:

$$f_X(x) = \int_{-\infty}^{\infty} f_{X,Y}(x,y)\,dy\qquad(4.12.3)$$

or

$$f_Y(x) = \int_{-\infty}^{\infty} f_{X,Y}(x,y)\,dx.\qquad(4.12.4)$$

Finally, we note that the joint *pdf* can be extended for finitely many random variables as in the discrete case.

As in bivariate random variable (discrete or continuous), it is important to notice the relationship among components of the vector. One particular important case is the one with no relationship at all. We define this case in the following definition.

Definition 4.12.2.

Let $\mathbf{X} = (X_1, X_2, \cdots, X_n)$ be a finite or denumerable random vector with joint *pdf* or *pmf* of $f_{\mathbf{X}}(x_1, x_2, \cdots, x_n)$. Let also denote the marginal *pdf* or *pmf* of X_i by $f_{X_i}(x_i)$. Then, $X_1, X_2, \cdots, X_n$ are called *mutually independent random variables* if for every $(x_1, x_2, \cdots, x_n)$ we have

$$f(x_1, x_2, \cdots, x_n) = f_{X_1}(x_1) \cdot f_{X_2}(x_2) \cdots \cdot f_{X_n}(x_n) = \prod_{i=1}^{n} f_{X_i}(x_i).\qquad(4.12.5)$$

In case the *pdf* or *pmf* is parametric, say with one parameter, θ, say $f(x|\theta)$, then with the same parameter value θ in each marginal *pdf* or *pmf*, the joint *pdf* or *pmf* will be

$$f(x_1, x_2, \cdots, x_n | \theta) = f_{X_1}(x_1|\theta) \cdot f_{X_2}(x_2|\theta) \cdots f_{X_n}(x_n|\theta) = \prod_{i=1}^{n} f_{X_i}(x_i|\theta).$$

(4.12.6)

We note that pairwise independence does not apply mutual independence. It is left as exercise to prove this statement by a counter example. We also not that if each one of the X_is is a vector, then we will have mutual random vector, as the result.

Example 4.12.1.

Let X and Y have the joint *pdf* as

$$f(x,y) = e^{-x-y}, \quad 0 < x, y < \infty.$$

(4.12.7)

Suppose

$$A = \{(x,y) : 0 < x < \infty, \ 0 < y < x/5\}.$$

Then, the probability that (X, Y) falls in A is

$$P\{(X,Y) \in A\} = \int_0^\infty \int_0^{x/5} e^{-x-y} \, dy \, dx = \int_0^\infty e^{-x} \left[-e^{-y}\right]_0^{x/5} dx$$

$$= \int_0^\infty e^{-x} \left[-\left(e^{-x/5} - 1\right)\right] dx = \int_0^\infty e^{-x} \left[1 - e^{-x/5}\right] dx$$

$$= \int_0^\infty \left[e^{-x} - e^{-6x/5}\right] dx = \left[-e^{-x} + \frac{5}{6} e^{-6x/5}\right]_0^\infty$$

$$= 0 + 0 - \left(-1 + \frac{5}{6}\right) = \frac{1}{6},$$

Example 4.12.2.

Let X and Y have the joint *pdf* as

$$f(x,y) = 5, \quad 0 \le x \le y < 1.$$

(4.12.8)

Let us find $f_X(x)$, $f_Y(y)$, $E(X)$, $E(Y)$ and $E(X^2)$.

Solution

From (4.12.8) we have

$$f_X(x) = \int_x^1 5 \, dy = 5y\big)_x^1 = 5 - x, \quad 0 \le x \le 1.$$

$$f_Y(y) = \int_0^y 5 \, dx = 5x\big)_0^y = 5y, \quad 0 \le y \le 1.$$

For the moments, we have:

$$E(X) = \int_0^1 \int_x^1 5x \, dy \, dx = \int_0^1 5x(1-x) \, dx = 5\left(\frac{x^2}{2} - \frac{x^3}{3}\right)_0^1$$

$$= 5\left(\frac{1}{2} - \frac{1}{3}\right) = \frac{5}{6}.$$

$$E(Y) = \int_0^1 \int_0^y 5y \, dx \, dy = \int_0^1 5y^2 \, dy = 5\frac{y^3}{3}\bigg)_0^1 = \frac{5}{3}.$$

$$E(X^2) = \int_0^1 \int_x^1 5x^2 \, dy \, dx = \int_0^1 5x^2(1-x) \, dx = 5\left(\frac{x^3}{3} - \frac{x^4}{4}\right)_0^1$$

$$= 5\left(\frac{1}{3} - \frac{1}{4}\right) = \frac{5}{12}.$$

4.13. FUNCTIONS OF A RANDOM VARIABLE

Definition 4.13.1.

We already know that a random variable, say X, associates the real number $X(\omega)$ to the outcome ω, where Ω is the sample space. Now suppose $\varphi(\cdot)$ is a function that associates real numbers onto real numbers. Then, the composite function $\varphi[X(\cdot)]$ is defined and with each outcome ω, $\omega \in \Omega$, it associates the real number $\varphi[X(\omega)]$. $Y(\omega) = \varphi[X(\omega)]$ *is called function of the random variable X.*

Definition 4.13.2.

A function of more than one variable can be defined similar to the one with one variable as in Definition 4.13.1. Thus, if $\mathbf{X} = (X_1, X_2, \cdots, X_n)$ is a random vector of n random variable that associates the sample space Ω to the space R^n of real n-tuples, then the function $\varphi(\cdot, \ldots, \cdot)$ on n real variables associates with each point in R_n a real number. Hence, we define $\varphi[\mathbf{X}] = \varphi[X_1(\cdot), X_2(\cdot), \cdots, X_n(\cdot)]$ for each $\omega \in \Omega$ as the real number $\varphi[X_1(\omega), X_2(\omega), \cdots, X_n(\omega)]$. $Y(\omega) = \varphi[X_1(\omega), X_2(\omega), \ldots, X_n(\omega)]$ is called *function of n, n* ≥ 1, *random variables*.

Example 4.13.1.

There are some common examples for a function of random variables. For instance, suppose X and Y are two random variables and a and b are constants. Then, the following are also random variables: $-X$, aX, $aX + bY$, X^a, max(X,Y). Proof of some of these are trivial and for some the reader is referred to Neuts (1973, pp. 108-109).

A question may now raise as if Y a random variable. Another question, but very important one is that how is the probability structure of the composite function related to its argument. In other words, if the answer to the first question is affirmative and *cdf* for X and Y are F_X and F_Y, respectively, how F_X and F_Y are related. The following theorem will give one property of a function of random variables.

Theorem 4.13.1.

If Y is a function of a discrete random variable X, say $Y = \varphi[X(\omega)]$, then

$$E(Y) = E\left[\varphi(X)\right] = \sum_{\omega \in \Omega} \varphi(\omega) p(\omega),$$ (4.13.1)

where, $p(\omega) = P(X = \omega)$.

Proof

The proof is left as an exercise.

Example 4.13.2.

Let X be a discrete random variable with *pmf* as follows:

ω	-3	-2	-1	0	1	2	3
$p(\omega)$	0.01	0.05	0.10	0.30	0.30	0.20	0.04

Suppose Y be a function of X defined by $Y = X^2 + 1$. It is obvious that Y is a random variable and its values, say y, are 1, 2, 5, 10. However, some these values are obtained from values of X in more than one way. Here they are: 10 by -3 and 3, 5 by -2 and 2, and 2 by -1 and 1. Thus, in calculating probabilities of these values probabilities will be added since the events are mutually exclusive. Hence, *pdf* for Y is:

y	1	2	5	10
$p(y)$	0.30	0.40	0.25	0.05

Thus, expected value of Y, directly from its *pmf* is:

$$E(Y) = (1)(.3)+(2)(.4)+(5)(.25)+(10)(.05) = .3 + .8 + 1.25 + .5 = 2.85.$$

However, we could find this value from Theorem 4.13.1, using $\Omega = \{-3, -2, -1, 0, 1, 2, 3\}$. Therefore,

$$E(Y) = E\left[3X^2 + 1\right] = \sum_{\omega \in \Omega}\left(3\omega^2 + 1\right)p(\omega).$$

Summarizing, we get

$y = 3\omega^2+1$	10	5	2	1	2	5	10
$p(\omega)$	0.01	0.05	0.10	0.30	0.30	0.20	0.04

Thus,

$$E(Y) = (10)(.01)+(5)(.05)+(2)(.10)+(1)(.30)+(2)(.30)+(5)(.20)+(10)(.04)$$
$$= .1 + .25 + .2 + .3 + .6 + 1 + .4 = 2.85.$$

Definition 4.13.3.

Let the random variables $X_1, X_2, \cdots, X_n$ be a sample of size n chosen from a population (of infinite size) in such way that each sample has the same chance to be selected, then the sampling is called *random sampling* (or *sampling from an infinite population*) and the result is called a *random sample*. If the population is finite of size N, then a sample of size n from this population such that each sample of the combination $\begin{pmatrix} N \\ n \end{pmatrix}$ would be referred to as a random sample.

We note that in case of a sample drawn from a finite population without replacement the random variables $X_1, X_2, \cdots, X_n$ are not mutually independent. It is left as exercise to show this fact.

Population may be known by its distribution $F(x)$. In that case, a random sample describes an experiment in which the variable of interest X has $F(X)$ as its *pdf*. Then, if the experiment is repeated and observation is made on X, the observations are represented by X_1, X_2, and so on, up to and including X_n. In other words, each X_i, $i = 1, 2, \cdots, n$, is an observation and each X_i has a marginal distribution $F(X)$. Each observation is taken such that its values have no effect or relationship with any other observation. In other words, $X_1, X_2, \cdots, X_n$ are mutually independent (i.e., knowing the value of one does not change the probabilities of the others).

Example 4.13.3.

Let $X_1, X_2, \cdots, X_n$ be a random sample from an exponential population with *pdf* $f_X(x|\mu)$. Let X_i represent lifetime of a certain type of light bulbs with an average of μ hours. We want to find the probability that the lifetime for all bulbs be more than 1500 hours.

Solution
From (4.11.10), the parametric *pdf* and *cdf* of X_i, respectively, are

$$f_X(x|\mu) = \begin{cases} \dfrac{1}{\mu} e^{-x/\mu}, & x \geq 0, \\ 0, & \text{elsewhere.} \end{cases} \tag{4.13.2}$$

and

$$F_X(x) = \begin{cases} 1 - e^{-x/\mu}, & x \geq 0 \\ 0 & \text{elsewhere,} \end{cases} \tag{4.13.3}$$

Hence, from (4.12.6), the joint *pdf* of the random sample is

$$f_{\mathbf{X}}(x_1, x_2, \cdots, x_n|\mu) = \prod_{i=1}^{n} f_{X_i}(x_i|\mu) = \prod_{i=1}^{n} \frac{1}{\mu} e^{-x_i/\mu} = \frac{1}{\mu^n} e^{-(x_1 + x_2 + \cdots + x_n)/\mu}. \tag{4.13.4}$$

Thus, to find the probability in question we have to integrate out $x_i s$ one at a time successively as

$$P\left(X_1 > 1500, \cdots, X_n > 1500\right) = \int_{500}^{\infty} \cdots \int_{500}^{\infty} \prod_{i=1}^{n} \frac{1}{\mu} e^{-x_i/\mu} dx_1 \cdots dx_n$$

$$= e^{-1500/\mu} \int_{1500}^{\infty} \cdots \int_{1500}^{\infty} \prod_{i=1}^{n} \frac{1}{\mu} e^{-x_i/\mu} dx_1 \cdots dx_n$$

$$= \left(e^{-1500/\mu}\right)^2 \int_{1500}^{\infty} \cdots \int_{1500}^{\infty} \prod_{i=2}^{n} \frac{1}{\mu} e^{-x_i/\mu} dx_2 \cdots dx_n$$

$$= \left(e^{-1500/\mu}\right)^3 \int_{1500}^{\infty} \cdots \int_{1500}^{\infty} \prod_{i=3}^{n} \frac{1}{\mu} e^{-x_i/\mu} dx_3 \cdots dx_n$$

$$\vdots$$

$$= \left(e^{-1500/\mu}\right)^n$$

$$= e^{-1500n/\mu}. \tag{4.13.5}$$

For example, if μ is large, then the probability of the execution time being greater than 1500 is 1, that is, we should be almost sure that the event will happen.

It is left as an exercise to find the same result as (4.13.5) directly by using independence property of the random sample.

As another example of a random sample, we state the following theorem.

Theorem 4.13.2.

Let $X_1, X_2, \cdots, X_n$ be a random sample from the normal distribution with mean μ and variance σ^2. Then, the distribution of the sample mean $\overline{X}$ is normal with

$$E(\overline{X}) = \mu, \text{ and } Var(\overline{X}) = \sigma^2/n. \tag{4.13.6}$$

Proof

The proof is left as an exercise. (Hint: Use moment generating function of normal distribution.)

Definition 4.13.4.

A measure of variability of the sampling distribution of the sample mean is called the *standard error of the sample mean*. If the population standard deviation, σ is known, then it is denoted by $\sigma_{\bar{X}}$ and is

$$\sigma_{\bar{X}} = \frac{\sigma}{\sqrt{n}}. \tag{4.13.7}$$

If the population standard deviation is not known, then denoting S as the sample standard deviation, then standard error of the sample mean, denoted by $S_{\bar{X}}$, is

$$S_{\bar{X}} = \frac{S}{\sqrt{n}}. \tag{4.13.8}$$

Now consider a population with mean and variance μ and σ^2, respectively. Passing to limit as n increases without bound, we see from (4.13.6) that variance of $\bar{X}$ decreases. Hence, distribution of $\bar{X}$ depends on the sample size n. This means, for different sizes of n we will have a sequence of distributions to deal with. It is interesting to see the limit of such a sequence as n increases without bound. Hence, let us consider the random variable W defined by

$$W = \frac{\bar{X} - \mu}{\sigma / \sqrt{n}}, \quad n = 1, 2, \cdots. \tag{4.13.9}$$

Then, the expected value and variance of W are

$$E(W) = E\left(\frac{\bar{X} - \mu}{\sigma / \sqrt{n}}\right) = \frac{E(\bar{X}) - \mu}{\sigma / \sqrt{n}} = \frac{\mu - \mu}{\sigma / \sqrt{n}} = 0$$

and

$$Var(W) = E\left(W^2\right) - \left(E(W)\right)^2 = E\left(W^2\right)$$

$$= E\left[\frac{\left(\bar{X} - \mu\right)^2}{\sigma^2 / n}\right] = \frac{E\left[\left(\bar{X} - \mu\right)^2\right]}{\sigma^2 / n} = \frac{\sigma^2 / n}{\sigma^2 / n} = 1.$$

Hence, W has a standard normal distribution for each positive integer n. Consequently, the limiting distribution of W is also standard normal. We now state a very important theorem that shows the distribution of W is standard normal regardless of distribution of the random sample.

Theorem 4.13.3. Central Limit Theorem

Let $X_1, X_2, \cdots, X_n$ be a random sample of size n from a distribution with finite mean μ and a finite positive variance σ^2. Then, the limiting distribution of $W = \dfrac{\overline{X} - \mu}{\sigma / \sqrt{n}} = $

$\dfrac{\sum\limits_{i=1}^{n} X_i - n\mu}{\sigma \sqrt{n}}$ is the standard normal (as $n \to \infty$).

Proof

See Hogg and Tanis (1993).

Example 4.13.4.

It is often difficult to find the exact distribution of W given by (4.13.9). One way to approximate it is by simulation. To do that a random sample of size n is taken repeatedly, say 1000 times, from a distribution with *pdf* $f(x)$ with mean μ and variance σ^2. Then, the value of W is computed for each sample leading to 1000 observations of values of W. For instance, let the sample size be 36 from the exponential distribution with mean 3. Then, probability that $\overline{X}$ is between 2.5 and 4 is approximately 0.8185. It is left as an exercise to conduct the simulation and verify the value given.

Example 4.13.5.

Consider a Poisson random variable, Y, with mean 10. The random variable Y can be thought of a random sample, $X_1, X_2, \cdots, X_{10}$, of size 10 from a Poisson distribution with mean 1, i.e., $Y = X_1 + X_2 + \cdots + X_{10}$. Thus, $W = \dfrac{\sum\limits_{i=1}^{10} X_i - (10)(1)}{(1)\sqrt{10}} = \dfrac{Y - 10}{\sqrt{10}}$ is approximately normal with mean 0 and variance 1 and Y is normally distributed with mean 10 and variance 10. Hence, for example, we have

$$P(15 < Y \leq 18) = P(15.5 \leq Y \leq 18.5) = P\left(\frac{15.5 - 10}{\sqrt{10}} < \frac{Y - 10}{\sqrt{10}} \leq \frac{18.5 - 10}{\sqrt{10}} \right)$$

$$= \Phi(2.690) - \Phi(1.739) = 0.9964 - 0.9591 = 0.0373.$$

Note that we have increased 15 to 15.5 because 15 is not included in the interval $15<Y\le18.5$. Now, if we directly use Poisson distribution with $\lambda = 10$, we will have

$$P(15<Y\le18) = P(Y=16)+P(Y=17)+P(Y=18)$$
$$= \frac{10^{16}e^{-10}}{16!}+\frac{10^{17}e^{-10}}{17!}+\frac{10^{18}e^{-10}}{18!}$$
$$= 0.0217+0.0128+0.0071 = 0.0416.$$

Of course, the difference between the values 0.0373 and 0.0416 is due to the approximation error.

EXERCISES

4.1. Introduction

4.1.1. Suppose we want to place two balls into two cells.

  (a) What are the possible outcomes?

  (b) What are the simple events?

4.1.2. Again consider Example 4.1.10. Assuming equiprobable property for each of the sample points,

  (a) what is the probability of (1, 5)?

  (b) what is the probability of (1, 6)?

4.1.3. In Example 4.1.10, suppose the machine in is to select two digits at random such that the sum of numbers selected is an even. What is the probability that both numbers are even?

4.1.4. In Example 4.1.10, suppose that we would let the first digit be an even number. What is the probability that both numbers are even?

4.1.5. Consider Exercise 4.1.3. Show that once the machine chooses the first number at random, the second cannot arbitrarily be chosen.

4.1.6. The probability of an event A and the probability of its complement A' always sum to

  A. 0   B. 1   C. 2   D. any positive value

4.1.7.   A CD player is playing a CD that has 8 songs on it plays the songs randomly. Two of the songs are the favorites. What is the probability that the second favorite is either the third or the sixth song that is playing?

4.1.8.   Three pairs of "before" and "after" pictures of three different peoples are given. Holding the "after" pictures and randomly match the "before" pictures, what is the probability of

(a)   all match?

(b)   none match?

4.1.9.   What is the number of ways to choose a set of 3 from a group of 10 without replacement?

4.1.10.   Suppose we are to choose 3 officers (chair, secretary, and treasurer) as 3 distinct members from a committee of 10 people. How many different ways can these officers be chosen?

4.1.11.   Suppose a machine selects digits at random from 1 to 9. We let the machine run four times. What is the probability of getting a 2 each time?

4.1.12.   Suppose a repairman has a diagnostic instrument which can identify problems with a machine three out of four times. Suppose he runs the diagnostic 8 times. What is the probability that he identifies the problem:

(a)   less than four times?

(b)   four times?

4.2. Counting Techniques

4.2.1.   Show that

(a) $_rC0 = 1$.
(b) $_rCr = 1$.
(c) $_rCk = _rCr\text{-}k$.
(d) $_rC1 = _rCr\text{-}1 = k$.

4.2.2.   Suppose there are 12 colored light bulbs with 3 indistinguishable yellow ones, 4 indistinguishable red, and 5 indistinguishable white. How many different ways can these bulbs be set in a box having twelve slots?

4.2.3.   Consider the set $\{1, 2, 3\}$.

(a) What is the permutation of the three digits taken all at a time?

(b) What is the permutation of the three digits taken one at a time?

4.2.4. Suppose a company manufactures four different items *A*, *B*, *C*, and *D*. The company wants to have two television advertisements showing one item on each show. Answer the following questions:

(a) From how many possibilities can the company chooses the items?

(b) How many of the ways involve the same item being used?

(c) What fraction of the ways involves the same item being used?

4.2.5. Suppose we have two urns *A* and *B* which we do not know which is which. Urn *A* contains 2 white marbles and *B* contains 3 greens. In order to draw a marble from the urns, we toss a fair coin. Suppose we drew the first marble and was green and we put it back into the urn it was drawn from. What is the probability that the second marble from the same urn drawn be also green?

4.2.6. Suppose we are to seat 3 people in 3 chairs at a table. How many different ways can this be done?

4.2.7. An experiment consists of three stages. There are five ways to accomplish the first stage, four ways to accomplish the second stage, and three ways to accomplish the third stage. The total number of ways to accomplish the experiment is

A. 60   B. 20   C. 15   D. 12

4.2.8. Suppose that a jar contains a fair and a two-headed coin. A coin is selected at random and shows heads when it is flipped. What is the probability that the coin drawn was the fair coin?

4.2.9. Suppose a box contains 48 colored light bulbs with 12 indistinguishable yellow, 12 indistinguishable red, 12 indistinguishable blue, and 12 indistinguishable white. Now we pick two bulbs without replacement.

(a) What is the probability that the second bulb is a yellow?
(b) What is the probability that the fifth bulb is a yellow?

4.2.10. Suppose the machine in the Example 4.1.10 were to select two digits at random. Use the sample space and counting the number of sample points to show that the probability of getting two digits is 1.

4.2.11. A committee of four is to be selected at random from a group of eight executives and four employees. Suppose that there must be at least one employee on the committee. What is the probability that exactly three employees are selected on the committee?

4.2.12.   **Birthdays.** In a group of k people, what is the probability that all have different birthday?

4.2.13.   **Birthdays.** Suppose that there are 30 people in a class.

(a)   What is the probability that no two of them have the same birthday?

(b)   If you had a dollar to bet "yes" or "no", on no two of them have the same birthday, which way would you bet?

(c)   For what number of people in a class would it be a fair bet either way?

4.2.14.   Consider a set of letters a, b, and c and their permutations. Assume equilikely probability for each of elements of the permutation. What is the expected value of the number of fixed letters among the elements of the permutations?

4.2.15.   What is the number of ways to choose a set of 3 from a group of 10?

4.2.16.   A "fair" coin is tossed two times.

(a)   What is the sample space?

(b)   What is the probability that two tails turn up?

4.2.17.   Three "fair" coins are tossed together.

(a)   What is the sample space?

(b)   What is the probability of obtaining two tails?

4.2.18.   Two "fair" dice are thrown.

(a)   What is the sample space?

(b)   Find the probability of an odd number turning up.

4.2.19.   Let $\Omega = \{A, B, C\}$. If P is a probability function on S such that $P(A) = .2$, and $P(C) = .5$, what is $P(B)$?

4.2.20.   Prove Stirling's Formula, formula (4.2.2).

4.2.21. Suppose a person goes to a movie or out to dinner on Friday nights. From the past experience he knows that the chance of going to a movie for him is 2/3, going to dinner is 4/9, and going to either is 7/9. What is the probability that he goes to:

(a) only one of the two?

(b) both?

(c) movie, but not to dinner?

4.2.22. Suppose a repairman tries three times to locate a problem with a unit. The probability that he locates the problem on the kth try is $1/(k + 2)$, $(k = 1, 2, 3)$. Find the probability that he locates the problem before the third try.

4.2.23. An urn contains 9 balls numbered from 1 to 9. Five balls are drawn without replacement. What is the probability that the largest number is 6?

4.2.24. An urn contains 20 balls of which 8 are yellow and 12 are red. If six balls are drawn without replacement, what is the probability that

(a) All six are of the same color?

(b) At least four are of the same color?

4.2.25. Suppose we have two urns, one contains two black and one white ball, the other contains one black and two white balls. We select an urn at random and choose two balls from it. Draw a tree diagram showing all possibilities when:

(a) the first ball is replaced before the second is drawn.

(b) the first ball is not replaced before the next is drawn.

4.2.26. A random number generator generates numbers at random from the unit interval [0,1]. Find:

(a) P(A), where $A = (1/3, 2/3]$.

(b) P(B), where $B = [0,1/2)$.

(c) the partition generated by A and B.

(d) the event algebra.

4.2.27. Suppose the pair (x,y) is the Cartesian coordinates of a point w chosen at random in a square with corners $(\pm 1, \pm 1)$.

(a) Find $P\left(x^2 + y^2 < a^2\right)$, $\mathbf{O} \leq \mathbf{a} \leq \mathbf{1}$; $1 \leq a \leq \sqrt{2}$, $a > \sqrt{2}$.

(b) For a given value of a, give the partition and event algebra generated by the event $\left\{x^2 + y^2 < a^2\right\}$.

4.2.28. For mutually exclusive events A_1, A_2, $\cdots$, A_n from a probability space $\left(\Omega, \mathcal{B}, P\right)$, show that: $P\left(\bigcup_i A_i\right) = \sum_i P\left(A_i\right)$.

4.2.29. Prove $P(\emptyset) = 0$.

4.2.30. If A^C is the complement of A, prove that $P(A^C) = 1 - P(A)$.

4.2.31. If A and B are events and A is a subset of B, prove that $P(B - A) = P(B) - P(A)$ and $P(A) \leq P(B)$.

4.2.32. If A and B are any two events, prove that
$P(A \cup B) = P(A) + P(B) - P(A \cap B)$.

4.2.33. For any two events A and B,

    A. $P(A \cup B) = P(A) + P(B)$
    B. $P(A \cup B) = P(\tilde{A})P(B)$
    C. $P(A \cup B) = P(A) + P(B) - P(A \cap B)$
    D. $P(A \cup B) = P(\tilde{A})P(B)$

4.2.34. If A and B are mutually exclusive events, then

    A. $P(A \cap B) = 0$
    B. $P(A \cup B) = 0$
    C. $P(A \cap B) = P(A \cup B)$
    D.   All of the above

4.2.35. If $P(A) = .35$ and $P(B) = .45$, then $P(A \quad B)$

    A.   is .10
    B.   is .80
    C.   is .20
    D.   Cannot be determined from the given information

4.2.36. Prove that if A, B, and C are any three events, then

$(A \cup B \cup C) = P(A) + P(B) + P(C) - P(A \cap B) - P(A \cap C) - P(B \cap C) + P(A \cap B \cap C)$.

4.2.37.   Toss a "fair" coin four times.

(a)   Write out the sample space.

(b)   Let A be the event that the first toss turns up a head, and let B be the event that the third toss turns up a head. Which one of {A, B}, {A, C}, {B, C}, {A, B, C} are independent?

4.2.38.   Suppose a repairman has two "trouble shooters" to locate a problem in a machine. The first instrument locates a problem one out of five times and the second does it four out of five times. What is the probability that a given problem be found if both A and B are used?

4.2.39.   Suppose the probability that an item passes a quality control test on the first attempt is 1/2, that it fails the first attempt and passes on the second attempt is 7/10, and that it fails the first two attempts and passes the third attempt is 8/10. If only three attempts are allowed for an item, what is the probability that it passes the test?

4.2.40.   A manufactured article that cannot be used if it is defective is required to pass through two inspections before it is permitted to be packaged. Experience shows that one inspector will miss 5% of the defective articles, whereas the second inspector will miss 4% of them. If good articles always pass inspection and if 10% of the articles turned out in manufacturing process are defective, what percentage of the articles that are produced and pass both inspections will be defective?

4.2.41.   Assume that there are equal numbers of male and female students in a high school and that the probability is 1/5 that a male student and 1/25 that a female student will choose science as major when attend college. What is the probability that:

(a)   a student at random will be a male science student?

(b)   a student selected at random will be a science student?

(c)   a science student selected at random will be a male student?

4.2.42.   Suppose we put keys of N door locks in a box. We mix the keys and randomly distribute them to the door locks so that each and all door locks has a key. Assume that the ith key is equally likely to fit any of the N locks. What is the expected number of keys that fit their locks?

4.2.43.   A machine is to choose a number at random between 1 and 100, inclusive. If the program is set such that the number chosen is a multiple of 3, what is the probability that the number selected is:

(a)   a multiple of 5?

(b)   a multiple of 5, but not of 9?

(c)   a multiple of exactly on of the integers 5 or 9?

(d)   there are no duplications in the first numbers chosen, i.e., all n numbers are different, for $n = 2, 3, \cdots$.

4.2.44.   Urn A has 3 white and 1 black balls and B has 2 whites and 4 blacks. An urn in selected at random and 2 balls are selected at random from it. Find the probability that:

(a)   both balls are white?

(b)   the urn B was chosen, given that both balls are white?

4.2.45.   "My Fun Fortune Cookies" are produced from a collection of 120 different fortunes. If n people go to dinner and each gets a fortune cookie, what is the probability that they are all different? Answer the question for $n = 3, 5, 10$, and 20.

4.3.   Tree Diagrams

4.3.1.   Let $A = \{a, b\}$, $B = \{1, 2, 3\}$, and $C = \{c, d\}$. Write the Cartesian product of these sets by a tree diagram.

4.3.2.   Suppose we have two urns, one contains one white and two black balls; the other contains two white and one black balls. An urn is chosen at random and, then two, balls are drawn from it without replacement. Draw the tree diagram for this experiment.

4.4.   Conditional Probability and Independence

4.4.1.   Suppose two repairmen A_1 and A_2 try independently to find a problem with a unit. The probability that A_1 finds the problem is 1/3 and the probability that A_2 finds it is 1/6. Find the probability that:

(a) both A_1 and A_2 find the problem.

(b) at least one will find the problem.

(c) neither will find the problem.

(d) only A_1 finds the problem.

4.4.2.   Consider Example 4.4.4. Let A and B be as described in that example. In addition, let D be: D = The second child is a girl. Show that pairwise independence does not imply that all three events are independent.

4.4.3.   If $P(A) = .40$, $P(B) = .30$, and $P(A \cap B) = .15$, then $P(A|B)$ is

A. 0.375   B. 0.12   C. 0.50   D.0.045

4.4.4.   If $P(A) = .60$, $P(B) = .40$, and $P(A|B) = .75$, then $P(A \cap B)$ is

A. 0.30   B. 0.45   C. 0.24   D. 0.80

4.4.5.   If events A and B are independent, which of the following statements are true?

A. A' and B are independent   B. A and B'are independent
C. A 'and B'are independent   D. All of the above are true

4.4.6.   If $P(A) = .30$, $P(B) = .60$, and $P(A \cap B) = .18$, then events A and B are

A. dependent   B. independent
C. mutually exclusive   D. complementary

4.4.7.   If $P(A) = .30$, $P(B) = .10$ and events A and B are independent, then $P(A^c B^c)$ is

A. 0.63   B. 0.70   C. 0.90   D. 0.40

4.4.8.   Suppose that A and B are independent events with $P(A) = .40$ and $P(B) = .70$, then $P(B^c|A^c)$ is

A. 0.28   B. 0.30   C.0.60   D. 1.10

4.5. The Law of Total Probability

4.5.1.   Let $x_1 = 2$, $x_2 = 5$, $x_3 = 6$, $x_4 = 4$, $x_5 = 7$, $x_6 = 10$, $x_7 = 3$. Find

(a) $\sum_{i=3}^{6} x_i$.

(b) $\sum_{i=1}^{7} x_1$

(c) $\sum_{i=1}^{7} x_i$.

(d) $\displaystyle\sum_{i=1}^{7} x_i^2$.

(e) $\displaystyle\sum_{i=1}^{7} x_i^2 - \left(\frac{1}{7}\sum_{i=1}^{7} x_i\right)^2$.

4.5.2.   An urn contains five white and ten red balls. Three balls are drawn at random one by one without replacement. Find the probability that:

(a) the first ball is white.

(b) the second ball is white given that the first was also white.

(c) the third ball is white given that both the first and the second were white.

4.5.3.   Given three events A, B, and C, with $P(A) = .5$, $P(B) = .4$ and $P[(A\cup B)'] = .2$ (where " E' " is the complement of the event E). Find the following probabilities:

(a) $P(A\cap B)$,

(b) $P(B|A)$, and

(c) $P(A|B)$.

4.5.4.   Consider all families with two children. Assume that boys and girls are equiprobable. Suppose a family with two children is chosen at random. It is found that the family has a girl.

(a) Write out the sample space for the original experiment before the girl was observed.

(b) Write out the conditional sample space given the observation.

(c) What is the probability that this family has also a boy?

4.5.5.   Consider Example 4.4.4. From the sample space of that example, we see that if we choose a family of two children, the probability that the second child is a girl will be 1/2.

(a) What if we know that the first child is also a girl?

(b) If we know that the family has a girl, what is the probability that both children are girls?

4.6. Discrete Random Variables

4.6.1.   Consider the set of all residents of the state of Texas as the sample space S. Group the residents of Texas according to three levels of education, namely: high school (H), undergraduates (U), and post-graduates (G). Define the sample space and a random variable on it for this case. What are the domain and range of your random variable?

4.6.2.   Prove Theorem 4.6.1.

4.6.3.   Generalize the problem stated in Exercise 4.2.8 to r coins in the jar. Suppose that the flip of the ith coin results a tail with probability i/r, $i = 1, 2, \cdots, r$. Suppose now one of the coins was randomly selected, flipped and the result was a tail. What is the probability that it was the 4th coin?

4.6.4.   A gambler has 3 coins in his pocket, two of which are fair and one is two headed. He selects a coin. If he tosses the selected coin

(a) and it turns up a head, what is the probability that the coin is a fair?

(b) 5 times, what is the probability that he gets five heads?

(c) 6 times, what is the probability of getting 5 heads followed by a tail?

4.7. Discrete Probability Distributions

4.7.1.   The probability mass function of a discrete random variable X is defined as $p(x)$ = $x/10$ for $x = 0,1,2,3,4$. Then, the value of the cumulative distribution function $F(x)$ at $x= 3$ is

A. 0.10   B. 030   C. 0.60   D. 0.90

4.7.2.   The cumulative distribution function $F(x)$ of a discrete random variable X is given by $F(0) = .30$, $F(1) = 0.70$, $F(2) = 0.90$, and $F(3) = 1.0$, then the value of the probability mass function $p(x)$ at $x = 1$ is

A. 0.30   B. 0.40   C. 0.20   D. 0.80

4.7.3.   Which of the following is true if $X \sim \text{Bin}(6, 0.3)$

A. $E(X) = 6.3$ B. $E(X) = 1.8$
C. $E(X) = 5.7$ D. None of the above is true

4.7.4.   Which of the following are true if $X \sim \text{Bin}(10, 0.75)$?

A. $V(X) = 7.5$ B. $V(X) = 9.25$ C. $V(X) = 1.875$

4.7.5.   If we perform a large number of independent binomial experiments, each with n = 10 trials, and $p = 0.60$, then the average number of successes per experiment will be close to

A. 10   B. 8   C. 6   D. 4

4.7.6.   Which of the following is (are) conditions of a binomial experiment?

A. There is a sequence of n trials, where n is fixed in advance of the experiment.
B. The trials are identical, and each trial can result in one of the same two possible outcomes, which we denote by success (S) or failure (F).
C. The trials are independent, so that the outcome of any particular trial does not influence theoutcome of any other trial.
D. The probability of success is the same (constant) from trial to trial; we denote this probability by p.
E. All of the above are conditions of a binomial experiment.

4.7.7.   The expected number of trials in 200 tosses of a fair coin is

A. 100   B. 200   C. 50   D. 120

4.7.8.   (***Binomial Stock Model***). Investing in equity stocks and other securities can be viewed as gambling. An obvious objective for gambling is to maximize fortune of the gambler. The act of gambling, like investing, can be summarized into two steps: (1) calculating possible payoffs and their respective probabilities; and (2) formulating this payoff information into an appropriate gambling (investing) strategy.

So suppose a gambler is able to bet any fraction of his money on the flip of a biased coin. With probability p he will win w dollars for every dollar bet and with probability $q = 1 - p$ he will lose his bet. The betting strategy involves finding a fraction β of the gambler's fortune ($0 \le \beta \le 1$), which if bet at each successive flip of the coin, will maximize his fortune. Find β as such.

4.7.9.   Let the random variable X be the number of calls per minute coming into a switchboard. Assume X is a Poisson random variable with mean 0.9 call per minute.

(a) Find the probability of no calls in a minute.

(b) Find the probability of three or more calls in a minute.

4.7.10. Which of the following statements is (are) true about a Poisson probability distribution with parameter λ?

A.   The mean of the distribution is λ
B.   The variance of the distribution is λ
C.   The parameter λ must be greater than zero
D.   All of the above statements are true

4.7.11. Assume that the approval rating of handling the job of the President of the United States in the year 2008 by the American public is 23%. We randomly choose 10 Americans and ask each of handling the job of the President. What is the probability that there are at least 2 approvers among the sample persons?

4.7.12. Suppose that telephone calls enter the university's switchboard, on the average, 2 every 5 minutes. Assuming that arriving calls has a Poisson *pmf* with parameter 4, what is the probability of less than 7 calls in 10 minutes period?

4.7.13. A material is known for 20% of breakage under stress. Five different samples of such materials are tested.

(a) What is the probability that the third sample resists the stress?

(b) What is the probability that the third and fourth samples resist the stress?

4.7.14. Suppose in a lottery game the chance of winning is 15%. If a person purchases 10 tickets, what is the probability of 3 winning tickets?

4.7.15. An electronic store is selling 10 boxes of AA size batteries at random. Two boxes contain some dead batteries. A customer is waiting on line to buy the sixth box. What is the probability that it contains dead batteries?

4.7.16. An engineer claims that a process should produce no more than 1 in 1000 defective items. We randomly select 100 of these items and find three defects. Is this consistent with the engineer's claim? Explain by using the Poisson approximation to the binomial.

4.7.17. Suppose a committee consists of seven members and we are interested in the number of members absent in each meeting. Let X denote the number of absentees in each meeting. Suppose also that the *cdf* of X is given by the following table:

x	< 0	0	1	2	3	4	5	6	7
F_X	0	0.10	0.60	0.70	0.75	0.85	0.95	0.99	1.00

Denote the probability of number of absents in each meeting p_X, $X = 0, 1, 2, \cdots, 7$. Find the *pmf*.

4.8.   Random Vectors

4.8.1.   Suppose X and Y have a joint distribution function given as

$$p_{XY} = \begin{cases} (1/24)(x^2 + y), & x = 1,\ 2;\ \ y = -1,\ 0,\ 2, \\ 0, & \textit{otherwise.} \end{cases}$$

Find the marginal distribution functions p_X and p_Y.

4.8.2.   Show that if a sample is drawn from a finite population without replacement, the random variables $X_1, X_2, \cdots, X_n$ are not mutually independent.

4.9.   Conditional Distribution and Independence

4.9.1.   Prove Theorem 4.9.1.

4.9.2.   Prove Theorem 4.9.4.

4.9.3.   Prove Corollary 4.9.1.

4.10. Discrete Moments

4.10.1.   In a game, a fair die is rolled. We win or lose a dollar amount equal to the number shown up, depending upon whether an even or odd number turned up, respectively.

(a) What is the expected win or loss in the game?

(b) What is the probability if change the order of winning and losing?

4.10.2.   Suppose that X is a discrete random variable whose values are 0, 1, 2, 3, 4, and 5, with probabilities .2, .3, .1, .1, .2, and .1, respectively. Find the mean and standard deviation of X.

4.10.3.   Let X be a discrete random variable with $V(X) = 8.6$, then $V(3X + 5.6)$ is

A. 77.4   B. 14.2   C. 83.0   D. 31.4

4.10.4. Prove Theorem 4.10.1.

4.10.5. Prove Theorem 4.10.2.

4.10.6. Prove Corollary 4.10.1. (Hint: Use mathematical induction.)

4.10.7. Prove Theorem 4.10.6.

4.10.8. Prove Theorem 4.10.7.

4.10.9. Prove Theorem 4.10.8.

4.10.10. Prove Theorem 4.10.9.

4.10.11. Prove Theorem 4.10.10.

4.10.12. Prove Theorem 4.10.11.

4.10.13. Suppose that 10 positive integers are taken at random with the same probability of being selected. Find the expected value of $X(X - 10)$, where X is the random variable representing the 10 positive integers.

4.10.14. Fifteen percent of the VCRs sold by a particular store will be returned for warranty service. Suppose the store sells 20 VCRs.

(a) Find the expected number, variance, and standard deviation of VCRs that will be returned.

(b) If more than five VCRs are returned, the store will lose money. Find the probability that the store loses money on the warranty service

4.10.15. Find expected value, variance and standard deviation of a Bernoulli random variable with probability $P(X = 1) = p$ and $P(X = 0) = 1 - p$. Show that

(a) $E(X) = p$,

(b) $E(X^2) = p$, and

(c) $VAR(X) = p(1 - p)$.

4.10.16. Consider the joint probability distribution of X and Y as below:

	$p(x,y)$	2	3	4
			X	
Y	-1	.10	.20	.05
	0	.27	.10	.15
	1	0	.03	.10

(a) Find $E(X)$, $E(Y)$, $E(X-Y)$, and $E(2X-3Y)$.

(b) Are X and Y independent?

(c) Show $E(XY) = E(X)E(Y)$.

4.10.17. Let X be the number of modules with programming errors in a piece of software. Let Y be the number of days it takes to debug the software. Suppose X and Y have the following joint probability mass function:

				X		
	$p(x,$	0	1	2	3	4
	0	.20	.08	.03	.02	.01
	1	0	.06	.09	.04	.01
Y	2	0	.04	.09	.06	.02
	3	0	.02	.06	.04	.03
	4	0	0	.03	.02	.02
	5	0	0	0	.02	.01

(a) Find $E(XY)$.

(b) Find the marginal probability mass function for each X and Y.

(c) Find the $Cov(X,Y)$

4.10.18. Suppose the cost of debugging the software in Example 4.10.15 is $1000 per defective module and $500 per day of debug time. If the cost function can be expressed as $W = 1000X + 500Y$, find the expected value, variance and standard deviation of the cost function.

4.10.19. Let X denote the number of hotdogs and Y the number of sodas consumed by an individual at a baseball game. Suppose X and Y have the following joint probability mass function:

	X			
p(x,	0	1	2	3
0	.06	.15	.06	.03
Y 1	.04	.20	.12	.04
2	.02	.08	.06	.04
3	.01	.03	.04	.02

(a) Find E(X), STD(X), E(Y), and STD(Y).

(b) Find Cov(X, Y) and Corr(X, Y).

(c) If hotdogs cost $3.00 each and soda cost $1.50 each, find an individual's expected value and standard deviation of total cost for sodas and hotdogs at a baseball game.

4.10.20. Prove that the mean and variance of a binomial random variable Y with parameters n and p are $E(Y)^{\cdot} = np$ and $Var(Y) = np(1-p)$, respectively.

4.10.21. Let Y denote the number of heads in 100 tosses of a fair coin. Find the expected value, variance and standard deviation of the number of heads.

4.10.22. Show that if $Y = a + bX$, then $E(Y) = a + bE(X)$.

4.11. Continuous Random Variables and Distributions

4.11.1. If X is a continuous random variable defined on the interval $[A, B]$, and the probability density function of X is $f(x; A, B) = 1/(A - B)$, $A < x < B$; and 0 otherwise, then X is said to have

    A. gamma distribution   B. normal distribution
    C. uniform distribution   D. Weibull distribution

4.11.2. If the probability density function of a continuous random variable X is f(x) = 0.5x 0< x< 2; and 0 otherwise. Then P (1<X< 1.5) is

    A. 0.5625   B.0.3125   C. 0.1250
    D. 0.4375

4.11.3. Which of the following standard normal probabilities are not correct?

    A. P(Z<- 1.25)= 0.8944   B. P (Z > 1.25) = 0.1056
    C. P (Z < 1.25) =0.8944   D. P(-1.25 < Z < 1.25) = .7888

4.11.4. If X is a normally distributed random variable with a mean of 10 and standard deviation of 4, then the probability that X is between 6 and 16 is

A. 0.9332   B. 0.7745   C. 0.6587
D. 0.0668

4.11.5. If X is a normally distributed random variable with a mean of 25 and a standard deviation of 8, then the probability that X exceeds 20 is approximately

A. 0.7357   B.0.2643   C. 0.6250   D. 0.3750

4.11.6. If X is a normally distributed random variable with a mean of 80 and a standard deviation of 12, then the $P(X = 68)$ is

A. 0.1587   B. 0.0000   C. 0.6587
D. 0.8413

4.11.7. Suppose a needle is spun on a circular dial. It will stop at a random angle, in standard position. Let X be a random variable denoting the position of the needle after each stop. Explain why X is a non-discrete random variable.

4.11.8. Suppose a chord is drawn at random in a circle. Let X denote the random variable measuring the length of chords drawn. Explain why X is a non-discrete random variable.

4.11.9. A number is to be chosen at random from the interval [0, 1]. What is the probability that the number is less than 2/3?

4.11.10. If X is a logistic random variable, show that $Y = \dfrac{1}{1+e^{-X}}$, $-\infty < X < \infty$, has a uniform distribution with mean 0 and variance equal to 1.

4.11.11. Show the following properties of Chi-squared random variable:
$\mu = r$,
$\sigma^2 = 2r$,
moment generating function, $M(t) = \left(1-2t\right)^{-r/2}$, $t < 1/2$, and
the Laplace transform of the $pdf = \left(1+2s\right)^{-r/2}$.

4.11.12. For Weibull random variable X, show that integrating (4.11.23) leads to the Weibull cumulative distribution function (cdf) as (4.11.24).

4.11.13. Prove that $\Gamma\left(\dfrac{1}{2}\right) = \sqrt{\pi}$.

4.11.14. Prove that $\Gamma(2\upsilon) = \dfrac{1}{\sqrt{\pi}} 2^{2\upsilon-1} \Gamma(\upsilon)\Gamma\left(\upsilon+\dfrac{1}{2}\right)$.

Hint: Use substitution $4(y-y^2) = s$ in $0 < y < \dfrac{1}{2}$.

4.11.15. The mean and variance of the Weibull random variable are

$$\mu_X = \frac{1}{\beta}\Gamma\left(1+\frac{1}{\alpha}\right) \text{ and } \sigma_X^2 = \frac{1}{\beta^2}\left\{\Gamma\left(1+\frac{2}{\alpha}\right) - \left[\Gamma\left(1+\frac{1}{\alpha}\right)\right]^2\right\},$$

respectively. When $1/\alpha$ in an integer, show that

$$\mu_X = \frac{1}{\beta}\left(\frac{1}{\alpha}\right)! \text{ and } \sigma_X^2 = \frac{1}{\beta^2}\left\{\left(\frac{2}{\alpha}\right)! - \left[\left(\frac{1}{\alpha}\right)!\right]^2\right\}.$$

4.12. Continuous Random Vector

4.12.1. Use a counter example to prove that pairwise independence does not apply mutual independence.

4.12.2. For Example 4.12.2, find A that was discussed in part (c) of Definition 4.12.1.

4.12.3. For Example 4.12.2, find $P\left(0 \leq X \leq \dfrac{1}{5}, 0 \leq Y \leq \dfrac{1}{5}\right)$.

4.12.4. For Example 4.12.2, find $f_X(x)$ and $f_Y(y)$.

4.12.5. For Example 4.12.2, find the first and second moments of X and Y.

4.13. Functions of a Random Variable

4.13.1. Prove Theorem 4.13.1.

4.13.2. Prove Theorem 4.13.2.

4.13.3. Find the numerical result under (4.13.5) directly using independence property of the random sample.

4.13.4. If Y is a function of a discrete random variable X, say $Y = \varphi[X(\omega)]$, then prove [formula (4.13.1)] that

$$E(Y) = E\big[\varphi(X)\big] = \sum_{\omega \in \Omega} \varphi(\omega) p(\omega),$$

where, $p(\omega) = P(X = \omega)$.

Chapter 5

STATISTICS

We start this chapter with a well-written article from The New York Times to show the value and applications of statistics from a non-academic and non-statistical professional view point. The article is just less a week old, the content is very much to date, and it is at the time of the annual meeting of the American Statistical Association in Washington, DC. A few names and sentences have been replaced by three dots. In our opinion, the omission would not change the intent of the article. We think it is a good starter for our students to appreciate the subject of statistics.

The New York Times   August 6, 2009

FOR TODAY'S GRADUATE, JUST ONE WORD: STATISTICS

By Steve Lohr

MOUNTAIN VIEW, Calif. — At Harvard, Carrie Grimes majored in anthropology and archaeology … But she was drawn to what she calls "all the computer and math stuff" that was part of the job. "People think of field archaeology as Indiana Jones, but much of what you really do is data analysis," she said. Now Ms. Grimes does a different kind of digging. She works at *Google,* where she uses statistical analysis of mounds of data to come up with ways to improve its search engine. Ms. Grimes is an Internet-age statistician, one of many who are changing the image of the profession as a place for dronish number nerds. They are finding themselves increasingly in demand — and even cool.

"I keep saying that the sexy job in the next 10 years will be statisticians," said Hal Varian, chief economist at Google. "And I'm not kidding." The rising stature of statisticians, who can earn $125,000 at top companies in their first year after getting a doctorate, is a byproduct of the recent explosion of digital data. In field after field, computing and the Web are creating new realms of data to explore — sensor signals, surveillance tapes, social network chatter, public records and more. And the digital data surge only promises to accelerate, rising fivefold by 2012, according to a projection by IDC, a research firm.

Yet data is merely the raw material of knowledge. "We're rapidly entering a world where everything can be monitored and measured," said Erik Brynjolfsson, an economist and director of the *Massachusetts Institute of Technology*'s Center for Digital Business. "But the big problem is going to be the ability of humans to use, analyze and make sense of the data." The new breed of statisticians tackle that problem. They use powerful computers and sophisticated mathematical models to hunt for meaningful patterns and insights in vast troves of data. The applications are as diverse as improving Internet search and online advertising, culling gene sequencing information for cancer research and analyzing sensor and location data to optimize the handling of food shipments.

Even the recently ended *Netflix* contest, which offered $1 million to anyone who could significantly improve the company's movie recommendation system, was a battle waged with the weapons of modern statistics. Though at the fore, statisticians are only a small part of an army of experts using modern statistical techniques for data analysis. Computing and numerical skills, experts say, matter far more than degrees. So the new data sleuths come from backgrounds like economics, computer science and mathematics. They are certainly welcomed in the White House these days. "Robust, unbiased data are the first step toward addressing our long-term economic needs and key policy priorities," *Peter R. Orszag*, director of the *Office of Management and Budget*, declared in a speech in May. Later that day, Mr. Orszag confessed in a *blog entry* that his talk on the importance of statistics was a subject "near to my (admittedly wonkish) heart." *I.B.M.*, seeing an opportunity in data-hunting services, created a Business Analytics and Optimization Services group in April. The unit will tap the expertise of the more than 200 mathematicians, statisticians and other data analysts in its research labs — but that number is not enough. I.B.M. plans to retrain or hire 4,000 more analysts across the company.

In another sign of the growing interest in the field, an estimated 6,400 people are attending the statistics profession's annual conference in Washington this week, up from around 5,400 in recent years, according to the American Statistical Association. The attendees, men and women, young and graying, looked much like any other crowd of tourists in the nation's capital. But their rapt exchanges were filled with talk of randomization, parameters, regressions and data clusters. The data surge is elevating a profession that traditionally tackled less visible and less lucrative work, like figuring out life expectancy rates for insurance companies.

Ms. Grimes, …, got her doctorate in statistics … in 2003 and joined Google later that year. She is now one of many statisticians in a group of 250 data analysts. She uses statistical modeling to help improve the company's search technology. For example, Ms. Grimes worked on an algorithm to fine-tune Google's crawler software, which roams the Web to constantly update its search index. The model increased the chances that the crawler would scan frequently updated Web pages and make fewer trips to more static ones. The goal, Ms. Grimes explained, is to make tiny gains in the efficiency of computer and network use. "Even an improvement of a percent or two can be huge, when you do things over the millions and billions of times we do things at Google," she said.

It is the size of the data sets on the Web that opens new worlds of discovery. Traditionally, social sciences tracked people's behavior by interviewing or surveying them. "But the Web provides this amazing resource for observing how millions of people interact," said Jon Kleinberg, a computer scientist and social networking researcher at *Cornell*. For example, in *research just published*, Mr. Kleinberg and two colleagues followed the flow of ideas across cyberspace. They tracked 1.6 million news sites and blogs during the 2008 presidential campaign, using algorithms that scanned for phrases associated with news topics like "lipstick on a pig." The Cornell researchers found that, generally, the traditional media leads and the blogs follow, typically by 2.5 hours. But a handful of blogs were quickest to quotes that later gained wide attention. The rich lode of Web data, experts warn, has its perils. Its sheer volume can easily overwhelm statistical models. Statisticians also caution that strong correlations of data do not necessarily prove a cause-and-effect link.

For example, in the late 1940s, before there was a polio vaccine, public health experts in America noted that polio cases increased in step with the consumption of ice cream and soft drinks, according to David Alan Grier, a historian and statistician at *George Washington University*. Eliminating such treats was even recommended as part of an anti-polio diet. It turned out that polio outbreaks were most common in the hot months of summer, when people naturally ate more ice cream, showing only an association, Mr. Grier said.

If the data explosion magnifies longstanding issues in statistics, it also opens up new frontiers.

"The key is to let computers do what they are good at, which is trawling these massive data sets for something that is mathematically odd," said Daniel Gruhl, an I.B.M. researcher whose recent work includes mining medical data to improve treatment. "And that makes it easier for humans to do what they are good at — explain those anomalies."

Andrea Fuller contributed reporting.

This chapter consists of three parts. They are: Part One: Descriptive Statistics (Section 1); Part Two: Inferential Statistics (Sections 2 – 7); and Part Three: Applications of Statistics (Sections 8 – 14).

PART ONE: DESCRIPTIVE STATISTICS

5.1. BASIC STATISTICAL CONCEPTS

Statistics is said to be a branch of mathematics dealing with gathering, analyzing and interpreting data, although some believe it as a discipline for itself such as any other science. Whichever, it is an important concept that its applications are now so vast and diverse that no area of science can do without. We start this chapter with defining some important terms that are widely used in statistics. Probability theory, as we discussed in the last chapter, allows us to model *randomness* in terms of *random experiments* that involves terms such as *sample space*, *event* and *distributions*, as we discussed. The area of statistics plays the role of bridging probability models to the real world.

Definition 5.1.1.

A statistical *population* is a collection or a set of all individuals, objects, or measurements of interest. A *sample* is a portion, subset or a part of the population of interest (finite or infinite number of them). The difference between the smallest and the largest sample points is called the *range*. A sample selected such that each element or unit in the population has the same chance to be selected is called a *random sample*. Study of the entire population is called a *census*.

Example 5.1.1.

To determine the average number of electronic devices per household in the United States, the totality of these devices per household constitutes the population for the study.

Example 5.1.2.

Suppose a manufacturing company with 10,000 employees is to determine productivity of its employees before and after casual dress during the summer time. Then, the entire number of employees will be the population to be studied.

Considering the population, however, may not, practically and economically, be feasible because it may be too much time consuming, and too costly, impossible to identify all members of it. Therefore, a sampling shows its necessity, except the desire to study the entire population. Of course, the sample must be *representative* of the entire population in order to make any prediction about the population. Random sample assures this condition. Symbolically, a random sample $\mathbf{X} = \left(X_1, X_2, \cdots, X_n \right)$ is defined as a sample consisting of n independent random variables with the same distribution. Each component of a random sample is a random variable representing observations. The term "random sample" is also used for a set of observed values $x_1, x_2, \cdots, x_n$ of the random variables. We should, however, caution that it is not always easy to select a random sample, particularly, when the population size is very large and the sample size is to be small. For instance, to select a sample of size 10 cartons of canned soup to inspect thousands of cartons in the storage, it is almost impossible to number all these cartons and then choose 10 at random. Hence, in cases like this, we do not have many choices; we have to do the best we can and hope that we are not seriously violating the randomness property of the sample.

In summary, the purpose of most statistical studies is to obtain information about the population from a random sample, by generalization. It is a common practice to identify a population and a sample with distribution of their values; *parameters* and *statistic*, respectively.

Definition 5.1.2.

A *parameter* is a number that describes a character of the population. A characteristic of a sample is called a *statistic*.

Example 5.1.3.

Numbers such as *mean* and *variance* of a *distribution* are examples of a *parameter*. Mean waiting time in a queue (waiting line) is a parameter of the population to obtain their newly marketed iPhone 3G made by Apple Inc., in early summer of 2008 (iPhone 3G is a small and lightweight handheld device that combines three products - a mobile phone, an iPod with touch controls, and an Internet communications device with desktop-class email, web browsing, maps, and searching. 3G technology gives iPhone fast access to the Internet and email over cellular networks around the world.)

A *statistic* is itself a random variable because the value of it is uncertain prior to gathering data. A *statistic*, say λ, is usually calculated based on a *random sample*. In other words, a *statistic* is a function of the *random vector* $\mathbf{x} = (x_1, x_2, \cdots, x_n)$, say, $\hat{\lambda}(\mathbf{x}) = f(x_1, x_2, \cdots, x_n)$.

5.1.1. Measures of Central Tendency

There are several ways to describe the center of a given n observations (*measures of central tendency*); the *arithmetic mean*, the *median* and the *mode* are three popular such ways. The arithmetic mean or more succinctly, the *mean*, (or the average) and to emphasize the randomness, the *sample mean* is defined below:

Definition 5.1.3.

Le $\mathbf{x} = (x_1, x_2, \cdots, x_n)$ be a random vector. Then, the statistic *sample mean* denoted by $\overline{x}$ (it is read as: x bar) is defined as

$$\overline{x} = \frac{1}{n} \sum_{i=1}^{n} x_i .$$ (5.1.1)

It is not always easy to find distribution of a statistic unless the statistic is "fairly simple" function of x and the population distribution has a "nice" form. We need to point out that the sample mean interprets as the balance point or center of sample points.

Example 5.1.4.

Suppose an electronic shop sells three units of an instrument IA, IB and IC at prices $20, $30, and $40 each, respectively. History shows that the percentages of sell of these items are: 30, 50 and 20, respectively. We want to find the distribution of the sample mean.

Solution

The probability distribution of revenue from a single randomly selected item is

Table 5.1.1.

	IA	IB	IC
x (in $)	20	30	40
$P(x)$	30% = 0.3	50% = 0.5	20% = 0.2

Hence, the average revenue (the expected value), μ, of a single randomly selected item is $\mu = (20)(.3) + (30)(.5) + (40)(.2) = 29$.

Now let X_1 and X_2 be the revenues from the first and second items sold, respectively. Assume that X_1 and X_2 are independent random variables and have the same distribution as listed in Table 5.1.1. We now pair up values of X_1 and X_2 as (x_1, x_2). Then, the probabilities of each component of a possible pair (given), joint probability of each pair (product of probabilities of each component), and sample means (average of both component, formula (5.1.1)) are summarized in Table 5.1.2.

Table 5.1.2.

x_1	x_2	$P(x_1, x_2)$	$\overline{x}$
20	20	0.09	20
20	30	0.15	25
20	40	0.06	30
30	20	0.15	25
30	30	0.25	30
30	40	0.10	35
40	20	0.06	30
40	30	0.10	35
40	40	0.04	40

Thus, the probability of each of the sample means can be calculated. For instance, probability of obtaining 30, is 0.06 + 0.25 + 0.06 = 0.37 (since 30 occurs three times (or with frequency 3) with probabilities 0.06, 0.25, and 0.06, and events are mutually exclusive). The word *frequency* refers to the number of times a data point is repeated. We can now calculate the rest of probabilities and obtain the distribution for $\overline{X}$ listed in Table 5.1.3.

Table 5.1.3. Distribution for $\overline{X}$

$\overline{x}$	20	25	30	35	40
Frequency	1	2	3	2	1
$P_{\overline{x}}(\overline{x})$	0.09	.15 + .15 = 0.30	.06 + .25 + .06 = 0.37	.10 + .10 = 0.20	0.04

Definition 5.1.4.

The statistic *sample median* (or simply median, also called the 2nd *quartile* or 50th *percentile*) of N (non-missing) observations $x_1, x_2, \cdots, x_n$ is loosely defined as the *middlemost* of the observed values. To find it, arrange $x_1, x_2, \cdots, x_n$ according to their values in ascending or descending order, then pick the middle value if N is odd, i.e., $(n+1)/2$, and the average (mean) of the middle two if n is even, i.e., $[(n/2 + (n+2)/2]/2$.

Example 5.1.5.

The *mean* and *median* of the set of observations 5, 7, 4, 6, 3, 10 and 9 are as follows:

$$\text{mean} = \bar{x} = \frac{5+7+4+6+3+10+9}{7} = 6.286,$$

and from 3, 4, 5, 6, 7, 9, 10, median = 6, since there are 7 (odd number) of observations. However, if we had 8 observations 5, 7, 4, 6, 3, 10, 3 and 9, then we would have had:

$$\text{mean} = \bar{x} = \frac{5+7+4+6+3+10+3+9}{8} = 5.875,$$

and from 3, 3, 4, 5, 6, 7, 9, 10, median = (5+6)/2 = 5.5, since there are 8 (even number) of observations.

Usually, the sample mean is preferred over the median, possibly because median does not carry all information of a sample. It may also be because the median is less sensitive to extreme data points than the mean. On the other hand, the median is sometimes preferred because of less affected by a few extreme values or "*outliers*".

Definition 5.1.5.

An *outlier* (or an extreme data point) is a sample point such that there is an unusually large gap between the largest or the smallest data point and this sample point. The median is less sensitive to extreme values than the mean. Thus, when data contain outliers, or are *skewed* (lack of symmetry), the median is used instead of the mean. We say a distribution is *skewed* if one tail extends farther than the other. Outliers cause skewness. An outlier on the far left will cause a *negative or left skew*; while on the far right will cause a *positive or right skew*.

We refer the reader for calculation of skewness to Wilks (1962, p.265).

Example 5.1.6.

Suppose an engineering firm has five engineers who are paid annual salaries as $70,000, $80,000, $85,000, $85,000, $90,000. However, the chief executive officer (CEO), who also is an engineer, earns $180,000. The sample mean in this case is $(70000 + 80000 + 85000 + 85000 + 90000 + 180000)/6 = 98,333.33$. But, by arranging the data points in ascending order, i.e., 70,000, 80,000, 85,000, 85,000, 90,000, 180,000, the median will be (85000+85000)/2 = $85,000, that is most representative of this group of engineers. As it can be seen this median will not change if the outlier, 180,000, is dropped, because we will then have 5 data points, the middle value of which is $85,000.

Definition 5.1.6.

The *mode* is most frequent of the observed values. A distribution may have more than one mode, in which case, the data set is called have *multi-modes*. When the number is two, it is called *bimodal*.

Note that, as in the case of other measures of central tendencies, mode of a data set loses some of information of data points.

5.1.2. Organization of Data

A set of data points may summarize or grouped in a simple way or into a suitable number of classes (or categories). However, as in measures of central tendencies, this grouping may cause loss of some information in the data. This is because, instead of knowledge of an individual data point, knowledge of its belonging to a group will be known.

Definition 5.1.7.

A *simple frequency distribution* is a grouping of data set according to the number of repetition of a data point. The ratio of a frequency to the total number of observations is called *relative frequency*. A relative frequency multiplied by 100 will yield a *percent relative frequency*.

Example 5.1.7.

Suppose a set of observations of daily emission (in tons) of sulfur from an industrial plant is given in Table 5.1.4:

Table 5.1.4. A Set of Observations (Raw Data

20.2	12.7	18.3	18.3	23.0
21.5	10.2	17.1	07.3	11.0
18.3	20.2	20.2	20.2	18.3
17.1	11.0	12.7	11.0	23.0
23.0	18.3	12.7	07.9	12.7
17.1	21.5	07.9	11.0	12.7

A *simple frequency distribution* for this data set is shown in Table 5.1.5.

Table 5.1.5. Simple Frequency Distribution

Data Point	Frequency	Relative Frequency	Percent Relative Frequency
07.3	1	1/30	03.33
07.9	2	2/30	06.67
10.2	1	1/30	03.33
11.0	4	4/30	13.33
12.7	5	5/30	16.67
17.1	3	3/30	10.00
18.3	5	5/30	16.67
20.2	4	4/30	13.33
21.5	2	2/30	06.67
23.0	3	3/30	10.00
Total	30	1	100

It can be seen from Table 5.1.5, this data set has two modes, namely, 12.7 and 18.3 since these are the values with highest frequencies. Thus, the distribution is bimodal.

For a simple frequency table, the sample mean (5.1.1) becomes:

$$\overline{x} = \frac{\sum_{i=1}^{k} f_i x_i}{n} , \qquad (5.1.2)$$

where x_i's represent data points with their frequencies.

Before discussing other types of grouping data, we need a few more terms. One of the simplest ways to assess dispersion in a data set is to compare the *minimum* and *maximum*.

Definition 5.1.8.

The *minimum* is the smallest value in a data set, and the *maximum* is the largest value. The *maximum – minimum* is called *range*, which is a statistic that is often used to describe dispersion in data sets.

Example 5.1.8.

In Example 5.1.7, the *minimum* is 7.3, the *maximum* is 23, and the *range* is $23 - 7.3 = 15.7$.

Another grouping of data is a *frequency distribution with classes* (or *intervals*). Here is how we construct it. Find the *range*. Decide how many *classes* you want to have. However, we don't want too many or too few *classes*. Classes may be chosen with equal lengths. If that is the case, divide the range by the number of classes selected and choose the rounded up number. This number will be the *length* (or *size*) of a *class*. Now to avoid overlaps, move 0.5 (if data is with integral digits and 0.05 if it is with no digits and starts with the tenth decimal place, .005 if it starts with the hundredth decimal place and so on), down and up from each end of the interval (*class boundaries*) and choose the interval half-open on the right end. If, however, the length of an interval is given, then divide the range by the length of an interval and choose the rounded up number to find the number of classes. After the intervals have been determined, count the number of data points in each class to find *frequency* for each class.

Example 5.1.9.

Suppose in grouping a set of data points with minimum 62 and maximum 125, i.e., range of 63, we decide to have 5 *class intervals*. Thus, the class sizes will be $63/5 \approx 13$. Hence, the 5 non-overlapping intervals are: (60.5, 73.5), (73.5, 86.5), (86.5, 99.5), (99.5, 112.5) and (112.5, 125.5). Class boundaries are 60.5, 73.5, 86.5, 99.5, 112.5, and 125.5.

Example 5.1.10.

We saw in Example 5.1.8 that range of the data points in Example 5.1.7 was 15.7. If we decide to have 5 *classes*, then size of a class will be $15.7/5 = 3.14 \approx 4$. Thus, the 5 non-overlapping intervals are: (6.8, 10.8), (10.8, 14.8), (14.8, 18.8), (18.8, 22.8) and (22.8, 26.8). See Table 5.1.6. On the other hand, if we had decided to have a *class length* to be 3, then the number of classes would have been $15.7/3 = 5.23 \approx 6$. Thus, the 6 non-overlapping intervals are: (6.8, 9.8), (9.8, 12.8), (12.8, 15.8), (15.8, 18.8), (18.8, 21.8), and (21.8, 24.8).

Table 5.1.6. Grouped Frequency Distribution

Class Interval	Frequency	Class Interval	Frequency
(06.8, 10.8)	04	(06.8, 09.8)	03
(10.8, 14.8)	09	(09.8, 12.8)	10
(14.8, 18.8)	08	(12.8, 15.8)	00
(18.8, 22.8)	06	(15.8, 18.8)	08
(22.8, 26.8)	03	(18.8, 21.8)	06
Total	30	(21.8, 24.8)	03
		Total	30

In order to calculate statistics such as mean, variance and standard deviation, using the class-interval grouping, the midpoints of each interval (some times called *class marks*) will be used for the x_i's in (5.1.1) and (5.1.2).

We cannot choose one or more data values as the mode or modes in this case as we did from Table 5.1.5. We could, on the other hand, find a class in which a mode is located. Such a class is called the *modal class*. Midpoint of the modal class is the mode. Hence, as we see from Table 5.1.6, the modal class is $10.8 - 14.8$, in case of 5 classes, and $15.8 - 18.8$, in case of 6 class. Thus, in case of 5 classes the mode is $(10.8+14.8)/2 = 12.8$, and in case of 6 classes it is $(15.8+18.8)/2 = 16.8$.

Using a statistical computer package, makes it is easy to find most basic information about a data set such as mean, median, standard deviation and more. Minitab is one of such software that is very user friendly and useful for classroom instruction. After entering the raw data from Example 5.1.7 in a column, say, C1, by just three clicks ("Basic Statistics", Display Descriptive Statistics", and the column number "C1") we obtain the following information on the Work Session:

Descriptive Statistics: C1

Variable	N	Mean	Median	TrMean	StDev	SE Mean
C1	30	15.880	17.100	15.969	4.923	0.899

Variable	Minimum	Maximum	Q1	Q3
C1	7.300	23.000	11.000	20.200

On the display, "N" is the number of data points (or the *sample size*), 30. In this case "*mean*" is the *sample mean* defined by formula (5.1.1). The mode or modes that are not given by Minitab through the descriptive statistics can be seen on a graphic representation of data.

Now let us sort the data in descending order (could have been in ascending order):

23.0, 23.0, 23.0, 21.5, 21.5, 20.2, 20.2, 20.2, 20.2, 18.3, 18.3, 18.3, 18.3, 18.3, 17.1, 17.1, 17.1, 12.7, 12.7, 12.7, 12.7, 12.7, 11.0, 11.0, 11.0, 11.0, 10.2, 7.9, 7.9, 7.3

Since there are 30 data items, the median has been calculated from Definition 5.1.4, i.e., the average of items 15 and 16, which is $(17.1+17.1)/2 = 17.1$.

Definition 5.1.9.

To avoid outliers, it is customary to *trim* the data. *The trimmed mean* is the mean of the *trimmed data*, i.e., of the remaining data set. Minitab denotes this mean by *TrMean*.

By *trimmed mean* it is meant to cut the data about 5 to 10 percent (rounded to the nearest integer) on each end (after sorted in ascending or descending order). Minitab uses 5%. Thus, taking a rounded 5% of each side drops 2 data items from each end. This leaves us with 26 data points as:

23.0, 21.5, 21.5, 20.2, 20.2, 20.2, 20.2, 18.3, 18.3, 18.3, 18.3, 18.3, 17.1, 17.1, 17.1, 12.7, 12.7, 12.7, 12.7, 12.7, 11.0, 11.0, 11.0, 11.0, 10.2, 7.9, 7.9.

Hence, applying formula (5.1.1), trimmed mean is 15.969, as shown in the Minitab session. By construction, trimmed mean, like median, is less sensitive to extreme values than the mean. Although this term is a bit higher level than basic descriptive statistics and is important in hypothesis testing, Minitab adds this information as part of the descriptive statistics, anyway.

The notation *SE Mean* (on MINITAB) is referred to the *standard error of the mean*. *SE Mean* is an estimate of the dispersion that we would observe in the distribution of sample mean, if we continue to take samples of the same size from the population. The value of *SE Mean* is the standard deviation divided by $\sqrt{N}$, i.e.,

$$standard\ error\ of\ the\ mean = \frac{standard\ deviation}{\sqrt{N}}, \qquad (5.1.3)$$

where N is the sample size. Notation $\sigma_{\bar{x}}$ is sometimes used for standard error of the mean.

Definition 5.1.10.

For a sample of size n, and a number p such that $0 < p < 1$, the $(100p)^{th}$ *sample percentile* is a sample point at which there are approximately np sample points below it and $n(1-p)$ sample point above it. The 25th *percentile* (also called the *first*, denoted by Q_1) and the 75th *percentile* (also called the *third*, denoted by Q_3) are the highest value for the lowest 25% of the data points and the lowest value for the highest 25% of the data points, respectively. Similarly, we could have *deciles* that are the percentiles in the 10th increments, such as the first decile as the 10th percentile, the fifth decile as the 50th percentile or median, and so on. The *interquartile range*, denoted by *IQR*, is the range of the middle 50% of the data points and is calculated as the difference between Q_3 and Q_1, i.e., $Q_3 - Q_1$. Below we will show how to calculate percentiles.

Example 5.1.11.

The *IQR* for our data set in Example 5.1.8 (the Minitab results on Example 5.1.7) is 20.2 - 11.0 = 9.2. We note that the *IQR* is relatively insensitive to extreme values. For example, the *IQR* would be 20.200 - 12.275 = 7.925, if we use 26 data points, after trimming the data set.

There are several graphic representations such as *Histogram, Dot Plot, Box Plot, Stem-and-Leaves*, and *Scatter Plot*. We will discuss the first and the last of these.

Stem-and-leaf displays a data set without loss of individual data points. John W. Tukey introduced this graphic method about three decades ago.

To display a data set in a stem-and-leaf form, take all digits (including decimal places) of a data point except the most right-hand digit as the stem and the most right-hand digit as the leaf. See example below for detail.

The following process is generally true for any ordered sample points. Other points, they do not have to be ordered according to the stem-and-leaf method.

To calculate a percentile, we consider three cases for the $(n+1)p$, namely, (1) $(n+1)p$ an integer, (2) $(n+1)p$ as an integer plus a proper fraction, and (3) $(n+1)p < 1$.

1. If $(n+1)p$ is an integer, then the $(100p)^{th}$ sample percentile is the $(n+1)p^{th}$ ordered data point in the ordered stem-and-leaf display.

2. If $(n+1)p$ is not an integer, but is equal to $r+a$, where r is the whole part and a is the proper fraction part of $(n+1)p$, then take the weighted average of the rth and $(r+1)^{st}$ ordered data points. That is, if $\tilde{\pi}_p$ denotes this quantity, it will be found as

$$\tilde{\pi}_p = D_r + a(D_{r+1} - D_r) = (1-a)D_r + aD_{r+1}, \text{ where } D_r = \text{ the } r^{th} \text{ ordered data}$$

   point and $D_{r+1} = (r+1)^{st}$ ordered data point.

3. If $(n+1)p < 1$, then the sample percentile is not defined.

Example 5.1.12.

Consider the following data set.

77	72	83	66	59
97	73	43	61	91
54	47	70	65	90
69	76	60	38	74
76	58	73	75	93

Construct the stem-and-leaf display. Find the following:

(a) Q_1, (b) Q_2, (c) Q_3, (d) 60th percentile, (e) 90th percentile, and (f) *IQR*.

Solution

Since the data points are double digits, as we read them from left to right, we record the left-hand digit as "stem" and the right hand digit as the leaf. Thus, we have:

Stem	Leaf	f
7	7 2 3 0 6 4 6 3 5	9
8	3	1
6	6 1 5 9 0	5
5	9 4 8	3
9	7 1 0 3	4
4	3 7	2
3	8	1

We now order them as follows:

Stem	Leaf	f	Cumulative f
3	8	1	1
4	3 7	2	3
5	4 8 9	3	6
6	0 1 5 6 9	5	11
7	0 2 3 3 4 5 6 6 7	9	20
8	3	1	21
9	0 1 3 7	4	25

This is called the *ranked stem-and-leaf display.*

We find the percentiles as follows:

(a) Q_1 is the 25th percentile. Thus, $p = 0.25$ and $(n+1)p = (26)(0.25) = 6.50 = 6 + 0.50$. Thus, $r = 6$ and $a = .50$. Hence, $\tilde{\pi}_{.25} = (1-.5)D_6 + .5D_7 = (.5)(59) + (.5)(60) = 59.5$.

(b) Q_2 is the 50th percentile. Thus, $p = 0.50$ and $(n+1)p = (26)(0.50) = 13$. Hence, the 13th ordered data point is 72 and $Q_2 = \tilde{\pi}_{50} = 72$.

(c) Q_3 is the 75th percentile. Thus, $p = 0.75$ and $(n+1)p = (26)(0.75) = 19.5 = 19 + 0.50$. Thus, $r = 19$ and $a = .50$. Hence,

$$\tilde{\pi}_{.75} = (1-.5)D_{19} + .5D_{20} = (.5)(76) + (.5)(76) = 76.$$

(d) For the 60th percentile, $p = 0.60$ and $(n+1)p = (26)(0.60) = 15.60 = 15 + 0.60$. Thus, $r = 15$ and $a = .60$. Hence, $\tilde{\pi}_{.60} = (1-.6)D_{15} + .4D_{16} = (.4)(73) + (.6)(74) = 73.6$.

(e) For the 90th percentile $p = 0.90$ and $(n+1)p = (26)(0.90) = 23.40 = 23 + 0.40$. Thus, r
= 23 and $a = .40$. Hence, $\tilde{\pi}_{.90} = (1-.4)D_{23} + .6D_{24} = (.4)(91) + (.6)(93) = 92.2$.

(f) $IQR = 76 - 59.5 = 16.5$.

We can interpret (a) through (e) as approximately 25%, 50%, 75%, 60% and 90% of the
sample points are less than 59.5, 72, 76, 73.6 and 92.2, respectively. The IQR in (f) tells us
that about 16.5% of data points fall between the first and the third quartiles.

Example 5.1.13.

Let us find the 60th percentile for the exponential distribution with mean 10.

In this case, $p = 0.6$, $f(x) = 10e^{-0.10x}$ and $F(x) = 1 - e^{-0.10x}$. Thus, using (5.1.4), we
need to solve $F(\pi_{.60}) = 0.60$ for $\pi_{.60}$. But, $F(\pi_{.60}) = 1 - e^{-0.10\pi_{.60}} = 0.60$ or
$e^{-0.10\pi_{.60}} = 0.40$. Hence, $\pi_{.60} = -10\ln(0.40) = 9.163$. That is, 60% of data set is below
9.163.

One of the most common graphic presentations of data set is a "*histogram*". A *histogram*
displays data that have been summarized into *class intervals* (such as in Table 5.1.6). It is a
graph that indicates the "shape" of a sample. Among other things, a *histogram* can be used to
assess the *symmetry* or *skewness* of the data. A *histogram* of a frequency class distribution is
constructed as adjacent rectangles. The bases of the rectangles represent the endpoints of the
class intervals and the heights represent the frequencies of the classes. To construct a
histogram, the horizontal axis is divided into equal intervals, and a vertical bar (or a strip) is
drawn at each interval to represent its frequency (the number of data points that fall within the
interval).

We may construct a *relative frequency histogram* using rectangles or strips with relative
frequency for each class as an estimate for the probability for that class as follows: the base of
each rectangle is the interval bounded by the class boundaries. The area of each rectangle i is
the relative frequency f_i / n for that interval, where n is the sample size (the total number of
observations), $i = 1, 2, \cdots, k$, and k is the number of class intervals.

Example 5.1.14.

Consider the set of observations of daily emission (in tons) of sulfur from an industrial
plant is given in Example 5.1.7. The ordered data points are as follows:

07.3	07.9	07.9	10.2	11.0
11.0	11.0	11.0	12.7	12.7
12.7	12.7	12.7	17.1	17.1
17.1	18.3	18.3	18.3	18.3
18.3	20.2	20.2	20.2	20.2
21.5	21.5	23.0	23.0	23.0

Now the median for this data set is Q_2, i.e., $p = 0.50$. Hence, $(n+1)p = (31)(.5) = 15.5$. Thus, $r = 15$ and $a = .5$. Therefore, Q_2 is the average of the 15th and the 16th data points, which is 17.1. On the other hand, for the 53rd percentile, $(31)(.53) = 16.43$, yielding $r = 16$ and $a = 0.43$. Hence,

$$\tilde{\pi}_{.53} = (1 - .43)D_{16} + .57D_{17} = (.57)(17.1) + (.43)(18.3) = 9.747 + 7.869 = 17.616.$$

Also, the 95th percentile is $\tilde{\pi}_{.95} = (1 - .45)D_{29} + .45D_{30} = (.55)(23) + (.45)(23) = 23$.

Using MINITAB, we obtain the following information as well as two histograms. Note that the 'Confidence Interval' will be discussed later in this chapter.

Descriptive Statistics: Observations

Variable	Mean	SE Mean	TrMean	StDev	Variance	CoefVar	Sum of Squares
Observations	15.880	0.899	15.969	4.923	24.231	31.00	8267.940

Variable	Minimum	Q1	Median	Q3	Maximum	Range	IQR	Skewness
Observations	7.300	11.000	17.100	20.200	23.000	15.700	9.200	-0.19

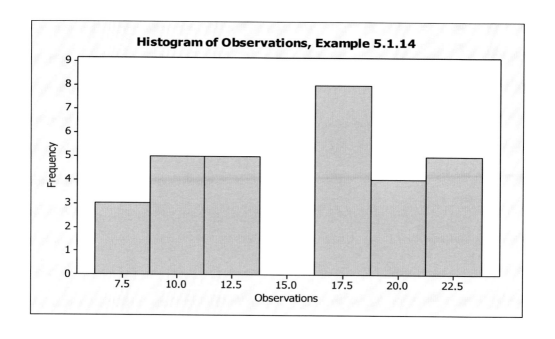

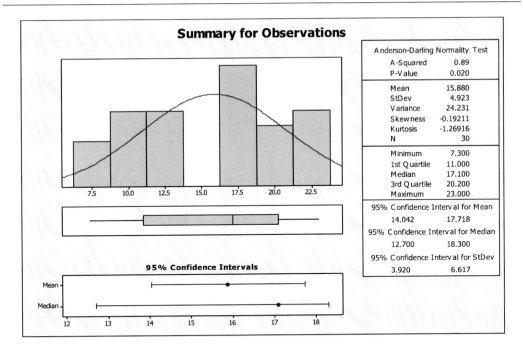

[Note that for a continuous random variable X with *pdf* as $f_X(x)$ and *cdf* as $F_X(x)$, that we will define and discuss later, the $(100p)^{th}$ percentile denoted by π_p is a number such that the area under the curve $f(x)$ to the left of π_p is p, i.e.,

$$p = \int_{-\infty}^{\pi_p} f(x)dx = F(\pi_p).]$$ (5.1.4)

Definition 5.1.11.

Now if we denote the boundaries of an interval by b_{i-1} and b_i, $i = 1, 2, \cdots, k$, then the *relative frequency histogram*, denoted by $h(x)$ is defined as

$$h(x) = \frac{f_i}{n(b_i - b_{i-1})}, \ i = 1, 2, \cdots, k.$$ (5.1.5)

Note that a sample point x will fall within one of the intervals $[b_{i-1}, \ b_i)$ or $(b_{i-1}, \ b_i]$, whether we want to take the half-open interval to be open from the left or from the right; we take it open from the right.

Definition 5.1.12.

The measure of sharpness of peak of a distribution is referred to as *kurtosis*.

Similar to skewness, positive or negative values of *kurtosis* will cause flatter than or sharper than the peak of the normal curve. For calculation of kurtosis the reader may see Wilks (1962, p. 265).

Example 5.1.15.

Consider Example 5.1.9. Let us assume that the data has been tally (or tabulated) and summarized as in Table 5.1.7.

Table 5.1.7. Data Grouping by Class Intervals. Example 5.1.15.

Class Intervals	Tally	Frequency f_i	Cumulative Frequency	Relative Frequency
(60.5, 73.5)	ЖҐ ЖҐ ЖҐ I	16	16	.16
(73.5, 86.5)	ЖҐ ЖҐ ЖҐ ЖҐ IIII	24	40	.24
(86.5, 99.5)	ЖҐ ЖҐ ЖҐ ЖҐ ЖҐ ЖҐ ЖҐ II	37	77	.37
(99.5, 112.5)	ЖҐ ЖҐ ЖҐ III	18	95	.18
(112.5, 125.5)	ЖҐ	5	100	.05
		100		1

Thus, for a data point x in each class the corresponding relative frequency histogram $h_i(x)$ is:

$$h_1(x) = \frac{16}{100 \cdot 13} = .01231, \ 60.5 < x \le 73.5$$

$$h_2(x) = \frac{24}{100 \cdot 13} = 01846, \ 73.5 < x \le 86.5$$

$$h_3(x) = \frac{37}{100.13} = .02846, \ 86.5 < x \le 99.5$$

$$h_4(x) = \frac{18}{100 \cdot 13} = 01385, \ 99.5 < x \le 112.5$$

$$h_5(x) = \frac{5}{100 \cdot 13} = 00385, \ 112.5 < x \le 125.5$$

As it can be seen, since we are taking each relative frequency as an area of the rectangle, the total area of all rectangles is equal to 1. Thus, if E is an event composed of a union of the first three class intervals, then the probability that a data point is less than 99.5 will be the sum of relative frequencies in the first three intervals, i.e., 0.77. This is the area under $h(x)$ from 0 to 99.5.

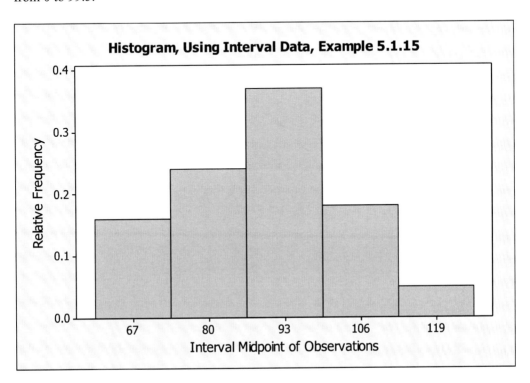

5.1.3. Measures of Variability

During the discussion above, we mentioned three measures of central tendencies. Now, we will discuss a measure of dispersion.

Definition 5.1.13.

Le $\mathbf{x} = (x_1, x_2, \cdots, x_n)$ be a random vector with sample mean $\bar{x}$. Then, the statistic *sample variance* (sometimes called *mean square*) denoted by s^2, is defined as

$$s^2 = \frac{1}{n-1}\sum_{i=1}^{n}(x_i - \bar{x})^2 .$$ (5.1.6)

Note use of the term $n - 1$ instead of n in the denominator. Although, logically, we should use n (and some authors in their publications by a couple of decades ago used n), it tends to underestimate the population variance σ^2. Since the primary use of the sample variance s^2 is to estimate the population variance σ^2, by replacing $n - 1$ and, thus, enlarging s^2, the tendency is corrected. The same correction is made in use of sample *standard deviation*, which is the positive square root of the sample variance. It is easy to see that (5.1.6) is equivalent to (5.1.7) below, which is used for practical calculation of sample variance,

$$s^2 = \frac{1}{n-1}\left[\sum_{i=1}^{n} x_i^2 - \frac{\left(\sum_{i=1}^{n} x_i\right)^2}{n}\right]. \tag{5.1.7}$$

In case that data is grouped in a simple frequency table, the sample variance would be calculated using

$$s^2 = \frac{\sum_{i=1}^{k} f_i x_i^2 - \left(\sum_{i=1}^{k} f_i x_i\right)^2 / n}{n-1}. \tag{5.1.8}$$

Example 5.1.16.

Returning to Example 5.1.7, we now show how to calculate the sample variance and sample standard deviation, denoted by in the Minitab Session as below.

Thus, using (5.1.7) we will have

$$s^2 = \frac{1}{30-1}\left[8267.94 - \frac{226957}{30}\right] = 24.2313,$$

and the sample standard deviation, StDev, $= \sqrt{24.2313} = 4.92253 \approx 4.923$.

Now from MINITAB, we obtain the same values as below:

Descriptive Statistics: C1

Variable	Mean	StDev	Variance
C1	15.880	4.923	24.231

X	X^2
20.2	408.04
21.5	462.25
18.3	334.89
17.1	292.41
23.0	529.00
17.1	292.41
12.7	161.29
10.2	104.04
20.2	408.04
11.0	121.00
18.3	334.89
21.5	462.25
18.3	334.89
17.1	292.41
20.2	408.04
12.7	161.29
12.7	161.29
7.9	62.41
18.3	334.89
7.3	53.29
20.2	408.04
11.0	121.00
7.9	62.41
11.0	121.00
23.0	529.00
11.0	121.00
18.3	334.89
23.0	529.00
12.7	161.29
12.7	161.29
476.4	8267.94

The following theorem states some properties of a random sample.

Theorem 5.1.1.

Let $X_1, X_2, \cdots, X_n$ be a random sample from a population with mean and variance μ and $\sigma^2 < \infty$, respectively. Then

a. $E(\bar{X}) = \mu$.

b. $Var(\bar{X}) = \dfrac{\sigma^2}{n}$.

c. $E(S^2) = \sigma^2$.

Proof

See Casella and Berger (1990, p. 208).

PART TWO: INFERENTIAL STATISTICS

As mentioned, since sampling is recommended in most cases over usage of population, the purpose of most statistical investigations is to generalize information contained in samples to the populations from which the samples were obtained. This is the basis for statistical inference. The term *inference* means a conclusion or a deduction.

Two major areas *hypotheses testing* and *estimation* (*point and interval*) are main components of the methods of *statistical inference*, in the classical approach. A third area is *localization*. An *estimate* (i.e., an educated guess) of a distribution based on a sample and drawing a conclusion about a parameter based on a sample is called a *statistical inference*.

5.2. ESTIMATION

We now discuss each one of the first two main areas. Different methods of estimation are available such as *point estimation*, *interval estimation* and *nonparametric estimation*. We will discuss the first two.

5.2.1. Point Estimation

Definition 5.2.1.

A *point estimate* of a parameter μ is the most plausible value of μ. The statistic that estimates this parameter is called the *point estimator* of μ and is denoted by $\hat{\mu}$ (read it as "mu" hat).

In other words, if the *pdf* is known but the parameter, μ, is unknown, first take a random sample, another random sample, and continue n times independently. Then, try to guess the value of the parameter μ. This means that we want to find a number $\hat{\mu}$ as a function of observations $(x_1, x_2, \cdots, x_n)$, i.e., $\hat{\mu} = Y(x_1, x_2, \cdots, x_n)$. The function Y is a statistic estimating μ, i.e., a *point estimator* for μ. We want the computed value $\hat{\mu} = Y(x_1, x_2, \cdots, x_n)$ to be closed to the actual value of the parameter μ. Note that Y is a random variable and, thus, has a *pdf* by its own.

Example 5.2.1.

What the Definition 5.2.1 is to say is that we make n observations of a random variable whose values are recorded as $(x_1, x_2, \cdots, x_n)$. For instance, we may toss an unfair coin 10 times. Hence, a 10-observation may be $(1, 0, 1, 0, 1, 0, 0, 0, 1, 0)$, where 0 and 1 stand for tails and heads, respectively. If p is the probability of obtaining a head, then the problem is to estimate p. Thus, the term *inference* in this case is to estimate the unknown probability of success, p, of the Bernoulli distribution. This example is generic. It could, for instance, be an example for an automatic switch to run between "off" and "on".

There are questions about a point estimator that needs to be answered. For instance, how good is an estimator? In other words, what should the properties for an estimator be for it to be chosen? How do we know that there is not a better choice of an estimator? We will address both of these questions below.

To answer the first question, it is known that a good estimator is the one that, on the average, is equal to the parameter (*unbiased*) and its *variance is as small as possible* (*minimum variance*). Hence, we have the following definitions:

Example 5.2.2.

Suppose a manufacturer has developed an item to improve the quality of the current item which has shown some defect. The new item has been tested n (say, 50) times. Let X be the number of tests that do not show the old defect. The manufacturer is interested in finding the proportion of all such tests that result in no defect. That is, we have to find the probability of no defect in a single item produced.

Solution

Let p denotes the proportion in question. If we observe that x (the number of favorite results of the tests, say, 30) tests show no defect, then an estimator for p, denoted by $\hat{p}$, is $\hat{p} = \dfrac{X}{n}$. Thus, the estimate of p is $\dfrac{x}{n} = 30/50 = .60$.

Definition 5.2.2.

A point estimator $\hat{\mu}$ of a parameter μ is *unbiased* if $E(\hat{\mu}) = \mu$, otherwise it is said to be *biased*. The *bias* of $\hat{\mu}$, denoted by $B(\hat{\mu})$, is defined by $B(\hat{\mu}) = E(\hat{\mu}) - \mu$.

Example 5.2.3.

Let $E(X) = \mu$ be the expected value of X. We want to investigate biasedness or unbiasedness of the sample mean, $\overline{X}_n$, and sample variance, S^2. It is easy to show that

$$E\left(\frac{1}{n}\sum_{i=1}^{n} X_i\right) = \frac{1}{n} E\left(\sum_{i=1}^{n} X_i\right). \tag{5.2.1}$$

Hence, from (5.1.1) and (5.2.1) we have:

$$E\left(\overline{X}_n\right) = \frac{1}{n}(n\mu) = \mu. \tag{5.2.2}$$

Equation (5.2.2) shows that the sample mean is unbiased. In other words, the sample mean is centered about the actual mean.

Now let us denote the sample variance S^2 by $Var\left(\overline{X}_n\right)$ and the population variance by σ_X^2. Then,

$$Var\left(\overline{X}_n\right) = Var\left(\frac{1}{n}\sum_{i=1}^{n} X_i\right) = \frac{1}{n^2} Var\left(\sum_{i=1}^{n} X_i\right) = \frac{\sigma_X^2}{n}. \tag{5.2.3}$$

Equation (5.2.3) shows that the sample variance is not unbiased.

Example 5.2.4.

Let $X_1, ..., X_n$ be a random sample of size n from a normal distribution with mean μ and variance σ^2. Let $\overline{X}$ and S^2 be the sample mean and sample variance, respectively. We will show that these statistics are both unbiased estimators.

It should be noted that the use of point estimation is limited. This is because point estimations are never exact. Hence, it is necessary to find how far off the true value of estimate is likely to be. One way to measure the distance is *MSE* defined below that should accompany an estimate.

Definition 5.2.3.

Let μ be a parameter and $\hat{\mu}$ an estimator of it. Then, the *mean square error* of $\hat{\mu}$, denoted by *MSE*, (and if there is a confusion, by $MSE(\hat{\mu})$) is defined by

$$MSE(\hat{\mu}) \equiv MSE = \left[E(\hat{\mu}) - \mu \right]^2 + Var(\hat{\mu}). \tag{5.2.4}$$

What (5.2.4) says is that *MSE* is the square of the bias plus the variance of the estimator, i.e.,

$$MSE(\hat{\mu}) = \left[Bias(\hat{\mu}, \mu) \right]^2 + Var(\hat{\mu}).$$

If the estimator, $\hat{\mu}$, is unbiased, then *MSE* is the variance of the estimator, $\hat{\mu}$, i.e.,

$$MSE = Var(\hat{\mu}) = E\left[(\hat{\mu} - \mu)^2 \right]. \tag{5.2.5}$$

In that sense, the *MSE* assesses the quality of the estimator in terms of its variation and unbiasness.

Example 5.2.4.

Let X be a binomial random variable with a sample of size n and unknown probability p. Let $\hat{p} = \dfrac{X}{n}$ be an estimator of p. We will see that $\hat{p}$ is unbiased and will find its *MSE*.

Solution
We know that $E(X) = np$ and $Var(X) = np(1-p)$. On the other hand,

$$E(\hat{p}) = E\left(\frac{X}{n} \right) = \frac{E(X)}{n} = \frac{np}{n} = p.$$

Thus, $E(\hat{p}) - p = p - p = 0$, i.e., $\hat{p}$ is unbiased.
Now the variance of $\hat{p}$ is

$$Var(\overline{p}) = Var\left(\frac{X}{n} \right) = \frac{Var(X)}{n^2} = \frac{np(1-p)}{n^2} = \frac{p(1-p)}{n}.$$

Thus, $MSE_{\hat{p}} = 0 + \dfrac{p(1-p)}{n} = \dfrac{p(1-p)}{n}$. This was anticipated because of (5.2.5).

Definition 5.2.4.

For a random sample $(x_1, x_2, \cdots, x_n)$ and a parameter μ, the statistic Y that minimizes $E\left[(Y - \mu)^2\right]$ is called the *minimum mean square error statistic*. If Y is unbiased and minimizes $E\left[(Y - \mu)^2\right]$, then it is called the *unbiased minimum variance estimator of* μ.

5.2.1.a. *Method of Moments*

There are different methods of point estimation, one common one is the *method of moments*. This is perhaps the oldest method of point estimation. It dates back to Karl Pearson in the late nineteenth century; although it always yields some type of estimate. However, in many cases, it may not be the best, i.e., it may be improved. The method is as follows:

Let $X_1, X_2, \cdots, X_n$ be a sample of size n from a population with *pdf* or *pmf* $f\left(x | \mu_1, \mu_2, \cdots, \mu_k\right)$, where $\mu_1, \mu_2, \cdots, \mu_k$ are parameters of the distribution. Then, estimators are obtained by equating the first k sample moments to the corresponding first k sample population and solving the system of k equations and k unknowns. Thus, we will have

$$
\begin{cases}
M_1 = E(X) = \bar{X} = \dfrac{1}{n}\sum_{i=1}^{n} X_i^1 \\[2mm]
M_2 = E(X^2) = \dfrac{1}{n}\sum_{i=1}^{n} X_i^2 \\[2mm]
\vdots \\[2mm]
M_k = E(X^k) = \dfrac{1}{n}\sum_{i=1}^{n} X_i^k.
\end{cases}
\tag{5.2.6}
$$

Estimators $\hat{\mu}_1, \hat{\mu}_2, \cdots, \hat{\mu}_k$ of $\mu_1, \mu_2, \cdots, \mu_k$ is the solution of (5.2.6). Let each of $M_1, M_2, \cdots, M_k$ be a function of $\mu_1, \mu_2, \cdots, \mu_k$, say $\varphi(\mu_1, \mu_2, \cdots, \mu_k)$, as they usually are. Then (5.2.6) can be written as

$$\begin{cases} M_1 = \varphi_1\left(\mu_1, \mu_2, \cdots, \mu_k\right) \\ M_2 = \varphi_2\left(\mu_1, \mu_2, \cdots, \mu_k\right) \\ \vdots \\ M_k = \varphi_k\left(\mu_1, \mu_2, \cdots, \mu_k\right). \end{cases} \tag{5.2.7}$$

Example 5.2.5.

Let $X_1, \ldots, X_n$ be a random sample of size n from a distribution with parameter μ and *pdf* as $f(x, \mu) = \mu x^{\mu-1}$, for $0 < x < 1$, $0 < \mu < \infty$. We want to estimate μ, using the method of moments.

Solution
For the expected value of the distribution, we have

$$E(x) = \int_0^1 x \mu x^{\mu-1}\, dx = \mu \int_0^1 x^{\mu} dx = \mu \frac{1}{\mu+1} x^{\mu+1} \Big|_0^1 = \frac{\mu}{\mu+1}.$$

Sine $\overline{X}$ is an unbiased estimator of μ, we write $\overline{X} = \dfrac{\mu}{\mu+1}$ and solve for μ. Hence,

$\mu = \mu\overline{X} + \overline{X} \Rightarrow \mu(1 - \overline{X}) = \overline{X}$ or $\hat{\mu} = \dfrac{\overline{X}}{1 - \overline{X}}$. Therefore, $\dfrac{\overline{X}}{1 - \overline{X}}$ is an estimator for μ.

5.2.1.b. Maximum Likelihood Estimator (MLE)

The maximum likelihood (together with some of its variants) is the most widely used method of estimation. Along with a list of its applications, *MLE* would cover practically the whole field of statistics. There are different methods to find out how unbiased an estimator is. The most popular method among statistician, especially when the sample size is large, is the *maximum likelihood* method. This method allows one to choose a value for the unknown parameter that most likely is the closest to the observed data.

Definition 5.2.5.

Let $X_1, X_2, \cdots, X_n$ be a random vector with a joint *pmf*

$$f\left(x_1,\ x_2,\ \cdots,\ x_n;\ \theta_1,\ \theta_2,\ \cdots,\ \theta_n\right), \tag{5.2.8}$$

where $\theta_1, \theta_2, \cdots, \theta_n$ are unknown parameters. When $x_1, x_2, \cdots, x_n$ are observed values of $X_1, X_2, \cdots, X_n$ and the function f given in (5.2.8) is taken as a function of the parameters, f is called the *likelihood function*. The values of $\theta_1, \theta_2, \cdots, \theta_n$ that maximize the likelihood function (5.2.8), denoted by $\hat{\theta}_1, \hat{\theta}_2, \cdots, \hat{\theta}_n$, are called the *maximum likelihood estimates* of $\theta_1, \theta_2, \cdots, \theta_n$ and briefly referred to as *MLE*'s. In other words, $\hat{\theta}_1, \hat{\theta}_2, \cdots, \hat{\theta}_n$ values that most likely produce the sample values $x_1, x_2, \cdots, x_n$. If $X_1, X_2, \cdots, X_n$ are taken instead of $x_1, x_2, \cdots, x_n$, then $\hat{\theta}_1, \hat{\theta}_2, \cdots, \hat{\theta}_n$ are referred to as the *maximum likelihood estimators* of $\theta_1, \theta_2, \cdots, \theta_n$.

Example 5.2.6. Procedure for Finding MLE by Example

We now give a procedure to find *MLE* by considering a special case. Let p be the probability of success in a sequence of Bernoulli trials. The probability p could be thought of as a proportion of a large population with a certain characteristic. We want to find the *MLE* for p. We will show that the answer is the relative frequency of successes.

Solution

Suppose that X is a Bernoulli random variable with parameter p, i.e., with *pmf*

$$b(x; p) = p^x (1-p)^{1-x}, \quad x = 0, 1; \quad p \in \Omega = \{p : 0 \le p \le 1\}, \tag{5.2.9}$$

where Ω is the *parameter space*, i.e., the set of all possible values of the parameter. We now take a random sample of size n, say $X_1, X_2, \cdots, X_n$. Let $x_1, x_2, \cdots, x_n$ denote the set of observed values of $X_1, X_2, \cdots, X_n$. Then, from (5.2.8) and (5.2.9), the probability that the sample $X_1, X_2, \cdots, X_n$ take the particular values $x_1, x_2, \cdots, x_n$ is the joint pdf of $X_1, X_2, \cdots, X_n$ evaluated at the observed values, i.e.,

$$P(X_1 = x_1, \ X_2 = x_2, \ \cdots, \ X_n = x_n) = \prod_{i=1}^{n} p^{x_i} (1-p)^{1-x_i} = p^{\sum_{i=1}^{n} x_i} (1-p)^{n - \sum_{i=1}^{n} x_i}, \tag{5.2.10}$$

which is the likelihood function. We regard (5.2.10) as a function of p and will find the value of p that maximizes this likelihood function. Since the trials are independent, denoting the likelihood function of p by $L(p)$, from (5.2.10) we will have

$$L(p) \equiv f\left(x_1, \ x_2, \ \cdots, \ x_n; p\right) = P\left(X_1 = x_1, \ X_2 = x_2, \ \cdots, \ X_n = x_n\right)$$
$$= f\left(x_1; p\right) f\left(x_2; p\right) \cdots f\left(x_n; p\right) \tag{5.2.11}$$
$$= p^{\sum\limits_{i=1}^{n} x_i} (1 - p)^{n - \sum\limits_{i=1}^{n} x_i}, \qquad 0 \le p \le 1.$$

To maximize $L(p)$, we take derivative of $L(p)$ with respect to p, and set it equal to zero.

$$\frac{dL(p)}{dp} = \left(\sum_{i=1}^{n} x_i\right) p^{\sum\limits_{i=1}^{n} x_i - 1} (1 - p)^{n - \sum\limits_{i=1}^{n} x_i} - \left(n - \sum_{i=1}^{n} x_i\right) p^{\sum\limits_{i=1}^{n} x_i} (1 - p)^{n - \sum\limits_{i=1}^{n} x_i - 1} = 0, \ 0 < p < 1.$$

Or

$$p^{\sum\limits_{i=1}^{n} x_i - 1} (1 - p)^{n - \sum\limits_{i=1}^{n} x_i} \left[\frac{\sum\limits_{i=1}^{n} x_i}{p} - \frac{n - \sum\limits_{i=1}^{n} x_i}{1 - p}\right] = 0, \qquad 0 < p < 1.$$

Or

$$\frac{\sum\limits_{i=1}^{n} x_i}{p} - \frac{n - \sum\limits_{i=1}^{n} x_i}{1 - p} = 0, \qquad 0 < p < 1. \tag{5.2.12}$$

Or

$$\sum_{i=1}^{n} x_i - np = 0,$$

from which we find

$$p = \frac{\sum\limits_{i=1}^{n} x_i}{n} = \overline{x}. \tag{5.2.13}$$

Relation (5.2.13) is the statistic that is the maximum likelihood estimator of p. We denote this by $\widehat{p}$. Thus,

$$\hat{p} = \frac{\sum_{i=1}^{n} x_i}{n} = \bar{x}, \tag{5.2.14}$$

where $\bar{x}$ is the sample mean.

We note that rather than directly taking derivative of the likelihood function, often it is easier to take derivative the natural logarithm of the likelihood function. Thus, we could have done as follows: from (5.2.11) we have

$$\ln L(p) = \left(\sum_{i=1}^{n} x_i \right) \ln p + \left(n - \sum_{i=1}^{n} x_i \right) \ln(1-p), \; 0 < p < 1,$$

and

$$\frac{d \left(\ln L(p) \right)}{dp} = \left(\sum_{i=1}^{n} x_i \right) \frac{1}{p} + \left(n - \sum_{i=1}^{n} x_i \right) \frac{-1}{1-p} = 0, \; 0 < p < 1. \tag{5.2.15}$$

But, (5.2.15) is the same as (5.2.12). Thus, the rest of steps would be the same, and hence, (5.2.14) will give $\bar{x}$ as the *MLE* for p.

Now we consider a specific case. A manufacturer takes a sample of 20 from a product to see if the product is defective. It is observed that the second (2nd), seventh (7th), fifteenth (15th), eighteenth (18th) and twentieth (20th) are defective. Let 0 and 1 represent, respectively, defectives and non-defective of an item checked. Then, the sample observed looks as follows: 1, 0, 1, 1, 1, 1, 0, 1, 1, 1, 1, 1, 1, 1, 0, 1, 1, 0, 1, 0. Let X_i, $i = 1, 2, \cdots$, represent the sample points, i.e., $X_i = 0$ or 1. Let also p be the probability that the item chosen is not defective and $1 - p$ if it is defective, i.e.,

$$P(X_i = 1) = p \text{ and } P(X_i = 0) = 1 - p, \; i = 1, 2, \cdots.$$

Then, the joint *pmf* of the sample chosen will, in general, looks like the following:

$$f(x_1, x_2, \cdots, x_{20}; p) =$$

$$\underbrace{p}_{1} \underbrace{(1-p)}_{2} \underbrace{pppp}_{3-6} \underbrace{(1-p)}_{7} \underbrace{ppppppp}_{8-14} \underbrace{(1-p)}_{15} \underbrace{pp}_{16-17} \underbrace{(1-p)}_{18} \underbrace{p}_{19} \underbrace{(1-p)}_{20}$$

$$= p^{15}(1-p)^5. \tag{5.2.16}$$

The relation (5.2.16) could actually be obtained from (5.2.10). The question, then, is for what value of p a sample may most likely look like the one observed. To answer this question, we must maximize the function given in (5.2.16) or equivalently, its logarithm. Hence,

$$\frac{d}{dp} \ln f\left(x_1, \ x_2, \ \cdots, \ x_{20}, \ p\right) = \frac{15}{p} - \frac{5}{1-p} = 0 \tag{5.2.17}$$

[as in (5.2.15)] implies that $p = 15/20$. Since the distribution is binomial, $15/20 = k/n$, where k is the observed number of successes (non-defectives). Therefore, $\hat{p} = 3/4$ is the maximum likelihood estimate for p. That is, $p = 3/4$ makes the chance of observing the particular data in hand ($x = 20$) as large as possible.

Note that the location of the defective items is not important. If we were told that, the number of defectives was 5. Then, the probability of having exactly 5 defective items among 20 would have been $\binom{20}{5} p^{15}(1-p)^5$. This expression will also be maximum when $p = 3/4$.

Example 5.2.7. Estimating Poisson Parameter by MLE.

Let $X_1, X_2, \cdots, X_n$ represent a random sample with its n independent observed values $x_1, x_2, \cdots, x_n$ from a Poisson random variable with parameter λ. We want to find the *MLE* for λ.

Solution
From the Poisson assumption, the probability of observing x_i events in the ith trial is:

$$p_X\left(x_i | \lambda\right) \equiv P\left(X_i = x_i | \lambda\right) = \frac{\lambda^{x_i} e^{-\lambda}}{x_i!}, \qquad i = 1, 2, \cdots, n; \quad x_i = 0, 1, 2, \cdots. \tag{5.2.18}$$

Then, from (5.2.10), the log likelihood function will be

$$\ln L(\lambda) = \sum_{i=1}^{n} \ln p_X(x_i | \lambda) = \sum_{i=1}^{n}\left(x_i \ln \lambda - \lambda - \ln(x_i!)\right)$$

$$= \ln \lambda \sum_{i=1}^{n} x_i - n\lambda - \sum_{i=1}^{n} \ln(x_i!). \tag{5.2.19}$$

Taking derivative with respect to λ of (5.2.19) and set it equal to zero, we obtain

$$\frac{d \ln L(\lambda)}{d\lambda} = \frac{1}{\lambda}\sum_{i=1}^{n} x_i - n = 0,$$

from which

$$\hat{\lambda} = \frac{1}{n}\sum_{i=1}^{n}x_i = \bar{x}. \tag{5.2.20}$$

Thus, *MLE* for λ is $\bar{x}$, the sample mean.

5.2.2.   Interval Estimation

We mentioned that use of point estimation is limited and when used, it should accompanied by an *MSE*. In practice, to avoid this problem, another method of estimation is used, called the *interval estimator* or *confidence interval*. A confidence interval is an interval with lower and upper limits constructed in such a way that the true valued of the estimate falls within this interval. To show how likely it is that the true value falls within the interval, a number called the *level of confidence* accompanies the confidence interval. We now formally define these terms.

Definition 5.2.6.

Let $X = (X_1,...,X_n)$ be a random sample from a distribution. Let $f(x_1,...,x_n)$ and $g(x_1,...,x_n)$ be a pair of functions of the sample such that $f(x) \leq g(x)$, for all $x \in X$. Suppose the inference $f(x) \leq \mu \leq g(x)$ is made on an observed value x. Then, the interval $[f(x), g(x)]$, $(f(x), g(x))$, $[f(x), g(x))$, or $(f(x), g(x)]$ is called an *interval estimator*. The proportion of all possible samples for which the confidence interval is covered is called the *confidence level*.

What the definition defines is that an interval estimator of a population parameter is an interval which is predicted to contain the parameter. The probability of such containment is the confidence. The functions f and g in Definition 5.2.6 are usually finite. However, infinite values of them are also some times of interest. For instance, if $f(x) = -\infty$, then we will have the half-open interval $(-\infty, g(x)]$. In such a case, we will consider $\mu \leq g(x)$, without mentioning of the lower bound. A similar case may be of interest for the upper bound. That is, $g(x) = \infty$, implying $[f(x), \infty)$ and $\mu \geq f(x)$.

Example 5.2.8.

Let us consider $X_1, X_2, \cdots, X_n$ as a sample of size n from a normal population with *known variance* σ^2. We want to use this sample to estimate the unknown mean, μ, and find a 95% confidence for it.

Solution

As we have seen, the sample mean (statistic) $\bar{X}$ is an unbiased point estimator of the population mean (parameter) μ. We saw in the previous chapter (Theorem 4.13.2) that sampling from $\Phi\left(\mu,\sigma^2\right)$ is normally distributed. We also saw that distribution of $\bar{X}$ has mean μ and variance σ^2/n. This implies that the standardized random variable Z from (4.11.25) for $\bar{X}$

$$Z = \frac{\bar{X}-\mu}{\sigma/\sqrt{n}}.$$
(5.2.21)

has standard normal distribution. Thus, if we desire the endpoints of the interval for which the value of the estimator falls within, with, say 95%, probability, then we need to look up z from the standard normal tables such that ($z_{\alpha/2} = z_{.025} = 1.96$)

$$P\left(-1.96 \le z \le 1.96\right) = 0.95.$$
(5.2.22)

Hence, substituting (5.2.21) in (5.2.22) we will have

$$P\left(-1.96 \le \frac{\bar{x}-\mu}{\sigma/\sqrt{n}} \le 1.96\right) = 0.95.$$
(5.2.23)

Now, we simplify the event $-1.96 < \dfrac{\bar{X}-\mu}{\sigma/\sqrt{n}} < 1.96$ from (5.2.23) as follows

$$-1.96 < \frac{\bar{X}-\mu}{\sigma/\sqrt{n}} < 1.96 \Rightarrow -1.96\left(\sigma/\sqrt{n}\right) < \bar{X}-\mu < 1.96\left(\sigma/\sqrt{n}\right)$$
$$\Rightarrow -\bar{X}-(1.96)\left(\sigma/\sqrt{n}\right) < -\mu < \bar{X}+(1.96)\left(\sigma/\sqrt{n}\right)$$
$$\Rightarrow \bar{X}-1.96\left(\sigma/\sqrt{n}\right) < \mu < \bar{X}+1.96\left(\sigma/\sqrt{n}\right).$$
(5.2.24)

Thus, from (5.2.23) and (5.2.24) we will have

$$P\left(\bar{X}-1.96\frac{\sigma}{\sqrt{n}} < \mu < \bar{X}+1.96\frac{\sigma}{\sqrt{n}}\right) = 0.95.$$
(5.2.25)

The event in (5.2.25) is the interval we are looking for. In other words, out of every 100 times we sample, 95 times μ will falls within the interval $\bar{X} \pm 1.96\dfrac{\sigma}{\sqrt{n}}$. The interval

$$\left[\bar{X}-1.96\frac{\sigma}{\sqrt{n}}, \bar{X}+1.96\frac{\sigma}{\sqrt{n}}\right]$$ is called a *95 percent confidence interval estimator of the*

population mean μ. For an observed value $\bar{x}$ of $\bar{X}$, the interval $\bar{x} \pm 1.96 \dfrac{\sigma}{\sqrt{n}}$ is called a *95*

percent confidence interval estimate of the mean μ. For instance, let X represent the lifetime of a light bulb. Suppose X is normally distributed with mean μ and standard deviation of 36. The manufacturer chose 27 light bulbs and lighted them to observe the lifetime until each burned out. The mean of this sample was 1478 hours. To estimate the mean of all such light bulbs, a 95% confidence interval is decided. Hence, we will have

$$\left[\bar{x} - z_{.025} \frac{\sigma}{\sqrt{n}}, \bar{x} + z_{.025} \frac{\sigma}{\sqrt{n}} \right] = \left[1478 - 1.96 \frac{36}{\sqrt{27}}, 1478 - 1.96 \frac{36}{\sqrt{27}} \right]$$

$$= \left[1464.42, 1491.58 \right].$$

Assigning normal distribution to a random variable may not be seen easily. In such case, the Central limit Theorem can be used to approximate the confidence interval. This is because when the sample size is large enough, the ratio $\dfrac{\bar{X} - \mu}{\sigma / \sqrt{n}}$ [(5.2.21)] is $N(0,1)$. Hence, a $100(1 - \alpha)\%$ confidence interval for μ, is

$$P\left(-z_{\alpha/2} \leq \frac{\bar{x} - \mu}{\sigma / \sqrt{n}} \leq z_{\alpha/2} \right) = 1 - \alpha, \tag{5.2.26}$$

that implies

$$\left[\bar{x} - z_{\alpha/2} \frac{\sigma}{\sqrt{n}}, \bar{x} + z_{\alpha/2} \frac{\sigma}{\sqrt{n}} \right]. \tag{5.2.27}$$

Other percent confidence intervals can be found similarly.

In case the *variance is unknown*, the sample variance, S^2, can be used. In this case, however, we can again use the quantity $(\bar{X} - \mu)/(\sigma / \sqrt{n})$ defined in (5.2.21) with the population standard deviation σ replace by the sample standard deviation, i.e.,

$$T \equiv \frac{\bar{X} - \mu}{S / \sqrt{n}}. \tag{5.2.28}$$

However, T defined in (5.2.28) is not normally distributed; it rather has *Student's t distribution* (or just simply, *t distribution*). Before introduction of the *t* distribution, the quantity $(\bar{X} - \mu)/(S / \sqrt{n})$, which is a random variable (since it depends upon S and $\bar{X}$), was treated as a standard normal until 1908. This is okay as long as the sample size is large, since in that case S is very close to σ with high probability.

Historically, t distribution was first found by W. S. Gosset in his paper Gosset (1908). He worked in brewing industry and lived from 1876 to 1937 in Dublin. He studied statistics under Karl Pearson at University College of the University of London. His statistical research papers were published under the pseudonym "Student".

Discussion of the confidence interval for the population mean so far has been based on large sample sizes (say, $n \geq 30$), regardless of the distribution. This is because of the Central Limit Theorem in case that the distribution is not being normal. We know that in that case $\bar{X} \sim N\left(\mu, \sigma^2 / n\right)$ that implies that $\left(\bar{X} - \mu\right) / \left(\sigma / \sqrt{n}\right) \sim N(0,1)$. In this case the sample variance is very close to the population variance. This is why the quantity $\left(\bar{X} - \mu\right) / \left(S / \sqrt{n}\right)$ is approximately normal with mean 0 and standard deviation 1.

When the sample size is small (say, $n < 30$), S may not be close to σ and so $\bar{X}$ may not be normally distributed. (We should caution, however, that if the sample is selected from a normal population, it will be normally distributed even when its size is small.) For a small sample size, no good general method for finding confidence interval is available. But, if the population is almost normal and, hence the sample standard deviation may not be close to the population standard deviation, the quantity $\left(\bar{X} - \mu\right) / \left(S / \sqrt{n}\right)$ may still be used. However, this quantity is not normally distributed any longer. The *Student's t distribution* will be used in this case.

Definition 5.2.7.

A random variable T is said to have a *Student's t distribution with d degrees of freedom* if its *pdf* is

$$f_T(t) = \frac{\Gamma\left(\dfrac{d+1}{2}\right)}{\Gamma\left(\dfrac{d}{2}\right)} \frac{1}{\sqrt{d\pi}} \frac{1}{\left(1 + \dfrac{t^2}{d}\right)^{(d+1)/2}}, \quad -\infty < t < \infty. \tag{5.2.29}$$

In this case we write $T \sim t_d$. In particular, let $X_1, X_2, \cdots, X_n$ be a random sample of size n from a normal distribution with mean μ and standard deviation σ. Then, the quantity $\left(\bar{X} - \mu\right) / \left(S / \sqrt{n}\right)$ has a *Student's t distribution with n − 1 degrees of freedom*.

It can be shown that $E(T) = 0$ for $d \geq 2$. Note also that if $d = 1$, then the t distribution is called *Cauchy distribution* and its mean and variance do not exist. Further note that, the variance of T exists only for $d \geq 3$, which is $d/(d-2)$.

Thus, for a $X_1, X_2, \cdots, X_n$ of size n, $n < 30$, from a normal population, a $100(1 - \alpha)\%$ confidence interval for another item, say Y, to be drawn from this population, using t-distribution, is given by

$$\left[\bar{x} - t_{n-1,\alpha/2} s \sqrt{1 + \frac{1}{n}}, \bar{x} + t_{n-1,\alpha/2} s \sqrt{1 + \frac{1}{n}} \right].$$
(5.2.30)

Hence, a 95% confidence for Y means

$$P\left[\bar{x} - t_{n-1,\alpha/2} s \sqrt{1 + \frac{1}{n}}, \; Y < \bar{x} + t_{n-1,\alpha/2} s \sqrt{1 + \frac{1}{n}} \right] = 0.95.$$
(5.2.31)

Example 5.2.29.

A sample of size 10 had a mean 1000 and variance 576. We want to find a 95% confidence interval for the population mean.

Solution

The sample size being 10 implies that we have 9 degrees of freedom. For a 95% confidence interval, $\alpha = 0.025$. From a *t-Table* in the Appendix, we have $t_{9,.025} = 2.262$. Hence, referring to (5.2.30), we will have

$$\left[1000 - t_{9,.025}(24) \sqrt{1 + \frac{1}{10}}, 1000 + t_{9,.025}(24) \sqrt{1 + \frac{1}{10}} \right]$$

$$= \left[1000 - (2.262)(24) \sqrt{1 + \frac{1}{10}}, 1000 + (2.262)(24) \sqrt{1 + \frac{1}{10}} \right]$$

$$= 1000 \pm 56.9377 = [943.0623, 1056.9377].$$

In case a 90% interval would have been desired, $\alpha = 0.05$, and we would have $t_{9,.05} = 1.833$. Hence, the interval would be

$$\left[1000 - (1.833)(24) \sqrt{1 + \frac{1}{10}}, 1000 + (1.833)(24) \sqrt{1 + \frac{1}{10}} \right]$$

$$= 1000 \pm 46.1392 = [953.8608, 1046.1392].$$

We now want to use an example to find a *confidence interval for the mean of a population* from which a large sample has been chosen. See Navidi (2010, p.189).

Example 5.2.30.

A study on life distribution of microdrills drilling a low-carbon alloy steel, shows that a drill should have a minimum lifetime (expressed as the number of holes drilled before failure) of 10 holes drilled before failure. Out of a sample of 144 microdrills 120 meet this requirement. Let p represent the proportion of microdrills in the population that satisfy such a requirement. We want to find a 95% confidence interval for p.

To do this, let X be a random variable representing the number of microdrills in the population that satisfy the requirement Thus, X is a binomial random variable with $n = 144$ with unknown parameter p. We have seen that $\hat{p} = X / n$ is an unbiased estimate for p. Thus, $\hat{p} = 120/144 = 0.833$. The sample size being 144 (large), using the Central Limit Theorem, implies that X is normal with mean np and variance $np(1 - p)$. Using $\hat{p} = X / n$, we see that $\hat{p}$ is normal with mean p and variance $[p(1 - p)/n]$. Hence, the standard error (5.1.3) of $\hat{p}$ is $\sigma_{\hat{p}} = \sqrt{p(1-p)/n}$. However, value of $\sigma_{\hat{p}}$ is not known since p is unknown. Hence, it cannot be used to form the confidence interval. To overcome this problem, it is tradition to use $\hat{p}$ for p. Thus, $\sigma_{\hat{p}} \approx \sqrt{\hat{p}(1-\hat{p})/n}$.

Now, $\hat{p}$ being approximately normal, critical value for a 95% confidence interval from the z-Table is 1.96. Therefore, a 95% confidence interval for p is $\hat{p} \pm 1.96\sqrt{\hat{p}(1-\hat{p})/n}$.

In general, the width of a confidence interval for a proportion, denoted by $\pm w$, is

$$\pm w = \pm z_{\alpha/2}\sqrt{\hat{p}(1-\hat{p})/n} \ . \tag{5.2.32}$$

Now solving $w = z_{\alpha/2}\sqrt{\hat{p}(1-\hat{p})/n}$ for n, we obtain

$$n = \frac{z_{\alpha/2}^2 \hat{p}(1-\hat{p})}{w^2} \ . \tag{5.2.33}$$

Equation (5.2.33) can be used to estimate the sample size to construct a level $100(1 - \alpha)\%$.

If no value of $\hat{p}$ is known, $\hat{p}$ may be taken as 0.5. This value maximizes $\hat{p}(1 - \hat{p})$ and n found in this case will not produce a width greater than w. In this case, we will have

$$n = \frac{z_{\alpha/2}^2}{4w^2} \ . \tag{5.2.34}$$

5.3. HYPOTHESIS TESTING

In the previous section, we saw how from a random sample we estimate the population parameters and we developed a range of values, confidence interval, within which we expected the parameters be located with some confidence level. In this section, we want to test statements about the parameters such as if the mean mileage driven on a brand of car's tire is more than 50,000 miles, or if the average lifetime of a light bulb is more that 2000 hours. Although using confidence interval, we are able to tell if the value of a statistic falls within two values with some probability, we cannot tell if the value can be less than or greater than a value within the interval. Such questions need to be hypothesized. We now define what we mean by a hypothesis.

Definition 5.3.1.

A *statistical hypothesis* is a conjecture (a statement, an idea, an assumption, a guess, or a theory) about the nature pf a population. It is often stated in terms of a parameter of the population. By *hypothesis testing* it is meant to determine the truth or falsity of a conjecture. The default position is called the *null hypothesis* and is denoted by H_0. The negation of H_0 is called the *alternative hypothesis* and is denoted by H_1.

What Definition 5.3.1 is saying is that the *null hypothesis* states that the effect of different values of the parameter is the same, while the *alternative hypothesis* states that different values of the parameter have different effects.

Definition 5.3.2.

A statistic (a characteristic of the sample such as sample mean) whose value is determined from the sample observations is called a *test statistic* used to decide to reject or accept the null hypothesis.

Definition 5.3.3.

The set of values of the *test statistics* for which H_0 is rejected is called the *critical region*. The point dividing the region where the null hypothesis is rejected and the region where it is not rejected is called the *critical value*.

Example 5.3.1.

In testing for the mean of a distribution, say μ, the test statistic is taken as the random variable Z, whose value is calculated by:

$$Z = \frac{\bar{X} - \mu}{\sigma / \sqrt{n}}, \tag{5.3.1}$$

[similar to (5.2.21) for small sample sizes], where $\bar{X}$ is the sample mean (with large sample size n, $n > 30$) that is normally distributed with mean and standard deviation as μ and $\sigma / \sqrt{n}$, respectively. Hence, the test is to find if the difference between the sample mean $\bar{X}$ and the population mean μ is statistically significant. This is done by finding the number of standard deviations the sample mean is from the population mean using (5.3.1).

Definition 5.3.4.

Testing a given null hypothesis causes two possible types of errors to occur: *Type I error* and *Type II error*. If H_0 is rejected while it is actually true, it is said that the *Type I error* has occurred. On the other hand, if H_0 is not rejected while it is actually not true, it is said that the *Type II error* has occurred.

When the null hypothesis, H_0, is rejected, it sends a strong message that there is a high chance that H_0 is not consistent with the observed value data. On the other hand, if H_0 is not rejected means that there is a small chance that it is consistent with the observed data. However, it is important to note that the purpose of testing H_0 is not to determine the truth or fallacy of it. It is rather to determine if truth of assumption of H_0 is consistent with the data collected. That is why it makes sense that H_0 should be rejected only if the sample data are not likely if H_0 is true. Hence, to design a test, in order to establish a hypothesis, the hypothesis should be stated as the alternative, H_1. On the other hand, to discard a hypothesis, it should be stated as H_0.

Definition 5.3.5.

In order to reject H_0, it is customary to require that whenever H_0 is true, the probability of being rejected is less than or equal to a small number α. The number α is called the *level of significance of the test* (or *significance level of the test*). α is pre-assigned and its value is often chosen as 0.01, 0.05, and 0.10. Choosing α in advance is the same as to choose the critical region in advance. α is the probability of a Type I error.

Definition 5.3.6.

Let β be the probability of a Type II error. Then, the probability of rejecting the tested hypothesis when it is false (i.e., an alternative hypothesis is true), denoted by $1 - \beta$, is called the *power of the test*. The ideal power function is $\beta = 0$. However, that is only possible in trivial cases. Hence, the practical hope is that power function near 1.

We summarize the types of errors in decision making in the Table 5.3.1.

Table 5.3.1. Error Types in Decisions

	Accept H_0	Reject H_0
H_0	Correct Decision	Type I error
H_1	Type I error	Correct Decision

(True Hypothesis)

Example 5.3.2.

Suppose X is a random variable with binomial distribution with $n = 5$ and p. Let the hypotheses be:

H_0: $p \le 1/2$, H_1: $p > 1/2$.

(a) The test rule is "reject H_0 if and only if all successes are observed". Hence, the power function for this test is: $\beta(p) = P(X = 5) = p^5$. The probability of a Type I error is $\beta(p) \le (1/2)^5 = 0.0312$, for all $p \le 1/2$, (very small), while since $\beta(p)$ is too small, for Type II will be too large for most $p > 1/2$. The probability of a Type II error is less than $1/2$ only if $p > (1/2)^{1/5} = 0.87$.

(b) To reduce the probability of Type II error in this case, we consider to "reject H_0 if $X = 3$, 4 or 5". The power function in this case is $\beta(p) \le P(X = 3, 4, \text{or } 5) = \binom{5}{3}p^3(1-p)^2 + \binom{5}{4}p^4(1-p)^1 + \binom{5}{5}p^5(1-p)^0$.

Definition 5.3.7.

Assume the null hypothesis, H_0, is true (in a trial, assumption of innocent of the defendant). Choose a random sample to test the hypothesis. The measure of the strength of disagreement between the observed values of the sample and H_0 that produces a number between 0 and 1 is called a *p-value* (or *observed* (or *achieved*) *significance level*). (Note that p here stands for probability. Hence, the p-value stands for *probability–value*.) Alternatively, p-value is the smallest significance level at which the data lead to rejection of H_0. The smaller the p-value (say, 0.05 or less), the stronger the evidence that H_0 is not true. Hence, when p-value is sufficiently small, we will state that *there is strong evidence to reject H_0*. On the other hand, if the p-value is large, it would mean that there is a strong evidence not to reject

H_0. In practice, the significance level is not given. It is rather desired to find the p-value, using the observations.

To *perform a hypothesis testing*, the following steps needs to be taken:

Step 1. State H_0 and H_1. (In case that the hypotheses include an inequality with equal sign "$\leq$" or "$\geq$", the equal sign "$=$" must be in the statement of the null and not in the alternative hypothesis.)

Step 2. State a level of significance, α, $0 < \alpha < 1$.

Step 3. Identify the test statistic, i.e., a value from the sample information to determine whether to accept or reject the null hypothesis, such as z, t, F, and χ^2 (chi-square).

Step 4. Formulate the decision rule, i.e., a specific condition under which the null hypothesis is accepted or rejected.

Step 5. Make a statistical decision (reject or accept H_0).

Step 6. Compute the p-value of the test statistic.

Step 7. Interpret the decision in term of the stated problem.

Note that the significance level α (the chance of incorrectly rejecting a true H_0) is needed for the approach of the rejection region. Accessing technology has caused statisticians to stay away from this approach and use the p-value approach. However, p and α are related as follows: Reject H_0 if the p-value is less than α; H_0 fails to be rejected if $p \geq \alpha$.

To *calculate the p-value for a large-sample hypothesis testing*, the following steps needs to be taken:

Step 1.   Calculate the value of the test statistic, z, from the set of observations.

Step 2.   Consider two cases: one-tailed and two-tailed. Hence, choose either Step 2a or Step 2b below:

Step 2a.   Case of one-tailed: the p-value is equal to the tail area greater than z in the same direction as the alternative hypothesis.
In this case, if H_1 is of the form "$>$", then the p-value is the area to the right of or above the observed value of z value.
In this case, if H_1 is of the form "$<$", then the p-value is the area to the left of or below the observed value of z value.

Step 2b.   Case of two-tailed: the p-value is equal to twice the tail area greater than z in the same direction as the sign of z.
In other words, there are two cases: $z > 0$ and $z < 0$. Thus, choose either Step 2b1 or Step 2b2 below:

Step 2b1. Case of $z > 0$. Then, the p-value is twice the area to the right of or above the observed z value.

Step 2b2.   Case of z < 0. Then, the p-value is twice the area to the left of or below the observed z value.

Example 5.3.3.

Suppose a manufacturer of a certain light bulb notices that about 5 percent of the products fail. The quality control department (QCD) recommends a procedure to improve the production. To test the new procedure, QCD suggests a sample of size 200 light bulbs be tested with the new procedure. Hence, for the test to show any improvement for the new procedure, the proportion of failure should be significantly less than 5 percent. Thus, we need a rule as when to accept the new procedure as an improvement.

Let p, n, and X denote the proportion of product failure, the sample size, and the number of failures in the sample, respectively. Then, $p = 0.05$, $n = 200$, and X is unknown. We want to test the ratio $X/200$ to be significantly less than 0.05, say, less than or equal to 0.02. Hence, we want to have the number of failure to be less than or equal to 5 in the 200 items tested. However, the percent of failure with this sampling may still be 5 while we are observing 5 or fewer failures in testing the 200 light bulbs. In other words, we are accepting the new procedure recommended by QCD as improvement while, in act, it is not. This incorrect decision is the Type I error in this example. But, it is possible that we observe 6 defective light bulbs in the 200 light bulbs tested ($x/200 = 0.3$, yet the new procedure is really an improvement with a failure of 1.5 percent. That is, we are not to accept the new procedure as an improvement, while, in fact, it is.

Let us consider the probabilities of the errors. Assume that trails in the sampling are independent and the chance of failure is the same throughout the sampling. Thus, X will be a binomial random variable with $n = 200$ and unknown p. We choose our unbiased estimator as $\hat{p} = X/200$. We want to test our "improvement" hypothesis. In other words, since we expect to have a change for improvement, we take H_0 as $p = 0.05$ (*simple null hypothesis*, completely specifies the distribution), and H_1 as $p < 0.05$ (*composite alternative hypothesis*, does not completely specifies the distribution). Our expectation, i.e., the rule of testing, is to reject H_0 and accept H_1 if $X \le 5$.

As we mentioned, if $X \le 5$ while $p = 0.05$, then we have made the Type I error. The probability of this type of error, as in Definition 5.2.27, α. Hence,

$$\alpha \cong P(X \le 5; p = 0.05) = \sum_{x=0}^{5} \binom{200}{x} (0.05)^x (1-0.05)^{200-x} = 0.067. \qquad (5.3.4)$$

Note that to find the value of this expression, not only we could find with available Binomial Tables or a new technology, since $p = 0.05$ is small relative to the sample size $n = 200$, we can use Poisson approximation with $\lambda = np = (200)(0.05) = 10$.

Now let us compute the probability of Type II error if we assume improvement of failure to 1.5 percent. But, this error occurs if $X > 5$ while, in fact, $p = 0.015$. Hence,

$$\beta \cong P(X > 5; p = 0.015) = \sum_{x=6}^{200} \binom{200}{x} (0.015)^x (1 - 0.015)^{200-x}. \qquad (5.3.5)$$

Repeating the same argument as above, computing (5.3.5) using Poisson $\lambda = np = (200)(0.015) = 3$, we will have

$$\beta \cong 1 - P(X \le 5) = 1 - \sum_{x=0}^{5} \frac{3^x e^{-3}}{x!} = 0.084. \qquad (5.3.6)$$

What (5.3.6) shows is probably a satisfaction for the QCD since with decreasing the chance of failure from 0.05 to 0.015, there is only about 8 percent chance of accepting failure probability of 0.05 and rejecting the improvement plan.

Example 5.3.4.

Suppose that packing department of a food company observes some weight irregularities for a particular item. Standard pack weight is 450 grams. The company is aware that the standard deviation for this type of packing is $\sigma = 10$ grams. The irregularity observation is to be tested with $\alpha = 0.05$.

Note that the irregularity could be either overweight or underweight. This type of test is called a *two-tail test*. Let the standard pack weight be denoted by μ. Then, we state our hypotheses as follows:

H_0: $\mu = 450$, H_1: $\mu \ne 450$.

We now want to find the region of rejection. Since the variance is known, the appropriate statistic is (5.3.1). The Company collects 30 packs from different locations and records the weights as follows:

430	448	441	436	451	457	439	458	438	435
443	450	443	430	438	456	448	459	449	431
438	437	456	453	459	438	459	455	453	451

Using MINITAB we obtain the sample mean and sample standard deviation, respectively, as $\bar{X} = 445.97$ and $S = 9.51$. Thus,

$$Z = \frac{\bar{X} - \mu}{\sigma / \sqrt{n}} = \frac{445.97 - 450}{9.51 / \sqrt{30}} = -2.32.$$

Hence, from the standard normal table, we have to find the area to the right of test statistic Z, i.e., $P(Z > 2.32) = 0.0102$. This is because $\Phi(-z) = 1 - \Phi(z)$.

When the sample size is large and the test statistic is negative, p-value is taken as twice the absolute value of the test statistic. Thus, p-value is 0.0204 which is less than $\alpha = 0.05$. Therefore, there is a strong evidence to the packing weights are different from the standard weight 450 grams. The rejection region is $Z > 1.96$ or $Z < -1.96$. It is clear that $z = -2.32 < -1.96$.

Example 5.3.5.

Consider a binary system of communication that sends messages in the form of digits 0 and 1. Signals emitted last no longer a fixed unit of time. They are often distorted and sometimes corrupted by random noise. Human error is also a common source of problems. That is, sending 0 instead of 1 and vice versa. Every fixed amount of time, the receiver has to decide which possible signal was transmitted. Because of such problems and errors, the decisions may be wrong.

Let p_0 and p_1 be probabilities that the signals emitted were 0's and 1's, respectively. If a 1 appears, a pulse of light is transmitted which lasts no longer that a fixed unites of time. If a 0 is received, no light pulse is transmitted. It is for this reason that this type of communication system is called an "on-off" system. The receiver has to make a decision on the basis of the following hypothesis:

H_0: The signals sent were 0's and 0's were transmitted.
H_1: The signals sent were 0's and some 1's were transmitted.

Note that p_0 and p_1 are actually *a priori* probabilities for the causes 0 and 1, respectively. Thus, probabilities $P\left(n|H_0\right)$ and $P\left(n|H_1\right)$ are probabilities of having $n, n = 0, 1, 2, \cdots$, signals sent, given that the emitted signals were 0's or 1's, respectively. Hence, if we look at the experiment of deciding which signal was sent, the outcomes of such an experiment could be denoted by $\left(H_i, n\right)$, $i = 0, 1$.

Now suppose we divide the nonnegative real axis into two disjoint parts R_0 and R_1 such that the union of R_0 and R_1 is the nonnegative real axis. Then if n turns out to be in R_0, it means that the receiver decides that no pulse was transmitted, i.e., 0 message was sent. If, on the other hand, n turns out to be in R_1, it means that the receiver decides a pulse was transmitted, i.e., 1 message was sent. Hence, a *correct decision* means that a pulse was sent and n is in R_1, or no pulse was sent and n is in R_0. The complementary event to this is called an *error*. Thus, a *wrong decision* means that no pulse was sent and n is in, or a pulse was sent and n is in R_1. Using the law of total probability, we can write the probabilities of these events as:

$$P(correctd\ decision) = \sum_{n=0}^{\upsilon} P(H_0, n) + \sum_{n=\upsilon+1}^{\infty} P(H_1, n),$$

$$P(wrong\ decision) = \sum_{n=0}^{\upsilon} P(H_1, n) + \sum_{n=\upsilon+1}^{\infty} P(H_0, n),$$

where υ is the divider number for the two parts of the axis. Note that the choice of is based on the maximizing and minimizing the probabilities of correct and wrong decisions, respectively. Thus, according to Bayes' theorem the conditional probabilities can be found as follows:

$$P(H_0|n) = \frac{P(H_0, n)}{P(n)}, \text{ and } P(H_1|n) = \frac{P(H_1, n)}{P(n)},$$

where $P(n) = P(H_0|n) + P(H_1|n)$.

Note again that $P(H_0|n)$ and $P(H_1|n)$ are *a posteriori* probabilities for H_0 and H_1, respectively, and $P(H_0|n) + P(H_1|n) = 1$.

Using p_0 and p_1, we can rewrite these conditional probabilities as:

$$P(H_0|n) = \frac{p_0 P(n|H_0)}{P(n)}, \text{ and } P(H_1|n) = \frac{p_1 P(n|H_1)}{P(n)}.$$

We saw a discussion of Bernoulli parameter. A standard procedure for testing the Bernoulli probability of success p for a specified value is as follows:

Let p_0 be a pre-assigned probability of success in a Bernoulli trial. To test this value, we write the simple null hypothesis as

H_0: $p = p_0$.

Use the following theorem that we state without proof:

Theorem 5.3.1.

Let X be the number of successes in n independent Bernoulli trials, then the random variable X/n has an approximate normal distribution, $N[p_0, (1-p_0)/n]$, where $N(\mu, \sigma^2)$ denotes the normal distribution with mean μ and variance σ^2, provided H_0: $p = p_0$ is true and n is large.

Choose
H_1: $p > p_0$.

Now consider the test H_0: $p = p_0$ against H_1: $p > p_0$ that rejects H_0 and accepts H_1 if and only if

$$Z = \frac{\dfrac{X}{n} - p_0}{\sqrt{\dfrac{p_0(1 - p_0)}{n}}} \geq z_\alpha.$$

In other words, reject H_0 and accept H_1 if X/n exceeds p_0 by z_α standard deviations of X/n and vice versa. Now, since under H_0, Z is approximately normal with mean 0 and variance 1, then the probability for the test to occur is the significance level α.

Note that had we taken H_1: $p < p_0$ instead of H_1: $p > p_0$, then $Z \leq -z_\alpha$ would be the appropriate test level.

5.4. INFERENCE ON TWO SMALL-SAMPLE MEANS

In the previous section we tested hypothesis with large and small samples. Now we want to see what happens if we have two populations. Recall that when sample is small, the Central Limit Theorem cannot be applied. So, we want to find the confidence interval for the difference between means of two different normal populations. We want to consider three cases.

Case 1. Population Variances not Necessarily Equal, but Unknown (Confidence Interval)

Let $X_1, X_2, \cdots, X_m$ and $Y_1, Y_2, \cdots, Y_n$ be samples from two independent normal populations with sizes n_X and n_Y and means μ_X and μ_Y, respectively. Then, it can been shown that the quantity

$$\frac{(\bar{X} - \bar{Y}) - (\mu_X - \mu_Y)}{\sqrt{\dfrac{S_X^2}{n_X} + \dfrac{S_Y^2}{n_Y}}} \sim Student's\ t \qquad (5.4.1)$$

with

$$\upsilon = \left[\frac{\left(\dfrac{S_X^2}{n_X} + \dfrac{S_Y^2}{n_Y} \right)^2}{\dfrac{\left(\dfrac{S_X^2}{n_X} \right)^2}{n_X - 1} + \dfrac{\left(\dfrac{S_Y^2}{n_Y} \right)^2}{n_Y - 1}} \right], \tag{5.4.2}$$

degrees of freedom, where [.] here means rounded down to the nearest integer. Then, from (5.4.1), a $100(1 - \alpha)\%$ confidence interval for the difference of means $(\mu_X - \mu_Y)$ is

$$\left[\bar{X} - \bar{Y} - t_{\upsilon,\alpha/2} \sqrt{\frac{S_X^2}{n_X} + \frac{S_Y^2}{n_Y}}, \; \bar{X} - \bar{Y} + t_{\upsilon,\alpha/2} \sqrt{\frac{S_X^2}{n_X} + \frac{S_Y^2}{n_Y}} \right]. \tag{5.4.3}$$

The confidence interval in (5.4.3) means that

$$P\left(\bar{X} - \bar{Y} - t_{\upsilon,\alpha/2} \sqrt{\frac{S_X^2}{n_X} + \frac{S_Y^2}{n_Y}} \le \mu_X - \mu_Y \le \bar{X} - \bar{Y} + t_{\upsilon,\alpha/2} \sqrt{\frac{S_X^2}{n_X} + \frac{S_Y^2}{n_Y}} \right) = 1 - \alpha. \tag{5.4.4}$$

We note that the sample variance is an unbiased estimate of the population variance. When it is from a sample of size, say, n, it has $n - 1$ degrees of freedom.

Example 5.4.1.

Two website designs, called standard and revised, are observed for the time browsers spend on them. Times spent by browsers are normally distributed with $\left(\mu_X, \sigma_X^2 \right)$, for the standard, and $\left(\mu_Y, \sigma_Y^2 \right)$, for the revised, as mean and variance, for each, where μ_X, σ_X^2, μ_Y, and σ_Y^2 are, respectively, sample mean (average time a browser spends observing standard website) and sample variance of the standard, and sample mean and sample variance of the modern sites. A sample of 10 from each design was taken and was observed that the amount of times spent on each by browsers had means (in minutes) and standard deviations as below

	Standard			Revised	
n_X	$\bar{X}$	S_X^2	n_Y	$\bar{Y}$	S_Y^2
10	46.10	91.81	10	35.30	63.27

We want to test equality of the means of these websites, i.e., if $\mu_X = \mu_Y$.

Solution

We define the null and alternative hypotheses as follows:

$$H_0: \ \mu_X - \mu_Y \le 0, \ H_1: \ \mu_X - \mu_Y > 0$$

From (5.4.1), our test statistic is

$$t = \frac{(\bar{X} - \bar{Y}) - 0}{\sqrt{\dfrac{S_X^2}{n_X} + \dfrac{S_Y^2}{n_Y}}} . \tag{5.4.5}$$

From the data given above and (5.4.5), we find our test statistic as

$$t = \frac{46.1 - 35.3}{\sqrt{\dfrac{91.81}{10} + \dfrac{63.27}{10}}} = \frac{10.8}{3.94} = 2.74$$

From (5.4.2), the number of degrees of freedom is

$$\upsilon = \left[\frac{\left(\dfrac{91.81}{10} + \dfrac{63.27}{10} \right)^2}{\dfrac{\left(\dfrac{91.81}{10} \right)^2}{9} + \dfrac{\left(\dfrac{63.27}{10} \right)^2}{9}} \right] = \left[\frac{240.498}{13.814} \right] = 17.41 \cong 17 \cdot$$

Now, using a *t*-Table for the 17 degrees of freedom, the values cutting off 1% and 0.5% in the right-hand tail are 2.567 and 2.898, respectively. Thus, the area in the right-hand tail corresponding to values greater than or equal to the observed value 2.74 is between 0.005 and .01. Therefore, the *p*-value will be between 0.005 and .01. That is, there is strong evidence that the difference of means is positive. In other words, with high probability the mean observed time on standard website is greater than the one of the modern one.

From (5.4.3), for the 1% cutting off in the right-hand tail, i.e., 98% confidence is $[0.686, 20.914]$, because

$$\left[\bar{X} - \bar{Y} - t_{\upsilon, \alpha/2} \sqrt{\frac{S_X^2}{n_X} + \frac{S_Y^2}{n_Y}}, \ \bar{X} - \bar{Y} + t_{\upsilon, \alpha/2} \sqrt{\frac{S_X^2}{n_X} + \frac{S_Y^2}{n_Y}} \right] = 46.10 - 35.30 \mp (2.567)(3.94).$$

Now, we use MINITAB to see if what we computed is confirmed.
95% Confidence interval:

Two-Sample T-Test and CI

Sample	N	Mean	StDev	SE Mean
1	10	46.10	9.58	3.0
2	10	35.30	7.95	2.5

Difference = mu (1) - mu (2)
Estimate for difference: 10.8000
95% lower bound for difference: 3.9516
T-Test of difference = 0 (vs >): T-Value = 2.74 P-Value = 0.007 DF = 17

99% Confidence interval:

Two-Sample T-Test and CI

Sample	N	Mean	StDev	SE Mean
1	10	46.10	9.58	3.0
2	10	35.30	7.95	2.5

Difference = mu (1) - mu (2)
Estimate for difference: 10.8000
99% lower bound for difference: 0.6947
T-Test of difference = 0 (vs >): T-Value = 2.74 P-Value = 0.007 DF = 17

We see that in both cases the p-value is 0.007, which is between 0.005 and 0.01. We also see that the t-value and the degrees of freedoms are the same. Thus, the values calculated by formulae are consistent with what the software shows.

Case 2. Population Variances Equal, but Unknown (Confidence Interval)

In case that variances of both populations are known to be equal ($\sigma_X^2 = \sigma_Y^2$), then a $100(1 - \alpha)\%$ confidence interval for the difference of means, $\mu_X - \mu_Y$, is

$$\left[\bar{X} - \bar{Y} - t_0 S_p \sqrt{\frac{1}{n_X} + \frac{1}{n_Y}}, \ \bar{X} - \bar{Y} + t_0 S_p \sqrt{\frac{1}{n_X} + \frac{1}{n_Y}} \right], \tag{5.4.6}$$

where

$$S_p^2 = \frac{(n_X - 1)S_X^2 + (n_Y - 1)S_Y^2}{n_X + n_Y - 2} \tag{5.4.7}$$

is defined as the *pooled estimator of the common population varience*, $n_X + n_Y - 2$ is the degrees of freedom and $t_0 = (n_X + n_Y - 2)t_{\alpha/2}$.

We note that equality of variances is rarely assumed and, hence, this method that was widely used before is rarely used these days.

Example 5.4.2.

Suppose test scores on a standardized test in two schools with large (X) and small (Y) ethnic populations are normally distributed with (μ_X, σ_X^2) and (μ_Y, σ_Y^2) as mean and (known to be equal, $\sigma_X^2 = \sigma_Y^2 = \sigma^2$) variance for each. Suppose that the scores have been organized and the following summary is given:

School with Large Ethnicity			School with Small Ethnicity		
n_X	$\overline{X}$	S_X^2	n_Y	$\overline{Y}$	S_Y^2
5	79.52	57.21	8	75.32	49.13

Now from (5.4.6), we have

$$S_p = \sqrt{\frac{(n_X - 1)S_X^2 + (n_Y - 1)S_Y^2}{n_X + n_Y - 2}} = \sqrt{\frac{4(57.21) + 7(49.13)}{5 + 8 - 2}} = 7.216.$$

From the *t*-Table, for a 98% confidence with 11 degrees of freedom, we have $t_{(.01)(11)} = 2.718$. Hence, from (5.4.6), we obtain

$$\overline{X} - \overline{Y} \pm t_0 S_p \sqrt{\frac{1}{n_X} + \frac{1}{n_Y}} = 79.52 - 75.32 \mp t_{(.01)(11)} S_p \sqrt{\frac{1}{5} + \frac{1}{8}} = 4.2 \pm (2.718)(7.216)(0.57).$$

Thus, the 98% confidence interval for the difference of means is $[-6.9795, 15.3795]$.

Using MINITAB, we obtain

Two-Sample T-Test and CI

Sample	N	Mean	StDev	SE Mean
1	5	79.52	7.56	3.4
2	8	75.32	7.01	2.5

Difference = mu (1) - mu (2)
Estimate for difference: 4.20000
98% CI for difference: (-6.98113, 15.38113)
T-Test of difference = 0 (vs not =): T-Value = 1.02 P-Value = 0.329 DF = 11
Both use Pooled StDev = 7.2158

Case 3. Population Variances Equal, but Unknown (Testing Null Hypothesis)

To test a null hypothesis on the difference of means in case of small-sample when it is not know if variances are equal, again we choose our samples as before with (mean and variance) for each sample as $\left(\mu_X, \sigma_X^2\right)$ and $\left(\mu_Y, \sigma_Y^2\right)$, respectively. Assume $\sigma_X^2 = \sigma_Y^2$. We choose the degrees of freedom as (5.4.2) and we compute the test statistic t from (5.4.1) as

$$t = \frac{(\bar{X} - \bar{Y}) - \Delta_0}{\sqrt{\dfrac{S_X^2}{n_X} + \dfrac{S_Y^2}{n_Y}}}, \tag{5.4.8}$$

where Δ_0 is the value of $\mu_X - \mu_Y$ specified by H_0. Then, we set the null hypothesis for the difference of means $(\mu_X - \mu_Y)$ as

$$H_0 : \mu_X - \mu_Y \leq \Delta_0, \ H_0 : \mu_X - \mu_Y = \Delta_0, \ \text{or} \ H_0 : \mu_X - \mu_Y \geq \Delta_0. \tag{5.4.9}$$

Now, we compute the p-value as an area under the Student's t curve with υ degrees of freedom. Degrees of freedom, υ, depend on H_1 as follows:

H_1:	p-value
$\mu_X - \mu_Y \neq \Delta_0$	Sum of the areas in the tails cut off by $-t$ and t
$\mu_X - \mu_Y < \Delta_0$	Area to the left of t
$\mu_X - \mu_Y > \Delta_0$	Area to the right of t

Once again, though rarely used, when population variances are known to be equal, we use (5.4.7) as the pooled standard deviation. Then, we compute

$$t = \frac{\left(\bar{X} - \bar{Y}\right) - \Delta_0}{S_p \sqrt{\dfrac{1}{n_X} + \dfrac{1}{n_Y}}}$$ (5.4.10)

and the p-value with the same degrees of freedom as before.

5.5. ANALYSIS OF VARIANCE (ANOVA)

Comparison of means of several populations is called *analysis of variance (ANOVA)*. By that we mean, estimating means of more than one population simultaneously. ANOVA is one of the most widely used statistical methods used. ANOVA is not about variance, it is rather about analyzing variations in means. In statistical design, ANOVA is used to determine how one can get the most information on the most populations with the fewer observations. Of course, this desire is extremely important in the industry due to minimizing the cost.

Definition 5.5.1.

The varying quantities in an experiment are called *factors* (variables) and the experiments containing factors are called *factorial experiments*. Values of the factors are called *levels* of factors or *treatments*. If the experimental units are assigned to treatments at random (will all possible assignments to be equally likely), then the experiment is called a *completely randomized experiment*.

Definition 5.5.2.

In a completely randomized experiment, each treatment represents a population and responses observed for the units assigned to that treatment represents a simple random sample from that population. The population means are called *treatment means*. If there are more than two samples (possibly with different sizes) to test for equality of their means, the method used is called *one-way analysis of variance*.

Usually, there are three assumptions to be satisfied in order to use ANOVA. These are:

(1) populations be normally distributed,

(2) populations have equal variances, and

(3) the samples are selected independently.

The subject of ANOVA involves one more distribution. It is called *Snedecor's F distribution* and is defined below. The *F* distribution, which is a ratio of two variances and its *pdf* is similar to the *t* distribution is named in honor of Sir Ronald Fisher. Whenever, three assumptions are met, *F* is used as the statistic. The information for test of equality of several means is usually summarized in *ANOVA table*. The table heading consists of "source", "sum of squares", 'degrees of freedom", "mean square", and "*F*-ratio", for example.

Definition 5.5.1.

A random variable *F* is said to have a *Snedecor's F distribution with d_1 and d_2 degrees of freedom* if its *pdf* is

$$f_F(x) = \frac{\Gamma\left(\dfrac{d_1+d_2}{2}\right)}{\Gamma\left(\dfrac{d_1}{2}\right) \times \Gamma\left(\dfrac{d_2}{2}\right)} \left(\frac{d_1}{d_2}\right)^{d_1/2} \frac{x^{(d_1/2)-1}}{\left[1+\left(\dfrac{d_1}{d_2}\right)x\right]^{(d_1+d_2)/2}}, \quad 0 < x < \infty. \quad (5.5.1)$$

In this case we write $X \sim F_{d_1,d_2}$. In particular, let $X_1, X_2, \cdots, X_n$ be a random sample of size *n* from a normal distribution with mean μ_X and standard deviation σ_X. Let also $Y_1, Y_2, \ldots, Y_m$ be a random sample of size *m* from an independent normal distribution with mean μ_Y and standard deviation σ_Y. Let $F = \left(S_X^2 / \sigma_X^2\right) / \left(S_Y^2 / \sigma_Y^2\right)$ be a random variable. Then, *F* has *Snedecor's F distribution with $n - 1$ and $m - 1$ degrees of freedom*.

We note that a variance ratio may have an *F* distribution even if the parent populations are not normal, Keller (1970).

We list a selected properties of the *F* distribution in the following theorem.

Theorem 5.5.1.

a. If $X \sim F_{d_1,d_2}$, then its reciprocal is also *F*, but with reverse degrees of freedom, i.e.,

$$\frac{1}{X} \sim F_{d_2,d_1}.$$

b. If $X \sim t_d$, then $X^2 \sim F_{1,d_2}$.

c.   If $X \sim F_{d_1, d_2}$, then $\dfrac{\dfrac{d_1}{d_2} X}{\left(1 + \dfrac{d_1}{d_2} X\right)} \sim beta\left(\dfrac{d_1}{2}, \dfrac{d_2}{2}\right)$, where *pdf* of *beta distribution*

with parameters α and β is

$$f(x|\alpha, \beta) = \frac{1}{B(\alpha, \beta)} x^{\alpha-1}(1-x)^{\beta-1}, \quad 0 < x < 1, \ \alpha > 0, \ \beta > 0, \tag{5.5.2}$$

where

$$B(\alpha, \beta) = \int_0^1 x^{\alpha-1}(1-x)^{\beta-1} dx. \tag{5.5.3}$$

Proof

See Casella and Berger (1990, p. 228).

For a $100(1 - \alpha)$% confidence interval for the ratio σ_1^2 / σ_2^2, we have

$$P\left[F_{\alpha/2} < \left(S_1^2 / \sigma_1^2\right) / \left(S_2^2 / \sigma_2^2\right) < F_{1-(\alpha/2)}\right] = 1 - \alpha. \tag{5.5.4}$$

Since $\left(S_1^2 / \sigma_1^2\right) / \left(S_2^2 / \sigma_2^2\right) = \left(S_1^2 / S_2^2\right)\left(\sigma_2^2 / \sigma_1^2\right)$, from (5.4.4) we have

$$\frac{F_{\alpha/2}}{S_1^2 / S_2^2} < \frac{\sigma_2^2}{\sigma_1^2} < \frac{F_{1-(\alpha/2)}}{S_1^2 / S_2^2} \implies \frac{S_1^2 / S_2^2}{F_{\alpha/2}} > \frac{\sigma_1^2}{\sigma_2^2} > \frac{S_1^2 / S_2^2}{F_{1-(\alpha/2)}}. \tag{5.5.5}$$

Therefore, from (5.5.5), by reversing the order, and from (5.5.4) we will have

$$P\left(\frac{S_1^2 / S_2^2}{F_{1-(\alpha/2)}} < \frac{\sigma_1^2}{\sigma_2^2} < \frac{S_1^2 / S_2^2}{F_{\alpha/2}}\right) = 1 - \alpha. \tag{5.5.6}$$

Thus, a $100(1 - \alpha)$% confidence interval for the ratio σ_1^2 / σ_2^2 is

$$\left[\frac{S_1^2 / S_2^2}{F_{1-(\alpha/2)}}, \frac{S_1^2 / S_2^2}{F_{\alpha/2}}\right]. \tag{5.5.7}$$

It should be noted that from part (a) of Theorem 5.5.1, if reverse of the random variable is to be used, not only that the degrees of freedom interchanges, the confidence levels will interchange as well. In other words,

$$F_{\alpha/2,(d_1,d_2)} = \frac{1}{F_{[1-(\alpha/2)],(d_2,d_1)}}.$$
(5.5.8)

Example 5.5.1.

Suppose we have sampled two brands of an electronic equipment, one with 5 items and another with 10. The following is the data observed:

Brand 1:   129 135 345 143 274
Brand 2:   186 124 246 125 146 58102 267 121 240

We want to compute a 95% confidence interval for the ratio of population variances.

Solution
Using MINITAB, we have the following information:

Descriptive Statistics: Brand 1, Brand 2

```
Variable    N    N*   Mean   SE Mean   StDev   Variance   Minimum     Q1   Median
Brand 1     5    0    205.2   44.1     98.6    9720.2     129.0    132.0   143.0
Brand 2     10   0    161.5   22.1     69.8    4878.3     58.0     116.3   135.5

Variable      Q3    Maximum   Kurtosis
Brand 1     309.5    345.0     -1.68
Brand 2     241.5    267.0     -1.16
```

Hence, we have $S_1^2 = 9720.2$, $S_2^2 = 4878.3$, $S_1^2/S_2^2 = 9720.2/4878.3 = 1.9925$, with $n = 5$, $m = 10$, $d_1 = 4$, and $d_2 = 9$. Thus, from the F-Table and (5.4.8), we have $F_{.025,(4,9)} = \dfrac{1}{F_{.975,(9,4)}} = \dfrac{1}{8.90} = 0.1124$. Now, from (5.5.7), for a 95% confidence interval for the ratio, we have

$$\left[\frac{S_1^2/S_2^2}{F_{1-(\alpha/2)}}, \frac{S_1^2/S_2^2}{F_{\alpha/2}}\right] = [1.9925/4.72, 1.9925/0.1124] = [0.4221, 17.7269].$$

Example 5.5.2.

To test strength of a material, supposed four samples are taken, one during the process and three after the process every six weeks. The results of observations are as follows (using appropriate units):

Sample 1:	410	620	730	640		
Sample 2:	900	780	1150	780		
Sample 3:	850	1100	1170	880	1050	
Sample 4:	890	885	910	690	670	590

We now use MINITAB for a one-way ANOVA with a 95% confidence on the null hypothesis:

H_0: all treatment means are equal

One-way ANOVA: Sample 1, Sample 2, Sample 3, Sample 4
ANOVA Table

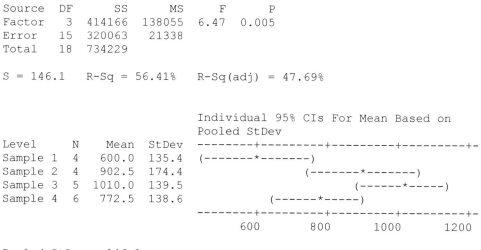

```
Source   DF      SS      MS      F      P
Factor   3    414166  138055  6.47   0.005
Error    15   320063   21338
Total    18   734229

S = 146.1   R-Sq = 56.41%   R-Sq(adj) = 47.69%


                             Individual 95% CIs For Mean Based on
                             Pooled StDev
Level       N     Mean  StDev  --------+---------+---------+---------+-
Sample 1    4    600.0  135.4  (-------*-------)
Sample 2    4    902.5  174.4                   (-------*-------)
Sample 3    5   1010.0  139.5                        (------*-----)
Sample 4    6    772.5  138.6           (------*-----)
                             --------+---------+---------+---------+-
                                 600       800      1000      1200

Pooled StDev = 146.1
```

As it can be seen, for the test of the null hypothesis "all treatment means are equal" the *p*-value is 0.005. Thus, we reject the hypothesis and conclude that not all treatment means are equal.

5.6. LINEAR REGRESSION

Often, there is an interest in knowing the relationship between two variables. Finding a relationship and degree of dependence are what regression analysis does. Regression is the most popular statistical methodology that utilizes the relationship among two or more variables to predict one variable in terms of the others. This tool (in different forms such as linear, non-linear, parametric, non-parametric, simple and multiple) is widely used in different disciplines such as engineering, biological sciences, social and behavioral sciences, and business. For instance, we might be interested in analyzing the relation between two variables such as strength of and stress on a beam in a building; or a student's grade point average and his/her grade in a statistics course. Analysis of regression not only is to show the relationship, it is also for prediction.

The relation we are referring to here is of two kinds: functional and statistical. Functional relation is based on the standard definition of a function in mathematics. However, statistical relation is not as rigor as the functional. That is because statistical relation is a relation between observed values of the variables. These observations do not necessarily fall on the curve of a relation.

Example 5.6.1.

Suppose a retail store sells an item. The company is interested in relation between the amount of money spent on advertising and the number of items sold. During the next six months, the Sales Department collected data on both of these items from the amount of money spent on advertising in different similar cities in different parts of the country and the number of items sold in those cities. Similarity is defined as almost same population, same average as income, and same education. Then, the statistical analyst of the company selected a random sample of size 10 from the data collected (10 cities as City 1, City 2, $\cdots$, City 10). This data is reported in Table 5.6.1. We want to portray this information in a scatter plot and observe the relation between the amount of money spent on advertising and the number of items sold.

Solution

To respond to the inquiries, we will refer to the amount of money spent on advertising as *independent* variable, denoted by X, and the number of items sold as *dependent* variable denoted by Y.

In general the variable that is being estimated or predicted is called *dependent variable* or *response variable*; and the variable that provides the basis for estimation is called the *independent* or *predictor* or *input variable*. As a common practice, when we are presenting data graphically, the independent or predictor variable is scaled on the X-axis (horizontal) and the dependent variable on the Y-axis (vertical).

Table 5.6.1. Advertising Spent and Amount Sold.

City #	Amount of Money Spent on Advertising (in $1000)	Number of Sales (in 100)
1	10	15
2	20	30
3	30	60
4	40	80
5	5	15
6	5	20
7	10	20
8	10	25
9	10	15
10	15	50

Now, we want to represent the data by a scatter plot. To do that, we use MINITAB. We start to re-represent the data using X and Y, as independent and dependent variables, respectively and re-write the Table 5.6.1 as Table 5.6.2. Then, graph, as Figure 5.6.1.

We can see from the scatter plot (a statistical relation) that as the amount of money spent on advertising, the number of items sold increase or at lease does not decrease. This, relation can be mathematically modeled as a line. Hence, we can find a line to fit the points. The linear equation we will obtain, using MINITAB, will have a functional relation.

As it can be seen from Figure 5.6.2, a relation (functional relation) between money spent on advertising and number sold is $Y = 445 + 0.184 X$. Note that the correlation coefficient can be obtained by MINITAB as 0.929, showing a very strong dependency between the variables.

Table 5.6.2. Money Spent on Advertising and Number Sold

City #	X	Y
1	10,000	1,500
2	20,000	3,000
3	30,000	6,000
4	40,000	8,000
5	5,000	1,500
6	5,000	2,000
7	10,000	2,000
8	10,000	2,500
9	10,000	1,500
10	15,000	5,000

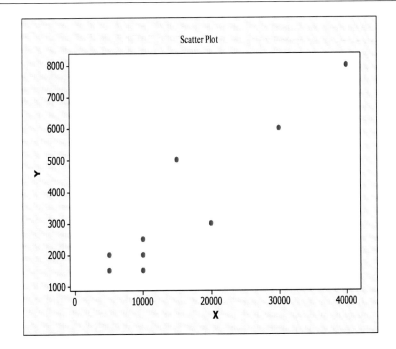

Figure 5.6.1. Scattered Plot.

The simplest relation between two variables is a straight line, say $Y = \beta_0 + \beta_1 x$. In a relation as such, if the constants (or parameters) β_0 and β_1 are given, then, for each value of x, the value of the response, Y, will be predicted. However, in practice, as in Example 5.6.1, such precision is almost impossible and, thus, subject to *random error*. Thus, in general we have the following definition:

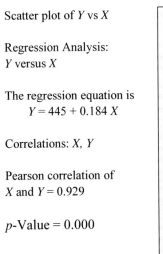

Scatter plot of Y vs X

Regression Analysis: Y versus X

The regression equation is
    $Y = 445 + 0.184\,X$

Correlations: X, Y

Pearson correlation of X and $Y = 0.929$

p-Value $= 0.000$

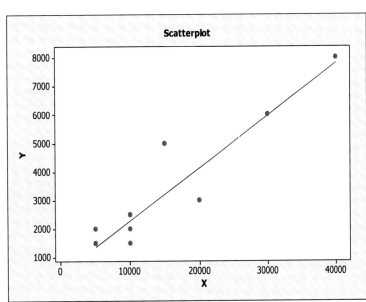

Figure 5.6.2. Functional Relation between Money Spent on Advertising and Number Sold.

Definition 5.6.1.

Let a relation between the predictor (or input) variable x and response variable Y be of the linear form

$$Y = \beta_0 + \beta_1 x + \varepsilon, \qquad (5.6.1)$$

where β_0 and β_1 are *parameters*, that are called *regression coefficients*, and ε is, what is called, the *random error* with mean 0. Then, the relation in (5.6.1) is called a *simple linear regression*.

Alternatively, since x is a variable (and not random variable), a *simple linear regression* is a relation between the predictor and response variables such that for any value of the predictor variable x, the response variable is a random variable, Y, with mean $E(Y)$ as

$$E(Y) = \beta_0 + \beta_1 x. \qquad (5.6.2)$$

The reason for (5.6.2) is that from (5.6.1), $E(\varepsilon) = 0$, and x being just a variable (in this case $\beta_0 + \beta_1 x$ plays the role of a constant), we have

$$E(Y) = E(\beta_0 + \beta_1 x + \varepsilon) = \beta_0 + \beta_1 x + E(\varepsilon) = \beta_0 + \beta_1 x.$$

Note that the response variable is denoted by an upper case Y, while the predictor is denoted by a lower case x. Reason for this is that the predictor, x, is chosen to be a constant with varying values. However, the response, Y, is taken as a random variable. Hence, the lower case y is used for an observed value of Y. Note also that from (5.6.1) and (5.6.2), the response variable Y comes from a *cdf* whose mean is (5.6.2).

In repeated trials, for $i = 1, 2, \cdots, n$, (5.6.1) and (5.6.2) may be written as

$$Y_i = \beta_0 + \beta_1 x_i + \varepsilon_i, \qquad (5.6.3)$$

and

$$E(Y_i) = \beta_0 + \beta_1 x_i, \qquad (5.6.4)$$

where, Y_i, $i = 1, 2, \cdots, n$, is the value of the response variable in the ith trial, x_i is the value of the predictor variable (a known constant) in the ith trial, and ε_i is a random error term with $E(\varepsilon_i) = 0$.

Note that from (5.6.3) we see that the response Y_i in the ith trial falls short or exceeds the value of the regression function by the error term ε_i. In other words, ε_i is the deviation of Y_i from its mean. It is assumed that the error terms ε_i have constant variance σ^2. Hence, it can be shown that responses Y_i have the same variance, i.e.,

$$Var\left(Y_i\right) = \sigma^2 .$$ (5.6.5)

Thus, the regression model (5.6.3) assumes the *cdfs* of Y have the same variance, regardless of the value of the predictor, x, in each trial.

We further note that the error terms, ε_i, are assumed to be uncorrelated. Thus, the outcome in any trial has no effect on the error term for any other trial. This implies that Y_i and Y_j, $i \neq j$, are uncorrelated.

Example 5.6.2.

Referring to the Example 5.6.1, the scatter plot suggests that, subject to random error, a straight- line relation between the amount of money spent on advertising and the number of units sold is appropriate. That is, we may choose a simple linear regression model in this case.

Example 5.6.3.

Suppose that regression model (5.6.3) applies to Example 5.6.1 as

$$Y_i = 445 + 0.184\, x_i + \varepsilon_i.$$

Suppose also that in the ith City, the company spends the amount of \$25,000 for advertising and sold 5000 items in that city. In other words, $x_i = 25,000$ and $Y_i = 5,000$. Hence, the amount of error in this particular city may be calculated as follows:

$$E(Y_i) = 445 + 0.184(25000) = 5045, \; Y_i = 5000 - 5045 = \text{-}45,$$

and, therefore, $\varepsilon_i = -45$.

Often, values of β_0 and β_1 in the regression equation (5.6.1) or (5.6.4) are not known and they need to be estimated. The observed values of the response Y_i corresponding to the predictor values $x_i, i = 1, 2, \cdots, n$, can be used to estimate the regression parameters β_0 and β_1. So, let B_0 and B_1 be the estimators of β_0 and β_1, respectively. Then, for the predictor value x_i, the estimator of the response random variable Y_i will be $B_0 + B_1 x_i$. Hence, the difference between the values of the response and its estimator, the random error, is

$$\varepsilon_i = Y_i - \left(B_0 + B_1 x_i\right).$$ (5.6.6)

Of course, we would have been most pleased if there were no error and we would have observed the exact values of the response. Since this is not practically possible, the smallest values of the deviation (error) ε_i is desirable. A method to accomplish this desire is called the *least square method* and the estimators are called *least square estimators*. The essence of the method is to minimize the sum of the squares of the errors. It is as follows:

Consider a pair of observations (x_i, Y_i), $i = 1, 2, \cdots, n$, the deviation of Y_i from its expected value given in (5.6.4), and the sum of the squares of the errors, $\sum_{i=1}^{n} \varepsilon_i^2$. We might question why sum of squares of errors rather than the sum of error. This is because since an error may be negative (as in Example 5.6.2) the sum of errors may be small, while there are large errors among the observations. So, the sum is minimum due to cancellation of positive and negative terms. However, this cannot happen with square of terms. We now define *least square estimators*.

Definition 5.6.2.

Given pairs of observations (x_i, Y_i), $i = 1, 2, \cdots, n$, the *least square estimators* of β_0 and β_1 are the values of B_0 and B_1 such that

$$\sum_{i=1}^{n} \varepsilon_i^2 = \sum_{i=1}^{n} \left(Y_i - B_0 - B_1 x_i \right)^2 \tag{5.6.7}$$

is minimum.

Now let $\hat{\beta}_0$ and $\hat{\beta}_1$ be point estimators of β_0 and β_1, respectively, that satisfy the least-square method criterion. ($\hat{y}$ reads y hat.) It can be shown (though, it is beyond the scope of this book) through solving of the so called *normal equations*

$$\begin{cases} \sum_{i=1}^{n} Y_i = n\hat{\beta}_0 + \hat{\beta}_1 \sum_{i=1}^{n} x_i \\ \sum_{i=1}^{n} x_i Y_i = \hat{\beta}_0 \sum_{i=1}^{n} x_i + \hat{\beta}_1 \sum_{i=1}^{n} x_i^2 \end{cases} \tag{5.6.8}$$

that

$$\hat{\beta}_1 = \frac{\sum_{i=1}^{n} \left(x_i - \bar{x} \right)\left(Y_i - \bar{Y} \right)}{\sum_{i=1}^{n} \left(x_i - \bar{x} \right)^2}, \tag{5.6.9}$$

and

$$\hat{\beta}_0 = \frac{1}{n}\left(\sum_{i=1}^{n} Y_i - \hat{\beta}_1 \sum_{i=1}^{n} x_i\right) = \overline{Y} - \hat{\beta}_1 \overline{x},\tag{5.6.10}$$

where $\overline{x}$ and $\overline{Y}$ are the means of the x_i and Y_i observations, respectively.

For practical purposes, if we denote

$$S_{xY} = \sum_{i=1}^{n}(x_i - \overline{x})(Y_i - \overline{Y}),\ S_{xx} = \sum_{i=1}^{n}(x_i - \overline{x})^2,\ \text{and}\ S_{YY} = \sum_{i=1}^{n}(Y_i - \overline{Y})^2,\tag{5.6.11}$$

then we can rewrite (5.6.9) and (5.6.10) as

$$\hat{\beta}_1 = \frac{S_{xY}}{S_{xx}}\ \text{and}\ \hat{\beta}_0 = \overline{Y} - \hat{\beta}_1 \overline{x}.\tag{5.6.12}$$

Note that S_{YY} is the total amount of variation in observed y values. It is called the *total sum of squares*, and sometimes denoted by *SST*. It should also be noted that the least square method is used to fit a line to a set of data. It is not an inference tool. Hence, the least square estimators are not really estimators, they are rather solutions.

Using the least square estimators, we define the relation between them as follows:

Definition 5.6.3.

Given sample estimators $\hat{\beta}_0$ and $\hat{\beta}_1$ of the regression parameters β_0 and β_1, respectively, estimate of the regression function is the line

$$\hat{y} \equiv \hat{\beta}_0 + \hat{\beta}_1 x,\tag{5.6.13}$$

that is called the *estimated regression line* (or *function*), where $\hat{\beta}_1$ is the slope and $\hat{\beta}_0$ is the intercept of the line.

It can be shown that the least square estimators found in (5.6.9) and (5.6.10) are unbiased and have the smallest (or minimum) variance among all unbiased linear estimators.

Note that from (5.6.4), $\hat{y}$ in (5.6.13) is a point estimator of the mean response, given the predictor x.

Definition 5.6.4.

The difference between the observed (actual) value of the response random variable Y_i and corresponding predicted response [estimated regression line, $\hat{y}$ (equation (5.6.13)], for each i, denoted by r_i, is called the ith *residual*.

From Definition 5.6.4 and (5.6.13), r_i is

$$r_i = Y_i - \hat{y} = Y_i - \hat{\beta}_0 - \hat{\beta}_1 x_i, \ i = 1, 2, \cdots, n. \tag{5.6.14}$$

Note that the sum of residuals is 0, i.e.,

$$\sum_{i=1}^{n} r_i = 0. \tag{5.6.15}$$

Also, note that the sum of squares of residuals, denoted by SS_r, also called the *residual sum of squares* or the *error sum of squares* denoted by SSE, is

$$SS_r = \sum_{i=1}^{n} \left(Y_i - \hat{\beta}_0 - \hat{\beta}_1 x_i \right)^2. \tag{5.6.16}$$

Example 5.6.4.

Let us consider Example 5.6.1 again. In Example 5.6.3 we found the error when comparing the observed value and the exact value of responses when the predictor and the response were given. Now for the same example, let us

  (a) find the estimated regression line,
  (b) predict the number of items sold when the amount of money spent for advertising in that City is 25,000,
  (c) calculate the residual for the first and the sixth cities, and
  (d) find SS_r.

Solution

To respond to (a) and (b), we can perform the calculations (1) using a computer program, (2) manually or (3) using a hand-calculator. We have already done (1) in Example 5.6.1. Below, we will do (2).

  (a) From Table 5.6.3, we have $\bar{x} = 15500$ and $\bar{Y} = 3300$. To use (5.6.13), we form Table 5.6.3 below.

Thus, from (5.6.12) we have

$$\hat{\beta}_1 = \frac{S_{xY}}{S_{xx}} = \frac{789500000}{3505250000} = 0.184221748 \tag{5.6.17}$$

and

$$\hat{\beta}_0 = \bar{Y} - \hat{\beta}_1 \bar{x} = 3300 - (0.184221748)(15500) = 444.562906. \tag{5.6.18}$$

Hence, from (5.6.13), the estimated regression line is

$$\hat{y} = \hat{\beta}_0 + \hat{\beta}_1 x = 444.562906 + 0.184221748\, x. \tag{5.6.19}$$

Table 5.6.3. Example 5.6.4.

i	x_i	Y_i	$x_i - \bar{x}$	$Y_i - \bar{Y}$	$(x_i - \bar{x})$ $\times (Y_i - \bar{Y})$	$(x_i - \bar{x})^2$	$(Y_i - \bar{Y})^2$	r_i	r_i^2
1	10,000	1,500	-5,500	-1,800	9900000	30,250,000	3,240,000	-786.780706	619023.8793
2	20,000	3,000	4,500	-300	-1350000	20,250,000	90,000	-1128.998506	1274637.627
3	30,000	6,000	14,500	2,700	39150000	210,250,000	7,290,000	28.783694	828.5010403
4	40,000	8,000	24,500	4,700	115150000	600,250,000	22,090,000	186.565894	34806.8328
5	5,000	1,500	-10,500	-1,800	18900000	110,250,000	3,240,000	134.328194	18044.0637
6	5,000	2,000	-10,500	-1,300	13650000	110,250,000	1,690,000	634.328194	402372.2577
7	10,000	2,000	-5,500	-1,300	7150000	30,250,000	1,690,000	-286.780706	82243.17333
8	10,000	2,500	-5,500	-800	4400000	30,250,000	640,000	213.219294	45462.46733
9	10,000	1,500	-5,500	-1,800	9900000	30,250,000	3,240,000	-786.780706	619023.8793
10	15,000	5,000	-500	1,700	-850000	250,000	2,890,000	1792.110394	3211659.664
Total	155000	33000	0	0	216000000 $= S_{xY}$	1172500000 $= S_{xx}$	46,100,000 $= S_{YY}$	-0.00496	6308102.345 $= SS_r$
Mean	15500	3300							

Note that the regression equation that, using MINITAB, we found in (5.6.19) is the same as what we found earlier in Example 5.6.1. Hence, the graph of this line is shown in Figure 5.6.2.

(b) From (5.6.17) and given $x = 25000$, we obtain the predicted member of sold item equal to

$$\hat{y} = 444.562906 + (0.184221748)(25000) = 5050.106616. \tag{5.6.20}$$

Note also that the response value found in (5.6.20) is the same as we were given in Example 5.6.3, except for the round off.

(c) From Table (5.6.3), we see that $x_1 = 10000$, $Y_1 = 1500$, $x_2 = 20,000$, and $Y_2 = 3,000$. Hence, from (5.6.14) and (5.6.18), we have

$$r_1 = Y_1 - \hat{\beta}_0 - \hat{\beta}_1 x_1 = 1500 - 444.562906 + 0.184221748\,(10000) = -786.78039.$$

$$r_5 = Y_5 - \hat{\beta}_0 - \hat{\beta}_1 x_5 = 2000 - 444.562906 + 0.184221748\,(5000) = 634.32835.$$

(d) From table 5.6.3, we see that $SS_r = 6308102.345$.

Using MINITAB, we will have the following results that they confirm the results we obtained manually above.

Regression Analysis: No. of Sale versus $ spent for Ad

```
The regression equation is
No. of Sale = 445 + 0.184 $ spent for Ad


Predictor             Coef   SE Coef      T      P
Constant             444.6     490.3   0.91  0.391
$ spent for Ad     0.18422   0.02593   7.10  0.000


S = 887.982     R-Sq = 86.3%     R-Sq(adj) = 84.6%

PRESS = 8269741     R-Sq(pred) = 82.06%


Analysis of Variance

Source             DF          SS          MS       F      P
Regression          1    39791898    39791898   50.46  0.000
Residual Error      8     6308102      788513
Total               9    46100000


Unusual Observations

          $
       spent    No.
         for     of
Obs       Ad    Sale    Fit   SE Fit   Residual   St Resid
  4    40000    8000   7813      695        187       0.34 X
 10    15000    5000   3208      281       1792       2.13R

R denotes an observation with a large standardized residual.
X denotes an observation whose X value gives it a large influence.
```

In this display (the ANOVA table), we should be noted that the value under "Coef", across "Constant", 444.6, is $\hat{\beta}_0$, and across "X" is $\hat{\beta}_1$. Also, the value "6308102" under "SS" and across "Residual Error" is SS_r.

Residual Plots for Number of Sales

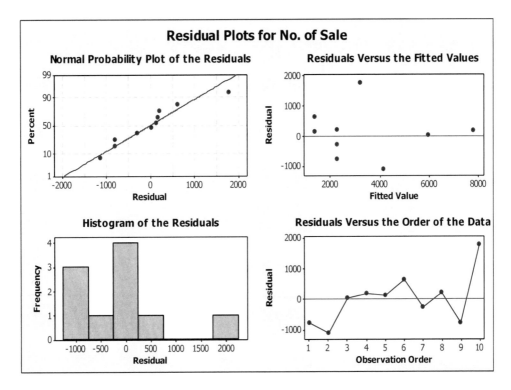

Let us go back to our linear regression model defined by (5.6.1) and (5.6.3). We have already seen that β_0 and β_1 are unknown parameters that have to be estimated by (5.6.12), for instance. We assume that all response random variables, Y_i, are independent. In other words, a response from the predictor value x_i will be assumed to be independent of the response from the predictor variable x_j, $i \neq j$.

We have already made some comments about the error random variable ε or ε_i, see comments above that include (5.6.2) – (5.6.5). Among different possible distributions for ε, we take normal with mean 0 and variance σ^2. In other words, for any predictor value x, our response random variable is normally distributed with mean $E(Y_i) = \beta_0 + \beta_1 x_i$ and $Var(Y_i) = \sigma^2$, $i = 1, 2, \cdots, n$, which is unknown and needs to be estimated.

To (point) estimate the variance of the response variable, σ^2, we observe values of the response variables Y_i, $i = 1, 2, \cdots, n$, corresponding to the predictor values x_i, $i = 1, 2, \cdots, n$. In Section 5.1, the sample variance was defined by (5.1.6). Hence, for Y_i, the sample variance

$$S^2 = \frac{1}{n-1} \sum_{i=1}^{n} (Y_i - \bar{Y})^2 ,$$

which, as we stated there, is an unbiased estimate for the population variance, σ^2.

Residuals are used to consider the appropriateness of the regression model. The residual sum of squares may be used, among other things, to estimator σ^2 as

$$\hat{\sigma}^2 = S^2 = \frac{SS_r}{n-2},$$

(5.6.21)

where $n - 2$ is the number of degrees of freedom associated with the estimator, equivalently, associated with error sum of squares.

The error sum of squares, SS_r, nay be interpreted as how much of variation in y cannot be attributed to a linear relationship between x and y. If $SS_r = 0$, there is no unexplained variation. Hence, the total sum of squares, S_{YY}, plays its roll as the measure of the total amount of variation in observed y. It is the sum of squared deviations about the sample mean of y. The ratio SS_r/S_{YY} is the proportion of total variation that cannot be explained by the simple regression model. Thus, $1 - SS_r/S_{YY}$, which is a number between 0 and 1, is the proportion of observed y variation explained by the regression model. This quantity is important and as we conclude this section, we define it as follows:

Definition 5.6.5.

The quantity $1 - SS_r/S_{YY}$, denoted by r^2, is called the *coefficient of determination*, i.e.,

$$r^2 = 1 - \frac{SS_r}{S_{YY}}.$$

(5.6.22)

Example 5.6.5.

Let us consider Example 5.6.4. From the MINITB display, we see that the value "46100000" under "SS" and across "Total" is S_{YY}. Also "R-Sq = 86.3%" is $100r^2$. This quantity is called the *goodness-of-fit statistic* r^2. To see how r^2 has been calculated, note that from Definition 5.6.5, we have

$$r^2 = 1 - \frac{SS_r}{S_{YY}} = \frac{S_{YY} - SS_r}{S_{YY}} = \frac{SSR}{S_{YY}},$$

(5.6.23)

where $SSR = S_{YY} - SS_r$ is the *regression sum of squares*, i.e., the ratio of explained variation to total variation.

From the MINITAB display in Example 5.6.4 (ANOVA table), the value '39791898" under "SS" and across "Regression" is SSR. Thus,

$$r^2 = \frac{SSR}{S_{YY}} = \frac{39791898}{4600000} = 0.863.$$

5. PART THREE: APPLICATIONS OF STATISTICS

So far, we have seen two parts of statistics, namely, descriptive and inferential. In the last sections of this chapter, we want to discuss two applications: (1) reliability and (2) up-and-down design. We will include some very recent studies on these subjects that may not be available in the existing textbooks of this kind, if at all.

5.7. RELIABILITY

The reliability of an item, or a product, is becoming the top priority for the third millennium and technically sophisticated customer. Manufacturers and all other producing entities are sharpening their tools to satisfy that customer. Estimation of the reliability has become a concern for many quality professionals and statisticians. In engineering, reliability analysis has become a branch of it that concerns with estimating the failure rate of components or systems. Thus, it is necessary to calculate measures that assess reliability of a system or a component of it.

A *system*, whose *reliability* is to be considered, consists of a number of *components*. These components may be installed in parallel, series, or a mixture of both. State of the system to be in working, partially-working, or not working condition, depends upon the same as for the components. Thus, let a system have n components, denoted by x_i, $i = 1, 2, \cdots, n$. Suppose that each component assumes only two states, namely, functioning, denoted by 1, and non-functioning, denoted by 0. Then

$$x_i = \begin{cases} 1, & \text{component } i \text{ is functioning,} \\ 0, & \text{component } i \text{ is non-functioning,} \end{cases} \qquad i = 1, 2, \cdots, n. \qquad (5.7.1)$$

Definition 5.7.1.

Consider a system with only two states, namely, functioning, denoted by 1, and non-functioning, denoted by 0. Denote the state of the system by Ψ. Suppose that the state of the system is determined completely by the states of its components. Then, Ψ will be a function of the components, i.e.,

$$\Psi = \Psi(\mathbf{x}), \text{ where } \mathbf{x} = \left(x_1, x_2, \cdots, x_n \right). \qquad (5.7.2)$$

The function $\Psi(x)$ in (5.7.2) is called the *structure function* of the system. The number of components, n, of the system is called the *order of the system*. The phrase *structure Ψ* is used in place of "*structure having structure function Ψ.*"

A *series structure* (*n-out-of-n*) functions if and only if each component functions, while a *parallel structure* (*1-out-of-n*) functions if and only if at least one component functions. Thus, for these two cases, structure functions are given by

$$\Psi(\mathbf{x}) = \prod_{i=1}^{n} x_i = \min(x_1, x_2, \cdots, x_n), \text{ series structure,} \qquad (5.7.3)$$

and

$$\Psi(\mathbf{x}) = \coprod_{i=1}^{n} x_i \equiv 1 - \prod_{i=1}^{n}(1 - x_i) = \max(x_1, x_2, \cdots, x_n), \text{ parallel structure,} \qquad (5.7.4)$$

respectively.

A *k-out-of-n structure* functions if and only if k out of n components function. In this case, the structure function is

$$\Psi(\mathbf{x}) = \begin{cases} 1, & \text{if } \sum_{i=1}^{n} x_i \geq k, \\ 0, & \text{if } \sum_{i=1}^{n} x_i < k. \end{cases} \qquad (5.7.5)$$

Relation (5.7.5) is equivalent to

$$\Psi(\mathbf{x}) = \prod_{i=1}^{n} x_i, \quad k = n, \text{ k-out-of-n structure,} \qquad (5.7.6)$$

while,

$$\Psi(\mathbf{x}) \equiv (x_1, x_2, \cdots, x_k) \amalg (x_1, x_2, \cdots, x_{k-1}, x_{k+1}) \amalg \cdots \amalg (x_{n-k+1}, x_2, \cdots, x_n)$$
$$= \max\{(x_1, x_2, \cdots, x_k), (x_1, x_2, \cdots, x_{k-1}, x_{k+1}), \cdots, (x_{n-k+1}, x_2, \cdots, x_n)\},$$
$$1 \leq k \leq n, \qquad (5.7.7)$$

where every choice of k out of the n components appears exactly one time.

Example 5.7.1.

An airplane with 3 engines such that it can function if and only if two of three engines function is an example of the 2-out-of-3 system. The structure function for this system is

$$\Psi(\mathbf{x}) = x_1 x_2 \amalg x_1 x_3 \amalg x_2 x_3$$
$$\equiv x_1 x_2 x_3 + x_1 x_2 (1 - x_3) + x_1 (1 - x_2) x_3 + (1 - x_1) x_2 x_3.$$

Example 5.7.2.

Consider a stereo hi-fi system with the following components: (1) FM tuner, (2) CD-player, (3) Amplifier, (4) Speaker 1, and (5) Speaker 2. Let us consider the system functioning when we can hear music through FM or CD-player. Hence, the structure function for this system is

$$\Psi(\mathbf{x}) = \left(x_1 \amalg x_2 \right) x_3 \left(x_4 \amalg x_5 \right).$$

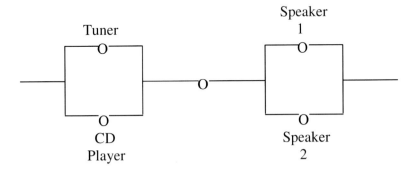

Figure 5.7.1. Hi-Fi, Example 5.7.2.

Denote by $(\cdot_i, x)$ as follows: $(1_i, x) \equiv (x_1, \cdots, x_{i-1}, 1, x_{i+1}, \cdots, x_n)$, $(0_i, x) \equiv (x_1, \cdots, x_{i-1}, 0, x_{i+1}, \cdots, x_n)$, $(\cdot_i, x) \equiv (x_1, \cdots, x_{i-1}, \cdot, x_{i+1}, \cdots, x_n)$.

Definition 5.7.2.

For a system, the ith component is called *irrelevant* to the structure Ψ, if Ψ is constant with respect to x_i, i.e., $\Psi(1_i, x) = \Psi(0_i, x)$ for all $(\cdot_i, x)$. Otherwise, the structure is called *relevant* to the structure.

Definition 5.7.3.

A system is called *coherent* if (1) its structure function is an increasing function and (2) each component is relevant.

We now consider the state of a component i as a random variable, X_i, whose indicator function was given by (5.7.1). We also assume that components of a system are statistically independent.

Definition 5.7.4.

Let

$$P(X_i = 1) \equiv r_i, \ i = 1, 2, \cdots, n. \tag{5.7.8}$$

The probability, r_i, that the component i functions is called the *reliability of component i*. Thus, by property of the indicator function, we have

$$E(X_i) = r_i, \ i = 1, 2, \cdots, n. \tag{5.7.9}$$

Definition 5.7.5.

As before, let

$$P\{\Psi(\mathbf{X}) = 1\} \equiv R, \ \mathbf{X} = (X_1, X_2, \cdots, X_n). \tag{5.7.10}$$

The probability, R, that the system functions is called the *reliability of the system*. Once again, by property of the indicator function, we have

$$E[\Psi(\mathbf{X})] = R. \tag{5.7.11}$$

Let us define r as the random vector $\mathbf{r} = (r_1, r_2, \cdots, r_n)$. Based on the independence assumption, the system reliability, R, will be a function of r, i.e.,

$$R = R(r). \tag{5.7.12}$$

If components are equi-reliable with reliability r, then the reliability of the system is dented by $R(r)$. Each $R(r)$ and $R(r)$ is called the *reliability function of the structure* Ψ.

We note that if the assumption of independence does not hold for a system, then (5.7.12) may not hold. That is, if the components are not independent, then the system reliability may not be a function of the component reliability alone.

Example 5.7.3.

From (5.7.3) and (5.7.4), the reliability function for series and parallel structures and the hi-fi of Example 5.7.2 are respectively as

$$\mathbf{R(r)} = \prod_{i=1}^{n} r_i \ , \text{ series structure}. \tag{5.7.13}$$

$$\mathbf{R(r)} = \prod_{i=1}^{n} r_i = 1 - \prod_{i=1}^{n} \left(1 - r_i\right), \text{ parallel structure.} \tag{5.7.14}$$

$$\mathbf{R(r)} = \left(r_1 + r_2 - r_1 r_2\right) p_3 \left(r_4 + r_5 - r_4 r_5\right), \text{ hi-fi Example 5.7.2.} \tag{5.7.15}$$

In case that components are equi-reliable, then the k-out-of-n system $\Psi(x) = 1$ if and only if $\sum_{i=1}^{n} x_i \geq k$ has reliability function

$$\mathbf{R}(r) = \sum_{i=1}^{n} \binom{n}{i} r^i (1-r)^{n-i}, \; k\text{-out-of-}n \text{ structure.} \tag{5.7.16}$$

For example, for the case of 2-out-of-3 of Example 5.7.1, the reliability function is

$$\mathbf{R}(r) = 3r^2(1-r) + r^3, \; \textit{2-out-of-3} \text{ structure.} \tag{5.7.17}$$

We now want to consider the reliability of a system as a function of time, rather than components.

Definition 5.7.6.

Let a random variable, say T, represent the lifetime of a component or a system with $f_T(t)$ and $F_T(t)$ as *pdf* and *cdf*, respectively. The *reliability function*, at a time t, denoted by $R(t)$, is the probability that the component or the system is still functioning at time t, i.e.,

$$R(t) = P(T > t). \tag{5.7.18}$$

From (5.7.18), it is clear that

$$R(t) = P(T > t) = 1 - P(T \leq t) = 1 - F_T(t). \tag{5.7.19}$$

From (5.7.19), we see that

$$R'(t) = -f_T(t). \tag{5.7.20}$$

When the number of components or systems is large, as we have seen before, relative frequency interprets as probability. Hence, $R(t)$, as in (5.7.19), is the fraction of components or systems that fails after time t.

We leave it as an exercise to show that for a continuous non-negative random variable X with *cdf* as $F_X(x)$, the expected value of X, $E(X)$ is

$$E(X) = \int_0^\infty \left[1 - F_X(t)\right] dt.$$

(5.7.21)

If X is a non-negative discrete random variable, then

$$E(X) = \sum_{x=0}^\infty P(X > x).$$

(5.7.22)

It can also be shown that from (5.7.19), the *mean functioning* time or *survival* is

$$E(T) = \int_0^\infty f_T(t) dt = \int_0^\infty R_T(t) dt.$$

(5.7.23)

Example 5.7.4.

Suppose the lifetime of a machine, denoted by X, has a continuous *cdf* as $F_X(x)$. We want to find the conditional probability that the machine will be working at time x, given that it is working at time t, $X > t$.

Solution

The question under consideration may be formulated as

$$F_X(x|X > t) = P(X \le x|X > t) = \frac{P\left[(X \le x) \cap (X > t)\right]}{P(X > t)}.$$

(5.7.24)

But,

$$(X \le x) \cap (X > t) = \begin{cases} \varnothing, & x < t, \\ t < X \le x, & x \ge t. \end{cases}$$

(5.7.25)

Hence, using (5.7.25) in (5.7.24), we obtain the *cdf*, $F_X(x)$ and *pdf* [derivative of $F_X(x)$ with respect to x] required, respectively, as

$$F_X(x|X > t) = \begin{cases} 0, & x \le t, \\ \dfrac{F_X(x) - F_X(t)}{1 - F_X(t)}, & x > t, \end{cases}$$

(5.7.26)

and

$$f_X(x|X > t) = \begin{cases} 0, & x \le t, \\ \dfrac{f_X(x)}{1 - F_X(t)}, & x > t. \end{cases}$$

(5.7.27)

Definition 5.7.7.

The *failure rate function* (or *force of mortality*), denoted by $r(t)$, is defined by

$$r(t) = f_T\left(x|T > t\right)_{x=t}.$$ (5.7.28)

The function $r(t)$ is an increasing function of t, i.e., the larger t (the older the unit) is the better is the chance of failure within a short interval of length h, namely $r(t)h$. From (5.7.20), we can see that (5.7.28) is equivalent to

$$r(t) = f_T\left(t|T > t\right) = \frac{-R'(t)}{R(t)}.$$ (5.7.29)

From (5.7.28), we may interpret $r(t)dt$ as the probability that a unit that has functioned up to time t will fail in the next dt units of time, i.e., between t and $t + dt$. In other words,

$$P\left(t < T \le t + dt|T > t\right) = f_T\left(t|T > t\right)dt = r(t)dt.$$ (5.7.30)

Example 5.7.5.

Suppose a unit, whose lifetime is represented by a random variable T has a constant failure rate function λ. We want to find the *pdf* and *cdf* of T.

Let $r(t)$ denote the failure rate of the unit. Then, from (5.7.27) we will have

$$r(t) = \frac{-R'(t)}{R(t)} = \lambda$$

or

$$\frac{R'(t)}{R(t)} = -\lambda.$$ (5.7.31)

Assuming the unit is working at the starting time, i.e., $R(0) = 1$, (5.7.31) will be a linear first order differential equation with initial condition $R(0) = 1$. Thus, solving (5.7.31), we obtain

$$-\int_0^t \lambda d\tau + c = \int_0^t \frac{R'(\tau)}{R(\tau)} d\tau = \ln R(t),$$

or $R(t) = e^{c-\lambda t}$, or $R(t) = Ce^{-\lambda t}$, where $C = e^c$. But, $R(0) = 1$ implies that $C = 1$, and thus,

$$R(t) = e^{-\lambda t}, \quad t > 0. \tag{5.7.32}$$

Hence, from (5.7.20)

$$f_T(t) = -R'(t) = \lambda e^{-\lambda t}, \quad t > 0. \tag{5.7.33}$$

In other words, if the random variable T has a constant failure rate function, then it is an exponential random variable and it is called the *exponential failure law*. The law implies that the mean failure rate for exponential failure law is $E(T) = 1/\lambda$.

It should be noted that the constant failure rate means it does not increase as the unit ages. In most cases, this property cannot relate to human populations. This is because, if that would have been true, it would have meant that an 85-years old person would have as much chance of living another year as would a 25-years old person. Thus, a constant failure rate is not a suitable function for most human populations and populations of manufactured components. In other words, the failure rates $r(t)$ is usually an increasing function of t.

Theorem 5.7.1.

In general, the failure rate function and the reliability are related by

$$R(t) = e^{-\int_0^t r(\tau)d\tau} \tag{5.7.34}$$

and

$$f_T(t) = r(t)e^{-\int_0^t r(\tau)d\tau}. \tag{5.7.35}$$

Proof

Left as an exercise.

Example 5.7.6.

The *Weibull failure law* has failure rate function given as

$$r(t) = \alpha\beta^{-\alpha}t^{\alpha-1}, \quad t \geq 0, \quad \alpha, \beta > 0. \tag{5.7.36}$$

From (5.7.34), the reliability is given by

$$R(t) = \left(\frac{t}{\beta}\right)^\alpha, \quad t \ge 0. \tag{5.7.37}$$

Hence, from (5.7.35) and (5.7.37), the *pdf* for *T* is:

$$f_T(t) = \frac{\alpha t^{\alpha-1}}{\beta^\alpha} e^{-(t/\beta)^\alpha}, \quad t \ge 0. \tag{5.7.38}$$

The Weibull failure rate function, for $\alpha > 1$, is a monotonically increasing function with no upper bound. This property gives an edge for the Weibull distribution over the gamma distribution, where the failure rate function is always bounded by $1/\beta$. Another property that gives an edge for the Weibull distribution over the gamma distribution is by varying the values of α and β, a wide variety of curves can be generated. Because of this, the Weibull distribution can be made to fit a wide variety of data sets.

In practice, often several units of a system (such as electronic devices in which current is to flow from a point A to point B) are connected in series, parallel, or a combination of. In such cases, the system fails if either one unit fails (the series case) or all fail (parallel case). For a case as described, we state the following theorem.

Theorem 5.7.2.

Suppose a system is composed of n units with reliability functions $R_1(t), R_2(t), \cdots, R_n(t)$. Assume that units fail independently of each other. Then, the reliability function of each of the series system and parallel systems, respectively denoted by $R_S(t)$ and $R_P(t)$, is, respectively, given by

$$R_S(t) = \prod_{i=1}^n R_i(t) \text{ and } R_P(t) = 1 - \prod_{i=1}^n (1 - R_i(t)). \tag{5.7.39}$$

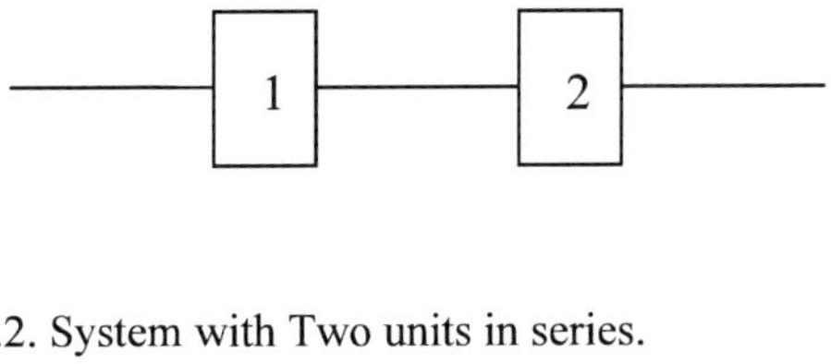

Figure 5.7.2. System with Two units in series.

Proof

See Higgens and Keller-McNulty (1995, p. 375).

Note that for a system with a combination of both series and parallel units, the reliability of the system is calculated by repeated applications of Theorem 5.7.2.

Example 5.7.7.

Suppose that the system consists of units 1 and 2 in series and function if both units function at the same time. See Figure 5.7.2. Assume that the units function independently. Let it be known that probability units 1 and 2 function are 0.97 and 0.90. We want to find the probability that the system functions.

To respond to our question, we use the probability of independent events since we want both units function together in order for the system to function. Hence,

Reliability of the system = P(system function)
$= P$(both units function)
$= P$(Unit 1 functions)$\times P$(Unit 2 functions)
$= (0.97)(0.90) = 0.873$.

Example 5.7.8.

Suppose that the time to failure of two units 1 and 2 of a system is exponentially distributed with parameters $\lambda_1 = 2$ and $\lambda_2 = 4$, for 1 and 2, respectively. From (5.7.34), the reliability functions of the units are $R_1(t) = e^{-0.5t}$ and $R_2(t) = e^{-0.4t}$, respectively. Assume that the units fail independently. Thus, from (5.7.38), the reliability functions for the series and parallel systems, respectively, are

$$R_S(t) = \prod_{i=1}^{n} R_i(t) = \left(e^{-0.5t}\right)\left(e^{-0.4t}\right) = e^{-0.9t},$$

and

$$R_P(t) = \prod_{i=1}^{n} R_i(t) = 1 - \left(1 - e^{-0.5t}\right)\left(1 - e^{-0.4t}\right) = 1 - \left(1 - e^{-0.4t} - e^{-0.5t} + e^{-0.9t}\right)$$

$$= e^{-0.4t} + e^{-0.5t} - e^{-0.9t}.$$

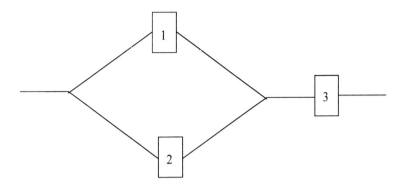

Figure 5.7.3. System with Two units in parallel and both in series with the third.

For instance, if a unit of time is a 1000 hours, then the probability that the series and parallel systems function more than a unit of time, each, is $R_S(1) = e^{-0.9} = 0.4066$ and $R_P(t) = e^{-0.4} + e^{-0.5} - e^{-0.9} = 0.8703$, respectively.

Example 5.7.9.

Suppose that a system consists of units 1 and 2 in parallel and then, unit 3 connected in series with both units 1 and 2. See Figure 5.7.3. Assume that the time to failures of the units of a system are exponentially distributed with parameters $\lambda_1 = 2$, $\lambda_2 = 4$, and $\lambda_3 = 5$. Further, assume that the units fail independently. Note that the units 1 and 2 form a parallel system with reliability function, using (5.7.39), as

$$R_{(1,2)P}(t) = e^{-0.4t} + e^{-0.5t} - e^{-0.9t}.$$

Then, the combined 1 and 2 form a series with 3. Hence, the reliability of the system is

$$R_{1,2,3}(t) = \left(e^{-0.4t} + e^{-0.5t} - e^{-0.9t}\right)\left(e^{-0.2t}\right) = e^{-0.6t} + e^{-0.7t} - e^{-1.1t}.$$

Thus, for instance, if a unit of time is 1000 hours, then the probability that the combined series and parallel system functions more than 3 units of time, each, is

$$R_{1,2,3}(3) = e^{-(0.6)(3)} + e^{-(0.7)(3)} - e^{-(1.1)(3)} = 0.7981.$$

5.8. ESTIMATION OF RELIABILITY

In Definition 5.7.6 [relation (5.7.18)], we defined the reliability of a unit as the probability of survival time of the unit beyond a time unit t, denoted by $R(t) = P(T > t)$. In that case, we were considering the lifetime of a unit compared with the failure time t. We now offer alternative vocabularies in defining the reliability, $R(t)$.

As we mentioned, the product reliability seems, finally, becoming the top priority for the third millennium and a technically sophisticated customer. Manufacturers and all other producing entities are sharpening their tools to satisfy such customer. Estimation of the reliability has become a concern for many quality control professionals and statisticians. When Y represents the random value of a stress (or supply) that a device (or a component) will be subjected to in service, and X represents the strength (or demand) that varies from product to product in the population of devices. The device fails at the instant that the stress applied to it exceeds the strength and functions successfully whenever $X > Y$. Then, the reliability (or the measure of reliability) R is defined as $P(Y < X)$, i.e., the probability that a randomly selected device functions successfully.

Definition 5.8.1.

Let the random variable X that represents the strength (that varies from item to item in the population of devices) takes place of the lifetime, T, of a device. Let also Y that represent the random value of a stress (that a device will be subjected to in service) takes place of the time of failure. Then, the *reliability* that a randomly selected devise survives under the stress exerted, denoted by R, is the probability that the device functions successfully, i.e.,

$$R = P(Y < X).$$ (5.8.1)

Note that there is no time factor in this alternative definition of reliability. In fact, we are considering a system with only one component. If the component fails, the system fails.

Chronologically, the stress-strength model originated in a nonparametric form by Wilcoxon (1945). A strength of a nonparametric approach was that there was no assumption on X and Y. However, a weakness of it was that the approach was too inefficient for practical purposes. Explicit consideration of the relevance of $R = P(Y < X)$ seems to have begun with Birnbaum (1956). The first attempt to study $R = P(Y < X)$, parametrically, under certain assumptions of X and Y, was done by Owen et al. (1964). They constructed confidence limits for $R = P(Y < X)$ when X and Y were assumed normally distributed. Subsequently, point estimation of R has been considered for several parametric families of distributions and under different assumptions. Johnson (1988), who specifically considers estimation in the Weibull case, when the stress and strength distributions have a common shape-parameter, presents a good overview. The studies lead in applications in many engineering problems under the umbrella of "reliability". The research continued to expand and get into the area of military, and particularly, medical-related problems in the last two decades. It seems that now, the clinical trial is the fasted growing area for application of the reliability. Applications of reliability have further expanded to contain psychology sociology, in order to compare random variables that represent state of affair of in two or more situations at different times.

The effect of the sample size and the values of the distribution parameters on the estimation of R was studied by Shayib (2005). He considered the effect of parameters ratio and sample sizes on the calculations of R. Haghighi and Shayib (2007) considered the preferred distribution function for estimating $P(Y < X)$ in the Burr type X, Gamma and Weibull cases. Haghighi and Shayib (2009) considered the case that the distributions of X and Y are independent Weibull, sharing the same known shape parameter α, but unknown and different scale parameters, i.e., $X \sim$ Weibull(α, β_1) and $Y \sim$ Weibull(α, β_2). (We will use the notation $X \sim$ Weibull(α, β) for the random variable X having a Weibull probability density function with parameters α and β.)

The Weibull *pdf* and *cdf* are given in (4.11.22) and (4.11.23), respectively, in the previous chapter. α is the shape parameter while $\beta^{1/\alpha}$ is the scale parameter of the distribution. Frequently, α is taken as $1 < \alpha < 5$, and we do the same. The expected value and the variance of X in this case, respectively, are

$$E(X) = \beta^{1/\alpha} \Gamma\left(1 + \frac{1}{\alpha}\right),$$ (5.8.2)

and

$$Var(X) = \beta^{2/\alpha} \left\{ \Gamma\left(1 + \frac{2}{\alpha}\right) - \left[\Gamma\left(1 + \frac{1}{\alpha}\right)\right]^2 \right\}.$$ (5.8.3)

The Weibull distribution is commonly used for as a model for life lengths and in the study of breaking strengths of materials because of the properties of its failure rate function. It is also an excellent distribution to describe the life of a manufactured component. Resistors used in the construction of an aircraft guidance system have lifetime lengths that follow a Weibull distribution with $\alpha = 2$, and $\beta = 10$, (with measurements in thousands of hours).

The value of R has been calculated, in the literature, under the assumed distributions and using different methods of estimations. In this section, we will discuss estimation of the reliability defined in (5.8.1). We will use estimation methods: *MLE*, *Shrinkage Estimation Procedures*, and *Method of Moments*. In this section, we are interested in the selection of the method of estimation that will give the highest value of R in the Weibull distribution.

5.8.1.   Estimation of the Reliability by MLE Method

To date, maximum likelihood estimate (*MLE*) is the most popular method for estimating reliability $R = P(Y < X)$. This is because *MLE* is flexible and general. A significance test for R has been introduced based on the *MLE* for R. Due to the existence of a unique solution of the maximum likelihood equations and since the shape parameter, α, is given, we will use the *MLE* method to find the estimate of the parameters β_1 and β_2 based on the simulated sample values with an assigned sample sizes. Thus, the *MLE*'s $\hat{\beta}_1$ and $\hat{\beta}_2$ (and at the same time the

MVUE) of β_1 and β_2 based on a sample sizes m and n, and the shape parameter α, are given by

$$\hat{\beta}_1 = \frac{1}{m}\sum_1^m x^\alpha, \quad \hat{\beta}_2 = \frac{1}{n}\sum_1^n y^\alpha. \qquad (5.8.4)$$

The *MLE* of *R* in this sequel will be denoted by $\hat{R}$. It can be shown that

$$R = P(Y < X) = \frac{1}{1+\dfrac{\beta_2}{\beta_1}} = \frac{1}{1+\rho}, \qquad (5.8.5)$$

where

$$\rho = \frac{\beta_2}{\beta_1}. \qquad (5.8.6)$$

Note that since from (5.8.5) we have

$$\hat{\rho} = \frac{\hat{\beta}_2}{\hat{\beta}_1}, \qquad (5.8.7)$$

Then,

$$\hat{R} = \frac{1}{1+\hat{\rho}}. \qquad (5.8.8)$$

Because *R* is monotonic in ρ, inference on ρ is equivalent to inference on *R*.

Example 5.8.1.

Let $X_1, X_2, ..., X_m$ be a random sample for *X*, and $Y_1, Y_2, ..., Y_n$ be the corresponding random sample for *Y*. To conduct a simulation, due to the involved and tedious calculations of the *MLE* of α, we choose $\alpha = 2, 3$, and 4, when the ratio of β_1/β_2 is 1, 1/2, and 2, with β_1 and β_2 as parameters for *X* and *Y*, respectively. We choose the ratio of the scale parameters to compare the effect measures of the reliability. We also choose $m = n$, and $m = 5(1)10$, (5 to 10 with increment 1). Based on 1000 runs, estimated values of the different estimators have been calculated and listed in Table 5.8.1. Additionally, included in Table 5.8.1 is the actual values of *R* calculated from (5.8.5).

5.8.2.   Estimation of Reliability by Shrinkage Procedures

Definition 5.8.2.

In some applications, an experimenter often possesses some knowledge of the experimental conditions based on the behavior of the system under consideration, or from the past experience, and is, thus, in a position to give an educated guess or an initial estimate of the parameter of interest. Given a prior estimate R_0 of R, we are looking for an estimator that incorporates this information. These estimators are, then, called "*shrinkage estimators*" as introduced by Thompson (1968).

In his paper, Thompson, considered Shrinkage of the minimum variance unbiased linear estimator (MVULE) as a method for lowering the mean square error toward origin in the parameter space. In a different terms, a shrinkage estimator is an estimator that, either explicitly or implicitly, incorporates the effects of shrinkage. In other words, a naive or raw estimate is improved by combining it with other information. A review on the shrinkage estimation is given in Schäfer and Strimmer (2005), for example. One general result is that many standard estimators can be improved, in terms of mean square error, by shrinking them towards zero (or any other fixed constant value). A well-known example arises in the estimation of the population variance based on a simple sample.

In this section, following Thompson (1968), we choose a guessed value for R_0, and compute a value of θ, denoted by θ_1, for the estimators of the form

$$\tilde{R} = \theta \, \hat{R} + (1 - \theta) R_0 , 0 \leq \theta \leq 1. \tag{5.8.9}$$

Then, using the *p*-value of the *likelihood ratio test* (*LRT*, defined below), we find three other values for θ denoted by θ_2, θ_3, and θ_4, by different methods. These estimators are given in the next two sections. The values of θ will be used to compute shrinkage estimators of R, denoted by R_1, R_2, R_3, and R_4. Then, we will compare these estimators with each other and with the *MLE* of R.

5.8.2.a. Shrinking towards a Pre-specified *R*

Here, we are looking for θ_1 in the estimator $\tilde{R} = \theta\hat{R} + (1-\theta)R_0$ that minimizes its mean square error *MSE*

$$MSE \left(\tilde{R}_1 \right) = E(\tilde{R} - R)^2 = E[\theta_1\hat{R} + (1 - \theta_1)R_0 - R]^2 . \tag{5.8.10}$$

The value of θ_1 that minimizes this *MSE* can be shown to be

$$\theta_1 = \frac{(R - R_0) E(\hat{R} - R_0)}{E(\hat{R}^2) - 2R_0 E(\hat{R}) + R_0^2},$$ (5.8.11)

subject to $0 \leq \theta_1 \leq 1$. Moreover this value depends on the unknown parameter R. Substituting $\hat{R}$ instead of R, we get

$$\theta_1 = \frac{(R - R_0) E(\hat{R} - R_0)}{E(\hat{R}^2) - 2R_0 E(\hat{R}) + R_0^2},$$ (5.8.12)

Hence, our first shrinkage estimator is

$$\tilde{R}_1 = \hat{\theta}_1 \hat{R} + (1 - \hat{\theta}_1) R_0 .$$ (5.8.13)

We now obtain approximate values of $E(\hat{R})$ and $Var(\hat{R})$. For this purpose, we state the following theorem:

Theorem 5.8.1.

Supposed $X \sim \text{Weibull}(\alpha, \beta_1)$ and $Y \sim \text{Weibull}(\alpha, \beta_2)$, as given in (4.11.22). Let U and T be independent identically distributed exponential random variables with parameters β_1 and β_2, and with *pdf* $h(u) = (1/\beta_1)e^{-u/\beta_1}$ and $h(t) = (1/\beta_2)e^{-t/\beta_2}$, respectively. Then, $V = U/T$ has probability density function

$$g(v) = \begin{cases} \beta_1\beta_2(\beta_2 v + \beta_1)^{-2}, & v > 0 \\ 0, & \textit{otherwise.} \end{cases}$$ (5.8.14)

Proof

The proof is easily shown by using the transformations of variables of the continuous type.

Example 5.8.2.

In the very special case, when $m = n$ and $\beta_2/\beta_1 = 1$, we see that $\hat{\rho} = v$, and $\hat{R} = 1/(1+v)$. Thus, it can easily be shown that $E(\hat{R}) = 1/2$, $E(\hat{R}^2) = 1/3$, and $Var(\hat{R}) = 1/12$. These values will be used in calculating $\hat{\theta}_1$.

Example 5.8.3.

For the ratio V in Theorem 5.8.1, we choose

$$V = \frac{\beta_1}{\beta_2}\left(\frac{\overline{Y}_n^\alpha}{\overline{X}_m^\alpha}\right) \sim F_{2n,2m}. \tag{5.8.15}$$

Noting that

$$\hat{R} = \frac{1}{1 + \dfrac{\beta_2}{\beta_1}V}, $$

we get

$$E(\hat{R}) = \left[1 + \frac{\beta_2}{\beta_1}E(V)\right]^{-1} + Var(V)\left[\left(\frac{\beta_2}{\beta_1}\right)^2\left(1 + \frac{\beta_2}{\beta_1}\right)E(V)\right]^{-3}, \tag{5.8.16}$$

and

$$Var(\hat{R}) = Var(V)\left(\frac{\beta_2}{\beta_1}\right)^2\left[1 + \frac{\beta_2}{\beta_1}E(V)\right]^{-2}, \tag{5.8.17}$$

where $E(V)$ and $Var(V)$ can be determined from the F-distribution. In these formulas β_1 and β_2 are further replaced by their *MLEs* respectively, for numerical computations. This will complete discussion on θ_1 given by (5.8.11).

5.8.2.b. Shrinking Using the p-value of the LRT

In section 5.8.2.a. we found the value for θ, namely θ_1, and that established the first shrinkage estimator. In this subsection, we will pursue another approach to find θ values in order to establish the other shrinkage estimators, namely, R_2, R_3, and R_4. Now, we will find the other three values of θ, which will give us the other estimators by finding the *p*-value of the hypothesis in the *Likelihood Ratio Test* (*LRT*) procedure.

For testing

$$H_0: R = R_0 \text{ vs. } H_1: R \neq R_0, \tag{5.8.18}$$

the *LRT* is of the form:

reject H_0 when $\dfrac{\overline{Y}^\alpha}{\overline{X}^\alpha} < C_1$ or $\dfrac{\overline{Y}^\alpha}{\overline{X}^\alpha} > C_2$,   (5.8.19)

where $\overline{Y}^\alpha$ is the sample mean of n value for Y^α, and similarly for $\overline{X}^\alpha$. In addition, the sample sizes are not necessary equal. This follows by noticing that (5.8.18) is equivalent to

$$H_0: \quad \beta_1 = \left[\frac{R_0}{1 - R_0}\right]\beta_2 \quad vs. \quad H_1: \quad \beta_1 \neq \left[\frac{R_0}{1 - R_0}\right]\beta_2. \qquad (5.8.20)$$

We now consider the following theorem:

Theorem 5.8.2.

Let $X \sim \text{Weibull}(\alpha, \beta_1)$ and $Y \sim \text{Weibull}(\alpha, \beta_2)$, as given in (4.11.22).

(i) $U = X^\alpha$ has an exponential distribution with parameter β_1 and *pdf*

$$h(u) = \begin{cases} \dfrac{e^{-u/\beta_1}}{\beta_1}, & u > 0, \\ 0, & otherwise. \end{cases}$$

(ii) Letting $\overline{U}$ be the mean of U, $\dfrac{2m\overline{U}}{\beta_1} \sim \chi^2_{2m}$.

(iii) $V = Y^\alpha$ has an exponential distribution with parameter β_2 and *pdf*

$$g(v) = \begin{cases} \dfrac{1}{\beta_2} e^{-v/\beta_2}, & v > 0 \\ 0, & otherwise, \end{cases}$$

(iv) Letting $\overline{V}$ be the mean of V, $\dfrac{2n\overline{V}}{\beta_2} \sim \chi^2_{2n}$,

and

v) $\dfrac{\beta_1\overline{V}}{\beta_2\overline{U}} \sim F_{2n,2m}$.

Proof

Cases (i) and (iii) are clear by using transformation $U = X^\alpha$ and $V = Y^\alpha$ in distributions of X and Y, respectively. For cases (ii), (iv) and (v) see Baklizi and Abu Dayyeh (2003).

Now, from (5.8.21) and Theorem 5.8.2, when $m = n$, we have

$$T \equiv \frac{R_0}{1-R_0} \frac{\bar{V}}{\bar{U}} \sim F_{2m,2m}.$$ (5.8.21)

Because of the skewness of the F-distribution and since we have a two-tailed test, the p-value (5.8.20) is

$$z = 2\ min\ [\ P_{H_0}\ (T > t),\ P_{H_0}\ (T < t)] = 2\ min\ [1 - F(t),\ F(t)],$$ (5.8.22)

where t is the observed value of the test statistic T that has an F distribution function under H_0. The p-value of this test indicates how strongly H_0 is supported by the data. A large p-value indicates that R is close to the prior estimate R_0 (Tse and Tso, 1996). Thus, from (5.8.9) we can use this p-value to form the shrinkage estimator

$$\tilde{R}_2 = \theta_2 \hat{R} + (1 - \theta_2)R_0,$$ (5.8.23)

where $\theta_2 = 1 - p$-value.

Based on the comments just made, to have the estimator even closer to R_0, we take the square root of the p-value. Thus, a third shrinking estimator is

$$\tilde{R}_3 = \theta_3 \hat{R} + (1 - \theta_3)R_0,$$ (5.8.24)

where $\theta_3 = 1 - \sqrt{p\text{-value}}$.

Now, to have the estimator far away from R_0, we take the square of the p-value such that

$$\tilde{R}_4 = \theta_4 \hat{R} + (1 - \theta_4)R_0,$$ (5.8.25)

where $\theta_4 = 1 - (p\text{-value})^2$, which is the forth shrinkage estimator.

5.8.3. Estimation of Reliability by the Method of Moments

In reference to (5.8.2) and (5.8.3), and since α is assumed given, we have two ways of estimating β_1 (or β_2) using the first and the second moments. Using the first moment of the random variable and the statistic of the sample we have

$$\hat{\rho} = \frac{\hat{\beta}_2}{\hat{\beta}_1} = \frac{\bar{Y}^\alpha}{\bar{X}^\alpha}, \tag{5.8.26}$$

since α is given, the expression in α for the gamma function disappears by finding the ratio. Moreover, by using the second moment of the random variable and the corresponding statistic from the sample we see that

$$\hat{\rho} = \frac{\hat{\beta}_2}{\hat{\beta}_1} = \frac{S_2{}^\alpha}{S_1{}^\alpha}, \tag{5.8.27}$$

where S is the standard deviation of the sample. Based on the above setting we have two more estimators to consider. These are, based on the first moment;

$$\hat{R}_5 = \frac{1}{1+\hat{\rho}} = \frac{1}{1+\dfrac{\bar{Y}_n^\alpha}{\bar{X}_m^\alpha}}, \tag{5.8.28}$$

and, based on the second moment, we have

$$\hat{R}_6 = \frac{1}{1+\hat{\rho}} = \frac{1}{1+\dfrac{S_2{}^\alpha}{S_1{}^\alpha}}. \tag{5.8.29}$$

The simulation will be run as above for the comparison.

Example 5.8.2.

A simulation study is conducted to investigate the performance of the estimators: $\tilde{R}_1, \tilde{R}_2, \tilde{R}_3$, and $\tilde{R}_4$ in relation to the *MLE* and using the θ values found by the 4 procedures outlined above. In addition to the 4 shrinkage estimators found above based on the *MLE* and the prior assigned value of R, there are two more that are based on the method of moments. These estimators are found by using the first and second moments of the distribution, respectively, and are denoted by R_5 and R_6.

The nomenclature of the simulation is as given below, for the case $m = n$, where

m: Number of X observations and is taken to be $m = 5(1)10$,

n: Number of Y observations and is taken to be $n = 5(1)10$.

R: The true value of $R = P(Y < X)$ and is taken to be 0.5, and

R_0: The initial estimate of R and is taken to be 0.45.

For each combination of m, n, R, and R_0, 1000 samples were generated for X and for Y taking ρ to be 1 and $R = 0.5$. This in turn corresponds to the ratio β_2 / β_1 to be 1. As it was mentioned earlier, since R is monotonic in ρ, the smaller the ρ is the larger R will be. The estimators are calculated and the biases and $MSEs$ of shrinkage estimators are obtained. The shrinkage estimators seem to perform better for small sample sizes than for large sample sizes. This is expected, as the sample size increases, the precision of the MLE increases, whereas shrinkage estimators are still affected by the prior guess R_0, which may be poorly made. This applies in the case of the exponential distribution. Further simulation is done for the Weibull distribution to check on the validity of the above statement.

Remarks

Industry is data rich and yet it is very often that decisions are made based on 2, and in most cases on less than 6, readings. The above discussion was done based on the Weibull distribution since in model fitting it is a strong competitor to the gamma distribution. Both the gamma and Weibull distributions are skewed, but they are valuable distributions for model fitting. Weibull distribution is commonly used as a model for life length because of the properties of its failure rate function which is unbounded for a shape parameter greater than 1.

Comparing estimators may be conducted using quite different criteria. In this section, we are comparing the estimators set above using the MSE values of those estimators that were obtained from the simulation. $MSEs$ of all the estimators are tabulated in Tables (5.8.1), (5.8.3), and (5.8.5). Tables (5.8.2), (5.8.4) and (5.8.6) display the ranking of the estimators based on the sample sizes used in the simulation. It is worth noting that the MLE of R, namely $\hat{R}$, is surpassed by the shrinkage estimators in almost all the cases. The estimators found using the method of moments came at the end of the list except in one case when $m = 8$. In addition, the estimator that was calculated based on the pre-assigned value of R, namely R_1, took the lead in all case.

Again, as it is completely perceived that the larger the sample the better is not a good practical assumption for evaluating the reliability of a system. Attention should be paid to the values of the parameters that control that distribution when it is assumed for the stress and strength data. In the case under consideration, the estimated values of the estimators came very close to the actual value calculated when the parameters' assumed values were used. It is completely clear that the insight knowledge for the underlined distribution and the ratio of its parameters will affect the calculating reliability of the item. Further work will be conducted with different ratios for the sample sizes and scale parameters of the distributions. In all cases to be considered latter, the shape parameter will be assumed given and need not to estimated.

Table 5.8.1. Weibull Distribution. The Estimated Average Valuesof the Estimators Standard Deviation and *MSE*

$m = n,\ \beta_1/\beta_2 = 1,\ \alpha = 2,\ R = 1/2,\ R_0 = .45$

Estimator			m			
	5	6	7	8	9	10
$\hat{R}$	0.4996	0.4989	0.4973	0.4692	0.5021	0.4992
	0.1515	0.1393	0.1284	0.1249	0.1127	0.1082
	0.0230	0.0194	0.0165	0.0166	0.0127	0.0117
R_1	0.4582	0.4564	0.4607	0.4552	0.4553	0.4552
	0.0271	0.0195	0.0142	0.0148	0.0139	0.0130
	0.0025	0.0023	0.0017	0.0022	0.0022	0.0022
R_2	0.4876	0.4875	0.4715	0.4877	0.4900	0.4924
	0.1280	0.1189	0.0976	0.1030	0.0996	0.0962
	0.0165	0.0143	0.0104	0.0108	0.0100	0.0093
R_3	0.4786	0.4779	0.4576	0.4781	0.4802	0.4628
	0.1010	0.0941	0.0748	0.0818	0.0801	0.0980
	0.0107	0.0093	0.0074	0.0072	0.0068	0.0110
R_4	0.4933	0.4940	0.4860	0.4940	0.4964	0.4986
	0.1437	0.1303	0.1157	0.1148	0.1100	0.1060
	0.0207	0.0177	0.0136	0.0132	0.0121	0.0112
R_5	0.4983	0.4967	0.4944	0.4964	0.4974	0.4975
	0.1565	0.1429	0.1353	0.1271	0.1188	0.1123
	0.0245	0.0204	0.0183	0.0162	0.0141	0.0126
R_6	0.4966	0.4967	0.4938	0.4981	0.4987	0.4995
	0.2230	0.2031	0.1916	0.1792	0.1714	0.1631
	0.0497	0.0412	0.0368	0.0321	0.0294	0.0266

Table 5.8.2. The ranking of the estimators with $m = n,\ m = 5(1)10$

Samples size	How they rank
5	$R_1, R_3, R_2, R_4, \hat{R}, R_5, R_6$
6	$R_1, R_3, R_2, R_4, \hat{R}, R_5, R_6$
7	$R_1, R_3, R_2, R_4, \hat{R}, R_5, R_6$
8	$R_1, R_3, R_2, R_4, R_5, \hat{R}, R_6$
9	$R_1, R_2, R_3, R_4, R_5, \hat{R}, R_6$
10	$R_1, R_2, R_3, R_4, \hat{R}, R_5, R_6$

Table 5.8.3. Weibull Distribution. The Estimated Average Values of the Estimators Standard Deviation and *MSE*

$m = n$, $\beta_1 / \beta_2 = 1/4$, $\alpha = 2$, $R = 0.2$, $R_0 = 0.25$

Estimator	5	6	m 7	8	9	10
$\hat{R}$	0.2201	0.2169	0.2105	0.2099	0.2107	0.2111
	0.0125	0.0978	0.0952	0.0881	0.0793	0.0748
	0.0006	0.0099	0.0092	0.0079	0.0064	0.0057
R_1	0.2249	0.2496	0.2489	0.2029	0.2487	0.2486
	0.1031	0.0006	0.0047	0.0187	0.0048	0.0044
	0.0125	0.0025	0.0024	0.0004	0.0024	0.0023
R_2	0.2261	0.2235	0.2177	0.2178	0.2184	0.2185
	0.0892	0.0816	0.0818	0.0783	0.0682	0.0642
	0.0086	0.0072	0.0070	0.0061	0.0050	0.0045
R_3	0.2319	0.2301	0.2252	0.2255	0.2263	0.2261
	0.0693	0.0640	0.0656	0.0621	0.0555	0.0519
	0.0058	0.0050	0.0049	0.0045	0.0038	0.0034
R_4	0.2223	0.2192	0.2129	0.2128	0.2133	0.2137
	0.1013	0.0920	0.0906	0.0842	0.0755	0.0712
	0.0108	0.0088	0.0084	0.0073	0.0059	0.0053
R_5	0.2201	0.2169	0.2107	0.2099	0.2107	0.2166
	0.1079	0.0978	0.0819	0.0881	0.0793	0.0784
	0.0120	0.0099	0.0068	0.0079	0.0064	0.0063
R_6	0.2482	0.2393	0.2210	0.2131	0.2214	0.2245
	0.1829	0.1640	0.0699	0.0696	0.0604	0.1161
	0.0358	0.0284	0.0053	0.0050	0.0041	0.0141

Table 5.8.4. The ranking of the estimators with $m = n$, $m = 5(1)10$

$m = n$, $\beta_1 / \beta_2 = 1/4$, $\alpha = 2$, $R = 0.2$, $R_0 = 0.25$

Samples size	How they rank
5	$\hat{R}, R_3, R_2, R_4, R_1, R_5, R_6$
6	$R_1, R_3, R_2, R_4, \hat{R}, R_5, R_6$
7	$R_1, R_3, R_6, R_5, R_2, R_4, \hat{R}$
8	$R_1, R_3, R_6, R_2, R_4, \hat{R}, R_5$
9	$R_1, R_3, R_6, R_2, R_4, \hat{R}, R_5$
10	$R_1, R_3, R_2, R_4, \hat{R}, R_5, R_6$

Table 5.8.5. Weibull Distribution . The Estimated Average Values of the Estimators. Standard Deviation and *MSE*

$m = n, \ \beta_1/\beta_2 = 4, \ \alpha = 2, R = 0.8, R_0 = 0.75$

				m		
Estimator	5	6	7	8	9	10
$\hat{R}$	0.7796	0.7839	0.7843	0.7873	0.7872	0.7883
	0.1129	0.1014	0.0954	0.0873	0.0834	0.0789
	0.0132	0.0105	0.0093	0.0078	0.0071	0.0064
R_1	0.7502	0.7507	0.7513	0.7560	0.7500	0.7705
	0.0027	0.0051	0.0076	0.0249	0.0025	0.0246
	0.0025	0.0026	0.0024	0.0026	0.0025	0.0015
R_2	0.7736	0.7765	0.7775	0.7803	0.7808	0.7821
	0.0950	0.0860	0.0818	0.0749	0.0718	0.0980
	0.0097	0.0079	0.0072	0.0060	0.0055	0.0099
R_3	0.7681	0.7701	0.7710	0.7731	0.7739	0.7751
	0.0750	0.0682	0.0653	0.0602	0.0575	0.0551
	0.0066	0.0055	0.0051	0.0043	0.0040	0.0037
R_4	0.7773	0.7808	0.7817	0.7847	0.7849	0.7862
	0.1066	0.0961	0.0909	0.0832	0.0796	0.0753
	0.0119	0.0096	0.0086	0.0072	0.0066	0.0059
R_5	0.7796	0.7839	0.7843	0.7873	0.7873	0.7883
	0.1129	0.1014	0.0954	0.0873	0.0834	0.0789
	0.0132	0.0105	0.0093	0.0078	0.0071	0.0064
R_6	0.7498	0.7552	0.7604	0.7695	0.7696	0.7711
	0.1907	0.1579	0.1501	0.1325	0.1326	0.1250
	0.0389	0.0289	0.0241	0.0185	0.0184	0.0165

Table 5.8.6. The ranking of the estimators with $m = n, m = 5(1)10$

$m = n, \ \beta_1/\beta_2 = 4, \ \alpha = 2, R = 0.8, R_0 = 0$

Samples size	How they rank
5	$R_1, R_3, R_2, \hat{R}, R_5, R_6, R_4$
6	$R_1, R_3, R_2, \hat{R}, R_5, R_6, R_4$
7	$R_1, R_3, R_2, \hat{R}, R_5, R_6, R_4$
8	$R_1, R_3, R_2, \hat{R}, R_5, R_6, R_4$
9	$R_1, R_3, R_2, \hat{R}, R_5, R_6, R_4$
10	$R_1, R_3, \hat{R}, R_5, R_2, R_6, R_4$

5.9. RELIABILITY COMPUTATION USING LOGISTIC DISTRIBUTIONS

As before, X and Y are assumed to be independent random variables. If $H_Y(y)$ and $f_X(x)$ are the *cdf* and *pdf* of Y and X, respectively, then it is well known that

$$R = P(Y < X) = \int_{-\infty}^{\infty} H_Y(z) f_X(z) dz .$$ (5.9.1)

The estimation of R is a very common concern in statistical literature in both distribution-free and parametric cases. We want to find an estimate for R and study the effect of the sample sizes and parameters' ratio on estimating R, when X and Y are assuming the logistic and extreme-value Type I distributions.

We consider the unknown-one-parameter logistic distributions for both X and Y, i.e., we assume that the local parameter α is 0. Hence, *cdf* and *pdf* are, respectively, as follow:

$$cdf:\ H_Y(y) = \frac{1}{1 + e^{-\frac{y}{\beta_1}}}, pdf:\ h_Y(y) = \frac{e^{-\frac{y}{\beta_1}}}{\beta_1 \left(1 + e^{-\frac{y}{\beta_1}} \right)^2}, \quad -\infty < y < \infty, \quad \beta_1 > 0 .$$ (5.9.2)

$$cdf:\ F_X(x) = \frac{1}{1 + e^{-\frac{x}{\beta_2}}}, pdf:\ f_X(x) = \frac{e^{-\frac{x}{\beta_2}}}{\beta_2 \left(1 + e^{-\frac{x}{\beta_2}} \right)^2}, \quad -\infty < x < \infty, \quad \beta_2 > 0 .$$ (5.9.3)

Theorem 5.9.1.

Assuming X and Y independent random variables with logistic probability distribution, we have

$$R(\beta_1, \beta_2) = \sum_{n=0}^{\infty} \sum_{m=0}^{\infty} \frac{(-1)^{n+m}}{1 + \dfrac{m}{n+1} \dfrac{\beta_2}{\beta_1}} .$$ (5.9.4)

Proof

Based on the assumption of the Theorem 5.9.1, it is easy to see that

$$R(\beta_1,\beta_2) = P(Y < X) = \int_{-\infty}^{\infty} \int_{-\infty}^{x} \frac{e^{-\frac{y}{\beta_1}}}{\beta_1\left(1+e^{-\frac{y}{\beta_1}}\right)^2} \frac{e^{-\frac{x}{\beta_2}}}{\beta_2\left(1+e^{-\frac{x}{\beta_2}}\right)^2} \, dy dx$$

$$= \int_{-\infty}^{\infty} \frac{e^{-\frac{x}{\beta_2}}}{\beta_2\left(1+e^{-\frac{x}{\beta_2}}\right)^2} \frac{1}{\left(1+e^{-\frac{y}{\beta_1}}\right)^x} \, dx \Bigg|_\infty$$

$$= \int_{-\infty}^{\infty} \frac{e^{-\frac{x}{\beta_2}}}{\beta_2\left(1+e^{-\frac{x}{\beta_2}}\right)^2} \frac{1}{\left(1+e^{-\frac{x}{\beta_1}}\right)} \, dx \ . \tag{5.9.5}$$

Let $z = e^{-\frac{x}{\beta_2}}$. Then, $dz = -\frac{1}{\beta_2} e^{-\frac{x}{\beta_2}} dx$, and $dx = -\beta_2 e^{\frac{x}{\beta_2}} dz = -\beta_2 \frac{1}{z} dz$. When

$x = -\infty$, $z = \infty$; when $x = \infty$, $z = 0$. Then, from (5.9.2) we will have:

$$R(\beta_1,\beta_2) = -\int_0^\infty \frac{z}{\beta_2(1+z)^2} \frac{1}{\left(1+z^{\frac{\beta_2}{\beta_1}}\right)} (-\beta_2 z^{-1} dz)$$

$$= \int_0^\infty \frac{1}{(1+z)^2} \frac{1}{\left(1+z^{\frac{\beta_2}{\beta_1}}\right)} \, dz \ . \tag{5.9.6}$$

Now,

$$\frac{1}{(1+z)^2} = -\frac{d}{dz}\frac{1}{1+z} = -\frac{d}{dz}\sum_{n=0}^{\infty}(-1)^n z^n = \frac{d}{dz}\sum_{n=0}^{\infty}(-1)^{n+1} z^n$$

$$= \sum_{n=0}^{\infty}(-1)^{n+1}\frac{d}{dz}z^n = \sum_{n=1}^{\infty}(-1)^{n+1} nz^{n-1}$$

$$= \sum_{n=1}^{\infty}(-1)^{n+1} nz^{n-1} = \sum_{n=0}^{\infty}(-1)^n (n+1)z^n .$$

Using the ratio test, we will have

$$\lim_{n\to\infty}\left|\frac{(-1)^{n+1}(n+2)z^{n+1}}{(-1)^n(n+1)z^n}\right|=|z|<1. \qquad (5.9.7)$$

Since for finite values of z we have $z=e^{-\frac{x}{\beta_2}}>0$, it follows from (5.9.7) that $0<z<1$.

Thus, the series $\displaystyle\sum_{n=0}^{\infty}(-1)^n(n+1)z^n$ converges for $0<z<1$.

Also let $u=z^{\frac{\beta_2}{\beta_1}}$. Then,

$$\frac{1}{\left(1+z^{\frac{\beta_2}{\beta_1}}\right)}=\frac{1}{(1+u)}=\sum_{n=0}^{\infty}(-1)^n u^n=\sum_{n=0}^{\infty}(-1)^n\left(z^{\frac{\beta_2}{\beta_1}}\right)^n=\sum_{n=0}^{\infty}(-1)^n z^{\frac{\beta_2}{\beta_1}n}.$$

However,

$$\left|z^{\frac{\beta_2}{\beta_1}}\right|<1\Rightarrow|z|^{\frac{\beta_2}{\beta_1}}<1\Rightarrow\frac{\beta_2}{\beta_1}\ln|z|<0\Rightarrow\ln|z|<0\Rightarrow0<z<1.$$

Therefore, for $0<z<1$ we have

$$\frac{1}{\left(1+e^{-\frac{x}{\beta_2}}\right)^2}\frac{1}{\left(1+e^{-\frac{x}{\beta_1}}\right)}=\sum_{n=0}^{\infty}(-1)^n(n+1)z^n\sum_{m=0}^{\infty}(-1)^m z^{\frac{\beta_2}{\beta_1}m}$$

$$=\sum_{m=0}^{\infty}(-1)^m z^{\frac{\beta_2}{\beta_1}m}-2z\sum_{m=0}^{\infty}(-1)^m z^{\frac{\beta_2}{\beta_1}m}+3z^2\sum_{m=0}^{\infty}(-1)^m z^{\frac{\beta_2}{\beta_1}m}+\cdots$$

$$=\sum_{m=0}^{\infty}(-1)^m z^{\frac{\beta_2}{\beta_1}m}-2\sum_{m=0}^{\infty}(-1)^m z^{1+\frac{\beta_2}{\beta_1}m}+3\sum_{m=0}^{\infty}(-1)^m z^{2+\frac{\beta_2}{\beta_1}m}+\cdots$$

Now let $\theta=\frac{\beta_2}{\beta_1}$. Then, from (5.9.2) we will have:

$$R(\theta) = \int_0^1 \left[\sum_{m=0}^{\infty} (-1)^m z^{\theta m} - 2\sum_{m=0}^{\infty} (-1)^m z^{1+\theta m} + 3\sum_{m=0}^{\infty} (-1)^m z^{2+\theta m} + \cdots \right] dz$$

$$= \sum_{m=0}^{\infty} \frac{(-1)^m z^{1+\theta m}}{1+m\theta} \bigg|_0^1 - 2\sum_{m=0}^{\infty} \frac{(-1)^m z^{2+\theta m}}{2+m\theta} \bigg|_0^1 + 3\sum_{m=0}^{\infty} \frac{(-1)^m z^{3+\theta m}}{3+m\theta} \bigg|_0^1 + \cdots$$

$$= \sum_{m=0}^{\infty} \frac{(-1)^m}{1+m\theta} - 2\sum_{m=0}^{\infty} \frac{(-1)^m}{2+m\theta} + 3\sum_{m=0}^{\infty} \frac{(-1)^m}{3+m\theta} + \cdots$$

$$= \sum_{n=0}^{\infty} (-1)^n (n+1) \sum_{m=0}^{\infty} \frac{(-1)^m}{(n+1)+m\theta} = \sum_{n=0}^{\infty} \sum_{m=0}^{\infty} \frac{(-1)^{n+m}}{1+\dfrac{m}{n+1}\theta}.$$

Hence,

$$R(\beta_1, \beta_2) = \sum_{n=0}^{\infty} \sum_{m=0}^{\infty} \frac{(-1)^{n+m}}{1+\dfrac{m}{n+1}\dfrac{\beta_2}{\beta_1}} \qquad (5.9.8)$$

and this proves Theorem 5.9.1.

Theorem 5.9.2. Computation of the MLE of R in case of the Logistic Distribution

The maximum likelihood estimate (*MLE*) for R in case of logistic distribution for X and Y is

$$\hat{R}\left(\hat{\beta}_1, \hat{\beta}_2\right) = \sum_{k=0}^{\infty} \sum_{l=0}^{\infty} \frac{(-1)^{k+l}}{1+\dfrac{l}{k+1}\dfrac{\hat{\beta}_2}{\hat{\beta}_1}}. \qquad (5.9.9)$$

Proof

To compute the *MLE* of R when X and Y are independent logistic random variables, we will obtain the *MLE* of each β_1 and β_2 (parameters of the one-parameter logistic distributions). Thus, suppose $Y_1, \cdots, Y_M$ and $X_1, \cdots, X_N$ are random samples from one-parameter-logistic distributions with parameters β_1 and β_2, respectively. We let $y_1, \cdots, y_m$ and $x_1, \cdots, x_n$ denote values of these random samples, respectively. Hence, the joint density function for each sample is the product of their marginal density functions, i.e.,

$$h_Y\left(y_1,\cdots,y_M;\beta_1\right) = \frac{e^{-\frac{y_1}{\beta_1}}}{\beta_1\left(1+e^{-\frac{y_1}{\beta_1}}\right)^2}\cdots\frac{e^{-\frac{y_M}{\beta_1}}}{\beta_1\left(1+e^{-\frac{y_M}{\beta_1}}\right)^2} = \frac{e^{-\frac{1}{\beta_1}\sum_{m=1}^{M}y_m}}{\beta_1^M\prod_{m=1}^{M}\left(1+e^{-\frac{y_m}{\beta_1}}\right)^2}. \quad (5.9.10)$$

$$f_X\left(x_1,\cdots,x_N;\beta_2\right) = \frac{e^{-\frac{x_1}{\beta_2}}}{\beta_2\left(1+e^{-\frac{x_1}{\beta_2}}\right)^2}\cdots\frac{e^{-\frac{x_N}{\beta_2}}}{\beta_2\left(1+e^{-\frac{x_N}{\beta_2}}\right)^2} = \frac{e^{-\frac{1}{\beta_2}\sum_{n=1}^{N}x_i}}{\beta_2^N\prod_{n=1}^{N}\left(1+e^{-\frac{x_n}{\beta_2}}\right)^2}.$$

$$(5.9.11)$$

Since the random variables Y and X are independent, their joint distribution is the product of their density functions. On the other hand the *MLE*'s of their parameters can be obtained by taking derivative of the natural logarithm of each with respect to their respective parameter being equal to zero. Therefore, from (5.9.9) and (5.9.10) we will have the followings:

$$\ln h_Y\left(y_1,\cdots,y_M;\beta_1\right) = \ln\left(e^{-\frac{1}{\beta_1}\sum_{m=1}^{M}y_m}\right) - \ln\left[\beta_1^M\prod_{m=1}^{M}\left(1+e^{-\frac{y_m}{\beta_1}}\right)^2\right]$$

$$= -\frac{1}{\beta_1}\sum_{m=1}^{M}y_m - M\ln\beta_1 - 2\sum_{m=1}^{M}\ln\left(1+e^{-\frac{y_m}{\beta_1}}\right). \quad (5.9.12)$$

Similarly

$$\ln f_X\left(x_1,\cdots,x_N;\beta_2\right) = -\frac{1}{\beta_2}\sum_{n=1}^{N}x_n - N\ln\beta_2 - 2\sum_{n=1}^{N}\ln\left(1+e^{-\frac{x_n}{\beta_2}}\right). \quad (5.9.13)$$

Hence,

$$\frac{\partial\ln h_Y}{\partial\beta_1} = \frac{1}{\beta_1^2}\sum_{m=1}^{M}y_m - \frac{M}{\beta_1} + \frac{2}{\beta_1^2}\sum_{m=1}^{M}\frac{y_m e^{-\frac{y_m}{\beta_1}}}{1+e^{-\frac{y_m}{\beta_1}}} = 0 \quad (5.9.14)$$

and

$$\frac{\partial\ln f_X}{\partial\beta_2} = \frac{1}{\beta_2^2}\sum_{n=1}^{N}x_n - \frac{N}{\beta_2} + \frac{2}{\beta_2^2}\sum_{n=1}^{N}\frac{x_n e^{-\frac{x_n}{\beta_2}}}{1+e^{-\frac{x_n}{\beta_2}}} = 0. \quad (5.9.15)$$

Now, we focus on the right hand side of (5.9.14), simply:

$$\frac{1}{\beta_1^2}\sum_{m=1}^{M} y_m - \frac{M}{\beta_1} + \frac{2}{\beta_1^2}\sum_{m=1}^{M} \frac{y_m e^{-\frac{y_m}{\beta_1}}}{1+e^{-\frac{y_m}{\beta_1}}} = 0. \tag{5.9.16}$$

Multiplying both side of (5.9.16) by $\dfrac{\beta_1^2}{M}$, we obtain:

$$\beta_1 = \bar{y} + \frac{2}{M}\sum_{m=1}^{M} \frac{y_m e^{-\frac{y_m}{\beta_1}}}{1+e^{-\frac{y_m}{\beta_1}}}$$

or

$$\beta_1 = \bar{y} + \frac{2}{M}\sum_{m=1}^{M} \frac{y_m}{1+e^{\frac{y_m}{\beta_1}}}, \tag{5.9.17}$$

where $\bar{y} = \dfrac{1}{M}\displaystyle\sum_{m=1}^{M} y_m$.

Equation (5.9.17) is a nonlinear equation in β_1. To solve it, we use iterative approximation. Hence, β_1 can be found as the solution of the equation

$$\beta_1 = f(\beta_1) , \tag{5.9.18}$$

where

$$f(\beta_1) = \bar{y} + \frac{2}{M}\sum_{m=1}^{M} \frac{y_m e^{-\frac{y_m}{\beta_1}}}{1+e^{-\frac{y_m}{\beta_1}}} .$$

Then,

$$f(\beta_{1(k)}) = \beta_{1(k+1)}, \tag{5.9.19}$$

where $\beta_{1(k)}$ means the kth iteration of $\hat{\beta}_1$. The iterating process may stop as $|\beta_{1(k+1)} - \beta_{1(k)}| < \varepsilon_1$, happens, where ε_1 is a small positive number. Hence,

$$\hat{\beta}_1 = \overline{y} + \frac{2}{M} \sum_{m=1}^{M} \frac{y_m}{1 + e^{\frac{y_m}{\hat{\beta}_1}}} \quad . \tag{5.9.20}$$

Similarly, we will have $g\left(\beta_{2(l)}\right) = \beta_{2(l+1)}$, where $\beta_{2(l)}$ means the lth iteration of $\hat{\beta}_2$. Thus,

$$\hat{\beta}_2 = \overline{x} + \frac{2}{N} \sum_{n=1}^{N} \frac{x_n}{1 + e^{\frac{x_n}{\hat{\beta}_2}}} \quad , \tag{5.9.21}$$

where $\overline{x} = \frac{1}{N} \sum_{n=1}^{N} x_n$.

The iterating process in this case may stop as $\left|\beta_{2(l)} - \beta_{2(l+1)}\right| < \varepsilon_2$ happens, where ε_2 is a small positive number. Having found $\hat{\beta}_1$ and $\hat{\beta}_2$, we will have

$$\hat{R}\left(\hat{\beta}_1, \hat{\beta}_2\right) = \sum_{k=0}^{\infty} \sum_{l=0}^{\infty} \frac{(-1)^{k+l}}{1 + \frac{l}{k+1} \frac{\hat{\beta}_2}{\hat{\beta}_1}}, \tag{5.9.22}$$

and this prove the Theorem 5.9.2.

The logistic distribution assumes that the scale parameter, β, is positive. However, if β approaches infinity, the *cdf* will be equal to 1/2, a degenerated distribution that is not desirable for us. Thus, we assume that both β_1 and β_2 are positive finite numbers.

Now we will consider infinite sample sizes for the *MLE* of β_1 and β_2, i.e., $M \to \infty$ and $N \to \infty$.

Theorem 5.9.3. *Asymptotic Behavior*

(1) For $M = \infty$, as $m \to M$ if y_m approaches either zero or infinity, then

$$\hat{\beta}_1 \to \overline{Y} \, . \tag{5.9.23}$$

(2) For $N = \infty$, as $n \to N$ if x_n approaches either zero or infinity, then

$$\hat{\beta}_2 \to \overline{X} \, . \tag{5.9.24}$$

(3) Under conditions stated in (1) and (2),

$$\widehat{R} = \sum_{k=0}^{\infty} \sum_{l=0}^{\infty} \frac{(-1)^{k+l}}{1 + \dfrac{l}{k+1} \dfrac{\overline{X}}{\overline{Y}}}. \tag{5.9.25}$$

Proof

We will focus on $\widehat{\beta}_1$ in (5.9.20). Similar results will be true for $\widehat{\beta}_2$ in (5.9.21). As we need to choose a random sample, no matter how large M is, it will be a finite number. Thus, from (5.9.20), $\lim\limits_{M\to\infty} \widehat{\beta}_1 = \overline{Y}$ if $\sum\limits_{m=1}^{\infty} \dfrac{y_m}{1 + e^{\frac{y_m}{\beta_1}}}$ is finite, i.e., if it is a convergent series. To be so, we need to show that

$$\lim_{m\to\infty} \left(\frac{y_m}{1 + e^{\frac{y_m}{\beta_1}}} \right) = 0. \tag{5.9.26}$$

Now, $\widehat{\beta}_1$ is finite because β_1 is. On the other hand, no matter what the value of m is, the value of y_m, although finite, is not under our control since it is chosen randomly. Therefore, the limit in (5.9.26) is zero either if y_m is closed to zero and $1 + e^{\frac{y_m}{\beta_1}}$ is not, or $1 + e^{\frac{y_m}{\beta_1}}$ is extremely larger than y_m to make the ratio zero. If y_m is closed to zero, then $1 + e^{\frac{y_m}{\beta_1}} > 0$ since $\widehat{\beta}_1 > 0$, and thus, the ratio is zero. On the other hand, if y_m is extremely large, then $1 + e^{\frac{y_m}{\beta_1}}$ will approach infinity. Thus, the ratio will approach zero. This completes part (1) of Theorem 5.9.1.

To prove part (2), we reason as we did for part (1). From (1) and (2), part (3) follows as a consequence. However, we have to note that (5.9.20) is a double alternating series. Since denominator increases as k and l, terms are decreasing. Thus, the series is convergent and $\widehat{R}$ is finite.

5.10. RELIABILITY COMPUTATION USING EXTREME-VALUE DISTRIBUTIONS

In this section we assume that both X and Y are independent and each has an unknown-one-parameter extreme-value type 1 (or double exponential or Gumbel-type) distributions.

Without loss of any generality, we assume that the location parameter is zero for each X and Y. Hence, the *cdf* and *pdf* for Y and X are, respectively, as follows:

$$F_Y(y;\theta_1) = e^{-e^{-y/\theta_1}}, \quad f(y;\theta_1) = \frac{1}{\theta_1} e^{-y/\theta_1} e^{-e^{-y/\theta_1}}, \quad -\infty < y < \infty, \quad \theta_1 > 0. \qquad (5.10.1)$$

$$H_X(x;\theta_2) = e^{-e^{-x/\theta_2}}, \quad h(x;\theta_2) = \frac{1}{\theta_2} e^{-x/\theta_2} e^{-e^{-x/\theta_2}}, \quad -\infty < x < \infty, \quad \theta_2 > 0. \qquad (5.10.2)$$

Theorem 5.10.1.

If

$$k\frac{\theta_2}{\theta_1} < 1, \quad k = 0,1,2,\cdots,$$

then,

$$R(\theta_1,\theta_2) = \sum_{k=0}^{\infty} \frac{(-1)^k}{k!} \Gamma\left(k\frac{\theta_2}{\theta_1}+1\right).$$

Proof

We assume that X and Y are independent random variables. Then, it is trivial that

$$R = P(Y < X) = \int_{-\infty}^{\infty} \int_0^x \frac{1}{\theta_1} e^{-\frac{y}{\theta_1}} e^{-e^{-\frac{y}{\theta_1}}} \frac{1}{\theta_2} e^{-\frac{x}{\theta_2}} e^{-e^{-\frac{x}{\theta_2}}} \, dy dx. \qquad (5.10.3)$$

To simplify (5.10.3), let $u = e^{-\frac{y}{\theta_1}}$. Then, $du = -\frac{1}{\theta_1} e^{-\frac{y}{\theta_1}} dy$, and

$dy = -\theta_1 e^{\frac{y}{\theta_1}} du = -\theta_1 \frac{1}{u} du$. Similarly, let $v = e^{-\frac{x}{\theta_2}}$. Then, $dv = -\frac{1}{\theta_2} e^{-\frac{x}{\theta_2}} dx$, and

$dx = -\theta_2 e^{\frac{x}{\theta_2}} dv = -\theta_2 \frac{1}{v} dv$.

Note that $v = e^{-\frac{x}{\theta_2}}$ implies that $\ln v = -\frac{x}{\theta_2}$ or $\frac{x}{\theta_2} = \ln\frac{1}{v}$. Also, when $x = -\infty$,

$v = \infty$; when $x = \infty$, $v = 0$; and when $y = -\infty$, $u = \infty$; when $y = x$,

$u = e^{-\frac{x}{\theta_1}} = e^{-\frac{\theta_2 x}{\theta_2 \theta_1}} = e^{-\frac{\theta_2 x}{\theta_1 \theta_2}} = e^{-\alpha\frac{x}{\theta_2}} = e^{-\alpha\ln\frac{1}{v}} = e^{\alpha\ln v} = e^{\ln v^\alpha} = v^\alpha$, where $\alpha = \frac{\theta_2}{\theta_1} > 0$. Thus,

$$R = P(Y < X) = -\int_0^\infty (-1) \int_{v^\alpha}^\infty \frac{1}{\theta_1} u e^{-u} \frac{1}{\theta_2} v e^{-v} \left(-\theta_1 \frac{1}{u} du\right)\left(-\theta_2 \frac{1}{v} dv\right)$$

$$= \int_0^\infty \int_{v^\alpha}^\infty e^{-u} e^{-v} du dv = \int_0^\infty -e^{-u}\Big)_{v^\alpha}^\infty e^{-v} dv = \int_0^\infty \left(-\left(0 - e^{-v^\alpha}\right)\right) e^{-v} dv$$

or

$$R(\alpha) = \int_0^\infty e^{-v^\alpha} e^{-v} dv . \tag{5.10.4}$$

Applying the series expansion of the exponential function, i.e., $e^{-u} = \sum_{k=0}^\infty \frac{(-1)^k u^k}{k!}$, on

the first part of the integrand in (5.10.4), we obtain

$$R(\alpha) = \int_0^\infty e^{-v} \left(\sum_{k=1}^\infty \frac{(-1)^k (v^\alpha)^i}{k!}\right) dv = \int_0^\infty e^{-v} \left(\sum_{k=1}^\infty \frac{(-1)^k v^{k\alpha}}{k!}\right) dv$$

$$= \int_0^\infty e^{-v} \left(1 - \frac{v^\alpha}{1!} + \frac{v^{2\alpha}}{2!} - \frac{v^{3\alpha}}{3!} + \frac{v^{4\alpha}}{4!} - \cdots + \frac{(-1)^k v^{i\alpha}}{k!} - \cdots\right) dv$$

$$= \int_0^\infty \left(\frac{e^{-v}}{0!} - \frac{v^\alpha e^{-v}}{1!} + \frac{v^{2\alpha}}{2!} e^{-v} - \frac{v^{3\alpha}}{3!} e^{-v} + \frac{v^{4\alpha}}{4!} e^{-v} - \cdots + \frac{(-1)^k v^{i\alpha}}{k!} e^{-v} - \cdots\right) dv \tag{5.10.5}$$

$$= \int_0^\infty e^{-v} dv - \int_0^\infty v^\alpha e^{-v} dv + \frac{1}{2}\int_0^\infty v^{2\alpha} e^{-v} dv - \frac{1}{3!}\int_0^\infty v^{3\alpha} e^{-v} dv + \cdots - \frac{1}{k!}\int_0^\infty v^{k\alpha} e^{-v} dv + \cdots .$$

Now, for $k = 0, 1, 2, \cdots$, let $k\alpha = \tau - 1$, i.e., $\tau = k\alpha + 1 = k\frac{\theta_2}{\theta_1} + 1$. $\alpha = \frac{\theta_2}{\theta_1} > 0$ implies

that $\tau > 1$ and

$$\int_0^\infty v^{k\alpha} e^{-v} dv = \int_0^\infty v^{\tau-1} e^{-v} dv = \Gamma(\tau) = \Gamma(k\alpha + 1) = \Gamma\left(k\frac{\theta_2}{\theta_1} + 1\right), \quad k = 0, 1, 2, \cdots . \tag{5.10.6}$$

Hence, using (5.10.6), (5.10.5) can be rewritten as

$$R(\theta_1, \theta_2) = \sum_{k=0}^{\infty} \frac{(-1)^k}{k!} \Gamma\left(k\frac{\theta_2}{\theta_1}+1\right), \tag{5.10.7}$$

as stated in the statement of the Theorem 5.10.3.

Now, since $0 \le R(\theta_1, \theta_2) \le 1$, the infinite series in (5.10.7) must converge. From the Euler formula for the gamma function we have

$$\Gamma(z) = \frac{1}{z} \Pi_{n=1}^{\infty} \frac{(n+1)^z}{n^{z-1}(n+z)}, \ |z| < 1. \tag{5.10.8}$$

Thus, letting

$$z = k\frac{\theta_2}{\theta_1} \tag{5.10.9}$$

in (5.10.7) and based on the difference equation satisfied by the gamma function, namely,

$$\Gamma(z+1) = z\Gamma(z), \tag{5.10.10}$$

we have

$$R(\theta_1, \theta_2) = \sum_{k=0}^{\infty} \frac{(-1)^k}{k!} \left[\Pi_{n=1}^{\infty} \frac{(n+1)^{k\frac{\theta_2}{\theta_1}}}{n^{k\frac{\theta_2}{\theta_1}-1}\left(n+k\frac{\theta_2}{\theta_1}\right)} \right]. \tag{5.10.11}$$

This is an alternative form for R stated in the Theorem 5.10.3. To complete the proof of Theorem 5.10.1, we note that from condition in (5.10.8) and substitution (5.10.10), for the series in (5.10.7) to converge, it suffices to choose θ_1 and θ_2 such that

$$k\frac{\theta_2}{\theta_1} < 1, \ k = 0, 1, 2, \cdots. \tag{5.10.12}$$

Theorem 5.10.4. Computation of the MLE of R in case of Extreme –Value Distribution

For $k\frac{\hat{\theta}_2}{\hat{\theta}_1} < 1$, $k = 0, 1, 2, \cdots$, we have:

$$\hat{R}\left(\hat{\theta}_1, \hat{\theta}_2\right) = \sum_{k=0}^{\infty} \frac{(-1)^k}{k!} \Gamma\left(k\frac{\hat{\theta}_2}{\hat{\theta}_1} + 1\right).$$

Proof

To compute the *MLE* of *R*, we will find the *MLE* of both θ_1 and θ_2. Based on a random sample for each *Y* and *X* of sizes *M* and *N*, respectively, we will have the following *MLE* estimators for θ_1 and θ_2, respectively,

$$\hat{\theta}_1 = \overline{Y} - \frac{\sum_{m=1}^{M} y_m e^{-y_m/\hat{\theta}_1}}{\sum_{m=1}^{M} e^{-y_m/\hat{\theta}_1}} \quad \text{and} \quad \hat{\theta}_2 = \overline{X} - \frac{\sum_{n=1}^{N} x_n e^{-x_n/\hat{\theta}_2}}{\sum_{n=1}^{N} e^{-x_n/\hat{\theta}_2}}. \tag{5.10.13}$$

Since the likelihood equations (5.10.13) do not admit explicit solution, the estimators $\hat{\theta}_1$ and $\hat{\theta}_2$ can be found numerically using methods such as iterative.

It is clear that both θ_1 and θ_2 being positive imply that both $\hat{\theta}_1$ and $\hat{\theta}_2$ have to be positive and, therefore, from (5.10.13) we must have $\dfrac{\sum_{m=1}^{M} y_m e^{-y_m/\hat{\theta}_1}}{\sum_{m=1}^{M} e^{-y_m/\hat{\theta}_1}} < \overline{Y}$ and

$\dfrac{\sum_{n=1}^{N} x_n e^{-x_n/\hat{\theta}_2}}{\sum_{n=1}^{N} e^{-x_n/\hat{\theta}_2}} < \overline{X}$.

Thus, from (5.10.7) and (5.10.13) we will have

$$\hat{R}\left(\hat{\theta}_1, \hat{\theta}_2\right) = \sum_{k=0}^{\infty} \frac{(-1)^k}{k!} \Gamma\left(k\frac{\hat{\theta}_2}{\hat{\theta}_1} + 1\right). \tag{5.10.14}$$

Similar condition to (5.10.12) holds for (5.10.14) as well. However, care should be taken since $\hat{\theta}_1$ and $\hat{\theta}_2$ are calculated rather than being chosen. Thus, if "infinity" is taken as a finite number, say 100, then, it suffices to choose θ_1 and θ_2 such that $\dfrac{\theta_2}{\theta_1} < \dfrac{1}{100}$ for (5.10.11) to make sense and (5.10.12) to give desirable value. This completes proof of Theorem 5.10.4.

Note that the bias and *MSE* of the estimation may be computed using (5.10.11) and (5.10.14).

Remark

In case $\hat{\theta}_1$ and $\hat{\theta}_2$ are, respectively, large compared with y_m and x_n, the right hand sides of each expression in (5.10.12), respectively, takes the form

$$\hat{\theta}_1 = \bar{Y}\left(1 - \frac{S_y^2}{\bar{Y}\hat{\theta}_1}\right) \text{ and } \hat{\theta}_2 = \bar{X}\left(1 - \frac{S_x^2}{\bar{X}\hat{\theta}_2}\right), \tag{5.10.15}$$

where $\bar{Y}$ and $\bar{X}$ are the sample means and S_y^2 and S_x^2 are sample variances for Y and X, respectively. Thus, for this case, from (5.10.14) and (5.10.15) we will have,

$$\hat{R}\left(\hat{\theta}_1, \hat{\theta}_2\right) = \sum_{k=0}^{\infty} \frac{(-1)^k}{k!} \Gamma\left(k \frac{\bar{X} - \dfrac{S_X^2}{\hat{\theta}_2}}{\bar{Y} - \dfrac{S_Y^2}{\hat{\theta}_1}} + 1\right). \tag{5.10.16}$$

5.11. SIMULATION FOR COMPUTING RELIABILITY USING LOGISTIC AND EXTREME-VALUE DISTRIBUTIONS

This is an example for both sections 14 and 15. We choose 7 different values of each β_1 and β_2 in pairs, i.e., (10, 01), (03, 1.5), (3.3, 03), (05, 05), (06, 07), (07, 14), (08, 20) such that the ratio β_2/β_1 are 0.10, 0.50, 0.9091, 1.0, 1.1667, 2.0, and 2.5. Computationally, we have chosen $\infty \equiv 1000$, and 2000 in infinite series. To solve the functional equations for β_1 and β_2 by iteration method, we use 1000 iterations. For the iteration processes, we, begin with the initial values of β_1 and β_2 as the means of the logistic random samples found in Step 6 below, denoted by $\bar{Y}$ and $\bar{X}$.

To find MLE's for β_1 and β_2, we solve the functional equations using iteration method. To assure the difference of consecutive terms approaching zero, we have chosen 1000 iterations that cause the difference less than $\varepsilon = 10^{-6}$.

We choose two random sample vectors, Y, X of small sizes (m, n), i.e., $Y = (Y_1, Y_2, \cdots Y_m)$ and $X = (X_1, X_2, \cdots X_n)$, such the ratio of m/n takes value 0.1, 0.5, 0.75, 0.9, 1, 1,1, 1.11, 1.33, 1.5, 1.9, 2, 2.5 and 3.

To choose random samples from logistic distribution we take the following steps (First for Y and similarly for X):

Step 1.   Choose a random sample of size, say M, from the uniform distribution (between 0 and 1), say $\{p_1, p_2, \cdots p_M\}$.

Step 2.   Set each point of the random sample found in Step 1 equal to the logistic distribution. If p is such a point, then write

$$p = \frac{1}{1 + e^{-\frac{y}{\beta_1}}}.$$ (5.11.1)

Step 3.   Solve (5.11.1) for y, i.e.,

$$p + pe^{-\frac{y}{\beta_1}} = 1, \quad pe^{-\frac{y}{\beta_1}} = 1 - p, \quad e^{-\frac{y}{\beta_1}} = \frac{1-p}{p},$$

$$-\frac{y}{\beta_1} = \ln\left(\frac{1-p}{p}\right), \quad y = -\beta_1 \ln\left(\frac{1-p}{p}\right),$$

and, thus

$$y = \beta_1 \ln\left(\frac{p}{1-p}\right).$$ (5.11.2)

Step 4.   The set of y's found in Step 3 form a random sample of size M from the logistic distribution with parameter β_1. However, some of the sample points may be negative and are not desirable.

Step 5.   Since β_1 (and β_2) must be positive, choose the maximum value from the set in Step 4.

Step 6.   Repeat Step 1 through Step 4, say L times, to generate a logistic random sample of size L.

Step 7.   Similarly, choose a random sample of size N to estimate β_2. Let q be the logistic point in Step 2 and find x, say, from (5.11.2), i.e.,

$$x = \beta_2 \ln\left(\frac{q}{1-q}\right).$$ (5.11.3)

x's found according to (5.11.3) will produce a logistic sample of size N. We follow Step 5 and Step 6 to have a logistic random sample of the same size, L, for estimating β_2.

The results of the computations are given in the tables below based on the ratio of M/N for both logistic and extreme-value distributions.

Table 5.11.1- LA: *Logistic, M/N < 1.0*

(β_1, β_2)	β_2 / β_1	$\hat{R}(\hat{\beta}_1, \hat{\beta}_2)$	Bias	MSE 1.0e-004*
\multicolumn{5}{c}{Given $M = 5$, $N = 50$ $M/N = 0.10$}
(10,01)	0.1000	0.7471	0.0001	0.0277
(03,1.5)	0.5000	0.7437	- 0.0004	0.0093
(3.3,03)	0.9091	0.7504	0.0008	0.0768
(05,05)	1.0000	0.7497	- 0.0009	0.0417
(06,07)	1.1667	0.7532	0.0008	0.1281
(07,14)	2.0000	0.7561	- 0.0013	0.0297
(08,20)	2.5000	0.7579	- 0.0007	0.0113

Table 5.11.2- LB: *Logistic, M/N < 1.0*

(β_1, β_2)	β_2 / β_1	$\hat{R}(\hat{\beta}_1, \hat{\beta}_2)$	Bias	MSE 1.0e-005*
\multicolumn{5}{c}{Given $M = 10$, $N = 20$ $M/N = 0.50$}
(10,01)	0.1000	0.7475	0.0004	0.0554
(03,1.5)	0.5000	0.7446	0.0005	0.1681
(3.3,03)	0.9091	0.7481	- 0.0015	0.6907
(05,05)	1.0000	0.7504	- 0.0003	0.4575
(06,07)	1.1667	0.7527	0.0004	0.1979
(07,14)	2.0000	0.7568	- 0.0005	0.1538
(08,20)	2.5000	0.7583	- 0.0003	0.1294

Table 5.11.3-EA: *Extreme-Value, M/N < 1.0*

(θ_1, θ_2)	θ_2 / θ_1	$\hat{R}(\hat{\theta}_1, \hat{\theta}_2)$	Bias	MSE 1.0e-005*
\multicolumn{5}{c}{Given $M = 5$, $N = 50$ $M/N = 0.10$}
(100,01)	0.0100	0.3703	- 0.2633	0.0791
(210,02)	0.0095	0.3696	- 0.2640	0.1007
(330,03)	0.0091	0.3699	- 0.2636	0.0464
(450,04)	0.0089	0.3700	- 0.2634	0.0623
(650,05)	0.0077	0.3701	- 0.2632	0.0255
(850,06)	0.0071	0.3695	- 0.2637	0.0982
(1000,7)	0.0070	0.3694	- 0.2638	0.0114

Table 5.11.4-EB: *Extreme-Value, M/N < 1.0*

			Given $M = 10$, $N = 20$ $M/N = 0.50$		
(θ_1, θ_2)	θ_2 / θ_1	$\hat{R}(\hat{\theta}_1, \hat{\theta}_2)$	Bias	MSE 1.0e-005*
(100,01)	0.0100	0.3700	- 0.2637	0.0303
(210,02)	0.0095	0.3703	- 0.2633	0.1756
(330,03)	0.0091	0.3695	- 0.2640	0.0443
(450,04)	0.0089	0.3702	- 0.2633	0.1329
(650,05)	0.0077	0.3696	- 0.2636	0.0399
(850,06)	0.0071	0.3692	- 0.2640	0.0138
(1000,7)	0.0070	0.3695	- 0.2637	0.0249

Table 5.11.5-LA: *Logistic, M/N < 1.0*

			Given $M = 15$, $N = 20$ $M/N = 0.75$		
(β_1, β_2)	β_2 / β_1	$\hat{R}(\hat{\beta}_1, \hat{\beta}_2)$	Bias 1.0e-003*	MSE 1.0e-005*
(10,01)	0.1	0.7466	- 0.4558	0.0346
(03,1.5)	0.5	0.7442	0.0490	0.0880
(3.3,03)	0.9091	0.7498	0.2838	0.4911
(05,05)	1.0	0.7513	0.6571	0.2924
(06,07)	1.1667	0.7516	- 0.7823	0.2803
(07,14)	2.0	0.7572	- 0.1470	0.1030
(08,20)	2.5	0.7587	0.0968	0.0287

Table 5.11.6-LB: *Logistic, M/N < 1.0*

			Given $M = 09$, $N = 10$ $M/N = 0.90$		
(β_1, β_2)	β_2 / β_1	$\hat{R}(\hat{\beta}_1, \hat{\beta}_2)$	Bias 1.0e-003*	MSE 1.0e-005*
(10,01)	0.1	0.7472	0.1790	0.2224
(03,1.5)	0.5	0.7439	- 0.2237	0.0802
(3.3,03)	0.9091	0.7495	- 0.0031	0.3674
(05,05)	1.0	0.7504	- 0.2459	0.3032
(06,07)	1.1667	0.7523	- 0.0476	0.1478
(07,14)	2.0	0.7571	- 0.2645	0.3440
(08,20)	2.5	0.7580	- 0.6000	0.1463

Table 5.11.7-EA: *Extreme-Value, M/N < 1.0*

(θ_1, θ_2)	θ_2 / θ_1	$\hat{R}(\hat{\theta}_1, \hat{\theta}_2)$	Bias	MSE 1.0e-005*
Given M = 15, N = 20 *M/N = 0.75*				
(100,01)	0.0100	0.3700	- 0.2636	0.0214
(210,02)	0.0095	0.3700	- 0.2636	0.0728
(330,03)	0.0091	0.3698	- 0.2637	0.0276
(450,04)	0.0089	0.3700	- 0.2635	0.1436
(650,05)	0.0077	0.3691	- 0.2641	0.0242
(850,06)	0.0071	0.3692	- 0.2640	0.0099
(1000,7)	0.0070	0.3695	- 0.2637	0.0359

Table 5.11.8-EB: *Extreme-Value, M/N < 1.0*

(θ_1, θ_2)	θ_2 / θ_1	$\hat{R}(\hat{\theta}_1, \hat{\theta}_2)$	Bias	MSE 1.0e-006*
Given M = 09, N = 10 *M/N = 0.90*				
(100,01)	0.0100	0.3700	- 0.2636	0.6181
(210,02)	0.0095	0.3699	- 0.2637	0.3951
(330,03)	0.0091	0.3700	- 0.2635	0.4026
(450,04)	0.0089	0.3697	- 0.2637	0.3487
(650,05)	0.0077	0.3696	- 0.2636	0.4046
(850,06)	0.0071	0.3694	- 0.2638	0.2906
(1000,7)	0.0070	0.3693	- 0.2639	0.6049

Table 5.11.9-LA: *Logistic, M/N = 1.0*

(β_1, β_2)	β_2 / β_1	$\hat{R}(\hat{\beta}_1, \hat{\beta}_2)$	Bias 1.0e-003*	MSE 1.0e-004*
Given M = 10, N = 10 *M/N = 1.0*				
(10,01)	0.1000	0.7473	0.2941	0.0073
(03,1.5)	0.5000	0.7447	0.6135	0.0319
(3.3,03)	0.9091	0.7503	0.7558	0.0620
(05,05)	1.0000	0.7497	- 0.8780	0.1255
(06,07)	1.1667	0.7518	- 0.5223	0.0385
(07,14)	2.0000	0.7576	0.2833	0.0087
(08,20)	2.5000	0.7583	- 0.2726	0.0136

Table 5.11.10-LB: *Logistic, M/N = 1.0*

		Given $M = 30, N = 30$ $M/N = 1.0$		
(β_1, β_2)	β_2 / β_1	$\hat{R}(\hat{\beta}_1, \hat{\beta}_2)$	Bias 1.0e-003*	MSE 1.0e-005*
(10,01)	0.1000	0.7470	- 0.0862	0.0596
(03,1.5)	0.5000	0.7443	0.1410	0.0477
(3.3,03)	0.9091	0.7493	- 0.2520	0.1633
(05,05)	1.0000	0.7511	0.4641	0.3147
(06,07)	1.1667	0.7525	0.0907	0.1735
(07,14)	2.0000	0.7575	0.1462	0.0818
(08,20)	2.5000	0.7580	- 0.5291	0.0843

Table 5.11.11-EA: *Extreme-Value, M/N = 1.0*

		Given $M = 10, N = 10$ $M/N = 1.0$		
(θ_1, θ_2)	θ_2 / θ_1	$\hat{R}(\hat{\theta}_1, \hat{\theta}_2)$	Bias	MSE 1.0e-005*
(100,01)	0.0100	0.3699	- 0.2638	0.0843
(210,02)	0.0095	0.3699	- 0.2637	0.0542
(330,03)	0.0091	0.3705	- 0.2630	0.2047
(450,04)	0.0089	0.3702	- 0.2633	0.0775
(650,05)	0.0077	0.3695	- 0.2638	0.0299
(850,06)	0.0071	0.3699	- 0.2633	0.1463
(1000,7)	0.0070	0.3693	- 0.2639	0.0354

Table 5.11.12-EB: *Extreme-Value, M/N = 1.0*

		Given $M = 30, N = 30$ $M/N = 1.0$		
(θ_1, θ_2)	θ_2 / θ_1	$\hat{R}(\hat{\theta}_1, \hat{\theta}_2)$	Bias	MSE 1.0e-006*
(100,01)	0.0100	0.3697	- 0.2639	0.4026
(210,02)	0.0095	0.3699	- 0.2637	0.7615
(330,03)	0.0091	0.3698	- 0.2637	0.4983
(450,04)	0.0089	0.3697	- 0.2637	0.4661
(650,05)	0.0077	0.3698	- 0.2635	0.2500
(850,06)	0.0071	0.3693	- 0.2639	0.0925
(1000,7)	0.0070	0.3693	- 0.2639	0.4617

Table 5.11.13-LA: *Logistic, M/N > 1.0*

(β_1, β_2)	β_2 / β_1	$\hat{R}(\hat{\beta}_1, \hat{\beta}_2)$	Bias	MSE 1.0e-004*
		*Given M = 10, N = 09 M/N = 1.11*		
(10,01)	0.1	0.7474	0.0003	0.0193
(03,1.5)	0.5	0.7441	- 0.0000	0.0109
(3.3,03)	0.9091	0.7502	0.0006	0.0657
(05,05)	1.0	0.7502	- 0.0004	0.1067
(06,07)	1.1667	0.7520	- 0.0004	0.0710
(07,14)	2.0	0.7563	- 0.0011	0.0244
(08,20)	2.5	0.7587	0.0001	0.0078

Table 5.11.14-LB: *Logistic, M/N > 1.0*

(β_1, β_2)	β_2 / β_1	$\hat{R}(\hat{\beta}_1, \hat{\beta}_2)$	Bias	MSE 1.0e-005*
		*Given M = 20, N = 15 M/N = 1.33*		
(10,01)	0.1	0.7471	0.0270	0.1153
(03,1.5)	0.5	0.7442	0.1018	0.0694
(3.3,03)	0.9091	0.7491	- 0.4303	0.2002
(05,05)	1.0	0.7506	- 0.0443	0.1751
(06,07)	1.1667	0.7522	- 0.1433	0.3111
(07,14)	2.0	0.7569	- 0.4218	0.1845
(08,20)	2.5	0.7581	- 0.4450	0.0333

Table 5.11.15-EA: *Extreme-Value, M/N > 1.0*

(θ_1, θ_2)	θ_2 / θ_1	$\hat{R}(\hat{\theta}_1, \hat{\theta}_2)$	Bias	MSE 1.0e-005*
		*Given M = 10, N = 09 M/N = 1.11*		
(100,01)	0.0100	0.3703	- 0.2634	0.1317
(210,02)	0.0095	0.3700	- 0.2636	0.1295
(330,03)	0.0091	0.3698	- 0.2637	0.0315
(450,04)	0.0089	0.3699	- 0.2636	0.0245
(650,05)	0.0077	0.3696	- 0.2637	0.0505
(850,06)	0.0071	0.3693	- 0.2639	0.0086
(1000,7)	0.0070	0.3693	- 0.2639	0.0322

Table 5.11.16-EB: *Extreme-Value, M/N > 1.0*

		Given M = 20, *N* = 15 *M/N* = 1.33		
(θ_1, θ_2)	θ_2 / θ_1	$\hat{R}(\hat{\theta}_1, \hat{\theta}_2)$	Bias	*MSE* 1.0e-005*
(100,01)	0.0100	0.3702	- 0.2634	0.0829
(210,02)	0.0095	0.3702	- 0.2633	0.0758
(330,03)	0.0091	0.3698	- 0.2637	0.0310
(450,04)	0.0089	0.3704	- 0.2630	0.1954
(650,05)	0.0077	0.3697	- 0.2636	0.1032
(850,06)	0.0071	0.3694	- 0.2638	0.0187
(1000,7)	0.0070	0.3697	- 0.2635	0.0448

Table 5.11.17-L: *Logistic, M/N > 1.0*

		Given M = 20, *N* = 10 *M/N* = 2.0		
(β_1, β_2)	β_2 / β_1	$\hat{R}(\hat{\beta}_1, \hat{\beta}_2)$	Bias	*MSE* 1.0e-005*
(10,01)	0.1	0.7469	- 0.0001	0.0872
(03,1.5)	0.5	0.7441	0.0000	0.0449
(3.3,03)	0.9091	0.7503	0.0007	0.4179
(05,05)	1.0	0.7508	0.0002	0.3734
(06,07)	1.1667	0.7522	- 0.0002	0.4297
(07,14)	2.0	0.7562	- 0.0012	0.2922
(08,20)	2.5	0.7585	- 0.0000	0.0505

Table 5.11.18-E: *Extreme-Value, M/N > 1.0*

		Given M = 20, *N* = 10 *M/N* = 2.0		
(θ_1, θ_2)	θ_2 / θ_1	$\hat{R}(\hat{\theta}_1, \hat{\theta}_2)$	Bias	*MSE* 1.0e-005*
(100,01)	0.0100	0.3698	- 0.2638	0.0670
(210,02)	0.0095	0.3702	- 0.2634	0.0377
(330,03)	0.0091	0.3706	- 0.2629	0.1088
(450,04)	0.0089	0.3699	- 0.2635	0.0699
(650,05)	0.0077	0.3693	- 0.2640	0.0292
(850,06)	0.0071	0.3696	- 0.2636	0.0608
(1000,7)	0.0070	0.3693	- 0.2639	0.0236

Comments

Industry is data rich and yet very often decisions are made based on a few, between 2 and 6, observations. It is the ratio between the distributions' scale parameters that mostly determines the value of R. The difference in values of the sample sizes for the stress and the strength, for a given ratio, does not affect the estimated value of R as long as the ratio of those sample sizes is kept fixed. Thus, this study shows that it is imperative to have the parameter of the stress much smaller than that of the strength despite that they share the same distribution.

It has been observed, through the computation, that when, in case of the logistic distribution, the ratio β_2 / β_1 (or θ_2 / θ_1 in case of extreme-value distribution) increases from 0.1 to 2.5, the estimated value of R does not fluctuate that much, barely 0.01. This small change has appeared and was not affected by the ratio of the sample sizes M and N, whether for small or large values of the sample sizes. Moreover, the same situation comes up again when M and N are, practically speaking, very large regardless of their ratio. The same argument goes on for the Bias and the MSE analysis for estimating R, as it is given in (5.9.5) and (5.10.7). This indicates that the R estimator is robust with regard to those variables, namely the ratio of the parameters or that of the sample sizes. The whole scenario applies on the extremer –value distribution as well.

Theoretically, the random variable under the logistic or extreme-value distribution can assume any value on the real. For estimating the positive parameters in each case, some restrictions on the sample values had to be applied. For simulation purposes, the maximum in the generated sample was taken in both cases, and estimating the parameters was done on those new samples. Through simulation, it has come clear that the ratio between the parameters has to taken very small. As mentioned above that ratio did not affect the calculation on R. In addition to that, in practice, the investigator needs to check on the data where doe is fit. Since checking is needed in order to decide, which distribution will be suitable to use when the realization values are large. This is required since the numerical calculations had shown the R value is small for the extreme value when compared with that for the logistic.

Tables 5.11.1-LA and 5.11.2-LB show the calculations for the logistic distribution, while Tables 5.11.3-EA and 5.11.4-EB show the calculations for the extreme-value when the ratio of M/N < 1. Tables 5.11.5-LA and 5.11.6-LB show the calculations for the logistic distribution, while Tables 5.11.7-EA and 5.11.8-EB show the calculations for the extreme-value when M/N = 1. Tables 5.11.9-LA, 5.11.10-LB 5.11.11-EA, 5.11.12-EB display the results when M/N > 1 for the logistic and extreme-value distributions respectively. Tables 5.11.13-LA, 5.11.14-LB, 5.11.15-EA and 5.11.16-EB tabulate the results in the asymptotic case, i.e., when M and N tend to infinity. For the extreme-value distribution, the results of the computations are given in Tables 5.11.17-L and 5.11.18-E.

It is not a good practical assumption for evaluating the reliability of a system, that as it is completely perceived, the larger the sample the better. Although increasing the sample size will have a great effect on the reduction of variation, it has been shown through small samples that the ratio is more important that the sample size for calculation of reliability of a system. Attention should be paid to the values of the parameters that control that distribution for the stress and strength data.

It is completely clear that the insight knowledge for the underlined distribution and the ratio of its parameters will affect the calculated reliability of that item. The *MSE* analysis shows that the assumed estimated values of *R* under the considered distribution are very close to the actual values when the parameters assume their known values.

5.12. BASICS CONCEPTS ABOUT UP-AND-DOWN DESIGN

In this section, we will discuss basic information on a statistical design from a family of *up-and-down rules* for the sequential allocation of dose levels to subjects in a dose-response study. Although we base the discussion on a medical case, it can identically be applied to engineering cases by changing the vocabularies.

Considering a sequence of experiments in which a *treatment* (*stimulus* or *stress*) is applied at a finite number of levels (or *dosages*), and the cumulative number of responses at each level is given. Then, an experimental design is completely specified by the rule that allocates treatments to subjects. *Up-and-Down* (*U-D*) designs are rules that specify the treatment level for the next trial, merely, to be one level higher, one level lower, or the same as the treatment level selected for the current trial. The U-D rules have been studied in the past four decades. The transition probabilities associated with changing treatment levels are made to depend on the outcome of the experiment that follows the treatment.

Note that the treatment level selection follows a *random walk* that will be discussed in chapter seven. Since the experimental outcome is a random variable, the random walk is said to be in a *random environment*. The design shows that the performance of a random walk rule can be improved by using the estimate of the probability of outcome calculated from all previous responses.

We start our discussion by defining some terms we will be using in the section. By a *treatment* (*stimulus* or *stress*) we mean applying a test (giving a drug such as polychlorinated biphenyls (PCB)) to a subject (such as a rat) and wait for a time period (say, 24 hours) for the subject's *response*. The response (say, *toxicity*) is a *reaction* (say, toxic) or a *no-reaction* (say, *non-toxic*). In other words, a treatment is defined as a Bernoulli random variable, say *Y*. Outcomes of such a trial is a reaction or no-reaction (toxic or no-toxic).

Let us consider a finite sequence of experiments, each experiment with only two possible outcomes for a response, namely, toxic and non-toxic. Hence, a treatment will be represented by a Bernoulli random variable, say *Y*, (for the nth subject it will be represented by $Y(n)$). That is, we let a finite number, *N*, (*N* = 30, for instance) of subjects (such as rats, or beams) undergoes treatments one at a time. For a sample of size *N*, treatment on the nth subject will be denoted by $Y(n), n = 0, 1, 2, \cdots, N$. We let $Y(n) = 1$ or $Y(n) = 0$, depending upon reaction or no-reaction, respectively, of the treatment response. That is, $Y(n)$ keeps track of the outcomes but not of the dose assignments. [In the binomial distribution sense, a toxic (reaction) connotes failure and non-toxic (no-reaction) connotes success, in the binomial distribution sense.] We also use a finite number of treatment levels, say *K*. The rule for *allocating treatment levels* (*concentration levels, dose level*, or *dosage*) to subjects produces a sequence of treatment levels, say $X(n)$, $n = 0, 1, 2, \cdots, N$, with $x_j(n)$ being the value of *X* at the jth level and at the nth trial, $0 \le j \le K$. Generally, we denote by

$\Omega_j = \{x_j, j = 1, 2, \cdots, K\}$ the set of dose levels. In practice the dose levels are often chosen equally spaced on the logarithmic scale. $n = 0$ is the initial trial. We assume the initial treatment level, $X(0)$, as a fixed number rather than a random variable. $X(n), n \geq 1$, is the nth treatment level after the initial trial. Arbitrary outcomes and treatments may be denoted by X and Y, respectively, without explicitly mention of their position in the sequence of trials. Let us denote the probability of a response at the level x by $Q(x)$. Hence,

$$Q(x) \equiv P\{Y = 1 | X = x\} \text{ of a toxic and } P(x) \equiv 1 - Q(x) \text{ of a non-toxic at } x. \qquad (5.12.1)$$

we assume that $Q(x_j)$ is an increasing function of the dose, but constant over all trials.

The *primary objective*, in the case of a sequence of treatments discussed, is to estimate the unknown dose level μ that has a probability of response equal to a fixed value Ψ, $0 \leq \Psi < 1$. The dose level μ is called *the target quantile*, and Ψ is called the *target probability of response*. Usually, dose levels with high probability of response are not desirable. Thus, we may concentrate on target probabilities of response that are below the median, i.e., $0 < \Psi < 0.5$. By symmetry, analogous results can be obtained for $0.5 < \Psi < 1$.

There are different rules for allocating treatment levels to subjects. Below, will discus two very closed and relatively recent methods of choosing the next trial's treatment level; they are *Biased Coin Design*. These rules were developed by Durham and Flournoy in 1993 and 1994.

Allocation rules should satisfy an ethical requirement of Phase I clinical trials by assigning patients to dose levels in the neighborhood of the target dose and, hence, reducing unnecessary assignments to the most toxic doses. *Biased Coin Designs* discussed below comply with this requirement.

Definition 5.12.1. Biased Coin Design, BCD I [Flournoy and Durham (1994)]

Suppose that the nth subject has been given dose level x_j. (Note that we are dropping the n from $x_j(n)$ since there won't be a confusion.) Assume that spacing between dose levels used is one unit, i.e., a finite dose levels x_j could have values such as $1, 2, \cdots, K$. Let Ψ be the probability of a response at a target quantile, μ, as described in Definition 5.12.1. Fix Ψ between 0 and 1. Assume that a treatment has been given at level x_j, $0 \leq j \leq K$. Assignment of a dose level to the next subject according to the following instruction is called *biased coin design, BCD I*:

(i) Toss a biased coin, with the probability of getting a head (H) and a tail (T) equal to b and $1 - b$, respectively, $0 \leq b \leq 1/2$, where

$$b \equiv \frac{\Psi}{1 + \Psi}, \quad 1 - b \equiv \frac{1}{1 + \Psi}, \quad 0 \leq b \leq 1/2. \qquad (5.12.2)$$

(ii) If heads (H) is observed, treat the next subject at level x_{j+1}, i.e.,

$$X(n+1) = x_{j+1} \big| \big(X(n) = x_j \text{ and } H \big).$$
(5.12.3)

(iii) If tails (T) is observed and the outcome of the treatment at the nth trial is a no-reaction, then treat the next subject at the current level x_j, i.e.,

$$X(n+1) = x_j \big| \big(X(n) = x_j, Y(n) = 0, \text{and } T \big).$$
(5.12.4)

(iv) If the outcome of the treatment at the nth trial is a reaction, then, move up to level x_{j+1}, i.e.,

$$X(n+1) - x_{j-1} \big| \big(X(n) = x_j \text{ and } Y(n) = 1 \big).$$
(5.12.5)

From (5.12.2) – (5.12.5), the transition probabilities (probabilities of moving from one level to another) in each trial are as follows:

$$p_{x_1, x_0} \equiv P\{X(n+1) = x_0(n+1) | X(n) = x_1(n), T, \text{and } Y(n) = 1\} = 0,$$
(5.12.6)

$$p_{x_1, x_1} \equiv P\{X(n+1) = x_1(n+1) | X(n) = x_1(n), T \text{ and } Y(n) = 0\} = 1 - b,$$

$$p_{x_1, x_2} \equiv P\{X(n+1) = x_2(n+1) | X(n) = x_1(n) \text{ and } H\} = b,$$

$$p_{x_j, x_{j-1}} \equiv P\{X(n+1) = x_{j-1}(n+1) | X(n) = x_j(n), T, \text{and } Y(n) = 1\} = (1-b)Q(x_j), \quad j = 2, 3, \cdots, N,$$

$$p_{x_j, x_j} \equiv P\{X(n+1) = x_j(n+1) | X(n) = x_j(n), T, \text{and } Y(n) = 0\} = (1-b)P(x_j), \quad j = 2, \cdots, N-1,$$

$$p_{x_j, x_{j+1}} \equiv P\{X(n+1) = x_{j+1}(n+1) | X(n) = x_j(n) \text{ and } H\} = b, \qquad j = 1, 2, \cdots, N-1,$$

$$p_{x_N, x_{N-1}} \equiv P\{X(n+1) = x_{N-1}(n+1) | X(n) = x_N(n), T, \text{and } Y(n) = 1\} = (1-b)Q(x_N),$$

$$p_{x_N, x_N} \equiv P\{X(n+1) = x_N(n+1) | X(n) = x_N(n), T, \text{and } Y(n) = 0\} = 1 - p_{N, N-1},$$

$$p_{x_N, x_N + 1} \equiv P\{X(n+1) = x_{N+1}(n+1) | X(n) = x_N(n) \text{ and } H\} = 0.$$

Definition 5.12.2. Biased Coin Design, BCD II [Durham and Flournoy (1993)]

Refer to Definition 5.12.2 with the following modification: Assignment of a dose level to the next subject according to the following instruction is called *biased coin design, BCD II*:

(i) Let

$$b \equiv \frac{\Psi}{1-\Psi}, \quad 1-b \equiv \frac{1-2\Psi}{1-\Psi}, \quad 0 \le b \le 1/2.$$
(5.12.7)

(ii) If response of the current treatment is a non-toxic and the coin toss yields heads (H), then treat the next subject with one level higher, level x_{j+1}, i.e.,

$$X(n+1) = x_{j+1}(n+1) \big| \big(X(n) = x_j(n), Y(n) = 0 \text{ and } H \big).$$ (5.12.8)

(iii) If response of the current treatment is a non-toxic and the coin toss yields tails (T), then treat the next subject with the same level x_j, i.e.,

$$X(n+1) = x_j(n+1) \big| \big(X(n) = x_j(n), Y(n) = 0 \text{ and } T \big).$$ (5.12.9)

(iv) If response of the current treatment is a toxic, then treat the next subject with one level lower, x_{j-1}, i.e.,

$$X(n+1) = x_{j-1}(n+1) \big| \big(X(n) = x_j(n) \text{ and } Y = 1 \big).$$ (5.12.10)

Thus, from (5.12.6) – (5.12.10), the transition probabilities in each trial, in this case, are as follows:

$$p_{x_1,x_0} \equiv P\{X(n+1) = x_0(n+1) | X(n) = x_1(n) \text{ and } Y(n) = 1\} = 0,$$ (5.12.11)

$$p_{x_1,x_1} \equiv P\{X(n+1) = x_1(n+1) | X(n) = x_1(n)\} = 1 - bP(x_0),$$

$$p_{x_1,x_2} \equiv P\{X(n+1) = x_2(n+1) | X(n) = x_1(n)\} = bP(x_0),$$

$$p_{x_j,x_{j-1}} \equiv P\{X(n+1) = x_{j-1}(n+1) | X(n) = x_j(n), T, \text{ and } Y(n) = 1\} = Q(x_j), \quad j = 1, 2, \cdots, N,$$

$$p_{x_j,x_j} \equiv P\{X(n+1) = x_j(n+1) | X(n) = x_j(n), T, \text{ and } Y(n) = 0\} = (1-b)P(x_j), j = 2, \cdots, N-1,$$

$$p_{x_j,x_{j+1}} \equiv P\{X(n+1) = x_{j+1}(n+1) | X(n) = x_j(n) \text{ and } H\} = bP(x_j), \quad j = 1, 2, \cdots, N-1,$$

$$p_{x_N,x_{N-1}} \equiv P\{X(n+1) = x_{N-1}(n+1) | X(n) = x_N(n), T, \text{ and } Y(n) = 1\} = Q(x_N),$$

$$p_{x_N,x_N} \equiv P\{X(n+1) = x_N(n+1) | X(n) = x_N(n), T, \text{ and } Y(n) = 0\} = P(x_N),$$

$$p_{x_N,x_N+1} \equiv P\{X(n+1) = x_{N+1}(n+1) | X(n) = x_N(n) \text{ and } H\} = 0.$$

Example 5.12.1. Procedure for Maximum Likelihood Estimator of the Target Quantile μ.

We first want to use *MLE* and conduct simulation to estimate the target quartile, μ, assuming the unknown treatment levels follow a logistic distribution with cumulative probability distribution function as in (4.11.25) that for our case is

$$Q(x_j) = \frac{e^{\frac{x_j-a}{b}}}{1+e^{\frac{x_j-a}{b}}}, \quad j = 1, 2, \cdots, K,$$ (5.12.12)

where K is the maximum dose level, and a and b are the parameters of the distribution.

Let $K = 10$. In other words, there are 10 possible dose levels available for a treatment in a dose level set, denoted by $\Omega_j = \{1, 2, \cdots, 10\}$. We choose $a = 5.5$ and $b = 1.5$ in (5.12.12) as the initial values. Although the parameters are fixed, the variable of the distribution, x, is still unknown. Remember that μ is one of the x_js that we are targeting.

In general, to estimate the target quantile, μ, that is a function of the parameters a and b, we somehow have to estimate the parameters a and b. We choose *MLE* for this purpose. So, let

$$\theta_1 = -\frac{a}{b} \text{ and } \theta_2 = \frac{1}{b}.$$

(5.12.13)

Then, (5.12.12) can be re-written as

$$Q(x_j) = \frac{e^{\theta_1 + \theta_2 x_j}}{1 + e^{\theta_1 + \theta_2 x_j}}.$$

(5.12.14)

We may simulate the maximum likelihood estimators of θ_1 and θ_2 using a standard logistic regression software package for any value of Ψ, the target probability of response. Hence, we start the simulation as follows:

Let the initial dose level (trial 0) be $x_i = 3$. We run samples of sizes $N = 10$, 25, and 50, and replicate (R) each 8000 times. The process records the initial dose level, that is, the level to be used in the first trial, as soon as simulation starts. Then, a subject goes under treatment with this dose level. Based on the allocation rules we choose, *BCD* defined in Definitions 5.12.2 and 5.12.3, dose level for the next trial is determined and will be recorded as the second replication starts.

We test values 0.10, 0.20, 0.25, 0.33, and 0.50 of Ψ. Then, we record the number of responses at level x_j (denote it by R_j) and the total number of responses at that level (denoted by S_j). Then, the proportion of toxic and no-toxic at level x_j will be $\dfrac{R_j}{S_j}$ and $1 - \dfrac{R_j}{S_j}$, respectively. Hence, the regression lines of D verses $\ln \dfrac{R_j}{S_j - R_j}$ for logistic, using proportions obtained from simulation would provide maximum likelihood estimates $\hat{\theta}_1$ and $\hat{\theta}_2$ for each response function.

Note that *MLE* exists only if there are at least two levels at which both toxic and non-toxic occur. We choose the initial values of θ_1 and θ_2 as defined in (5.12.13) with $a = 5.5$ and $b = 1.5$.

To find the *true* value of μ, replace x_i in (5.12.14) by μ, let it equal to Ψ, and solve for μ. Hence, we have

$$Q(\mu) = \frac{e^{\theta_1 + \theta_2 \mu}}{1 + e^{\theta_1 + \theta_2 \mu}} = \frac{1}{1 + e^{-(\theta_1 + \theta_2 \mu)}} = \Psi \Rightarrow \mu = Q^{-1}(\Psi),$$

$$\frac{1}{1 + e^{-(\theta_1 + \theta_2 \mu)}} = \Psi \Rightarrow 1 + e^{-(\theta_1 + \theta_2 \mu)} = \frac{1}{\Psi} \Rightarrow e^{-(\theta_1 + \theta_2 \mu)} = \frac{1 - \Psi}{\Psi},$$

$$e^{-(\theta_1 + \theta_2 \mu)} = \frac{1 - \Psi}{\Psi} \Rightarrow e^{\theta_1 + \theta_2 \mu} = \frac{\Psi}{1 - \Psi} \Rightarrow \ln \frac{\Psi}{1 - \Psi} = \theta_1 + \theta_2 \mu,$$

$$\ln \frac{\Psi}{1 - \Psi} = \theta_1 + \theta_2 \mu \Rightarrow \mu = \frac{-\theta_1}{\theta_2} + \frac{1}{\theta_2} \ln \frac{\Psi}{1 - \Psi}. \tag{5.12.15}$$

Thus, from (5.12.13) and (5.12.15) we have

$$\mu = Q^{-1}(\Psi) = a + b \ln\left(\frac{\Psi}{1 - \Psi}\right). \tag{5.12.16}$$

Therefore, having estimates $\hat{\theta}_1$ and $\hat{\theta}_2$, using (5.12.13), we will have estimates of a and b and, thus, obtain estimate of μ from (5.12.16).

The following are steps summarizing the simulation discussed above:

Step 1.   Set $R = 8000$.

Step 2.   Choose $x_j = 3$ (initial value).

Step 3.   Take $N = 10$ (then, 25 and 50).

Step 4.   Choose $\Psi = 0.10$ (then, 0.20, 0.25, 0.33, and 0.50).

Step 5.   Let $b = \dfrac{\Psi}{1 + \Psi}$.

Step 6.   Choose $Q(x_j) = \dfrac{e^{\theta_1 + \theta_2 x_j}}{1 + e^{\theta_1 + \theta_2 x_j}}$, where $\theta_1 = -\dfrac{a}{b}$ and $\theta_2 = \dfrac{1}{b}$, with $a = 5.5$ and $b = 1.5$.

Step 7.   Record x_j ($R = 1$, $x_j = 3$, continue to the end of $R = 1$. Then, go to $R = 2$ with $x_j = 3$ again) by recorder N_j.

Step 8.   Test for a toxic or non-toxic. If $Q(x_j) \leq$ random number, then there is no toxic respond, otherwise there is a toxic response.

Step 9.   Record toxic responses at level j by recorder R_j and non-toxic responses at that level by recorder S_j.

Step 10.   Toss a biased coin (with bias $= b$) by choosing a random number R_n. If $R_n \leq b$, there is a head otherwise a tail.

Step 11.   Decide for moving up, down, or stay. If you have a head move up to x_{j+1}, $x_j = \min(x_j, N)$. If there is a tail and there is a no-response then stay put at level d_k. If there is a tail and there was a toxic response, then move down to x_{j-1}, $x_j = \max(x_j, 1)$.

Step 12.   The log likelihood function is the natural logarithm of L (the likelihood function), i.e., $\ln L = \sum_{j=1}^{N} \left[R_j (\theta_1 + \theta_2 d_j) - S_j \ln \left(1 + e^{\theta_1 + \theta_2 x_j} \right) \right]$.

Step 13.   Find *MLE* of θ_1 and θ_2, call them $\hat{\theta}_1$ and $\hat{\theta}_2$. (In calculating *MLE*, use $\theta_1 = -\dfrac{a}{b}$ and $\theta_2 = \dfrac{1}{b}$ with $a = 5.5$ and $b = 1.5$ as initial values for $\hat{\theta}_1$ and $\hat{\theta}_2$.

Step 14.   Calculate $\hat{a} = -\dfrac{\hat{\theta}_1}{\hat{\theta}_2}$ and $\hat{b} = \dfrac{1}{\hat{\theta}_2}$.

Step 15.   Find $\hat{\mu} = \hat{a} + \hat{b} \ln \left(\dfrac{\Psi}{1 - \Psi} \right)$.

This completes Example 5.12.1.

5.13. UP-AND-DOWN DESIGN

In this section, we want to discuss extend the discussion of basics in the previous section. We will conclude that the new method, *up-and-down rules*, which is originally due to Narayana, is superior to *random walk designs*, exist in the literature, especially for moderate to large samples, in the sense that the dose assignments are more tightly clustered around the target dose. In fact, as the sample size gets large, the probability of assignment goes to zero for dose levels not among the two closest to the target. In addition to targeting precision, we will find that, this has the benefit of avoiding dose levels with high toxicity for large sample sizes. We, finally, find that the isotonic regression estimator is superior to other estimators for small to moderate sample sizes.

The importance of *quantile estimation* is widely accepted in area of biomedical studies. For example, in Phase I clinical trials it is of interest to estimate the dose level at which a desired probability of toxic response is obtained. Sequential designs which, on the basis of prior responses, restrict allocation in the region of desired response and thereby limit exposure to high toxicity levels are appealing.

Let us consider an experiment in which subjects are sequentially assigned to various dose levels of a drug, the response is binary (toxic/nontoxic), and the set of possible dose levels is fixed. *The goal is to target the dose level, μ, which has a prescribed probability, Ψ, $0 \leq \Psi \leq 0.5$, of toxicity.* The target, μ, is the quantile of interest. We will use notations and notions discussed in the previous section.

It is well known (Chernoff, 1979) that the locally optimal design for this problem is a one-point design at μ, as long as Ψ is in the midrange of the unit interval. Mats, Rosenberger, and Flournoy (1998) noted that for $0.18 < \Psi < 0.82$, the optimal design is a one-point design for a wide variety of response functions. Of course, it is unlikely that μ actually is one of dose levels chosen for the experiment. In this case the optimal design is concentrated on the two points closest to μ. A certain amount of targeting can be accomplished using one of up-and-down designs since such a design can be made to produce a distribution of trials with the highest concentration around μ. However, none of these are two point distributions, even in the limit. Narayana, in his unpublished dissertation, (1953) proved that, in the case $\Psi = 0.5$, after a large number of trials all subsequent observations fall on the two dose levels closest to the median (or three, if the median is exactly on a dose level x_j and $j = 2, \ldots, K-1$).

Narayana introduced the following rule to estimate μ for $\Psi = 0.5$. Let $Y(n)$, $n = 0, 1, 2, \ldots, N$, be 1 or 0 according to the outcome of the nth trial, a *reaction (toxicity)* or a *no-reaction (no toxicity)*, respectively, that is $Y(n)$ keeps track of the outcomes but not of the dose assignments. Let $X_j(n)$ and $N_j(n)$, respectively, be the number of responses (toxic) and number of trials at the dose d_j up to and including the nth trial. Thus, the ratio $X_j(n)/N_j(n)$ gives an estimate for the probability of toxicity at the dose d_j. It is assume that $X_j(n)/N_j(n) = 0$, if $N_j(n)$ is zero. Hence, assign the next subject to level d_{j-1} if $X_j(n)/N_j(n) > 0.5$ and $Y(n) = 1$; level d_{j+1} if $X_j(n)/N_j(n) < 0.5$ and $Y(n) = 0$; otherwise assign to level d_j.

He proposed an improvement to the up-and-down design of Dixon and Mood (1948) targeting $\Psi = 0.5$ under which the trials distribution converge to two points closest to μ_{50}. This is accomplished by incorporating the entire history of responses on the current dose in the decision rule.

Flournoy and Durham (1993b) provided conditions on the transition probabilities to assure that the stationary (i.e., when sample size is very large) treatment distribution will be *unimodal*. In this example, the first subject is assigned to dose level x_1. A *start-up rule*, which is, usually, a function of the target probability of toxicity, is used to conserve resources by bringing the starting point of the primary design closer to the target.

We can find *MLE* of a target quantile, μ, as in Example 5.12.1, using the two-parameter logistic model (5.12.14). To ensure the existence of the *MLEs* of a and b, the data are augmented by adding two observations, one with toxic response and the other with a non-toxic response. The two observations are distributed over the dose levels in proportion to the actual sample size at each dose level. For example, if 5, 7, and 3 subjects were assigned to the first three dose levels, we add 10/15, 14/15, and 6/15 observations at those doses with 5/15,

7/15, and 3/15 toxicities correspondingly. The *MLE* of μ, based on the augmented data, is given by

$$\hat{\mu} = \left\{ \ln \left[\Psi / (1 - \Psi) \right] - \hat{a} \right\} / \hat{b},$$ (5.13.1)

where $\hat{a}$ and $\hat{b}$ are parameters estimates.

Note that the *MLE* is not reliable even for large sample sizes when dose levels are too close together.

When the target quantile falls on one of the dose levels, say x_m, the optimal allocation will put all the subjects on x_m. In the case, where the target dose level is in between dose levels x_m and x_{m+1}, i.e., $Q(x_m) < \Psi < Q(x_{m+1})$, the asymptotic variance of $\hat{\mu}$ [given in (5.13.1)] can be calculated from the variance covariance matrix of $\hat{a}$ and $\hat{b}$, using the *delta method*. (The *delta method* is a method for deriving an approximate probability distribution for a function of an asymptotically normal statistical estimator from knowledge of the limiting variance of that estimator. More broadly, the delta method may be considered a fairly general central limit theorem.)

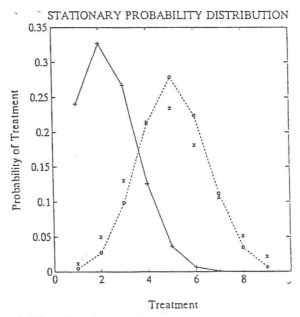

Legend: The stationary probability of receiving each treatment is given at integer value x = 1, 2, ..., 9 on the horizontal axis give the response function Q(x) is logistic 1/[1+exp(3.596-0.548x)]. These probabilities are connected by lines to improve their visibilities.

'+': target Ψ = 0.10 $\Rightarrow$ μ = 2.50 for BCD I & II

'o': target Ψ = 0.33 $\Rightarrow$ μ = 6.50 for BCD I

'x': target Ψ = 0.33 $\Rightarrow$ μ = 2.50 for BCD II

Figure 5.13.1. Stationary probability of treatment.

In case some other means of estimation are used, we can compare the designs by looking at how close the dose assignments are to the target dose level, i.e., how tightly the trails tend to concentrate around μ. For each trial this is measured by the mean squared error of the dose assignments calculated as the average of squared deviations of the dose assignment from the target dose level μ. In this regards, we should note that that the *isotonic regression estimator* is more efficient than *MLE* for small and moderate sample sizes, i.e., $N < 40$. (The concept of *isotonic regression estimator* is quite simple. Estimate $Q(x_j)$ by $\hat{Q}(x_j) = X_j(n)/S_j(n)$ for $j = 1, \cdots, K$. Let m be such that $\hat{Q}(x_m) \leq \Psi \leq \hat{Q}(x_{m+1})$. Then, estimate μ by logistic interpolating between x_m and x_{m+1}.)

EXERCISES

5.1.   Basics Statistical Concepts

5.1.1.   When desired information is available for all objects in the population, we have what is called a

A. population   B. sample   C. census   D. stem-and-leaf display

5.1.2.   Which of the following statements regarding histograms are correct?

A.   A unimodal histogram is one that rises to a single peak and then declines, whereas a bimodal histogram is one that has two different peaks.
B.   A unimodal histogram is positively skewed if the right or upper tail is stretched out compared to the left or lower tail.
C.   A unimodal histogram is negatively skewed if the left or lower tail is stretched out compared to the right or upper tail.
D.   A histogram is symmetric if the left half is a mirror image of the right half.
E.   All of the above

5.1.3.   A histogram that is negatively skewed is

A.   Skewed to the right
B.   Skewed to the left
C.   A histogram with exactly two peaks
D.   Symmetric

5.1.4.   Is the sample median always the center point of the distribution? If yes, explain why; if not, give an example.

5.1.5.   Is the sample median always between the sample mean and sample mode? If yes, explain why; if not, give an example.

5.1.6.   True or false: is the sample mean the most frequent data point?

5.1.7.   Find a sample size for which the median will always equal to one of the sample points.

5.1.8.   Which of the following is not a measure of location?

A. The mean   B. The median C. The mode   D. The variance

5.1.9.   Which of the following is not a measure of variability?

A. Variance   B. Standard deviation
C. Mean absolute deviation   D. Median

5.1.10. Given that $n = 10$, $\sum_{i=1}^{n} X_i = 25$, and $\sum_{i=1}^{n} X_i^2 = 512$, then the sample standard deviation is

A. 7.012   B. 6.704   C. 7.067
D. None of the above answers is correct

5.1.11. The sample space for the experiment in which a six-sided die is thrown twice consists of how many outcomes?

A. 6   B. 36   C. 12   D. 24

5.1.12. Is it possible for the standard deviation be 0? If so, give an example; if not, explain why not.

5.1.13. A sample of 100 married women was taken. Each was asked how many children she had. Responses were tallied as follows:

Number of Children	0	1	2	3	4	5 or more
Number of Women	25	22	29	13	7	4

a.   Find the average number of children in the sample.
b.   Find the sample standard deviation of the number of children.
c.   Find the sample median.
d.   What is the first quartile of the number of children?
e.   What proportion of the women had more than the average number of children?
f.   For what proportion of women was the number of children more than one standard deviation greater than the mean?
g.   For what proportion of the women was the number of children within one standard deviation of the mean?

5.1.14. Let X be a continuous random variable with pdf $f(x) = 1 - |x - 1|$, $0 \le x \le 2$.

(a) Find cdf of X and graph it.
(b) Find the first decile.
(c) Find 20th and 95th percentiles and show them on the graph of cdf of X.
(d) Find the IQR.

5.1.15. Consider the following data:

93	67	52	84	63
75	84	81	51	87
34	71	79	57	83
58	82	71	81	69
68	85	45	42	65

Construct a stem-and-leaf display.

5.1.16. There are 10 employees in a certain department of a company. Mean, median, and standard deviation of salaries are $70,000, $55,000, and $60,000, respectively. The highest salary is $100,000; however, it is incorrectly reported as $1,000,000. With this error, what are the mean, median and standard deviation?

5.1.17. In a post office, the weights of packages are converted from pound to kilogram (1 kg $\approx$ 2.2 lb). How this change does affect the mean of the packages mailed?

5.1.18. Suppose two catalysts A and B control the yield of an undesirable side product. A particular reaction was run several using A and B catalysts. Percentages yields for 24 runs of A and 20 runs of B are as follows:

Catalyst A					Catalyst B			
4.4	4.4	2.6	3.8		3.4	1.1	2.9	5.5
4.9	4.6	5.2	4.7		6.4	5.0	5.8	2.5
4.1	2.6	6.7	4.1		3.7	3.8	3.1	1.6
3.6	2.9	2.6	4.0		3.5	5.9	6.7	5.2
4.3	3.9	4.8	4.5		6.3	2.6	4.3	3.8
4.4	3.1	5.7	4.5					

a. Construct a histogram for the yields of each catalyst.
b. Construct a stem-and-leaf display for each catalyst.
c. Find mean and variance for each catalyst.
d. Use MINITAB to do a, b, and c, and compare your results.
e. Construct a simple group frequency table for each catalyst.
f. Construct a group frequency table for each catalyst with 3 class intervals.

5.2.   Estimation

5.2.1.   Suppose a random variable X represents a large population consisting of three measurements 0, 3, and 12 with the following distribution:

X	0	3	12
P_X	1/3	1/3	1/3

Write every possible sample of size 3.

5.2.2.   Assuming equiprobable property for every possible sample of size 3 in Exercise 5.2.1, what is the property of each?

5.2.3.   For Exercise 5.2.1, find the sample mean, $\overline{X}$ for each possible sample.

5.2.4.   For Exercise 5.2.1, find the sample median, M_m, for each possible sample.

5.2.5.   For Exercise 5.2.1, find the sampling distribution of the sample mean, $\overline{X}$.

5.2.6.   For Exercise 5.2.1, find the sampling distribution of the sample median, M_m.

5.2.7.   For Exercise 5.2.1, show that $\overline{X}$ is an unbiased estimator of the population parameter μ, in this case.

5.2.8.   For Exercise 5.2.1, show that M_m is a biased estimator of the population parameter μ, in this case.

5.2.9.   Let $X_1, X_2, \cdots, X_n$ represent a random sample with its n independent observed values $x_1, x_2, \cdots, x_n$ from a normal random variable with mean μ and variance σ^2. Find the *MLE* of $\theta = (\mu, \sigma^2)$.

5.2.10. Suppose $X_1, X_2, \cdots, X_n$ is a random sample of size n with *pdf* as indicated. Find the *MLE* for the parameter λ.

(a) $f(x; \lambda) = \lambda x^{\lambda - 1}$, $0 \leq x \leq 1, \ \lambda > 0$.

(b) $f(x; \lambda) = \dfrac{1}{\lambda} e^{-x/\lambda}$, $x \geq 0, \ \lambda > 0$.

(c) $f(x; \lambda) = \dfrac{1}{\lambda}$, $0 \leq x \leq \lambda, \ \lambda > 0$.

5.2.11. Find the maximum likelihood estimate of the parameter p in the *negative binomial distribution* with probability mass function as

$$f(x;p) = P(X = x) = \binom{x-1}{k-1} p^k (1-p)^{x-k}, \qquad x = k, k+1, \cdots, \quad 0 < p \le 1.$$

5.2.12. Find the maximum likelihood estimate of the parameter μ in the *Cauchy distribution* with probability density function as

$$f(x;\mu) = \frac{\mu}{\pi(x^2 + \mu^2)}, \qquad \mu > 0.$$

5.2.13. Find $z_{\alpha/2}$ for each of the following values of α:

(a) 0.10   (b) 0.05   (c) 0.01   (d) 0.20

5.2.14. A random sample of size 70 from a normally distributed population results a sample mean and standard deviation of 26.2 and 4.1, respectively. Find a 95% confidence interval for the population mean μ.

5.2.15. A random sample of 50 shoppers at a automotive part store showed that they spent an average of \$20.5 with variance of \$39.2. Find a 95% confidence interval for the average amount spent by a shopper at the store.

5.2.16. The time it take for a manufacturer to assemble an electronic instrument is a normal random variable with mean 1.2 hours and a variance 0.04 hours. To reduce the assembly time, the manufacturer implements a new procedure. Then, a random sample of size 35 is taken and it shows the mean assembly time as 0.9 hour. Assuming the variance remains unchanged, form a 95% confidence interval for the mean assembly time under the new procedure.

5.2.17. Find the value of $t_{n-1,\alpha/2}$ to construct a two sided confidence interval of the given level with the indicated sample size:

(a) 90% level, $n = 9$.   (b) 95% level, $n = 2$.

5.2.18. Find the value of $t_{n-1,\alpha}$ to construct a two sided confidence interval of the given level with the indicated sample size:

(a) 95% level, $n = 5$.
(b) 99% level, $n = 29$.

5.2.19. A heat transfer model from a cylinder immersed in a liquid predicts that the heat transfer coefficient for the cylinder becomes constant at very low heat rates of the fluid. Result of a sample of 10 measurements in W/m^2 K, are:

12.0 14.1 13.7 13.1 13.1 14.4 11.9 14.1 12.2 11.8

Find a 95% confidence interval for the heat transfer coefficient.

5.2.20. During a drought, a water utility of a particular town sampled 100 residential water bills and fond that 73 of residents had reduced their water consumption compared with last year.

(a) Find a 95% confidence interval for the proportion of residences, who reduced their water consumption.
(b) Find a 99% confidence interval for the proportion of residences, who reduced their water consumption.
(c) Find the sample size necessary for a 95% confidence interval to declare the proportion to within ±0.05.
(d) Find the sample size necessary for a 99% confidence interval to declare the proportion to within ±0.05.

5.2.21. Leakage from underground fuel tanks has been a source of water pollution. A sample of 87 gasoline station was taken and found 13 with at least one leaking underground tank.

(a) Find a 95% confidence interval for the proportion of gasoline stations with at least one leaking underground tank.
(b) Find a 90% confidence interval for the proportion of gasoline stations with at least one leaking underground tank.
(c) How many stations are necessary to be sampled for a 95% confidence interval to declare the proportion to within ±0.04?
(d) How many stations are necessary to be sampled for a 90% confidence interval to declare the proportion to within ±0.04?

5.3.   Hypothesis Testing

5.3.1. H_0 is rejected at the 5% level. (a), (b), (c), true or false?

(a) The result is satisfactorily significant at the 10% level.
(b) The result is satisfactorily significant at the 5% level.
(c) The result is satisfactorily significant at the 1% level.

5.3.2.   If p-value is 0.01, which is the best conclusion?

a. H_0 is definitely false.
b. H_0 is definitely true.
c.   There is a 50% chance that H_0 is true.
d. H_0 is possible, and H_1 is false.

e.   Both H_0 and H_1 are plausible.

5.3.3.   In a process that manufacturers tungsten-coat silicon wafers, the target resistance for a wafer is 85 mΩ. A random sample of 50 wafers was taken and resulted mean resistance of 0.5 mΩ, and standard deviation of 0.5 mΩ. Let μ be the mean resistance of the wafers manufactured by this process. A quality engineer is testing H_0: $\mu = 85$ versus H_1: $\mu \neq 85$. Find the p-value.

5.3.4.   Fill volume, in oz, of a large number of beverage cans was measured, with sample mean, $\overline{X} = 11.98$ and $\sigma_{\overline{X}} = 0.02$. Use this information to find the p-value for testing H_0: $\mu = 12.0$ versus H_1: $\mu \neq 12.0$.

5.3.5.   Suppose that at a company it is known that over the passed a few years, employees sick days averaged 5.4 days per year. To reduce this number, the company introduces telecommuting (allowing employees to work at home on their computers). After implementing the new policy, the Human Resource Department chose a random sample of 80 employees at the end of the year and found an average of 4.5 sick days with standard deviation of 2.7 days. Let μ be the mean sick days of all employees of the company. Find the p-value for testing hypothesis H_0: $\mu \geq 5.4$ versus H_1: $\mu < 5.4$.

5.3.6.   A pollster conducts a survey of a random sample of voters in a community to estimate the proportion who support a measure on public health insurance. Let P be the proportion of the population who support the measure. The pollster takes a sample of 200 voters and tests H_0: $P = 0.50$ versus H_1: $P \neq 0.50$ at 5% level. What is the power of the test if the true value of P is 0.55?

5.3.7.   For Exercise 5.3.6, how many voters must be sampled so that the power will be 0.8?

5.4.   Inference on Two Small-Sample Means

5.4.1.   In a manufacturing plant, an experiment was performed by making 5 batches of a chemical using the standard method A and the same for a new method B. The result was as follows:

Standard					Revised				
77.0	69.1	71.5	73.0	73.7	78.5	79.6	76.1	76.0	78.5

Find a 90% confidence interval for the difference in the mean product between the two means.

5.4.2.   In a comparison of the effectiveness of online learning with the traditional in-classroom instruction, 12 students enrolled in a business course online and 14 enrolled in a traditional in-classroom. The final exam scores were as follows:

Classroom	80	77	74	64	71	80	68	85	83	59	55	75	81	81
Online	64	66	74	69	75	72	77	83	77	91	85	88		

Does the mean score differ between the two types of course?

5.4.3.   Let $\sigma_1^2 = \sigma_2^2 = \sigma^2$.Find the pooled estimator of σ^2 for each of the following:

(a) $s_1^2 = 120, \; s_2^2 = 100, n_1 = n_2 = 25.$

(b) $s_1^2 = 12, \; s_2^2 = 20, n_1 = 20, n_2 = 10.$

(c) $s_1^2 = 3000, \; s_2^2 = 2500, n_1 = 16, n_2 = 17.$

5.4.4.   In a state, there are areas that cable television companies are prohibited from monopolizing an area and there are those areas that companies are not demanded to allow competition.   To check basic cable charges these two categories of company impose, a group is taking the responsibility to conduct an analysis. The group randomly samples basic rates for six cable television companies that have no competition and six for companies that face competition (but not from each other). The results are as follows:

No Competition	$14.44	26.88	22.78	25.78	23.34	27.52
Competition	$18.95	23.74	17.25	20.14	18.98	20.14

(a) What are the appropriate null and alternative hypotheses for this test?
(b) For the test in (a), conduct the test use $\alpha = 0.05$.

5.5.   Analysis of Variance (ANOVA)

5.5.1.   Two independent samples of sizes $n_1 = 6$ and $n_2 = 5$ are taken from two normally distributed populations 1 and 2, respectively.

Sample 1	3.1	4.4	1.2	1.7	0.7	3.4
Sample 2	2.3	1.4	3.7	8.9	5.5	

(a) Do these data provide sufficient evidence to indicate a difference between the population variances?
(b) Find and interpret the approximate observed significance level for the test.

5.5.2.   A manufacturer produces its products with two different procedures that, on the average, both approximately products the same number of products per day. A consultant for the manufacturer recommends that out of the two procedures, the one with the smaller variance in the output product per day stays permanently. To test the recommendation of the consultant, two independent random samples one from product of each procedure were taken and the following data were observed:

Procedure 1	
n_1	s_1^2
21 days	1432

Procedure 2	
n_2	S_2^2
21 days	3761

(a) Do these data provide sufficient evidence at $\alpha = 0.10$ to conclude that the variances in the number of units produced per day differ for the two procedures?

(b) If the answer to part (a) is affirmative, which procedure should the manufacturer choose?

(c) If the answer to part (a) is negative, what should the manufacturer do?

5.6.   Linear Regression

5.6.1.   Consider the following data:

X	8	5	4	6	2	5	3
Y	1	3	6	3	7	2	5

(a) Sketch a scattered plot for the data.

(b) Predict the relationship between the variables X and Y from the plot.

(c) Given the following information, calculate the least square estimates of β_0 and β_1.

X	
$\bar{X}$	SS_{XX}
4.7143	23.4286

Y	
$\bar{Y}$	SS_{XY}
3.8571	- 23.2857

(d) Plot the least square line on the scatter plot.

(e) Does the least square line in part (d) appear to fit the data well? Explain.

5.6.2.   A car dealership company is interested in modeling the relationship between the number of cars sold by the company each week and the average number of sales consultants on the floor each day during the week. The company believes that the relationship between the two items is linear. The following sample was provided by the company.   (Denote by Y, the number of cars sold by the company each week, and denote by X, the average number of sales consultants on the floor each day during the week).

Week of	Y	X
January 30	20	6
February 7	11	3
March 2	10	4
June 29	18	6
October 26	6	2

(a) Sketch a scattered plot for the data.
(b) Assume the relationship between the two variables X and Y is linear. Use the least square method to estimate the y-intercept and the slope of the line.
(c) Plot the least square line on the scatter plot.
(d) Interpret the least square estimate of s of the y-intercept and the slope of the line of means.
(e) Are the interpretations in part (d) meaningful?
(f) According to your least squares, approximately how many cars should the company should expect to cell in a week, if an average of 5 sales consultants is on the floor each day?

5.6.3.   From a graph in an article "A New Measurement Method of Diesel Engine Oil Consumption Rate" (J. Society Auto Engr., 1985, pp. 28 – 33), we read the following data

x	4	5	8	11	12	16	17	20	22	28	30	31	39
y	5	7	10	10	14	15	13	25	20	24	31	28	39

(a) Assume that x and y are related by the simple linear regression model. Using a significance level $\alpha = 0.05$, test H_0: $\beta_1 = 1$ versus $\beta_1 \neq 1$.

(b) Calculate the value of the sample correlation coefficient for this data.

5.6.4.   To study the relationship between speed (ft/sec) and stride rate (number of steps taken/sec) among female marathon runners, an investigation was carried out with a sample of 11 female runners ($n = 11$), the result of which are as follows:

$$\sum(\text{speed}) = 205.4, \sum(\text{speed})^2 = 388.08,$$
$$\sum(\text{rate}) = 35.16, \sum(\text{rate})^2 = 112.681, \text{ and}$$
$$\sum(\text{speed})(\text{rate}) = 660.130.$$

(a) Calculate the equation of the least squares line to be used to predict stride rate from speed.
(b) Calculate the equation of the least squares line to be used to predict speed from stride rate.
(c) Calculate the coefficient of determination for the regression of stride rate on speed of part (a).
(d) Calculate the coefficient of determination for the regression of stride rate on speed of part (b).
(e) How are the results in parts (c) and (d) related?

5.6.5. A recent article in *Business Week* listed "Best Small Companies". To determine the current company's sales and earnings, a random sample of 12 companies was chosen. As a result, the sales and earnings, in millions of US dollars, are summarized in the table bellow:

Company	Sales ($ millions)	Earnings ($ millions)
Papa John's International	89.2	4.9
Applied innovations	18.6	4.4
Integracare	18.2	1.3
Wall Data	71.7	8.0
Davidson Accessories	58.6	6.6
Chico's Fas	46.8	4.1
Checkmate Electronics	17.5	2.6
Royal Grip	11.9	1.7
M-Wave	19.6	3.5
Serving-N-Slide	51.2	8.2
Daig	28.6	6.0
Cobra Golf	69.2	12.8

Let sales and earning be the independent and dependent variables, respectively.

(a) Draw a scatter plot.
(b) Compute the coefficient of correlation.
(c) Compute the coefficient of determination.
(d) Interpret your findings in parts (b) and (c).
(e) Find the regression equation.
(f) For a company with $50,000,000 in sale, estimate the earnins.

5.6.6. Show that in a regression function, the responses Y_i have the same variance, i.e.,
$$Var(Y_i) = \sigma^2.$$
Hint: Show that $Var(Y_i) = Var(\beta_0 + \beta_1 x_i + \varepsilon_i) = Var(\varepsilon_i) = \sigma^2.$

5.7.   Reliability

5.7.1.   Derive relations (5.7.13), (5.7.14), (5.7.15), (5.7.16) and (5.7.17).

5.7.2.   If X is a non-negative discrete random variable, then

$$E(X) = \sum_{x=0}^{\infty} P(X > x).$$

5.7.3.   From (5.7.19), show that the *mean functioning* time or *survival* is

$$E(T) = \int_0^{\infty} f_T(t)dt = \int_0^{\infty} R_T(t)dt.$$

5.7.4.   From (5.7.20), show that (5.7.28) is equivalent to

$$r(t) = f_T(t|T > t) = \frac{-R'(t)}{R(t)}.$$

5.7.5.   Prove Theorem 5.8.1. (Hint: Use Example 5.7.5.)

5.7.6.   Suppose that a system consists of units 1 and 2 in series and function if both units function at the same time. See Figure 5.7.E1. Assume that the units function independently.

Figure 5.7.E1. System with Two units in series

(a) Suppose it is known that probability that unit 1 fails is 0.05 and of unit 2 is 0.03. Find the probability that the system functions.

(b) Suppose that probability that each of units 1 and 2 fail is p. Find the value p such that the probability that the system functions is 0.90.

5.7.7.   In Exercise 5.7.2, suppose three units are connected in series and each has probability p of failure. Find the value p such that the probability that the system functions is 0.90.

5.7.8.   In Exercise 5.7.2, suppose the units are connected in parallel and function if either unit functions.

    (a) Suppose it is known that probability that unit 1 fails is 0.08 and of unit 2 is 0.12. Find the probability that the system functions.

    (b) Suppose that probability that each of units 1 and 2 fail is p. Find the value p such that the probability that the system functions is 0.99.

5.7.9.   In Exercise 5.7.4, suppose three units are connected in parallel and each has probability p of failure. Find the value p such that the probability that the system functions is 0.99.

5.7.10. In Exercise 5.7.4, suppose each unit has a probability 0.5 of failure. What is the minimum number of units that must be connected parallel so that the probability that the system functions is at least 0.90.

5.8. Reliability Computation using Logistic and Extreme-Value Distributions

5.8.1.   Supposed $X \sim \text{Weibull}(\alpha, \beta_1)$ and $Y \sim \text{Weibull}(\alpha, \beta_2)$, as given in (4.11.24). Let U and T be independent identically distributed exponential random variables with parameters β_1 and β_2, and with pdf $h(u) = (1/\beta_1)e^{-u/\beta_1}$ and $h(t) = (1/\beta_2)e^{-t/\beta_2}$, respectively. Prove that $V = U/T$ has probability density function

$$g(v) = \begin{cases} \beta_1\beta_2(\beta_2 v + \beta_1)^{-2}, & v > 0 \\ 0, & \text{otherwise.} \end{cases}$$

5.8.2.   Show that

$$R = P(Y < X) = \cfrac{1}{1 + \cfrac{\beta_2}{\beta_1}} = \frac{1}{1 + \rho} \ ,$$

where $\rho = \dfrac{\beta_2}{\beta_1}$.

5.8.3.   Show that the value of θ_1 that minimizes MSE given in (5.8.10) is

$$\theta_1 = \frac{(R - R_0)\, E(\hat{R} - R_0)}{E(\hat{R}^2) - 2R_0\, E(\hat{R}) + R_0^2},$$

subject to $0 \le \theta_1 \le 1$.

5.8.4.   Prove Theorem 5.8.1.   (Hint: use the transformations of variables of the continuous type)

5.8.5.   Prove cases (i) and (iii) of Theorem 5.8.2.

5.9. Reliability Computation using Logistic Distributions

5.9.1.   Prove Theorem 5.9.1.

5.9.2.   For the logistic distribution, assumes that the scale parameter, β, is positive. Show that if β approaches infinity, the *cdf* will be equal to 1/2.

5.9.3.   In Theorem 5.9.3, show that parts (1) and (2) imply part (3).

5.10.   Reliability Computation using Extreme-Value Distributions

5.10.1.   Assume that X and Y are independent random variables.   In Theorem 5.10.1 show that

$$R = P(Y < X) = \int_{-\infty}^{\infty}\int_{0}^{x} \frac{1}{\theta_1} e^{-\frac{y}{\theta_1}} e^{-e^{-\frac{y}{\theta_1}}} \frac{1}{\theta_2} e^{-\frac{x}{\theta_2}} e^{-e^{-\frac{x}{\theta_2}}}\, dy\,dx \ .$$

5.10.2.   Give detail steps arriving (5.10.11).

5.10.3.   Show that for the series in (5.10.7) to converge, it suffices to choose θ_1 and θ_2 such that $k\dfrac{\theta_2}{\theta_1} < 1$, $k = 0, 1, 2, \cdots$.

5.10.4. In proof of Theorem 5.10.4, show that both θ_1 and θ_2 being positive imply that both $\hat{\theta}_1$ and $\hat{\theta}_2$ have to be positive

5.10.5. In proof of Theorem 5.10.4, drive (5.10.14) from (5.10.7) and (5.10.13).

5.11. Simulation for computing Reliability using Logistic and Extreme-Value Distributions

5.11.1. (Project). Repeat the numerical example used in section 5.11 with Weibull distribution.

5.12. Basics Concepts About Up-and-Down Design

5.12.1. (Project). Repeat Example 5.12.1 with extreme-value distribution.

5.13. Up-and-Down Design

5.13.1. (Project). Repeat the numerical example used in section 5.13 with extreme-value distribution.

Chapter 6

DIFFERENTIAL AND DIFFERENCE EQUATIONS

In recent years, there has been an increasing interest in the calculus of difference and differential-difference equations. We list three reasons for this. First, the advent of high-speed computers has led to a need for fundamental knowledge of this subject. Second, there are numerous applications of difference equations to engineering, sciences (such as physics, chemistry, biology, probability and statistics), economics and psychology. And third, the mathematical theory is of interest in itself, especially in view of the analogy of the theory to differential equations.

One main difference between difference equations and differential equations is that difference equations are equations that involve discrete changes of an unknown function, while differential equations involve instantaneous rates of changes of an unknown function. Difference equations are the discrete analogs of differential equations. They appear as mathematical models in situations where the variable takes only a discrete set of values. The theory and solutions of difference equations in many ways are parallel to the theory and solutions of differential equations.

6.1. INTRODUCTION

Definition 6.1.1.

A *difference equation* over a set S is an equation of the form

$$F(k, y_k, y_{k+1}, ..., y_{k+n}) = 0, \tag{6.1.1}$$

where F is a given function, n is some positive integer and k is in S. A *solution* of the difference equation (6.1.1) is a sequence $\{y_k\}$, which satisfies (6.1.1) for all values of k in S. A *general solution* is one, which involves exactly n arbitrary constants. A *particular solution* is a solution obtained from the general solution by assigning values to these constants.

Definition 6.1.2.

A difference equation is said to be *linear* over a set S, if it can be written in the form

$$f_0(k)y_{k+n} + f_1(k)y_{k+n-1} + \ldots + f_{n-1}(k)y_{k+1} + f_n(k)y_k = f(k), \tag{6.1.2}$$

where each of $f_0, f_1, \ldots, f_n$ and f is a function of k defined for all values of $k \in S$. If both f_0 and f_n are different from zero at each point of S, then the difference equation (6.1.2) is said to be of *order* n over S. If $f(k) = 0$, then the difference equation is called a *homogeneous difference equation*.

Example 6.1.1.

The following difference equations with order as listed are linear over S as indicated:

$8y_{k+1} - 7y_k = 0$, (order 1) over $S = \{0, 1, 2, 3, \cdots\}$.

$y_{k+2} + 5y_{k+1} + 6y_k = 2k$, (order 2) over $S = \{0, 1, 2, 3, \cdots\}$.

$y_{k+3} - 3y_{k+2} + 6y_{k+1} - 4y_k = -2k + 5$, (order 3) over $S = \{0, 1, 2, 3, \cdots\}$.

$ky_{k+2} - 7y_{k+1} + 3y_k = 0$, (order 2) over $S = \{1, 2, 3, \cdots\}$.

$y_{k+2} + 6y_k = 3k - 2$, (order 2) over $S = \{0, 1, 2, 3, \cdots\}$.

Example 6.1.2.

The following difference equations are nonlinear.

$$5y_{k+2} - y_k^3 = 0$$

$$y_k y_{k+3} - y_{k+1} = 4k^2 + 5k - 1$$

$$\frac{4}{7 + y_{k+1}^2} + y_{k+2} = k.$$

Example 6.1.3.

The sequence $\{y_k\} = \{3^k\}$, $k = 0, 1, 2, \cdots$, is a solution of the difference equation $y_{k+1} - 3y_k = 0$. This is because $y_{k+1} - 3y_k = 3^{k+1} - 3 \cdot 3^k = 3^{k+1} - 3^{k+1} = 0$ $k = 0, 1, 2, \cdots$.

Example 6.1.4.

We want to show that the function y given by

$$y_k = 1 - \frac{2}{k}, \quad k = 1, 2, 3, \cdots \tag{6.1.3}$$

is a solution of the first-order difference equation

$$(k+1)y_{k+1} + ky_k = 2k - 3, \quad k = 1, 2, 3, \cdots. \tag{6.1.4}$$

Solution

To do this, we substitute the values of y into (6.1.4) and see that an identity results for all positive integer values of k. From (6.1.3) we know that $y_{k+1} = 1 - \dfrac{2}{k+1}$. So, (6.1.4) becomes

$$(k+1)\left(1 - \frac{2}{k+1}\right) + k\left(1 - \frac{2}{k}\right) = 2k - 3.$$

Simplifying the left-hand side of this equation, we obtain

$$(k+1) - 2 + k - 2 = 2k - 3.$$

Thus, the function y does satisfy the difference equation (6.1.4).

Example 6.1.5.

We want to solve the nonlinear equation

$$y_{k+1} - y_k + ky_{k+1}y_k = 0, \quad y_1 = 2. \tag{6.1.5}$$

Solution

Dividing both sides of (6.1.5) by $y_{k+1}y_k$ we obtain

$$\frac{1}{y_{k+1}} - \frac{1}{y_k} = k \tag{6.1.6}$$

Now, letting $\dfrac{1}{y_k} = v_k$, (6.1.5) can be rewritten as

$$v_{k+1} - v_k = k, \ v_1 = \frac{1}{2} \tag{6.1.7}$$

It is easy to show that $v_k = \dfrac{k^2 - k + 1}{2}$ is a solution of (6.1.7). Therefore, the solution of

(6.1.5) is $y_k = \dfrac{2}{k^2 - k + 1}$.

Definition 6.1.4.

For the homogeneous difference equation

$$y_{k+n} + c_1 y_{k+n-1} + \cdots + c_{n-1} y_{k+1} + c_n y_k = 0, \tag{6.1.8}$$

an equation of the form

$$r^n + c_1 r^{n-1} + \cdots + c_{n-1} r + c_n = 0, \tag{6.1.9}$$

which is an algebraic equation of degree n, is called *characteristic equation* (or *auxiliary equation*) of the equation (6.1.8).

Theorem 6.1.1.

For the homogenous difference equation

$$y_{k+2} + b y_{k+1} + c y_k = 0 \tag{6.1.10}$$

with constant coefficients b and c, $c \neq 0$, let r_1 and r_2 be two roots of the characteristic equation

$$r^2 + br + c = 0. \tag{6.1.11}$$

Then, the following three cases are possible:

Case 1.

If r_1 and r_2 are two distict real roots such that $r_1 \neq r_2$, then the general solution of (6.1.10) is

$$Y_k = c_1 r_1^k + c_2 r_2^k .\qquad(6.1.12)$$

Case 2.

If r_1 and r_2 are real roots such that $r_1 = r_2$, then the general solution of (6.1.10) is

$$Y_k = (c_1 + c_2 k) r_1^k .\qquad(6.1.13)$$

Case 3.

If r_1 and r_2 are complex conjugate roots, in their polar form as $\rho(\cos\theta \pm i\sin\theta)$, then the general solution of (6.1.10) is

$$Y_k = A\rho^k \cos(k\theta + B),\qquad(6.1.14)$$

where A and B are constants.

Proof

See Goldberg (1961, p.141).

Example 6.1.6.

We want to find the general solution of the so called *Fibonacci equation*

$$y_{k+2} = y_{k+1} + y_k .\qquad(6.1.15)$$

Solution

The characteristic equation of Equation (6.1.15) is

$$r^2 - r - 1 = 0 .\qquad(6.1.16)$$

Equation (6.1.16) has two distinct real roots as $r_1 = \left(1 - \sqrt{5}\right)/2$ and $r_2 = \left(1 + \sqrt{5}\right)/2$.
Hence, the general solution of (6.1.15) is

$$Y_k = c_1 \left(\frac{1 - \sqrt{5}}{2}\right)^k + c_2 \left(\frac{1 + \sqrt{5}}{2}\right)^k .\qquad(6.1.17)$$

Example 6.1.7.

Let us find the general solution of the homogenous difference equation

$$9y_{k+2} + 6y_{k+1} + y_k = 0.$$

Solution

The characteristic equation for this difference equation is $9r^2 + 6r + 1 = 0$, which has a double real root $r_1 = r_2 = -1/3$. Hence, the general solution is

$$Y_k = (c_1 + c_2 k)\left(-\frac{1}{3}\right)^k.$$

Example 6.1.8.

Consider the difference equation

$$y_{k+2} + 4y_k = 0.$$

The characteristic equation for this difference equation is $r^2 + 4 = 0$, which has two complex conjugate solutions $r_1 = 2i$ and $r_2 = -2i$. Now, for the polar coordinate of the roots, we use the facts that

$$a + bi = \rho(\cos\theta + i\sin\theta), \quad \rho = \sqrt{a^2 + b^2}, \quad \sin\theta = b/\rho, \quad \cos\theta = a/\rho, \quad \text{and}$$
$$-\pi < \theta < \pi.$$

Hence, since $2i = 0 + 2i$ and $-2i = 0 - 2i$, we have $\rho = 2$ and $\theta = \pi/2$. Therefore, the general solution is

$$Y_k = A2^k \cos\left(\frac{k\pi}{2} + B\right).$$

6.2. LINEAR FIRST ORDER DIFFERENCE EQUATIONS

We now consider a linear first order difference equation of the form

$$y_{k+1} = ay_k + b, \quad k = 0, 1, 2, \cdots \tag{6.2.1}$$

with initial condition $y_0 = d$, where a, b, and d are constants. From Theorem 6.1.1, the characteristic equation of (6.2.1) is $r - a = 0$. Therefore, there is only one characteristic root and that is $r_1 = a$. Hence, the solution of the homogeneous equation part of (6.2.1), i.e.,

$$y_{k+1} = ay_k, \quad k = 0, 1, 2, \cdots \tag{6.2.2}$$

is

$$y_k = C_1(a)^k, \quad k = 0, 1, 2, \cdots . \tag{6.2.3}$$

For the particular solution of (6.2.1), there are two cases:

Case 1: $a = 1$

Let us try to find a particular solution of the form

$$y_k^* = Ak . \tag{6.2.4}$$

Hence, substituting (6.2.4) in (6.2.2), we have

$$\begin{aligned}
y_{k+1}^* - ay_k^* &= A(k+1) - a(Ak) \\
&= A\big[1 + (1-a)k\big].
\end{aligned} \tag{6.2.5}$$

Since $a = 1$, (6.2.5) becomes

$$y_{k+1}^* - y_k^* = A . \tag{6.2.6}$$

In order to satisfy (6.2.1), we must choose $A = b$. Therefore,

$$y_k^* = bk \tag{6.2.7}$$

is the desired particular solution. Thus, the general solution of (6.2.1) is

$$y_k = C_1 + bk . \tag{6.2.8}$$

Using the initial condition, $y_0 = d$, we can find C_1 as

$$y_0 = d = C_1 a^0 + 0 = C_1 . \tag{6.2.9}$$

Substituting C_1 from (6.2.9) in (6.2.8), our general solution with the initial condition will be

$$y_k = d + bk, \ k = 0, 1, 2, \cdots. \tag{6.2.10}$$

Case 2: $a \neq 1$

In this case, we try a particular solution of (6.2.1) of the form

$$y_k^* = A. \tag{6.2.11}$$

Substituting (6.2.11) in (6.2.1) and letting result equal to b, we have

$$y_{k+1}^* - ay_k^* = A - aA = b. \tag{6.2.12}$$

Thus, from (6.2.12), we have

$$A = \frac{b}{1-a}. \tag{6.2.13}$$

Therefore, using (6.2.13), the general solution of (6.2.1), in this case, is

$$y_k = C_1 a^k + \frac{b}{1-a}. \tag{6.2.14}$$

With initial condition, $y_0 = d$, from (6.2.14), we can find C_1 as

$$y_0 = d = C_1 a^0 + \frac{b}{1-a},$$

and, hence,

$$C_1 = d - \frac{b}{1-a}. \tag{6.2.15}$$

Thus, substituting (6.2.15) in (6.2.14), the general solution with the initial condition, in this case, is

$$y_k = \left[d - \frac{b}{1-a} \right] a^k + \frac{b}{1-a}, \ k = 0, 1, 2, \cdots. \tag{6.2.16}$$

Example 6.2.1.

Let us consider the *Malthusian* law (in biology) of population growth that assumes the rate of change of a population is proportional to the population present. There are two cases: continuous time and discrete time. In the continuous case, letting $N(t)$ denote the size of a population at time t and assuming it is a differentiable function with respect to t, we will have

$$N'(t) = pN(t),$$

where p is the constant of proportionality. However, in this example we will use the discrete case. This is more realistic for use in many applications such as biology. In such cases, the size N of the population is a step function. For instance, in some insect populations, one generation dies out before the next generation hatches. Thus, we address the following discrete population growth model. In such models, it is assumed that size of population from one generation to the next increases and the increase is proportional to the size of the former generation.

Let N_k, $k = 0, 1, 2, \cdots$, denote the size of the population of the kth generation and N_0 the initial size of the population. Then, the growth of the population from the kth generation to the $(k + 1)^{st}$ generation, $N_{k+1} - N_k$, will be proportional to the current generation N_k, i.e.,

$$N_{k+1} - N_k = \alpha N_k, \tag{6.2.17}$$

where α is the constant of proportionality. Equation (6.2.17) can be written as

$$N_{k+1} - (1 + \alpha)N_k = 0, \tag{6.2.18}$$

which is a linear homogeneous difference equation of order 1. It is ease to verify from (6.2.16) that the solution is

$$N_k = N_0(1 + \alpha)^k. \tag{6.2.19}$$

Note that equation (6.2.17) shows that the size of the current population depends upon the size of the previous one. In some cases, it may be more realistic to model the $(k+2)^{nd}$ generation population to depend upon the previous two populations, i.e.,

$$N_{k+2} + pN_{k+1} + qN_k = 0. \tag{6.2.20}$$

In such a case to solve equation (6.2.20) the Theorem 6.1.1 may be applied.

6.3. BEHAVIOR OF SOLUTIONS OF THE FIRST ORDER DIFFERENCE EQUATIONS

Given the general equation of a first order difference equation

$$y_{k+1} = ay_k + b,$$

with initial condition $y_0 = d$ and certain conditions, we want to analyze the behaviors of the solution sequence and the variations of the graphs. The graph of the solutions of difference equation is a set of pairs (k, y_k), $k = 0, 1, 2, \cdots$. The resulting points are joined by straight-line segments in order aid in seeing the way in which these values vary as k increases. Let

$$L = \frac{b}{1-a}.$$

Case 1: $a = 1$, $b = 0$

For this case, we have a homogeneous difference equation,

$$y_{k+1} - y_k = 0, \ k = 0, 1, 2, \cdots, \tag{6.3.1}$$

with initial condition

$$y_0 = d. \tag{6.3.2}$$

The characteristic equation corresponding to this difference equation is

$$r - 1 = 0. \tag{6.3.3}$$

From (6.3.3), we have $r = 1$. Hence, the general solution of (6.3.1) is

$$y_k = C_1. \tag{6.3.4}$$

Using (6.3.2), the constant C_1 in (6.3.4) can be found as

$$y_0 = d = C_1. \tag{6.3.5}$$

Substituting (6.3.5) in (6.3.4), the solution of (6.3.1) is

$$y_k = d. \tag{6.3.6}$$

It is trivial that graph of solution (6.3.6) is just a horizontal line that passes through $(0, d)$. Hence, the solution sequence is constant, converges to d, and $\lim_{k \to \infty} \{y_k\} = d$.

Example 6.3.1.

Solve the difference equation $y_{k+1} - y_k = 0$ with initial condition $y_0 = 2$.

Solution

From (6.3.6) the solution is $y_k = 2, \ k = 0,1,2,3,\ldots$ The graph of the solution sequence is a horizontal line that passes through $(0, 2)$. The solution sequence is constant, converges to 2, and $\lim_{k \to \infty} \{y_k\} = 2$.

Case 2: $a = 1, b > 0$

For this case, we have a difference equation,

$$y_{k+1} - y_k = b, \ k = 0,1,2,3,\ldots \tag{6.3.7}$$

with initial condition

$$y_0 = d. \tag{6.3.8}$$

The characteristic equation corresponding to this difference equation is

$$r - 1 = 0. \tag{6.3.9}$$

From (6.3.9), we have $r = 1$. Hence the general solution of corresponding to (6.3.7) homogeneous equation is

$$y_k = C_1. \tag{6.3.10}$$

Let us try to find a particular solution of (6.3.7) of the form

$$y_k^* = Ak. \tag{6.3.11}$$

Now we have that

$$y_{k+1}^* - y_k^* = A(k+1) - (Ak) = Ak + A - Ak = A.$$

In order to satisfy (6.3.7), we must choose $A = b > 0$. Therefore $y_k^* = bk$ is the desired particular solution. Hence, the general solution of (6.3.7) is

$$y_k = C_1 + bk. \tag{6.3.12}$$

Using (6.3.8), the constant C_1 in (6.3.12) can be found as

$$C_1 = d. \tag{6.3.13}$$

Substituting (6.3.13) in (6.3.12), the solution of (6.1.7) is

$$y_k = d + bk. \tag{6.3.14}$$

Therefore, the graph of the solution sequence is a line with a positive slope that passes through $(0, d)$. The solution sequence is increasing, diverges, and $\lim_{k \to \infty}\{y_k\} = \infty$.

Example 6.3.2.

Solve the difference equation $y_{k+1} - y_k = 3$ with initial condition $y_0 = 2$.

Solution

From (6.3.14), the solution is $y_k = 2 + 3k$, $k = 0, 1, 2, 3, \ldots$ .The graph of the solution sequence is a line with a positive slope of 3 that passes through $(0, 2)$. The solution sequence is increasing, diverges, and $\lim_{k \to \infty}\{y_k\} = \infty$.

Case 3: *a* = 1, *b* < 0

For this case, we have a difference equation (6.3.7). The general solution is given by the formula (6.3.14), but $b < 0$.Therefore, the graph of the solution sequence is a line with a negative slope that passes through $(0, d)$. The solution sequence is decreasing, diverges, and $\lim_{k \to \infty}\{y_k\} = -\infty$.

Example 6.3.3.

Solve the difference equation $y_{k+1} - y_k = -3$ with initial condition $y_0 = 1$.

Solution

From (6.3.14), the solution is $y_k = 1 - 3k$, $k = 0,1,2,3,\cdots$.The graph of the solution sequence is a line with a negative slope of -3 that passes through (0, 1). The solution sequence is decreasing, diverges, and $\lim\limits_{k\to\infty}\{y_k\} = -\infty$.

Case 4: $a = -1, d \neq L$

For this case, we have a difference equation,

$$y_{k+1} + y_k = b, \ k = 0,1,2,3,\cdots, \tag{6.3.15}$$

with initial condition

$$y_0 = d. \tag{6.3.16}$$

The characteristic equation corresponding to this difference equation is

$$r + 1 = 0. \tag{6.3.17}$$

From (6.3.17), we have $r = -1$. Hence, the general solution of corresponding to (6.3.15) homogeneous equation is

$$y_k = C_1(-1)^k, \ k = 0,1,2,3,\cdots. \tag{6.3.18}$$

Let us try to find a particular solution of (6.3.15) of the form

$$y_k^* = A, \tag{6.3.19}$$

where the constant coefficient A is undetermined. Now we have that

$$y_{k+1}^* + y_k^* = A + A = 2A.$$

In order to satisfy (6.3.15), we must choose $A = \dfrac{b}{2}$. Therefore, $y_k^* = \dfrac{b}{2}$ is the desired particular solution. Hence, the general solution of (6.3.15) is

$$y_k = C_1(-1)^k + \frac{b}{2}. \tag{6.3.20}$$

Using (6.3.16), the constant C_1 in (6.3.20) can be found as

$$C_1 = d - \frac{b}{2}.$$

(6.3.21)

Substituting (6.3.21) in (6.3.20), the solution of (6.3.15) is

$$y_k = (d - \frac{b}{2})(-1)^k + \frac{b}{2},$$

(6.3.22)

where

$$d \neq L = \frac{b}{2}.$$

Therefore, the solution sequence is alternating series that passes through $(0, d)$, and $\lim_{k \to \infty} \{y_k\}$ does not exists. The solution sequence diverges because $\lim_{k \to \infty} | y_k | \neq 0$ even though the series alternates regularly.

Example 6.3.4.

Solve the difference equation $y_{k+1} + y_k = 2$ with initial condition $y_0 = 4$.

Solution

From (6.3.22), the solution is $y_k = 3(-1)^k + 1$, $k = 0, 1, 2, 3, \cdots$. The solution sequence is alternating series that passes through $(0, 4)$, and $\lim_{k \to \infty} \{y_k\}$ does not exists.

Case 5: $a > 1, d > L$

For this case, we have a difference equation,

$$y_{k+1} - ay_k = b, \ k = 0, 1, 2, 3, \cdots,$$

(6.3.23)

with initial condition

$$y_0 = d.$$

(6.3.24)

The characteristic equation corresponding to this difference equation is

$$r - a = 0. \tag{6.3.25}$$

From (6.3.25), we have $r = a$. Hence, the general solution of the corresponding to (6.3.23) homogeneous equation is

$$y_k = C_1(a)^k. \tag{6.3.26}$$

Let us try to find a particular solution of (6.3.23) of the form

$$y_k^* = A, \tag{6.3.27}$$

where the constant coefficient A is undetermined. Now we have that

$$y_{k+1}^* - a y_k^* = A - aA = A(1-a).$$

In order to satisfy (6.3.23), we must choose $A = \dfrac{b}{1-a}$. Therefore, $y_k^* = \dfrac{b}{1-a}$ is the desired particular solution. Thus, the general solution of (6.3.23) is

$$y_k = C_1(a)^k + \frac{b}{1-a}. \tag{6.3.28}$$

Using (6.3.24), the constant C_1 in (6.3.28) can be found as

$$C_1 = d - \frac{b}{1-a}. \tag{6.3.29}$$

Substituting (6.3.29) in (6.3.28), the solution of (6.3.23)

$$y_k = (d - \frac{b}{1-a})a^k + \frac{b}{1-a}, \tag{6.3.30}$$

where

$$d > L = \frac{b}{1-a}.$$

Therefore, the graph of the solution sequence is an exponential curve that passes through $(0,d)$. The solution sequence is increasing, diverges, and $\lim_{k \to \infty}\{y_k\} = \infty$.

Example 6.3.5.

Solve the difference equation $y_{k+1} - 3y_k = -2$ with initial condition $y_0 = 6$.

Solution

From (6.3.30), the solution is $y_k = 5(3)^k + 1$, $k = 0, 1, 2, 3, \cdots$. The graph of the solution sequence is an exponential curve that passes through (0, 6). The solution sequence is increasing, diverges, and $\lim_{k \to \infty} \{y_k\} = \infty$.

Case 6: $a > 1, d < L$

For this case, we have a difference equation (6.3.23). The general solution is given by the formula (6.3.30), but $d < L = \dfrac{b}{1-a}$. Therefore, the graph of the solution sequence is an exponential curve that passes through (0, d). The solution sequence is decreasing, diverges, and $\lim_{k \to \infty} \{y_k\} = -\infty$.

Example 6.3.6.

Solve the difference equation $y_{k+1} - 2y_k = -3$, with initial condition $y_0 = -2$.

Solution

From (6.3.30), the solution is $y_k = -5(2)^k + 3$, $k = 0, 1, 2, 3, \cdots$. The graph of the solution sequence is an exponential curve that passes through $(0, -2)$. The solution sequence is decreasing, diverges, and $\lim_{k \to \infty} \{y_k\} = -\infty$.

Case 7: $0 < a < 1, d > L$

For this case, we have a difference equation (6.3.23). The general solution is given by the formula (6.3.30), but $d > L = \dfrac{b}{1-a}$. Therefore, the graph of the solution sequence is an exponential curve and steadily getting closer to $y = L$. The graph of the solution sequence passes through (0, d). The solution sequence is decreasing, converges to L, and $\lim_{k \to \infty} \{y_k\} = L$.

Example 6.3.7.

Solve the difference equation $y_{k+1} - \dfrac{1}{2}y_k = 2$ with initial condition $y_0 = 5$.

Solution

From (6.3.30), the solution is $y_k = \left(\dfrac{1}{2}\right)^k + 4$, $k = 0,1,2,3,\cdots$. The graph of the solution sequence is an exponential curve and steadily getting closer to $y = 4$. The graph of the solution sequence passes through (0, 5). The solution sequence is decreasing, converges to 4, and $\lim\limits_{k \to \infty}\{y_k\} = 4$.

Case 8: $0 < a < 1, d < L$

For this case, we have a difference equation (6.3.23). The general solution is given by the formula (6.3.30), but $d < L = \dfrac{b}{1-a}$. Therefore, the graph of the solution sequence is an exponential curve and steadily getting closer to $y = L$. The graph of the solution sequence passes through (0, d). The solution sequence is increasing, converges to L, and $\lim\limits_{k \to \infty}\{y_k\} = L$.

Example 6.3.8.

Solve the difference equation $y_{k+1} - \dfrac{1}{2}y_k = 3$ with initial condition $y_0 = 1$.

Solution

From (6.3.30), the solution is $y_k = -5\left(\dfrac{1}{2}\right)^k + 6$, $k = 0,1,2,3,\cdots$. The graph of the solution sequence is an exponential curve and steadily getting closer to $y = 6$. The graph of the solution sequence passes through (0, 1). The solution sequence is increasing, converges to 6, and $\lim\limits_{k \to \infty}\{y_k\} = 6$.

Case 9: $-1 < a < 0, d \neq L$

For this case, we have a difference equation (6.3.23). The general solution is given by the formula (6.3.30). Therefore, the graph of the solution sequence is an alternating series that is steadily getting closer to $y = L$. The graph of the solution sequence passes through $(0, d)$. The solution sequence converges to L, and $\lim_{k \to \infty}\{y_k\} = L$.

Example 6.3.9.

Solve the difference equation $y_{k+1} + \dfrac{1}{2} y_k = 6$ with initial condition $y_0 = 5$.

Solution

From (6.3.30), the solution is $y_k = \left(-\dfrac{1}{2}\right)^k + 4$, $k = 0, 1, 2, 3, \cdots$. The graph of the solution sequence is an alternating series that is steadily getting closer to $y = 4$. The graph of the solution sequence passes through $(0, 5)$. The solution sequence converges to 4, and $\lim_{k \to \infty}\{y_k\} = 4$.

Case 10: $a < -1, d \neq L$

For this case, we have a difference equation (6.3.23). The general solution is given by the formula (6.3.30). The graph of the solution sequence is an alternating series that is alternating farther away from zero. The graph of the solution sequence passes through $(0, d)$. The solution sequence does not converge, and $\lim_{k \to \infty}\{y_k\}$ does not exist.

Example 6.3.10.

Solve the difference equation $y_{k+1} + 2y_k = 3$ with initial condition $y_0 = 0$.

Solution

From (6.3.30), the solution is $y_k = 1 - (-2)^k$, $k = 0, 1, 2, 3, \cdots$. The graph of the solution sequence is an alternating series that is alternating farther away from zero. The graph of the solution sequence passes through $(0, 0)$. The solution sequence does not converge, and $\lim_{k \to \infty}\{y_k\}$ does not exist.

Case 11: $a \neq 1, d = L$

For this case, we have a difference equation (6.3.23). The general solution is given by the formula (6.3.30), where $d = L = \dfrac{b}{1-a}$. So, the solution is

$$y_k = \frac{b}{1-a} = L. \tag{6.3.31}$$

Therefore, the graph of the solution sequence is a horizontal line that passes through $(0, d)$. The solution sequence is constant, converges to L, and $\lim\limits_{k \to \infty}\{y_k\} = L$.

Example 6.3.11.

Solve the difference equation $y_{k+1} - 2y_k = -4$ with initial condition $y_0 = 4$.

Solution

From (6.3.31) the solution is $y_k = 4$, $k = 0, 1, 2, 3, \cdots$. Therefore, the graph of the solution sequence is a horizontal line that passes through $(0, 4)$. The solution sequence is constant, converges to 4, and $\lim\limits_{k \to \infty}\{y_k\} = 4$.

6.4. GENERATING FUNCTIONS AND THEIR APPLICATIONS IN DIFFERENCE EQUATIONS

Differential equations may be solved numerically using difference equations. Example 6.4.1 is a numerical solution of differential equation by the *Euler's method*:

Example 6.4.1.

The *Euler method* approximate the differential equation

$$\frac{dy}{dx} = f(x, y), \quad y(x_0) = y_0 \tag{6.4.1}$$

by the difference equation

$$y_{k+1} = y_k + hy'_k, \tag{6.4.2}$$

where $y(x_k) = y_k$, $y'(x_k) = y'_k = f(x_k, y_k)$, $h = \Delta(x) = x_{k+1} - x_k$.

Solution

Apply this method to solve the differential equation

$$y' + 2y = 2 - e^{-4x}, \quad y(0) = 1. \tag{6.4.3}$$

This is a simple linear differential equation so we can find that the exact solution is

$$y(x) = 1 + \frac{1}{2}e^{-4x} - \frac{1}{2}e^{-2x}. \tag{6.4.4}$$

Use Euler's Method with a step size of $h = 0.1$ to find approximate values of the solution at x = 0.1, 0.2, 0.3, 0.4 and 0.5. In order to use Euler's Method we first need to rewrite the differential equation into the form given in (6.4.1).

$$y' = 2 - e^{-4x} - 2y = f(x, y).$$

Note that $x_0 = 0$ and $y_0 = 1$. Then

$$y_1 = y_0 + hy'_0 = 1 + 0.1f(0,1) = 0.9$$
$$y_2 = y_1 + hy'_1 = 0.9 + 0.1f(0.1, 0.9) = 0.8530$$
$$y_3 = y_2 + hy'_2 = 0.8374$$
$$y_4 = y_3 + hy'_3 = 0.8398$$
$$y_5 = y_4 + hy'_4 = 0.8517$$

The exact value is $y(0.5) = 0.8837$. Hence, the first digit is correct.

As we mentioned earlier, the solution of a difference equation defined over a set of values is a sequence. Applying a *generating function* on a set of difference equations will result either in an algebraic or in a functional equation. Thus, generating functions plays extremely useful role in solving difference equations.

Definition 6.4.1.

Suppose $\{a_n\}$ is a sequence of real numbers such that $a_n = 0$ for $n < 0$. If the infinite series

$$G(z) \equiv \sum_{n=0}^{\infty} a_n z^n , \tag{6.4.5}$$

where z is a dummy variable, converges for $|z| < R$, where R is the radius of convergence, then $G(z)$ is called the *generating function* of the sequence $\{a_n\}$.

Note that if the sequence $\{a_n\}$ is bounded, then the series (6.4.5) converges for at least $|z| < 1$.

Example 6.4.2.

Generating function of the sequence $1, 1, 1, \cdots$ is $G(z) = \dfrac{1}{1-z}$ since division of 1 by (1

$- z$) yields $1 + z + z^2 + \dots + z^n + \cdots, \; |z| \le 1$.

As a property of generating functions, consider the product of two generating functions $G(z)$ and $H(z)$. So, using the notation introduced in (6.4.1) we have:

$$G(z)H(z) = \left(\sum_{i=0}^{\infty} a_i z^i \right)\left(\sum_{j=0}^{\infty} b_j z^j \right) = \sum_{i=0}^{\infty} \sum_{j=0}^{\infty} a_i b_j z^{i+j} . \tag{6.4.6}$$

By letting $k = i + j$, the double series on the right hand side of (6.2.6) can be written in a single power series as follows:

$$G(z)H(z) = \sum_{k=0}^{\infty} c_k z^k , \tag{6.4.7}$$

where

$$c_k = \sum_{i=0}^{k} a_i b_{k-i} . \tag{6.4.8}$$

Example 6.4.3.

Generating function of the sequence $1, 2, 3, \cdots$ is $G(z) = \dfrac{1}{(1-z)^2}$ since division of 1 by

$(1 - z)^2$ yields $1 + 2z + 3z^2 + \dots + (n+1)z^n + \cdots, \; |z| \le 1$.

Definition 6.4.2.

The sequence $\{c_k\}$ defined in (6.4.8) is called the *convolution* of the two sequences $\{a_i\}$ and $\{b_j\}$.

We now extend the property of generating function to the theory of probability by defining the generating function of a random variable X, representing the sequence $\{x_n\}$ of real numbers denoting values of the random variable as $G(z) = \sum_{n-0}^{\infty} x_n z^n$. From the definition of expectation of a discrete random variable X with probability mass function (*pmf*) p_x as $E(X) = \sum_{x=0}^{\infty} x p_x$, we can now define the probability generating function.

Definition 6.4.3.

Let X be a discrete random variable and k a nonnegative integer. Consider the random variable z^X with *pmf* p_k. Then the *probability generating function* or (if there is no fear of confusion) simply the *generating function* of X, denoted by $G(z)$, is defined as:

$$G(z) \equiv E(z^X) = \sum_{k=0}^{\infty} p_k z^k .$$
(6.4.9)

Note:
1. When X takes k as its value, z^X takes z^k as its value.

2. The idea of applying generating function to probability is intended to encapsulate all the information about the random variable. For instance from (6.4.9) we have the following:

(a) If $z = 1$, then $G(1) = \sum_{k=0}^{\infty} p_k = 1$.

(b) $\dfrac{d^k G(z)}{dz^k} = \sum_{k=n}^{\infty} k! p_k \dfrac{z^{k-n}}{(n-k)!}.$

Thus,

$$\left. \frac{d^k G(z)}{dz^k} \right)_{z=1} = \sum_{k=n}^{\infty} \frac{k! p_k}{(n-k)!} = E\big[X(X-1)\cdots(X-n+1) \big].$$
(6.4.10)

Therefore, (6.4.9) yields all the moments of X via (6.4.10). For instance, if $n = 1$, from (6.4.10) we will have the expectation of X. If $k = 2$, we will have $E(X(X-1)) = E(X^2) - E(X)$. Thus, variance can be obtained as:

$$Var(X) = E(X^2) - [E(X)]^2 = E(X(X-1)) + E(X) - [E(X)]^2. \text{ (6.4.11)}$$

(c) If we expand (6.4.9) by Taylor theorem, then

$$n! p_k = \frac{d^k G(z)}{dz^k}\bigg)_{z=0}.$$

Example 6.4.4.

If X is a Poisson random variable, then the generating function of X will be:

$$G(z) = \sum_{k=0}^{\infty} \frac{e^{-\lambda} \lambda^k}{k!} z^k = e^{-\lambda} \sum_{k=0}^{\infty} \frac{(\lambda z)^k}{k!} = e^{-\lambda} e^{\lambda z} = e^{-\lambda(1-z)}. \qquad (6.4.12)$$

Hence, from (6.4.10) and (6.4.12) we have:

$$E(X) = \frac{dG(z)}{dz}\bigg)_{z=1} = \lambda e^{-\lambda(1-z)}\bigg)_{z=1} = \lambda.$$

Thus, the parameter of the Poisson distribution is, in fact, the expectation of X.
To obtain the variance, from (6.4.10) - (6.4.11) we have:

$$E[X(X-1)] = \frac{d^2 G(z)}{dz^2}\bigg)_{z=1} = \lambda^2 + \lambda,$$

and

$$Var(X) = \lambda^2 + \lambda - \lambda^2 = \lambda.$$

Example 6.4.5.

Suppose that X is a binomial random variable with parameters n and p. We denote by $P_{k,n}$ the probability of exactly k successes in n repeated independent Bernoulli trials with probability of success in each trials as p (and, thus, of failure $q = 1 - p$) that lead to the binomial distribution, i.e.,

$$P_{k,n} = P(X = k), \; k = 0, 1, 2, \cdots, n,$$ (6.4.13)

with $\sum_{k=0}^{n} P_{k,n} = 1$.

Since the parameter n is fixed, $P_{k,n}$ is defined for the set of k-values $0, 1, 2, \cdots, n$. To find $P_{k,n}$, we derive a difference equation and then solve it by generating function method. We then will find the mean and variance from the generating function.

The event that $(k + 1)$ successes occur in $(n+1)$ trials is possible only in the following two mutually exclusive ways:

(1) by having $(k + 1)$ successes after n trials and then a failure in the $(n+1)^{st}$ trial, or

(2) by having k successes after n trials and then a success in the $(n+1)^{st}$ trial.

The probability of the event in case (1) is the product of the probability of $(k+1)$ successes in the first n trial and the probability of a failure in the $(n+1)^{st}$. Hence, the probability of the event in this case is $qP_{k+1,n}$. Similarly, the probability of the event in case (2) is $pP_{k,n}$. Therefore,

$$P_{k+1,n+1} = qP_{k+1,n} + pP_{k,n}, \; k = 0, 1, 2, \cdots, n, \;\; n = 0, 1, 2, \cdots.$$ (6.4.14)

Definition 6.4.4.

Equation (6.4.14) is a *partial difference equation* for the distribution function $P_{k,n}$.

Continuing our example, the event that zero successes occur in $(n+1)$ trials is possible only by having zero successes after n trials and then a failure in the $(n+1)^{st}$ trial. Thus,

$$P_{0,n+1} = qP_{0,n}, \; n = 0, 1, 2, \cdots.$$ (6.4.15)

Noting the fact that in 0 trials it is impossible to have any number of successes other than zero, we will have the initial conditions

$$P_{0,0} = 1 \text{ and } P_{k,0} = 0, \text{ for } k > 0.$$ (6.4.16)

The three equations (6.4.14) – (6.4.16) is the system of difference equations we said we will drive. We now want to solve the system for $P_{k,n}$. Thus, we define the generating function $G_n(z)$ for the sequence $\{P_{k,n}\}$ as follows:

$$G_n(z) = P_{0,n} + P_{1,n}z + P_{2,n}z^2 + \cdots + P_{k,n}z^k + \cdots = \sum_{k=0}^{\infty} P_{k,n}z^k \, , \, n = 0,1,2,\cdots. \quad (6.4.17)$$

Note that there is one such generating function for each value of n. To apply (6.4.17) on (6.4.14), we multiply (6.4.14) through by z^{k+1}, sum over k and obtain

$$\sum_{k=0}^{\infty} P_{k+1,n+1}z^{k+1} = q\sum_{k=0}^{\infty} P_{k+1,n}z^{k+1} + p\sum_{k=0}^{\infty} P_{k,n}z^{k+1} \, , \, n = 0,1,2,\cdots. \quad (6.4.18)$$

From (6.4.17), (6.4.18) can be rewritten as:

$$G_{n+1}(z) = (q + pz)G_n(z) + (P_{0,n+1} - qP_{0,n}) , \, n = 0,1,2,\cdots. \quad (6.4.19)$$

Using conditions (6.4.15), (6.4.19) may be yet be rewritten as:

$$G_{n+1}(z) = (q + pz)G_n(z), \, n = 0,1,2,\cdots, \quad (6.4.20)$$

which is a first-order difference equation for the generating function $G_n(z)$.

To solve (6.4.20), let $Y^n = G_n(z)$. Then, (6.4.20) becomes

$$Y^{n+1} = (q + pz)Y^n .$$

Since $Y^n \neq 0$ (otherwise there will not be any equation) we will have $Y = q + pz$ or

$$Y^n = (q + pz)^n ,$$

or

$$G_n(z) = (q + pz)^n, \, n = 0,1,2,\cdots. \quad (6.4.21)$$

Using the well-known *Binomial Theorem*, from (6.4.21) we have:

$$G_n(z) = (q + pz)^n = \sum_{k=0}^{n} \binom{n}{k}(pz)^k q^{n-k}$$

or

$$G_n(z) = \sum_{k=0}^{n} \binom{n}{k} p^k q^{n-k} z^k.$$ (6.4.22)

The coefficients of (6.4.22) are the distribution function we were looking for, i.e.,

$$P_{k,n} = P(X = k) = \binom{n}{k} p^k q^{n-k}, \; k = 0, 1, 2, \cdots, n.$$ (6.4.23)

Note that the right hand side of (6.4.23), indeed, defines the binomial probability distribution with parameters n and p. In other words, the generating function for the binomial distribution is given by (6.4.21).

To find the mean and variance, from (6.4.10) and (6.4.21) we have $G_n'(z) = n(q + pz)^{n-1} p$. Since $p + q = 1$, we will have

$$E(X) = G_n'(1) = n(q + p)^{n-1} p = np.$$

Also,

$$G_n''(z) = n(n-1)(q + pz)^{n-2} p^2.$$

Thus,

$$E[X(X-1)] = n(n-1)(q+p)^{n-2} p^2 = n(n-1)p^2.$$

On the other hand, $(E(X))^2 = (np)^2$. Hence,

$$Var(X) = E(X^2) - [E(X)]^2 = E[X(X-1)] + E(X) - [E(X)]^2$$
$$= n(n-1)p^2 + np - (np)^2 = np - np^2 = np(1-p) = npq.$$

Example 6.4.7.

A retail store has a list of items displayed. Items picked by customers are recorded so as the number of customers. We say an item is in *state* k on the nth trial if the item has been picked exactly k times among n customers entered the store, $k, n = 0, 1, 2, \cdots$. We denote the probability that an item is in state k on trial n by $p_{k,n}$. Let s_k denote the probability that an item in state k is picked. s_k is defined to be independent of the number of trials. The numbers

$s_0, s_1, s_2, \cdots$ can be estimated from the experimental data and so we assume that they are known. The unknown probabilities $p_{k,n}$ are to be determined.

Solution

The first step to determine $p_{k,n}$ is the derivation of a partial difference equation satisfied by these probabilities as we did in Example 6.4.5. The event that an item is in state $k + 1$ on trial $n + 1$ can occur in only two mutually exclusive ways: either (1) an item is in state $k + 1$ on trial n and is not picked on trial $n + 1$, or (2) the item is in state k on trial n and is picked on trial $n + 1$.

We assume that successive trials are statistically independent. Then, the probability of event (1) is the product of $p_{k+1,n}$ and $(1 - s_{k+1})$, the probability that this item is not picked on the following trail. Similarly, event (2) has as its probability the produce of $p_{k,n}$ and s_k. Hence,

$$p_{k+1,n+1} = (1 - s_{k+1}) p_{k+1,n} + s_k p_{k,n}, \quad k = 0, 1, 2, \cdots, \quad n = 0, 12, \cdots. \tag{6.4.29}$$

The above argument does not apply to $p_{0,n+1}$ since an item can be in state 0 (i.e., never be picked) on trial $n+1$ only if it is in state 0 on trial n and not picked on trial $n+1$. Thus, we have

$$p_{0,n+1} = (1 - s_0) p_{0,n}, \quad n = 0, 1, 2, \cdots. \tag{6.4.30}$$

We may assume a set of initial conditions expressing the fact that no item is picked before the first trial, i.e.,

$$p_{0,0} = 1, \text{ and } p_{k,0} = 0 \text{ for } k = 1, 2, \cdots. \tag{6.4.31}$$

Equations (6.4.29) and (6.4.30) constitute a system of difference equations with initial conditions (6.4.31). We will use the method of generating functions to solve the system and determine $p_{k,n}$.

Let k be fixed. Consider $p_{k,n}$, as the general term of the sequence $p_{k,0}, p_{k,1}, p_{k,2} \cdots$. We define the generating function of $p_{k,n}$ as follows:

$$G_k(z) = \sum_{n=0}^{\infty} p_{k,n} z^n, \quad k = 1, 2, \cdots. \tag{6.4.32}$$

We now apply (6.4.32) on (6.4.30). Thus, we will have

$$\sum_{n=0}^{\infty} p_{0,n+1} z^{n+1} = (1-s_0) \sum_{n=0}^{\infty} p_{0,n} z^{n+1}$$

or

$$\sum_{n=1}^{\infty} p_{0,n} z^{n} = (1-s_0) z \sum_{n=0}^{\infty} p_{0,n} z^{n}$$

or

$$G_0(z) - p_{0,0} = (1-s_0) z G_0(z)$$

or

$$[1-(1-s_0)z] G_0(z) = p_{0,0}. \tag{6.4.33}$$

From the initial condition (6.4.31) and (6.4.33) we will have

$$G_0(s) = \frac{p_{0,0}}{1-(1-s_0)z} = \frac{1}{1-(1-s_0)z}. \tag{6.4.34}$$

In a similar way, we may apply the generating function on (6.4.29). Thus,

$$\sum_{n=0}^{\infty} p_{k+1,n+1} z^{n+1} = (1-s_{k+1}) \sum_{n=0}^{\infty} p_{k+1,n} z^{n+1} + s_k \sum_{n=0}^{\infty} p_{k,n} z^{n+1}$$

or

$$\sum_{n=1}^{\infty} p_{k+1,n} z^{n} = (1-s_{k+1}) z \sum_{n=0}^{\infty} p_{k+1,n} z^{n} + s_k z \sum_{n=0}^{\infty} p_{k,n} z^{n}$$

or

$$G_{k+1}(z) - p_{k+1,0} = (1-s_{k+1}) z G_{k+1}(z) + s_k z G_k(z), \quad k = 0, 1, 2, \cdots$$

or

$$[1-(1-s_{k+1})z] G_{k+1}(z) = s_k z G_k(z), \quad k = 0, 1, 2, \cdots$$

or

$$G_{k+1}(z) = \frac{s_k z}{[1-(1-s_{k+1})z]} G_k(z), \quad k = 0, 1, 2, \cdots. \tag{6.4.35}$$

Equation (6.4.34) is a homogeneous first-order difference equation, which can be solved recursively as

$$G_k(s) = G_0(z) \prod_{i=0}^{k-1} \frac{zs_i}{1 - z(1 - s_{i+1})}, \quad k = 0, 1, 2, \cdots. \tag{6.4.36}$$

Substituting $G_0(z)$ from (6.4.34) into (6.4.36) we obtain the generating function of the probability distribution as:

$$G_k(s) = \frac{1}{1 - z(1 - s_0)} \prod_{i=0}^{k-1} \frac{zs_i}{1 - z(1 - s_{i+1})}, \quad k = 0, 1, 2, \cdots. \tag{6.4.37}$$

To find the probabilities $p_{k,n}$ we need to invert the generating functions found in (6.4.34) and (6.4.37). To do this note that $\dfrac{1}{1 - z(1 - s_0)} = \sum_{n=0}^{\infty} [(1 - s_0)z]^n$. Hence, from (6.4.34) we have

$$p_{0,n} = (1 - s_0)^n, \quad n = 0, 1, 2, \cdots. \tag{6.4.38}$$

Now, for $k = 1$ from (6.4.37) we have

$$G_1(z) = \frac{s_0 z}{[1 - s(1 - s_0)][1 - z(1 - s_1)]}. \tag{6.4.39}$$

Using the method of partial fractions, (6.4.39) can be written as

$$G_1(s) = \frac{s_0}{s_1 - s_0} \left[\frac{1}{1 - z(1 - s_0)} - \frac{1}{1 - z(1 - s_1)} \right], \quad s_1 \neq s_0. \tag{6.4.40}$$

As before, inverting generating functions in (6.4.40) when $s_1 \neq s_0$, i.e., writing (6.4.40) in power series and choosing coefficients yields

$$p_{1,n} = \frac{s_0}{s_1 - s_0} \left[(1 - s_0)^n - (1 - s_1)^n \right], \quad n = 0, 1, 2, \cdots. \tag{6.4.41}$$

Similarly, all probabilities $p_{k,n}$ for $k = 2, 3, \cdots$ can be explicitly found. We leave this for readers to find.

Often information may be directly obtained from the generating function without the necessity of finding explicit solutions. For instance, suppose that we are looking for the probability that an item is picked k times and then is never picked again in the future trials. The product $p_{k,n} s_k$ is the probability that an item is picked k times before the nth trial and then is picked again. Summing this probability over n, denoting the result by π_k, and using (6.4.32) we will have

$$\pi_k = \sum_{n=0}^{\infty} p_{k,n} s_k = s_k \sum_{n=0}^{\infty} p_{k,n} = s_k G_k(1) \tag{6.4.42}$$

From (6.4.34) we find $G_0(1) = \dfrac{1}{s_0}$, $s_0 \neq 0$. Thus, from (6.4.42) we have

$$\pi_0 = s_0 G_0(1) = 1. \tag{6.4.43}$$

Similarly, assuming $s_0 \neq 0$, $s_1 \neq 0, \ldots, s_k \neq 0$, from (6.4.35) and (6.4.43) we have

$$G_k(1) = G_0(1) \prod_{i=0}^{k-1} \frac{s_i}{1-(1-s_{i+1})} = \frac{1}{s_0} \prod_{i=0}^{k-1} \frac{s_i}{s_{i+1}} = \frac{1}{s_0} \frac{s_0}{s_1} \frac{s_1}{s_2} \frac{s_2}{s_3} \cdots \frac{s_{k-2}}{s_{k-1}} \frac{s_{k-1}}{s_k} = \frac{1}{s_k}.$$

Therefore, if all s_k ($k = 0,1,2,\cdots$) are positive, we will have

$$\pi_k = 1, \quad k = 0,1,2,\cdots, \tag{6.4.44}$$

i.e., an item in state k will surely move to state $k+1$ if the customers keep arriving. It follows that as the number of trials increases indefinitely, the probability of any finite number of picks is zero.

Example 6.4.8.

To study behavior changes under certain experimental conditions, we think of a sequence of events starting with the perception of a stimulus, followed by the performance of a response (pressing bar, running maze, etc.), and ending with the occurrence of an environmental event (presentation of food, electric shock, etc.).

Let the probability p that the response will occur during some specified time interval after the sequence is initiated measure the behavior. In other words, p denotes the subject's level of performance. This level is increased or decreased after each occurrence of the response according as the environmental factors are reinforcing or inhibiting. This design is referred to as an *adaptive design*.

We consider an experiment in which a subject is repeatedly exposed to a sequence of stimulus-response-environmental events. We divide the experiment into stages, each stage being a trial during which the subject is run through the sequence. The subject's level of performance is then a function of the trial number, denoted by n. Let p_n be the probability of the response (during the specified time interval following the stimulus) in the nth trial run. $n = 0$ denotes the initial trial that the subject is first introduced to the experiment proper. We let p_0 denote the initial value describing the disposition of the subject toward the response when he is first introduced to the experiment proper. We also denote $p_n, 0 \le p_n \le 1$, $n = 0, 1, 2, \cdots$, as the probability of a response at the nth trial, i.e., the probability that the subject's level of performance at trial number n is given by p_n. Note that $1 - p_n$ is the maximum possible increase in level and $-p_n$ is the maximum possible decrease in moving to trial $n+1$. This follows since 1 and 0 are the largest and smallest values of p_{n+1}.

Suppose that the subject's performance in trial $n+1$, although dependent upon his level of behavior in the preceding trial (as measured by p_n), is independent of the past record of performance leading to trial n, this is, posses the Markov property. We further assume that this dependence of p_{n+1} upon p_n is linear. Thus, the slope-intercept form of the equation of this straight line is

$$p_{n+1} = a + mp_n, \ n = 0, 1, 2, \cdots, \tag{6.4.45}$$

where a is the intercept (the value of p_{n+1} when p_n is 0) and m is the slope of the line (the amount by which p_{n+1} changes per unit increase of p_n).

To re-write (6.4.45) in the "*gain-loss*" form, we introduce the parameter b defined by equation

$$m = 1 - a - b, \tag{6.4.46}$$

so that

$$p_{n+1} = a(1 - p_n) - bp_n, \ n = 0, 1, 2, \cdots. \tag{6.4.47}$$

Equation (6.4.47) may be interpreted as the change in performance level (denoted by $\Delta p_n = p_{n+1} - p_n$) is proportional to the maximum possible gain and maximum possible loss, with a and b as constants of proportionality. It is for this reason that (6.4.47) is named the "*gain-loss*" form. The parameter a measures those environmental events which are reinforcing (e.g., punishing the subject). From (6.4.47) since $p_n = 0$ implies that $p_{n+1} = a$, and $0 \le p_{n+1} \le 1$, we must have $0 \le a \le 1$. On the other hand, if $p_n = 1$, from (6.4.45) and (6.4.46) we will have $p_{n+1} = 1 - b$, so that $0 \le 1 - b \le 1$, or $0 \le b \le 1$. It can easily be shown that the necessary and sufficient condition for $0 \le p_{n+1} \le 1$ is that $0 \le a \le 1$ and $0 \le b \le 1$. Therefore, $a = 0$ describes a situation in which no reward is given after the response occurs, b

$= 0$ describes a no-punishment trial, and $a = b$ implies that the measures of reward and punishment are equal.

Numerically, suppose $a = 0.4$ and $b = 0.1$. Then, $p_{n+1} = p_n + 0.4(1 - p_n) - 0.1p_n$ or

$$p_{n+1} = 0.5p_n + 0.4, \quad n = 0, 1, 2, \cdots. \tag{6.4.48}$$

If $p_0 = 0.2$, we may successively compute $p_1 = 0.5(0.2 + 0.4 = 0.5, \ p_2 = 0.5(0.5) + 0.4 = 0.65, \cdots$.

To solve (6.4.48) for p_n for every n, with a general term, we note that equation (6.4.47) (that can be re-written as $p_{n+1} = a - (a + b)p_n$) is a linear difference equation whose unique solution can easily be found (refer to Section 6.2) as:

$$p_n = \begin{cases} m^n\left(p_0 - \dfrac{a}{1-m}\right) + \dfrac{a}{1-m}, & \text{if } m \neq 1, \\ p_0 + an, & \text{if } m = 1. \end{cases}$$

Thus, with $a = 0.4, b = 0.1$ (i.e., $m = 1 - 0.4 - 0.1 = 0.5$) and $p_0 = 0.2$, we will have

$$p_n = (0.5)^n(0.2 - 0.8) + 0.8p_n, \text{ or } p_n = -0.6(0.5)^n + 0.8, \quad n = 0, 1, 2, \cdots.$$

Back to the general case, we note that (6.4.48) using (6.4.49) may be rewritten in the standard form

$$p_{n+1} = (1 - a - b)p_n + a \quad n = 0, 1, 2, \cdots, \tag{6.4.49}$$

which is a linear first-order difference equation with constant coefficients. Thus, we have the solution

$$p_n = \begin{cases} (1 - a - b)^n\left(p_0 - \dfrac{a}{a+b}\right) + \dfrac{a}{a+b}, & \text{if } a + b \neq 0, \ n = 1, 2, \cdots, \\ p_0, & \text{if } a + b = 0. \end{cases} \tag{6.4.50}$$

Since $0 \leq a \leq 1$ and $0 \leq b \leq 1$, we have $-1 \leq -a - b \leq 1$. End-points of this interval attained only if a and b are either both 1 or both 0. If $a = b = 1$, then $\{p_n\}$ oscillates finitely between the two values p_0 and $1 - p_0$. But, in all other cases the sequence $\{p_n\}$ converges, to the limit p_0 if $a = b = 0$, and to the limit $\dfrac{a}{a+b}$ otherwise. If $0 < a + b < 1$, then

$0 < 1 - a - b < 1$ and $\{p_n\}$ is monotonically decreases to $\dfrac{a}{a+b}$ if $p_0 > \dfrac{a}{a+b}$,

monotonically increases to $\dfrac{a}{a+b}$ if $p_0 < \dfrac{a}{a+b}$, and always equal to $\dfrac{a}{a+b}$ if $p_0 = \dfrac{a}{a+b}$.

If $1 < a + b < 2$, then $-1 < 1 - a - b < 0$ and $\{P_n\}$ is a damped oscillatory sequence with

limit $\dfrac{a}{a+b}$. The special case $a + b = 0$ yields a constant sequence (with value p_0); $a + b = 1$

produces a sequence each of whose element is a.

Two special cases of interest are when $a = 0$ and $a = b$ as discussed below.

Case 1.

$a = 0$, i.e., no reward is given after the response occurs.

Then, the difference equation (6.4.45) becomes $p_{n+1} = (1 - b)p_n$ with the solution

$p_n = (1 - b)^n p_0, n = 0, 1, 2, \cdots$. This is an equation which describes the steady (as $n \to \infty$)

decrease in response probability from the initial probability p_0. By plotting p as a function of

n we obtain a curve of experimental extinction.

Case 2.

$a = b$, i.e., the measures of reward and punishment are equal.

Ignoring the extreme subcases $a = b = 0$ and $a = b = 1$, the quantity

$(1 - a - b)^n$ as $n \to \infty$ and (6.2.46) show that $p_n \to \dfrac{a}{a+b}$, which is equal to 0.5. That is,

ultimately the response tends to occur (in the specified time interval after the stimulus is
presented) in half the trials. The balancing of reward and punishment produces, in the long-
run, a corresponding symmetry in performance.

We remark that modifying this example could be used for engineering where the amount
of stress is to be adaptive according to a desirable tolerance.

Theorem 6.4.1.

The generating function of sum of n independent random variables is the product of the
generating function of each.

Proof

Let X_1, X_2, $\cdots$, X_n be n independent random variables with corresponding generating functions $G_1(z) = E(Z^{X_1})$, $G_2(z) = E(Z^{X_2})$, $\cdots$, $G_n(z) = E(Z^{X_n})$. Let also $X = X_1 + X_2 + \cdots + X_n$. Then,

$$E(Z^X) = E(z^{X_1 + X_2 + \cdots + X_n}) = E(z^{X_1} z^{X_2} \cdots z^{X_n})$$
$$= E(z^{X_1}) E(z^{X_2}) \cdots E(z^{X_n})$$
$$= G_1(z) G_2(z) \cdots G_n(z).$$

Note that in the probability generating function defined in Definition 6.4.5, if we replace z by e^t, we will obtain the moment generating function of a discrete random variable defined in Definition 4.10.8.

Example 6.4.9.

Let us consider the following system of first order difference equations

$$\begin{cases} ap_0 = bp_1 \\ (a+b)p_k = ap_{k-1} + bp_{k+1}, \quad k = 1, 2, \cdots, \end{cases} \qquad (6.4.51)$$

where a and b are positive constants and p's are probabilities, i.e., $\sum_{k=0}^{\infty} p_k = 1$.

Solution

To solve for p_k, we apply the generating function method as follows. So, let

$$G(z) = \sum_{k=0}^{\infty} p_k z^k \qquad (6.4.52)$$

be defined as the generating function for the sequence $\{p_k, k = 1, 2, \cdots\}$. Multiplying both sides of the second equation of (6.4.51), summing over k, and applying (6.4.52) yields

$$(a+b)\sum_{k=1}^{\infty} p_k z^k = a \sum_{k=1}^{\infty} p_{k-1} z^k + b \sum_{k=1}^{\infty} p_{k+1} z^k$$

or

$$(a+b)\left[\sum_{k=0}^{\infty} p_k z^k - p_0\right] = az \sum_{k=0}^{\infty} p_k z^k + \frac{b}{z} \sum_{k=2}^{\infty} p_k z^k$$

or

$$(a+b)\left[G(z)-p_0\right]=azG(z)+\frac{b}{z}\left[G(z)-p_0-p_1z\right]$$

or

$$\left(a+b-az-\frac{b}{z}\right)G(z)=(a+b)p_0-\frac{b}{z}p_0-bp_1 . \tag{6.4.53}$$

Now adding the first equation of (6.4.51) to (6.4.53) and simplifying, we obtain

$$\frac{(a+b)z-az^2-b}{z}G(z)+ap_0=(a+b)p_0+bp_1-\frac{b}{z}p_0-bp_1$$

or

$$\left[(a+b)z-az^2-b\right]G(z)=b(z-1)p_0$$

or

$$G(z)=\frac{b(z-1)}{(a+b)z-az^2-b}p_0 . \tag{6.4.54}$$

From (6.4.54) we have

$$G(1)=1=\lim_{z\to 1}\frac{b(z-1)}{(a+b)z-az^2-b}p_0=\frac{b}{b-a}p_0$$

or

$$p_0=\frac{b-a}{b}=1-\frac{a}{b} . \tag{6.4.55}$$

Substituting (6.4.55) in (6.4.54) yields

$$G(z)=\frac{(b-a)(z-1)}{(a+b)z-az^2-b}=\frac{(b-a)(z-1)}{(z-1)(-az+b)}=\frac{b-a}{b-az} . \tag{6.4.56}$$

Applying the Maclaurin expansion of $G(z)$, i.e., $\sum_{n=0}^{\infty}\frac{G^{(k)}(0)}{k!}z^n$, on (6.4.56) we will have

$$G(z) = \left(1 - \frac{a}{b}\right)\left[1 + \frac{a}{b}z + \left(\frac{a}{b}\right)^2 z^2 + \cdots\right].$$

(6.4.57)

Therefore, $p_0 = 1 - \dfrac{a}{b}$, $p_1 = \left(1 - \dfrac{a}{b}\right)\dfrac{a}{b}$, $p_2 = \left(1 - \dfrac{a}{b}\right)\left(\dfrac{a}{b}\right)^2$, $\cdots$,

or

$$p_n = \left(1 - \frac{a}{b}\right)\left(\frac{a}{b}\right)^n, \ n = 0,1,2,\cdots,$$

(6.4.58)

which is the solution we are looking for.

6.5. DIFFERENTIAL-DIFFERENCE EQUATIONS

Definition 6.5.1.

In addition to differences in equations, there may also be derivatives or integrals. In such case, we refer to the equations as *differential-difference equation* or *integral-difference equation*. These equations can sometimes be solved by various special techniques such as generating functions, Laplace transforms, etc.

Example 6.5.1.

Solve the differential-difference equation by the method of generating functions.

$$y'_{k+1}(t) = y_k(t), \ k = 0,1,2,3,\cdots,$$

(6.5.1)

with initial conditions

$$y_0(t) = 1, \ y_k(0) = \delta_{k,o} = \begin{cases} 1, & \text{if } k = 0, \\ 0, & \text{if } k \neq 0. \end{cases} \ k = 0,1,2,3,\cdots.$$

(6.5.2)

Solution
We introduce the generating function

$$G(t,z) = \sum_{k=0}^{\infty} y_k(t)z^k .$$

(6.5.3)

Then, from the differential-difference equation (6.5.1) we have

$$\sum_{k=0}^{\infty} y'_{k+1}(t)z^k = \sum_{k=0}^{\infty} y_k(t)z^k$$

or

$$\frac{1}{z}\sum_{k=0}^{\infty} y'_{k+1}(t)z^{k+1} = \sum_{k=0}^{\infty} y_k(t)z^k . \tag{6.5.4}$$

The equation (6.5.4) can be written in terms of generating function $G(t,z)$ as

$$\frac{1}{z}\left[\frac{\partial}{\partial t}G(t,z) - y'_0(t)\right] = G(t,z). \tag{6.5.5}$$

Since $y_0(t) = 1$, we have from (6.5.5)

$$\frac{\partial G(t,z)}{\partial t} - zG(t,z) = 0. \tag{6.5.6}$$

The last equation (6.5.6) is linear first order differential equation with respect to t and therefore the general solution is:

$$G(t,z) = Ce^{tz}. \tag{6.5.7}$$

From (6.5.2), (6.5.3), and (6.5.7) we have

$$G(0,z) = C = \sum_{k=0}^{\infty} y_k(0)z^k = 1$$

Hence, $C = 1$ and the solution of the problem (6.5.1), (6.5.2) is

$$G(t,z) = e^{tz} = \sum_{k=0}^{\infty} \frac{(tz)^k}{k!} = \sum_{k=0}^{\infty} \frac{t^k}{k!}z^k \tag{6.5.8}$$

Since the coefficient of z^k in the expansion is $y_k(t)$, we find the solution

$$y_k(t) = \frac{t^k}{k!}, \quad k = 0,1,2,3,\cdots.$$

EXERCISES

6.1.   Introduction

A. Find the general solution of each of the following homogeneous difference equations:

6.1.1. $y_{k+2} - y_k = 0$.

6.1.2. $y_{k+2} - 5y_{k+1} + 6y_k = 0$.

6.1.3. $y_{k+2} + 2y_{k+1} + y_k = 0$.

6.1.4. $4y_{k+2} - 6y_{k+1} + 3y_k = 0$.

6.1.5. $4y_{k+2} - 4y_{k+1} + y_k = 0$.

6.1.6. $y_{k+2} + y_{k+1} + y_k = 0$.

6.1.7.   For each of the equations in Problems 6.1.1,…, 6.1.6, find a particular solution satisfying the initial conditions $y_0 = 0$ and $y_1 = 1$.

B. Find particular solution of the following second order difference equations by the method of undetermined coefficients.

6.1.8. $y_{k+2} - 5y_{k+1} + 6y_k = 3$.

6.1.9. $y_{k+2} - 3y_{k+1} + 2y_k = 1$.

6.1.10. $y_{k+2} - y_{k+1} - 2y_k = k^2$.

6.1.11. $y_{k+2} - 2y_{k+1} + y_k = 3 + 4k$.

6.1.12. $y_{k+2} - 2y_{k+1} + y_k = 2^k(k+1)$.

6.1.13. $8y_{k+2} - 6y_{k+1} + y_k = 5\sin\dfrac{k\pi}{2}$.

6.2. Linear First Order Difference Equations

Find the solution of the first order difference equations with the given initial condition and write out the first seven values of y_k in sequence form.

6.2.1. $y_{k+1} = y_k + 2, \quad y_0 = 2$.

6.2.2. $y_{k+1} + y_k - 3 = 0, \quad y_0 = 1$.

6.2.3. $y_{k+1} = 3y_k - 1, \quad y_0 = 2$.

6.2.4. $y_{k+1} + 3y_k = 0, \quad y_0 = 2$.

6.2.5. $3y_{k+1} = 2y_k + 4, \quad y_0 = -4$.

6.2.6. $2y_{k+1} - y_k = 4, \quad y_0 = 3$.

6.3. Behavior of Solutions of the First Order Difference Equations

Describe the behavior of the solution sequence $\{y_k\}$ of the following difference equations.

6.3.1. $y_{k+1} = y_k + 2, \quad y_0 = 2$.

6.3.2. $y_{k+1} + y_k - 3 = 0, \quad y_0 = 1$.

6.3.3. $y_{k+1} = 3y_k - 1, \quad y_0 = 2$.

6.3.4. $y_{k+1} + 3y_k = 0, \quad y_0 = 2$.

6.3.5. $3y_{k+1} = 2y_k + 4, \quad y_0 = -4$.

6.3.6. $2y_{k+1} - y_k = 4, \quad y_0 = 3$.

6.4.   Generating Functions and their Applications in Difference Equations

Find the generating function $G(z)$ of the sequence $\{a_n\}$:

6.4.1. $a_k = 1 + 3k$.

6.4.2. $a_k = \left(\dfrac{1}{3}\right)^k \dfrac{2}{3}$.

6.4.3. $a_k = (k+1)(k+2)$.

6.4.4. $a_k = \dfrac{k(k+1)}{2}$.

Find the sequence $\{a_n\}$ having the generating function $G(z)$:

6.4.5. $G(z) = \dfrac{2}{1-z} + \dfrac{1}{1-2z}$.

6.4.6. $G(z) = \dfrac{z}{(1-z)^2} + \dfrac{z}{1-z}$.

6.4.7. $G(z) = \dfrac{z^2}{1-z}$.

6.4.8. $G(z) = \dfrac{z}{1-2z}$.

6.4.9.   Solve the partial difference equation by the method of generating function:

$$2P_{k,n} = P_{k-1,n} + P_{k,n-1},$$

with initial conditions $P_{k,0} = 0$, $P_{0,n} = 1$ if $n = 1,2,3,...,$ $P_{0,0} = 0$.

Chapter 7

STOCHASTIC PROCESSES AND THEIR APPLICATIONS

The goal of this chapter is to utilize what we have discussed in the previous chapters, as much as possible, in their applications in stochastic processes. This approach is not new since the theory of stochastic process originated based on needs of physics in studying a random phenomena changing with time. Before doing so, we start with some basic vocabularies we need throughout the chapter.

7.1. INTRODUCTION

Definition 7.1.1.

Let $X(t)$ be a random variable, where t is a parameter (index, time point, or epoch) belonging to a set T (linear, denumerable, or nondenumerable). The family (or sequence) of random variables $\{X(t), t \in T\}$ is called a *stochastic process*. If T is countable (or denumerable), then the process is called a *discrete-time* (or *discrete-parameter*), while if it is continuous, then the process is called *continuous-time* (or *continuous-parameter*) *stochastic process*. Using a nonnegative index set, the discrete-time process is denoted by $\{X_n, n = 0, 1, \cdots\}$, while the continuous-time is denoted by $\{X(t), t \geq 0\}$. The set of all possible values of random variable $X(t)$, denoted by S, is called the *state space* of the process. The state space may be finite, countable or noncountable. The starting value of the process is called the *initial state* and is denoted by $X(0)$. The index set T is called the *parameter space*. We could, hence, say that a stochastic process is a sequence or a family of random variables indexed by a parameter.

Definition 7.1.1 allows a stochastic process to be any one of the following four categories:

1. Discrete-time and discrete state space, DD (or chain).
2. Discrete-time and continuous state space, DC.
3. Continuous-time and discrete state space, CD (or chain).
4. Continuous-time and continuous state space, CC.

Example 7.1.1.

Consider the net number of sales and returns of an item in a store at time t within 24 hours. For this stochastic process, the index set T is the set $\{0,1,2,\cdots,24\}$ and the state space is $\{0,\pm 1,\pm 2,\cdots\}$.

Example 7.1.2.

In Example 7.1.1., consider only arrival of customers at the store. Then, the number of customers arrived during the interval $[0, t]$ is a stochastic process in which $T = [0,\infty)$ and the state space $S = \{0,1,2,\cdots\}$. This is also an example of a pure birth process that will be considered in detail later.

Example 7.1.3.

Consider a small industrial firm with five computers and a technician to handle service of these computers, who can repair one computer per day. If there is a backlog of request for service, a failed computer must get in a waiting line (queue) until the technician repairs the failed computers ahead of it. History has shown that each computer has a 20% chance of failing during any day it is in operation. The failure of a computer is independent of the failure of another. If the backlogs at time t are the quantities of interest, then the set of states of this process is $\{0, 1, 2, 3, 4\}$, since if the fifth computer fails, after the technician completes repair of one, there will be a maximum of 4 computers waiting for service.

Example 7.1.4.

Suppose we have a devise that sends a massage in time and there is a noise present either unintentionally or intentionally (as in war-time). It is desired to filter the noise and have the true signal (Beekman 1974, p. 17). An example of a noise is the shot effect in vacuum tube. The random fluctuation in the intensity of flow of electrons from cathode to the anode causes the shot effect. This is an evolution of a physical phenomenon through time we mentioned that historically a stochastic process originated.

To see how this problem can be modeled as a stochastic process, we assume that when an electron arrives at the anode at time zero ($t = 0$) it produces an effect, say $f(t)$, at some point in time in the output circuit. We further assume that the output circuit is such that the effects of the various electrons add linearly. Thus, if the kth electron arrives at time t_k, then the total effects at time t, denoted by $X(t)$, may be represented as

$$X(t) = \sum_{\text{all}\, t_k < t} f(t - t_k),$$

where the series is assumed to be convergent. Assuming that electrons arrive independently, $X(t)$ may be assumed asymptotically a normal random variable. In that case, $\{X(t), 0 \le t \le 1\}$ is a *normal stochastic process*.

Example 7.1.5.

Let $X_1, X_2, \cdots, X_n$ be a random sample of size n from a population with an unknown distribution function $F(x)$. Then, its *empirical distribution function*, denoted by $F_n(x)$, is defined as

$$F_n(x) = \frac{1}{n}\left(\text{Number of } X_i\text{'s} \le x\right) = \frac{1}{n}\sum_{i=1}^{n} \Psi_x\left(X_i\right), \qquad -\infty < x < \infty, \qquad (7.1.1)$$

where Ψ_x is the indicator function,

$$\Psi_X = \begin{cases} 1, & \text{if } x \le X, \\ 0, & \text{if } x > X. \end{cases}$$

The function $F_n(x)$ being a random variable for each x, makes $\{F_n(x), -\infty < x < \infty\}$ a stochastic process.

Definition 7.1.2.

A sequence of real numbers $\{\tau_1, \tau_2, \cdots, \tau_n, \cdots, \tau_1 \ge 0\}$ is called a *point process* if $\tau_1 < \tau_2 < \cdots$ and $\lim_{n \to \infty} \tau_n = +\infty$. The numbers τ_n, $n = 1, 2, \cdots$, are called *event times* or *epochs*.

Definition 7.1.2 implies that a point process is a strictly increasing sequence of real numbers, which does not have a finite limit point. Examples of such process are numerous: arrival time points of tasks at service stations such as supermarkets, failure time points of machines, and time points of traffic accidents. In these examples, τ_n are times at which events occur, i.e., the arrival times.

We note that τ_n s do not necessarily have to be points of time. They could be points other than times such as location of potholes in a road. However, since the number of potholes is finite, in the real world, to satisfy the conditions stated in Definition 7.1.2 we need to consider finite sample from a point process. The arrival times are of less interest than the number arrived in an interval of time.

Definition 7.1.3.

Let the number of events occurred in an interval $(0, t]$, $t > 0$, at event times τ_n be denoted by $X(t)$, $t > 0$. Then, $X(t) = \max\{n, \tau_n \le t\}$. The sequence $\{X(t), t \ge 0\}$ is called the *counting process* belonging to the point process $\{\tau_1, \tau_2, \cdots, \tau_n, \cdots, \tau_1 \ge 0\}$. If it is assumed that no more than one event can occur at a time, the process is called *simple counting process*. When the time points τ_n are random variables, then $\{\tau_1, \tau_2, \cdots, \tau_n, \cdots, \tau_1 \ge 0\}$ will be called *random point process* with $P\left\{\lim_{n \to \infty} \tau_n = +\infty\right\} = 1$. If the time points are of different types, such as arrival of different types of jobs to a service station, the process is called a *marked point process*.

Note that the number of events that occur in an interval $(s, t]$, $s < t$, denoted by $X(s, t)$, is $X(s, t) = X(t) - X(s)$.

Definition 7.1.4.

Let $\theta_i = \tau_i - \tau_{i-1}$, $i = 1, 2, \cdots$, $\tau_0 = 0$. If $\{\theta_i, i = 1, 2, \cdots\}$ is a sequence of independent identically distributed (iid) random variables, then the random point process $\{\tau_1, \tau_2, \cdots\}$ is called *recurrent point process*.

Note that Poisson and renewal processes are two most important examples of recurrent point process. We will discuss these two processes below.

Definition 7.1.5.

Let $\theta_i = \tau_i - \tau_{i-1}$, $i = 1, 2, \cdots$, $\tau_0 = 0$. When random variables θ_i, $i = 1, 2, \cdots$, are non-negative, the recurrent point process $\{\theta_1, \theta_2, \cdots\}$ is called an *ordinary renewal process*. The intervals θ_i, $i = 1, 2, \cdots$, are called *renewal periods* or *renewal cycle length*.

Definition 7.1.6.

Let $\theta_i = \tau_i - \tau_{i-1}$, $i = 1, 2, \cdots$, $\tau_0 = 0$. Assume that θ_i, $i = 1, 2, \cdots$, are non-negative independent random variables. Further assume that

$$F_1(t) \equiv P\{\theta_1 \le t\}$$

and θ_i, $i = 2, 3, \cdots$, are identically distributed as

$$P\{\theta_i \le t\} \equiv F(t), \quad i = 2, 3, \cdots, \quad F(t) \ne F_1(t).$$

Then, the sequence $\{\theta_1, \theta_2, \cdots\}$ is called a *delayed renewal process*. The random time point at which the nth renewal occur is given by

$$T_n = \sum_{i=1}^{n} \theta_i, \quad n = 1, 2, \cdots. \tag{7.1.2}$$

The random point process $\{T_1, T_2, \cdots\}$ is called the *time points of renewal process*.

Example 7.1.6.

Suppose that in an ordinary renewal process, the periods $T \equiv \theta_i$ are normally distributed with parameter μ and σ. Although, the renewal theory has been extended to negative periods, for us, to make sure that T is non-negative, we assume that $\mu > 3\sigma$. It is well know that sum of independent normally distributed random variables is again normally distributed. The parameters of the sum is the sum of the parameters [see Beichelt (2006), Example 1.24, section 1.7.2]. Then, distribution of T_n, defined in (7.1.2), is

$$P(T_n \le t) = \Phi\left(\frac{t - n\mu}{\sigma\sqrt{n}}\right), \quad t \ge 0, \tag{7.1.3}$$

where $\Phi(x)$ is the standard normal distribution. Since T_n is the sum n iid random variables, formula (7.1.2) can be interpreted as follows: if n is sufficiently large (say 20 or higher), then T_n is approximately normal with parameters $n\mu$ and $n\sigma^2$.

Definition 7.1.7.

For τ_n, $n = 1, 2, \cdots$, as random variables, let $X(t) = \max\{n, \tau_n \le t\}$ be the random number of events occurring in the interval $(0, t]$. A continuous-time stochastic process $\{X(t), t \ge 0\}$ with state space $\{0, 1, 2, \cdots\}$ is called the *random counting process* belonging to the random point process $\{\tau_1, \tau_2, \cdots, \tau_n, \cdots\}$ if it satisfies the following three properties:

i. $X(0) = 0$;

ii. $X(\tau_1) \leq X(\tau_2)$ for $\tau_1 \leq \tau_2$; and

iii. for any τ_1 and τ_2 such that $0 \leq \tau_1 \leq \tau_2$, the number of events occurred in $(\tau_1, \tau_2]$, denoted by $X(\tau_1, \tau_2)$, is equal to $X(\tau_2) - X(\tau_1)$. $X(\tau_1, \tau_2)$ is called the *increment*. Note that $X(t) = X(0, t)$.

Note that every stochastic process $\{X(t), t \geq 0\}$ in continuous time having properties i – iii is the counting process of a certain point process $\{\tau_1, \tau_2, \cdots, \tau_n, \cdots\}$. Thus, statistically speaking, the stochastic processes $\{\tau_1, \tau_2, \cdots, \tau_n, \cdots\}$, $\{t_1, t_2, \cdots, t_n, \cdots\}$, and $\{X(t), t \geq 0\}$ are equivalent.

7.2. Markov Chain and Markov Process

In examples we presented in Section 7.1.1, we assumed independent events. In other words, the outcome of an experiment at one time is not affected by the outcome of other times. Tossing a fair coin is such an example. However, this should not give us the impression that it is always true. For instance, choosing random whole numbers between 1 and 100 without replacement is a dependent event since choosing without replacement increases chances for next numbers to be chosen.

Definition 7.2.1.

A discrete-time stochastic process $\{X_n, n = 0, 1, 2, \cdots\}$ is called *discrete-time Markov chain* or simply a *Markov chain* with a state space S if for a given sequence of values, state space, $x_i \in \{1, 2, \cdots\}$, $i = 0, 1, 2, \cdots$, the following property holds:

$$P\{X_n = x_n \mid X_0 = x_0, X_1 = x_1, \cdots, X_{n-1} = x_{n-1}\} = P\{X_n = x_n \mid X_{n-1} = x_{n-1}\}. \quad (7.2.1)$$

If the state space, S, is finite, the chain is called a *finite state Markov chain* or a *finite Markov chain*. Equation (7.2.1) is the *Markov property* for the Markov chain.

Essentially, a process with Markov property "forgets" the history except its current state and the current time. This property is called the *"forgetfulness"* property of a Markov chain or *Markov property* or *stationary Markov property*.

The Markov property (7.2.1) states that given that the process is in state x, the probability for its next transition has nothing to do with the history of the process prior to the current time. In other words, the conditional probability of transition from state x to state y on the next trial depends only on x and y and not on the trial number nor any previously visited states. That is, if the left hand side of (7.2.1) exists, then (7.2.1) shows dependence between the random variables X_n; and implies that given present stat of the system, the future is independent of the past.

A historical note: Andrei A. Markov was a Russian probabilist (1856 – 1922), who contributed much to the theory of stochastic processes.

Example 7.2.1.

A fair coin "forgets" how many times it has visited the "tails" state and the "heads" state at the current toss. For instance, P("heads" on this toss | "tails on the previous 5 tosses) is still 1/2.

Example 7.2.2. (Non-Markovian Stochastic Process)

Consider a man who changes his house often, but will have one house at a time. His purchasing a house depends upon when he purchased last. Hence, whether he buys a new house will surely depend on when he last purchased a house. Let $\{X(n), n = 0, 1, 2, 3, 4, 5\}$ denote this man's house buying history for the past 6 years. Assume $X(n)$ to be an indicator function defined as:

$$X(n) = \begin{cases} 1, & \text{if a house was bought in year } n, \\ 0, & \text{otherwise.} \end{cases}$$

Let us see how this process violates the Markov property, (7.2.1). If (7.2.1) were satisfied, we would have to have the following:

$$P\big[X(5)=1\big|X(4)=0\big] = P\big[X(5)=1\big|X(4)=0, X(3)=1, X(2)=0, X(1)=1, X(0)=0\big]$$

$$= P\big[X(5)=1\big|X(4)=0, X(3)=0, X(2)=1, X(1)=1, X(0)=0\big]. \quad (7.2.2)$$

But, (7.2.2) means that the chance that the man changes his house every other year is the same as skipping two years. This is unlikely for this man. Thus, it seems that the Markov property (7.2.1) does not hold for this case.

There are cases as in Example 7.2.2 that it is not clear if the Markov property holds in a stochastic process. For instance, would the history of real estate prices in a neighborhood is an indicative for predicting future prices in that neighborhood? This is why in some cases the Markov property is assumed to hold for a stochastic process to make it a Markov process.

Definition 7.2.2.

The right hand side of (7.2.1), $P\{X_n = x_n | X_{n-1} = x_{n-1}\}$, is called a *one-step* (*one time unit* or *one jump*) *transition probability* from state x_{n-1} to state x_n. In general, the conditional probability $P\{X_n = y | X_{n-1} = x\}$, $x, y \in S$, is called the *transition probability* from state x to state y, denoted by p_{xy}, i.e.,

$$p_{xy} = P\{X_n = y | X_{n-1} = x\}, \ x, y \in S. \tag{7.2.3}$$

Note that p_{xy} in some books and other literatures is for the transition from y to x rather than from x to y that we defined. See for example Karlin and Taylor (1975) and Kemeny and Snell (1960). Note also that p_{xx} is the probability that Markov chain stays in state x for another step. Further note that p_{xy} is non-negative and sums to 1 along each row, i.e.,

$$0 \le p_{xy} \le 1, \ \forall x, y \in S, \text{ and } \sum_{y \in S} p_{xy} = 1, \ \forall x \in S. \tag{7.2.4}$$

Definition 7.2.4.

For a discrete-time Markov chain $\{X_n, n = 0, 1, 2, \cdots\}$ with space state $S = \{1, 2, \cdots\}$, the matrix $\mathbf{P}(t)$ whose elements are the one-step transition probabilities of a Markov chain, $p_{xy}(t)$, is called the *matrix of transition probabilities* or *transition matrix*. That is,

$$\mathbf{P}(t) = \begin{bmatrix} p_{11}(t) & p_{12}(t) & p_{13}(t) & \cdots \\ p_{21}(t) & p_{22}(t) & p_{23}(t) & \cdots \\ p_{31}(t) & p_{32}(t) & p_{33}(t) & \cdots \\ \vdots & \vdots & \vdots & \vdots \end{bmatrix}. \tag{7.2.5}$$

With two properties of the matrix $\mathbf{P}(t)$, namely, $0 \le p_{xy}(t) \le 1$ and $\sum_{y \in S} p_{xy}(t) = 1$, $\forall x \in S$, as in (7.2.4), the transition matrix $\mathbf{P}(t)$ is called a *stochastic matrix*. In case that the state space S is finite with N elements, then $\mathbf{P}(t)$ is an $N \times N$ matrix. If columns also sum to 1, then the transition matrix $\mathbf{P}(t)$ is called *doubly stochastic*.

Example 7.2.3.

Here is a simple 2-state discrete-time Markov chain with a 2×2 transition matrix P as

$$\mathbf{P} = \begin{bmatrix} 1-\alpha & \alpha \\ \beta & 1-\beta \end{bmatrix}, \tag{7.2.6}$$

where $0 \le \alpha, \beta \le 1$. If $\alpha = \beta = 0$, then P in (7.2.6) will be the 2×2 identity matrix $\begin{bmatrix} 1 & 0 \\ 0 & 1 \end{bmatrix}$. On the other hand if $\alpha = \beta = 1$, then P will be the anti-diagonal matrix $\begin{bmatrix} 0 & 1 \\ 1 & 0 \end{bmatrix}$.

If $\alpha = \beta = 1/2$, then P will be $\begin{bmatrix} 1/2 & 1/2 \\ 1/2 & 1/2 \end{bmatrix}$. Note that a chain with identity transition matrix stays in the initial case forever; while with the anti-diagonal matrix flips state each time. In the last case, the chain may stay put or change with equal probability. Of course, P in all three cases is stochastic matrix.

Example 7.2.4.

Here is a chain with 4 states and a transition matrix P as

$$\mathbf{P} = \begin{bmatrix} 0 & 1/3 & 1/3 & 1/3 \\ 1/2 & 1/2 & 1/2 & 1/2 \\ 1/4 & 1/4 & 1/4 & 1/4 \\ 0 & 0 & 0 & 1 \end{bmatrix}. \tag{7.2.7}$$

Definition 7.2.3.

If the one-step transition probabilities are independent of time, the Markov chain is called *homogeneous*, denoted by p_{xy}, i.e., if

$$p_{xy} = P\{X_{n+m} = y | X_{n+m-1} = x\}, \ x, y \in S, \ m = 0, \pm 1, \cdots, \pm(n-1), \cdots. \tag{7.2.8}$$

For the general case, the *n-steps transition probability* from state x to state y in n steps is denoted by $p_{xy}^{(n)}$, i.e.,

$$p_{xy}^{(n)} = P\{X_{n+m} = y | X_m = x\}, \ x, y \in S, \ n = 0, 1, \cdots, \tag{7.2.9}$$

where $p_{xy}^{(1)} = p_{xy}$ and by convention we have

$$p_{xy}^{(0)} = \begin{cases} 1, & y = x \\ 0, & y \neq x. \end{cases} \tag{7.2.10}$$

The n-step transition matrix whose elements are $p_{xy}^{(n)}$ is denoted by $\mathbf{P}^n$.

Example 7.2.5.

Consider the transition matrix given in (7.2.6) in Example 7.2.2. Of course, the chain represented by such matrix is homogeneous. Without loss of generality, we call the states of this chain as 0 and 1. Hence, $p_{00} = 1 - \alpha$, $p_{01} = \alpha$, $p_{10} = \beta$, and $p_{11} = 1 - \beta$. We want to calculate the n-step transition probability matrix for this chain.

Solution

We can find the n-step transition probability matrix directly, using iterative method. To do so, we write

$$\mathbf{P}^n = \mathbf{P}^{n-1}\mathbf{P}. \tag{7.2.11}$$

Hence, from (7.2.11) elements of $\mathbf{P}^n$ can be written as follows:

$$\begin{cases} p_{00}^{(n)} = p_{00}^{(n-1)}(1-\alpha) + p_{01}^{(n-1)}\beta, \\ p_{01}^{(n)} = p_{00}^{(n-1)}\alpha + p_{01}^{(n-1)}(1-\beta), \\ p_{10}^{(n)} = p_{10}^{(n-1)}(1-\alpha) + p_{11}^{(n-1)}\beta, \\ p_{11}^{(n)} = p_{10}^{(n-1)}\alpha + p_{11}^{(n-1)}(1-\beta). \end{cases} \tag{7.2.12}$$

Let us calculate $p_{00}^{(n)}$. Remembering $p_{00}^{(0)} = 1$, $p_{00}^{(1)} = 1 - \alpha$, and $p_{01}^{(n)} = 1 - p_{00}^{(n)}$ (complement probability), from the first equation of (7.2.12) we have:

$$\begin{aligned} p_{00}^{(n)} &= p_{00}^{(n-1)}(1-\alpha) + p_{01}^{(n-1)}\beta \\ &= p_{00}^{(n-1)}(1-\alpha) + \left(1 - p_{00}^{(n-1)}\right)\beta \\ &= (1-\alpha-\beta)p_{00}^{(n-1)} + \beta. \end{aligned}$$

Recursively, the solution is

$$p_{00}^{(n)} = \begin{cases} \dfrac{\beta}{\alpha+\beta} + \dfrac{\alpha}{\alpha+\beta}(1-\alpha-\beta)^n, & \alpha+\beta>0, \\ 1, & \alpha=\beta=0. \end{cases} \qquad (7.2.13)$$

$p_{01}^{(n)}$ can be found from the fact that the matrix is stochastic, i.e., $p_{01}^{(n)} = 1 - p_{00}^{(n)}$, and formula (7.2.11). Hence,

$$p_{01}^{(n)} = \begin{cases} \dfrac{\alpha}{\alpha+\beta}\left[1-(1-\alpha-\beta)^n\right], & \alpha+\beta>0, \\ 0, & \alpha=\beta=0. \end{cases} \qquad (7.2.14)$$

For $p_{11}^{(n)}$, we can use the last equation of (7.2.12), the method we used to find $p_{00}^{(n)}$, and swapping α and β. $p_{10}^{(n)}$ can be found as the complement of $p_{11}^{(n)}$. Hence,

$$p_{11}^{(n)} = \begin{cases} \dfrac{\alpha}{\alpha+\beta} + \dfrac{\beta}{\alpha+\beta}(1-\alpha-\beta)^n, & \alpha+\beta>0, \\ 1, & \alpha=\beta=0, \end{cases} \qquad (7.2.15)$$

and

$$p_{10}^{(n)} = \begin{cases} \dfrac{\beta}{\alpha+\beta}\left[1-(1-\alpha-\beta)^n\right], & \alpha+\beta>0, \\ 0, & \alpha=\beta=0. \end{cases} \qquad (7.2.16)$$

Definition 7.2.5.

If the stochastic process in Definition 7.2.1 is of continuous-time, i.e., $\{X(t),t\geq 0\}$, then it is called a *Markov process*.

Similar definitions and formulas for Markov process follow from Markov chain by replacing n by t. Thus, (7.2.2) – (7.2.5) become

$$P\{X(t_n)\leq x_n \,|\, X(0)=x_0, X(t_1)=x_1,\cdots,X(t_{n-1})=x_{n-1}\}$$
$$= P\{X(t_n)\leq x_n \,|\, X(t_{n-1})=x_{n-1}\}. \qquad (7.2.17)$$

$$p_{xy}(t)=P\{X(t_n)=y \,|\, X(t_{n-1})=x\}, \; x,y\in S. \qquad (7.2.18)$$

$$p_{xy}(t) = P\{X(t_{n+m}) = y \,|\, X(t_{n+m-1}) = x\}, \, x, y \in S,$$

$$t_m = t_0, \pm t_1, \cdots, \pm t_{(n-2)}, \pm t_{(n-1)}, \cdots . \quad (7.2.19)$$

$$p_{xy}^{(n)}(t) = P\{X(t_{n+m}) = y \,|\, X(t_m) = x\}, \, x, y \in S, \, t \geq 0. \quad (7.2.20)$$

The *transition probability* from x to y in a time interval of length t, will be denoted by $p_{xy}(t)$. This means that, similar to (7.2.3) we have

$$p_{xy}(t) = P\{X(s+t) = y \,|\, X(s) = x\}. \quad (7.2.21)$$

Note that the distinction between discrete and continuous-time Markov chains is that in discrete-time chains there is a "jump" to a new state at times $1, 2, \cdots$, while in continuous-time chains a new state may be entered at any time $t \geq 0$.

We denote the probability mass function (*pmf*) associated with random variables X_n by $\{p_i(n)\}$, $i = 0, 1, 2, \cdots$, i.e., $p_i(n) = P\{X_n = i, i = 0, 1, 2, \cdots\}$. Similarly, we will have the probability density function (*pdf*) for a Markov process.

Definition 7.2.6.

The probability that a Markov process is in state x at time t, denoted by $p_x(t)$ (i.e., $p_x(t) = P\{X(t) = x\}$) is called the *absolute state probability* at time t. $\{p_x(t), x \in Z\}$, where Z is the set of integers, is said to be the (one-dimensional) *absolute distribution* of the Markov chain at time t, while $\{p_x(0), x \in Z\}$ is called the *initial probability distribution* of the Markov process.

Note that the absolute probability distribution of the Markov chain at time t according to the law of total probability satisfies the system of linear difference equations

$$p_y(t) = \sum_{x \in Z} p_x(0) p_{xy}(t), \qquad y \in Z. \quad (7.2.22)$$

To explain the Markov property, suppose we observe at time t_1, a Markov process is in a state x. At a later time t_2, we see that the process is still in state x. It is interesting to note that, under the Markov assumptions, the process is no more likely to make a transition to a state y at $t_2 + \Delta t$ than it was at time $t_1 + \Delta t$, since $P(X(t_1 + \Delta t) = y | X(t_1) = x) = P(X(t_2 + \Delta t) = y | X(t_2) = x)$. The process ignores or "forgets" the time it has already spent in state x, namely $t_2 - t_1$. It is as if transition probabilities are determined "from scratch" at each instant of time. (This forgetfulness in birth-death processes will later dictate the specific distribution of the random variable, T_x, the time spent in state x, the so-called *holding time*.)

This independence of time leads to the fact that the holding time, T_x, is an exponential random variable, and like all other exponential random variables T_x has the *forgetfulness* property. Formally, we define:

Definition 7.2.7.

A continuous random variable T is *forgetful* if

$$P\left(T > s + t \mid T > t\right) = P\left(T > s\right), \ \forall s, t \geq 0 . \tag{7.2.23}$$

Example 7.2.6.

If the notion of forgetfulness seems troublesome, consider that a fair coin "forgets" how many times it has visited the "tails" state and the "heads" state at the current toss. For example, P("heads" on this toss |"tails" on the previous 9 tosses) is still 1/2.

Example 7.2.7.

If we think of T as the lifetime of some instrument (or, holding time in an "up" state until transition to a "down" state), then the equation (7.2.23) may be viewed as, given that the instrument has "lived" for at least t hours ($T > t$), the probability it survives for another s hours is identical to the probability it survives s hours from time zero. In other words, the instrument "forgets" that it has been in use for t units of time and has the same prospects for longevity as when it was new.

We emphasize again that the exponential distribution is, indeed, forgetful. We formally state this fact in the following theorem:

Theorem 7.2.1.

If T is an exponential random variable with mean $1/\mu$ (and in particular if T is holding time for some state in a birth-death process), then T is forgetful.

Proof

We remind the reader that a random variable is exponential with parameter μ if its probability distribution function is given by:

$$P(X \leq x) = \begin{cases} 1 - e^{-\mu x}, & x \geq 0 \\ 0, & x < 0 \end{cases}$$

Using the conditional probability property

$$P(B|A) = \frac{P(A \cap B)}{P(A)}, \quad P(A) \neq 0, \text{ for the events } A \text{ and } B,$$

we have

$$P(T > s+t|T > t) = \frac{P(T > s+t \text{ and } T > t)}{P(T > t)} = \frac{P(T > s+t)}{P(T > t)}$$

$$= \frac{e^{-\mu(s+t)}}{e^{-\mu t}} = e^{-\mu s} = P(T > s).$$

Remark 7.2.1.

The exponential distribution function can be shown to be the only continuous probability distribution having the forgetfulness property and conversely a continuous probability distribution function having such a property is exponential. For discrete probability mass functions, however, only the geometric mass function has this property.

Definition 7.2.8.

For a Markov process with state space S, the equations

$$p_{xy}(t+\tau) = \sum_r p_{xr}(t) p_{ry}(\tau), \quad x, r, y \in S, \ t, \tau \geq 0 \tag{7.2.24}$$

or in matrix form

$$\mathbf{P}(t+\tau) = \mathbf{P}(t)\mathbf{P}(\tau), \ \forall t, \tau \in [0, \infty) \tag{7.2.25}$$

are called the *Chapman-Kolmogorov equations*. In this case, we say that the transition probabilities satisfy the Chapman-Kolmogorov equations.

Equations similar to (7.2.24) and (7.2.25) can be defined for discrete-time Markov chains.

Definition 7.2.9.

A Markov process whose transition probabilities $p_{xy}(t)$ do not depend on time t, is called *stationary* or *time-homogeneous* or simply *homogeneous*. Thus, for any times t and t_1 we have

$$p_{yx}(\Delta t) = P\{X(t+\Delta t) = y | X(t) = x\} = P\{X(t_1 + \Delta t) = y | X(t_1) = x\}. \qquad (7.2.26)$$

In other words, no matter what the current time t is, if the process is in state x, the probability of transition to a state y at time $t + \Delta t$ is $p_{yx}(\Delta t)$.

Definition 7.2.10.

The probability $\{\pi_x = p_x(0), x = 0, \pm1, \pm2, \cdots\}$ is called *stationary initial probability distribution* if

$$\pi_x = p_x(t), \ \forall t \geq 0, x = 0, \pm1, \pm2, \cdots. \qquad (7.2.27)$$

Example 7.2.8. Counting Particle

Radium 226 has a half-life of approximately 1600 years and its rate of decay is approximately constant over time. In particular, for a sample of 100 picograms, radium 226 emits alpha particles at an average rate of $\alpha = 3.7$ particles per second. Suppose we have a particle detector which counts sequentially, $X(t)$, the number of emitted alpha particles (from a sample of 100 picograms) by time t. For any elapsed time t, $X(t)$ may be considered a discrete random variable with values in the set of non-negative integers $\{0,1,2,3,\cdots\}$.

We begin our experiment by counting emissions from time $t = 0$ with the counter set to zero ($X(0) = 0$). For this process, we define a probability mass function $p_x(t)$, for each $t \geq 0$, as $p_x(t) = P(X(t) = x) = P(x$ particles are counted by time t), for $x = 0, 1, 2, \cdots$, where $p_0(0) = P\{X(0) = 0\} = 1$.

Now suppose that our particle counter registers 100,000 at 8:00 a.m. The probability that it counts 5 more particles in one second is

$$P(X(8:00:01) = 100,005 | X(8:00:00) = 100,000),$$

and this is equal to

$$P(X(10:00:01) = 100,005 | X(10:00:00) = 100,000),$$

the probability it counts 5 more particles in one second if the counter had registered 100,000 at 10:00:00 a.m.

7.3. CLASSIFICATION OF STATES OF A MARKOV CHAIN

We now classify the states of a Markov chain that will lead to classify the chain.

Definition 7.3.1.

A state y is *accessible* from state x, denoted by $x \rightarrow y$, if there is a positive probability, say $p_{xy}^{(n)} > 0$ for some $n \geq 0$. If y is accessible from state x ($x \rightarrow y$) and x is accessible from state y ($y \rightarrow x$), then we say x and y *communicate*, or they are *in the same class*. This relationship is denoted by $x \leftrightarrow y$. In other words, there are nonnegative integers n and m such that $p_{xy}^{(n)} > 0$ and $p_{yx}^{(m)} > 0$.

Definition 7.3.2.

In a Markov chain, a state i from the state space S is called *recurrent* if it will be visited infinitely many times, i.e., $P\big(X(n) = x \text{ for infinitly many } n\big) = 1$. The state x is called *transient* if it will not be visited infinitely many times, i.e., $P\big(X(n) = x \text{ for infinitly many } n\big) = 0$. In other words, $P\big(X(n) = x \text{ for finitly many } n\big) = 1$.

Definition 7.3.3.

Let

$$f_{xy}^{(n)} = P\big\{X(n) = y, X(k) \neq y, k = 1, 2, \cdots, n-1 \big| X(0) = x, n \geq 1\big\}. \qquad (7.3.1)$$

In (7.3.1), we defined F_{xy} as the probability that the chain, starting at state x at time 0, visits state y for the first time at the time point n. Hence, $f_{xy}^{(n)}$ defined in (7.3.1) is called the *first-passage time probabilities*.

Note that f_{xx} is the probability that state x ever returns to itself. Note also that the difference between $f_{xy}^{(n)}$ and $p_{xy}^{(n)}$ (that was defined earlier) is that $p_{xy}^{(n)}$ is the probability that the Markov chain starting from state x will reach the state y in n steps, but it may have visited state y in along the way, while in case of $f_{xy}^{(n)}$, the chain visits y for the first time.

From the law of total probability, it can be shown that a relation between $p_{xy}^{(n)}$ and $f_{xy}^{(n)}$ is

$$p_{xy}^{(n)} = \sum_{k=1}^{n} f_{xy}^{(k)} \, p_{yy}^{(n-k)} , \tag{7.3.2}$$

where $p_{yy}^{(0)} = 1, \ \forall y \in Z$.

Thus, the first passage time may be calculated recursively from

$$f_{xy}^{(n)} = p_{xy}^{(n)} - \sum_{k=1}^{n-1} f_{xy}^{(k)} \, p_{xy}^{(n-k)}, \quad n = 2, 3, \cdots . \tag{7.3.3}$$

If F_{xy} denotes the first-passage time random variable with *cdf* as $\left\{ f_{xy}^{(n)}, n = 1, 2, \cdots \right\}$, then expected first –passage time, denoted by μ_{xy}, is

$$\mu_{xy} = E(F_{x=y}) = \sum_{n=1}^{\infty} n f_{xy}^{(k)} . \tag{7.3.4}$$

Suppose a Markov chain starts from a state x. Let f_{xy} denote the probability that the chain ever makes a transition into y. Then,

$$f_{xy} = \sum_{n=1}^{\infty} n f_{xy}^{(n)} . \tag{7.3.5}$$

The state x is called *positive recurrent*, if $\mu_{xx} < 1$ and is called *null-recurrent*, if $\mu_{xx} = \infty$, and $f_{xx} < 1$.

We can redefine the transient state as in the following definition.

Definition 7.3.4.

A state x for which $p_{xx} = 1$ (so that $p_{xy} = 0$ for $y \neq x$) is called an *absorbing state* (or an absorbing barrier). Note that P(leaving an absorbing state) = 0. If a state is a non-absorbing state, it is called a *transient state*.

The relationship $\leftrightarrow$ has the following properties of *equivalence relation* (see Karlin and Taylor, 1975, and Allen, 2003):

1. Reflexive: $x \leftrightarrow x$.
2. Symmetry: $x \leftrightarrow y$ implies $y \leftrightarrow x$.
3. Transitivity: $x \leftrightarrow y$ and $y \leftrightarrow z$ implies $x \leftrightarrow z$.

The equivalence relation $\leftrightarrow$ on the states of the Markov chain defines a set of equivalence classes of the Markov chain.

Definition 7.3.5.

The sets of equivalence classes in a Markov chain are called the *classes of the Markov chain*.

Note that if every state of a Markov chain is accessible from every other state, then there is only one class for the chain. This relation partitions the state space S into disjoint, but not necessarily closed classes in such a way that two states x and y belong to the same class if and only if they communicate.

Definition 7.3.6.

A Markov chain with only one communication class is called an *irreducible* Markov chain. However, if there is more than one class, then it is called *reducible*.

Definition 7.3.7.

In a Markov chain, a set of states is called *closed* if it is impossible to access any state outside of that class from inside by only one transition.

What Definition 7.3.7 is saying is that a subset C of the state space of a Markov chain is closed if $\sum_{x \in C} p_{xy} = 1$ for all $y \in C$. In other words, if a Markov chain is in a closed state, it cannot leave that state since $p_{xy} = 0$ when $x \in C$ and $y \notin C$.

Example 7.3.1.

The class of an absorbing state x consists of only the state x and such a class is closed. Note that if x is not absorbing and it is the only state in its class, then the class is open and x is visited only one time, i.e., the chain will never revisit such class after it visited x for the first time.

7.4. RANDOM WALK PROCESS

Random walks serve as models for a variety of useful problems in applied probability. We begin the discussion of random walks with some examples including an example of a classic problem called "*Gambler's Ruin.*"

Example 7.4.1.

Suppose a particle starting from a point, say i, moves along the real axis one step forward to $i+1$ or one step backward to $i-1$, with equal probability. Let X_n represent the position of the particle after n steps. Thus, the starting point i will be the position X_0. Thus, the sequence $\{X_0, X_1, \cdots, X_n, \cdots\}$ is a discrete-time Markov chain with state space $S = \{0, \pm 1, \pm 2, \cdots\}$ and one-step transition probabilities

$$p_{ij} = \begin{cases} 1/2, & \text{for } j = i+1 \text{ or } j = i-1, \\ 0, & \text{otherwise.} \end{cases}$$

Now suppose we know that the initial step was on the real line somewhere between 1 and 4, inclusive. Suppose also that if the particle reaches at point 0 or 6, it will not be able to get out of it. In other words, states 0 and 6 are *absorbing states*. Thus, for this revised Markov chain, $\{X_0, X_1, \cdots\}$, the state space is $S = \{0, 1, 2, 3, 4, 5, 6\}$ and its one-step transition probabilities are

$$p_{ij} = \begin{cases} 1/2, & j = i+1 \text{ or } j = i-1, \text{ and } 1 \le i \le 5, \\ 1, & i = j = 0 \text{ or } i - j - 6, \\ 0, & \text{otherwise.} \end{cases}$$

The one-step transition matrix, in this case, is

$$\mathbf{P} = \begin{bmatrix} 1 & 0 & 0 & 0 & 0 & 0 & 0 \\ 1/2 & 0 & 1/2 & 0 & 0 & 0 & 0 \\ 0 & 1/2 & 0 & 1/2 & 0 & 0 & 0 \\ 0 & 0 & 1/2 & 0 & 1/2 & 0 & 0 \\ 0 & 0 & 0 & 1/2 & 0 & 1/2 & 0 \\ 0 & 0 & 0 & 0 & 1/2 & 0 & 1/2 \\ 0 & 0 & 0 & 0 & 0 & 0 & 1 \end{bmatrix}.$$

Example 7.4.2. Gambler's Ruin Problem

Two gamblers, designated by G1 and G2 (Gambler 1 and Gambler 2), play a simple gambling game. On each round (trial) of the game a coin is tossed and if H = "Heads" results then G1 wins \$1 from G2. On the other hand, if T = "Tails" results from the coin toss, then G2 wins \$1 from G1. The probabilities associated with "Heads" and "Tails" on the coin are unknown to the players so that both players feel free to continue the rounds. Rounds continue until one of the players loses all his money and is thereby "ruined".

Note that for this game to be "fair" the coin used in the model should have $P(H) = P(T) = 1/2$. On the other hand, if we wished to model a more general series of success/failure trials, where levels of skill are taken into account, we could use a biased coin. For example, a game where G2 is three times as skilled as G1 might use a model with $P(H) = 1/4$ and $P(T) = 3/4$.

For this example, suppose G1 has \$3 and G2 has \$1 before the games begin. Further suppose that (unknown to the players) $P(H) = 1/3$ and $P(T) = 2/3$. We can study the game by keeping track of the money that G1 has at the end of each round since G2 will have \$4 minus that amount. Over the course of the game, G1 will have a fortune ranging from \$0 to \$4. Figure 7.3.1 illustrates the possible dollar amounts of G1's fortune as well as the probabilities for his moving from one fortune value to the next during the process.

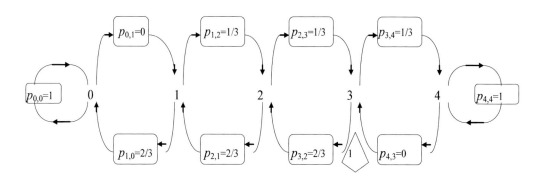

Figure 7.4.1. Transition Probability Diagram for Example 7.4.1.

In Figure 7.4.1, the values 0, 1, 2, 3, 4 are the *states* of the process and these represent possible dollar amounts (fortunes) that G1 might have during the course of the trials. Arrows between states of the form $x \rightarrow (p_{x,y}) \rightarrow y$ indicate that $p_{x,y}$ is the probability that G1 moves from \$x to \$y (given that he has \$x) on the next round of the game. For example, $2 \leftarrow (p_{3,2} = 2/3) \leftarrow 3$ means that if G1 has \$3 (state 3), then the probability that he loses a round and moves to \$2 (state 2) is 2/3. The arrowhead beneath state 3 contains the initial probability of beginning at state 3 and this probability is 1 since G1 begins with \$3.

Consider state 0. Here, G1 has \$0 and the game has stopped. Thus, we write $p_{0,0} = 1$ since there are no subsequent rounds. Similarly, we indicate game's end at state 4 (G1 has all \$4) by $p_{4,4} = 1$. If the game ends at state 0 (G1 loses all his money), we say that the process has been *absorbed* into state 0 since state 0 cannot be exited once entered. Similarly, state 4 is an *absorbing* state.

Note that for each state, the sum of the probabilities leaving that state is 1.

This transition probability diagram illustrates an example of what is called a *random walk*. In this process we "walk" back and forth (or up and down) from state to state and the direction of a step (transition) is determined in a random fashion by the coin toss.

Definition 7.4.1.

A (*simple*) *random walk* is described as follows:

- It consists of a subset X of the non-negative integers, called the *state space*, where each integer in X is called a *state*.

- Each state $k \in X$ is assigned an *initial probability*, $p_k(0)$, where $\sum_{k \in X} p_k(0) = 1.$

- Each state k has associated with it, a random experiment which results in an outcome of k, $k-1$, or $k+1$ with respective probabilities $p_{k,k-1}$, $p_{k,k}$ and $p_{k,k+1}$, with $p_{k,k+1} + p_{k,k-1} + p_{k,k} = 1$. Some, but not all of $p_{k,k-1}$, $p_{k,k}$ and $p_{k,k+1}$ may be zero. These *transition probabilities* are:
  (i) $p_{k,k-1} = P$(transition from state k to state $k-1$),
  (ii) $p_{k,k} = P$(transition from state k back to state k), and
  (iii) $p_{k,k+1} = P$(transition from state k to state $k+1$).

Example 7.4.3. (Gambler's Ruin Problem Continues)

Let us go back to Example 7.4.2. Remember we had five states 0, 1, 2, 3, 4. Let us generalize the number of states and assume that the state space has $N+1$ elements, i.e., $S = \{0, 1, 2, \cdots, N\}$. Recall that states in the Gambler's Problem represent the amount of money (fortune) of one gambler. The gambler bets \$1 per round and wins or loses. He is ruined if he reaches state 0. The probability is assumed to be $p > o$ and of losing is $q > 0$ with $p + q = 1$. In terms of transition probabilities, this assumption means that $p_{k,k+1} = p$ and $p_{k,k-1} = q$, for $1, 2, \cdots, N-1$. We further assume that $p_{oo} = p_{NN} = 1$, i.e., p_{oo} and p_{NN} are *absorbing boundaries*. All other transition probabilities are assumed to be zero. Note that in this case as in Example 7.4.2, transition of a sate to itself is not allowed, i.e., $p_{kk} = 0$ for all k except $k - 0$ and $k = N$. The transition matrix for this random walk is as follows:

$$\mathbf{P} = \begin{bmatrix} 1 & q & 0 & \cdots & 0 & 0 \\ 0 & 0 & q & \cdots & 0 & 0 \\ 0 & p & 0 & \cdots & 0 & 0 \\ 0 & 0 & p & \cdots & 0 & 0 \\ \vdots & \vdots & \vdots & \cdots & \vdots & \vdots \\ 0 & 0 & 0 & \cdots & q & 0 \\ 0 & 0 & 0 & \cdots & 0 & 0 \\ 0 & 0 & 0 & \cdots & p & 1 \end{bmatrix}. \tag{7.4.1}$$

It should be clear that there are three communication classes for this Markov chain. They are $\{0\}$, $\{1, 2, \cdots, N-1\}$, and $\{N\}$. The sets of states $\{0\}$ and $\{N\}$ are closed, but the set of states $\{1, 2, \cdots, N-1\}$ is not. The Markov chain in this case is reducible. The states 0 and N are absorbing states and all other states are transient. Hence, this is an example of a random walk with absorbing boundaries.

Example 7.4.4.

Consider a sequence of n independent Bernoulli trials, each with $P(\text{success}) = P(S) = p$ and $P(\text{failure}) = 1 - p = q$. Let the random variable $X(n)$ = the number of successes accumulated through n trials, that is, $X(n)$ represents the number of accumulated successes over n trials.

This process may model cases such as the following: (1) $X(n)$ might be the number of patients (out of n patients) who respond favorably to some particular medical treatment with a known success rate of p; $X(n)$ could represent the number of sales made in n sales calls where $P(\text{making a sale on a given call}) = p$; (2) $X(n)$ could represent the number of times, out of n trials, a beam survives application of a stress (i.e., amount of strength exceeds the amount of stress).

We will consider a third example, i.e., a random walk that is described as follows: A steps forward is denoted by 1 and staying put is denoted by 0. On any trial the probability of a success (a transition forward) is p, while the probability of staying put (a trial without a success) is $1 - p = q$, $p + q = 1$. We let a state, $X(n)$, be the position of the walker at the nth trial.

Suppose we start the process at trial 0 with no success, i.e., $X(0) = 0$. We also let $p_x(n)$ denote the probability that the process is at position x at the nth trial, i.e., $p_x(n) = P(X(n)) = x$. Then a difference equation describing the process is as follows:

$$p_x(n+1) = \sum_{y \in X} P_{yx} p_y(n) = p_{x-1,x} p_{x-1}(n) + p_{x,x} p_x(n). \tag{7.4.7}$$

From (7.4.7), we can write the transition matrix $P(n)$ (with infinite states) and a probability recursion table generated from it:

$$P(n) = \begin{array}{c} \\ 0 \\ 1 \\ 2 \\ 3 \\ 4 \\ \vdots \end{array} \begin{bmatrix} \begin{array}{cccccc} 0 & 1 & 2 & 3 & 4 & \cdots \end{array} \\ \begin{array}{cccccc} 1 & 0 & 0 & 0 & 0 & \cdots \\ q & p & 0 & 0 & 0 & \cdots \\ q^2 & 2pq & p^2 & 0 & 0 & \cdots \\ q^3 & 3q^2p & 3pq^2 & p^3 & 0 & \cdots \\ \vdots & \vdots & \vdots & \vdots & \vdots & \vdots \\ \vdots & \vdots & \vdots & \vdots & \vdots & \vdots \end{array} \end{bmatrix}.$$

Table 7.4.2. Probability Recursion Table for Example 7.4.4.

$p_x(n)$ / Trial (n)	State (x)					
	0	1	2	3	4	$\cdots$
0	1	0	0	0	0	$\cdots$
1	q	p	0	0	0	$\cdots$
2	q^2	$2pq$	p^2	0	0	$\cdots$
3	q^3	$3q^2p$	p^3	0	0	$\cdots$
4	$\vdots$	$\vdots$	$\vdots$	$\vdots$	$\vdots$	$\vdots$
$\vdots$	$\vdots$	$\vdots$	$\vdots$	$\vdots$	$\vdots$	$\vdots$

Note that each row n of the table contains the terms of the binomial expansion and consequently, the probability mass function for $X(n)$, is the Binomial distribution.

For example, if walker steps forward with probability .75, then the probability that he/she is at location 4 in 5 trials is

$$p_4(5) = \binom{5}{4} (.75)^4 (.25)^{5-4} = 0.3955.$$

Example 7.4.5. (Failure-Repair Model (FR))

Suppose a computer in a certain office has completely fails to work with probability, q, each time it is turned on. The probability that the machine functions without breakdown on a given trial is assumed to be $1 - q = p$. Each time the computer fails, a repairperson (an IT personnel) is called. The computer cannot be used until it is completely repaired.

Solution

It is clear that the process is a random walk. Let $\{0, 1\}$ denote the state space of this Markov chain. We define the random variable $X(n)$ to be the number of computers "up" (working properly) at the end of the nth trial. In this example, values of $X(n)$ are 0 or 1 since there is only one computer in the office. We also assume that the process begins in state 1 (that is "up"). Hence, $p_1(0) = 1$ and $p_0(0) = 0$. Thus,

$$p_1(1) = p_{01}p_0(0) + p_{11}p_1(0) = (1)(0) + (p)(1) = p = 1 - q$$

and

$$p_0(1) = p_{10}p_1(0) = (q)(1) = q.$$

Proceeding to trial 2 we have

$$p_1(2) = p_{01}p_0(1) + p_{11}p_1(1) = (1)(q) + (1 - q)(1 - q) = 1 - q + q^2$$

and

$$p_0(2) = p_{10}p_1(1) = (1 - q)q = q - q^2.$$

Continuing in this way, we can write our transition matrix and generate the probability recursion table as follows:

$$
P(n) = \begin{matrix} & 0 & 1 \\ 0 \\ 1 \\ 2 \\ 3 \\ 4 \\ \vdots \end{matrix}
\begin{bmatrix}
0 & 1 \\
q & 1-q \\
q-q^2 & 1-q+q^2 \\
q-q^2+q^3 & 1-q+q^2-q^3 \\
q-q^2+q^3-q^4 & 1-q+q^2-q^3+q^4 \\
\vdots & \vdots
\end{bmatrix}.
$$

Table 7.4.3. Probability Recursion Table for Example 7.4.5.

Trial (n) \ $p_x(n)$	State (x)	
	0	1
0	0	1
1	q	1-q
2	$q-q^2$	$1-q+q^2$
3	$q-q^2+q^3$	$1-q+q2-q3$
4	$q-q^2+q^3-q^4$	$1-q+q^2-q^3+q^4$
$\vdots$	$\vdots$	$\vdots$

The table reveals that for any trial n

$$p_1(n) = 1 + (-q) + (-q)^2 + \cdots + (-q)^n = \sum_{k=0}^{n} (-q)^k = \frac{1-(-q)^{n+1}}{1+q}$$

and, thus,

$$p_0(n) = 1 - \frac{1-(-q)^{n+1}}{1+q} = \frac{q+(-q)^{n+1}}{1+q}.$$

So if, for example, the machine breaks down 5% of the times it is used, then the probability that the machine is "up" after trial 5 is $p_1(5) = (1 - (-0.05)^6)/(1 + .05) = 0.952$.

Definition 7.4.5.

In a Markov chain, the greatest common divisor of $n \geq 1$, say d_i, such that $p_{ij}^{(n)} > 0$, the d_i is called the *period* of state i. If $p_{ii}^{(n)} = 0, \forall n > 0$, then the period of i is defined as infinite, i.e., $d_i = \infty$. If $d_i = 1$, then the state i is called *aperiodic*. An aperiodic and positive recurrent state is called *ergotic*.

It can be shown that for a finite irreducible aperiodic Markov chain, the rate of convergence of $p_{ij}^{(n)}$ to π_j is geometric. In other words, for a number r and for some $k \geq 1$

$$p_{ij}^{(k)} \geq r, \text{ for all states } i \text{ and } j. \tag{7.4.8}$$

Theorem 7.4.1.

If P is a finite irreducible and aperiodic, then for all states i and j we have

$$\left| p_{ij}^{(m)} - \pi_j \right| \leq (1-r)^{(m/k)-1}, \tag{7.4.9}$$

where k and r are as in (7.4.8).

Proof

See Suhov and Kelbert (2008, p.74)

Example 7.4.6. (Geometric Algebra of Markov Chain)

Let n be a positive integer. We place n points 0, 1, ..., $n-1$ around a unit circle at the vertices of a regular n-gon. Now, suppose a particle starts at one of these points and moves on to the nearest neighbor point with probability 1/2. Thus, we have a random walk. The transition matrix for this Markov chain is

$$
P(n) = \begin{array}{c} \\ 0 \\ 1 \\ 2 \\ \vdots \\ n-2 \\ n-1 \end{array}
\begin{array}{cccccc}
0 & 1 & 2 & \cdots & n-2 & n-1 \\
\begin{bmatrix} 0 & 1/2 & 0 & \cdots & 0 & 1/2 \\ 1/2 & 0 & 1/2 & \cdots & 0 & 0 \\ 1/2 & 0 & 0 & \cdots & 0 & 0 \\ \vdots & \vdots & \vdots & \vdots & \vdots & \vdots \\ 1/2 & 0 & 0 & \cdots & 0 & 0 \\ 1/2 & 0 & 0 & \cdots & 1/2 & 0 \end{bmatrix}
\end{array}.
\tag{7.4.10}
$$

Note that if n is even, the entire vertices partitions into two subsets of odd, say $S_o = \{1, 3, \cdots, n-1\}$, and even, say $S_e = \{0, 2, \cdots, n-2\}$. These subsets are periodic because if the particle starts at a vertex from the even subset, it will commute between the odd and even subsets. At all odd pint times it will be in the odd subset and at all even point times it will be in the even subset. Hence, the Markov chain is periodic when n is even. We leave it to the reader to discuss the case when n is odd.

It can be seen from the transition matrix $P(n)$ given in (7.4.10) that the Markov chain, in this case, is irreducible. Thus, if n is odd, the right-hand side of (7.4.8) can be written as

$$
\left(1 - \frac{1}{2^{n-1}}\right)^{m/(n-1)} \approx e^{\left(-\frac{m}{2^{n-1}(n-1)}\right)}.
\tag{7.4.11}
$$

Now, if we want uniformity in convergence when both n and m approach infinity, then we must ensure that the ratio $m/(2^n n)$ will approach infinity. In other words, it seems we must show that m must grow faster than $(2^n n)$. To show that this statement is true, we do as follows:

The matrix (7.4.10) is a Hermitian matrix, i.e., $P^T = P$. Thus, it has n orthonormal eigenvectors. These eigenvectors for a basis in the n-dimensional real Euclidean space R^n and its eigenvalues are all real. The eigenvectors can be found using a discrete Fourier transform. Hence, we consider

$$
\Psi_l(j) = \frac{1}{\sqrt{n}} e^{2\pi i l \frac{i}{n}}, \quad j, l = 0, 1, 2, \cdots, n-1.
\tag{7.4.12}
$$

For complete discussion see Suhov and Kelbert (2008, p.101).

7.5. Birth-Death (B-D) Processes

Definition 7.5.1.

A continuous-time Markov chain with state space as the set of non-negative integers, $Z = \{0,1,2,\cdots\}$, and transition probabilities

$$\begin{cases} p_{x,x+1} = \lambda_x, & x = 0,1,2,\cdots, \\ p_{x,x-1} = \mu_x, & x = 1,2,\cdots, \\ p_{x,y} = 0, & x = 0,1,2,\cdots, \quad y \neq x, \quad y \neq x \pm 1, \\ p_{x,x} = r_x, & x = 0,1,2,\cdots, \\ p_x = \lambda_x + \mu_x, & x = 0,1,2,\cdots, \quad \mu_0 = 0, \\ \lambda_x + \mu_x + r_x = 1 \end{cases} \qquad (7.5.1)$$

is called

(1) a *pure birth process*, if $\mu_x = 0$, for $x = 1,2,\cdots$,

(2) a *pure death process*, if $\lambda_x = 0$, for $x = 0,1,2,\cdots$, and

(3) a *birth – death process*, sometimes denoted by *B-D process*, $\lambda_x > 0$ and $\mu_x > 0$.

Note that no B-D process allows transitions of more than one step. Transitions such as $x \rightarrow x + 2$, $x - 3 \leftarrow x - 1$, etc., are not allowed. In other words, the probability of a transition of more than one step is considered to be negligible. Note also that a birth and death process may be called a *continuous-time random walk*. In that case, we can say that if the walker takes only forward steps of the fashion $0 \rightarrow 1 \rightarrow 2 \rightarrow 3 \rightarrow \cdots$, the chain is called a *pure birth process*. In this case, only transition probabilities λ_x are positive. If staying put is also allowed, then the process is called a *birth process*. In this case, only $\lambda_x > 0$ and $r_x > 0$. On the other hand, if only backward transitions $0 \leftarrow 1 \leftarrow 2 \leftarrow 3 \leftarrow \cdots$ are allowed, the process is called a *pure death process*. In this case, only transition probabilities μ_x are positive. If staying put is also allowed, then the process is called a *death process*. In this case only $\lambda_x = 0$. If only both forward and backward steps are allowed, the chain is called a *pure birth – death process*. In that case, $r_x = 0$. If staying put is allowed, the process is called a *birth – death process*.

Example 7.5.1.

Simple examples of B-D processes are: (1) the number of visitors (by time t) to Houston Museum of Art since the museum opened, for a pure birth; (2) the number of 1957 Chevrolet automobiles with original paint (at time t), for a pure death; and (3) the number of customers in a certain grocery store (at time t), for a birth and death.

Example 7.5.2.

Radium 226 has a half-life of approximately 1600 years and its rate of decay is approximately constant over time. In particular, for a sample of 100 picograms, radium 226 emits alpha particles at an average rate of $\lambda = 3.7$ particles per second. Suppose we have a particle detector which counts sequentially the number of emitted alpha particles (from a sample of 100 picograms) by time t. We let the random variable $X(t)$ represent such a number. For any elapsed time t, $X(t)$ may be considered a discrete random variable with values in the set of non-negative integers $\{0, 1, 2, \cdots\}$. We start the counting emissions from time $t = 0$ with the counter set to zero, that is $X(0) = 0$. For each $t \geq 0$, we define a probability mass function,

$p_x(t) = P\{x$ particles are counted by time $t\}$, for $x = 0, 1, 2, \ldots,$

where $p_0(0) = 1$. The set of non-negative integers $\{0, 1, 2, \cdots\}$ for values of the random variables $X(t)$ will be the state space in this case. Hence, $p_x(t) = P($the process is in state x at time $t)$.

Now suppose that over time, the process counts particles in the fashion $0 \rightarrow 1 \rightarrow 2 \rightarrow 3 \rightarrow \cdots$, i.e., moving forward from one state x to the next higher consecutive state $x + 1$, but never backward to a lower valued state. Thus, the process allows only arrivals (births) but no departures (deaths), and as such, it is an example of a *pure birth process*.

For a Markov process, the Chapman-Kolmogorov equation (7.2.15), can be rewritten as

$$p_{xy}(\Delta t + t) = \sum_r p_{xr}(\Delta t) p_{ry}(t) = \sum_{r \neq x} p_{xr}(\Delta t) p_{ry}(t) + p_{xx}(\Delta t) p_{xy}(t). \qquad (7.5.2)$$

Subtracting $p_{xy}(\Delta t)$ from both sides of (7.5.2), factoring and dividing both sides by Δt, we obtain

$$\frac{p_{xy}(\Delta t + t) - p_{xy}(t)}{\Delta t} = \sum_{r \neq x} \frac{p_{xr}(\Delta t)}{\Delta t} p_{ry}(t) + \left(\frac{p_{xx}(\Delta t) - 1}{\Delta t} \right) p_{xy}(t). \qquad (7.5.3)$$

Now passing to the limit as $\Delta t \rightarrow 0$, using properties of limit and summation, we will have

$$\lim_{\Delta t \to 0} \frac{p_{xy}(\Delta t + t) - p_{xy}(t)}{\Delta t} = \sum_{r \ne x} \left[\lim_{\Delta t \to 0} \frac{p_{xr}(\Delta t)}{\Delta t} \right] p_{ry}(t) + \left(\lim_{\Delta t \to 0} \frac{p_{xx}(\Delta t) - 1}{\Delta t} \right) p_{xr}(t).$$

$$(7.5.4)$$

Let us define

$$q_{xy} \equiv \lim_{\Delta t \to 0} \frac{p_{xy}(\Delta t)}{\Delta t}, \quad x \ne y, \ q_{xx} \equiv \lim_{\Delta t \to 0} \frac{p_{xx}(\Delta t) - 1}{\Delta t} \text{ and } q_x \equiv -q_{xx}. \qquad (7.5.5)$$

Note that q_{xy}, defined in (7.5.5), is always finite. Also note that q_x, which is non-negative, always exists and is finite when the state space S is finite. However, q_x may be infinite when S is denumerably infinite.

Now, when Δt is very small, from (7.5.5) it follows that

$$\begin{cases} p_{xy}(\Delta t) = \Delta t \, q_{xy} + o(\Delta t), & x \ne y, \\ p_{xx}(\Delta t) = \Delta t \, q_x + o(\Delta t), \end{cases} \qquad (7.5.6)$$

where $o(\Delta t)$ is the "little o" of Δt, i.e., $\lim_{\Delta t \to 0} \dfrac{o(\Delta t)}{\Delta t} = 0$. Thus, $\sum_y p_{xy}(\Delta t) = 1$, or

$\sum_{y \ne x} p_{xy}(\Delta t) + p_{xx}(\Delta t) - 1 = 0$. Hence, $\sum_{y \ne x} q_{xy} + q_{xx} = 0$, or

$$\sum_{y \ne x} q_{xy} = q_x. \qquad (7.5.7)$$

Now, from definition of derivative, relations (7.5.5) and (7.5.6), equation (7.5.4) can be written as

$$p'_{xy} = \sum_{r \ne x} q_{xr} p_{ry}(t) + q_x p_{xy}(t). \qquad (7.5.8)$$

Equation (7.5.8) is called *Chapman-Kolmogorov backward equation*. In matrix form it can be written as

$$\mathbf{P}'(t) = \mathbf{Q}\,\mathbf{P}(t), \qquad (7.5.9)$$

where, Q is a matrix whose elements are q_{xy}, and it is called the *transition density matrix, rate matrix* or *Q-matrix*.

Similarly, by referring to (7.2.15), we may write (7.5.2) as

$$p_{xy}(t+\Delta t) = \sum_r p_{xr}(t)p_{ry}(\Delta t) = \sum_{r\neq x} p_{xr}(t)p_{ry}(\Delta t) + p_{xy}(t)p_{xx}(\Delta t). \tag{7.5.10}$$

From (7.5.10), using similar steps as leading to (7.5.8) and (7.5.9), we will have

$$p'_{xy} = \sum_{r\neq x} p_{xr}(t)q_{ry} + q_x p_{xy}(t), \tag{7.5.11}$$

That is called *Chapman-Kolmogorov forward equation*, and

$$\mathbf{P}'(t) = \mathbf{P}(t)\mathbf{Q}. \tag{7.5.12}$$

7.6. PURE BIRTH PROCESS – POISSON PROCESS

Before considering a specific pure birth processes, we discuss a general pure birth process.

Definition 7.6.1.

A counting process $\{X(t), t \geq 0\}$ is called a *pure birth process with transition rate $\lambda(x)$*, $\lambda(x) > 0$, provided that the following hold:

(i)　　The process begins in state m, the minimum of the state space, i.e.,

$$P\{X(0) = m\} = 1. \tag{7.6.1}$$

(ii)　　The probability that the count increases by 1 (1 "arrival") in a small time interval $[t, t + \Delta t]$ is approximately proportional to the length, Δt, of the time interval. The transition rate, $p_{x,x+1} = \lambda(x)$, is assumed to be a function of the state x and may vary from state to state, i.e.,

$$P\{X(t+\Delta t) = x+1 | X(t) = x\} = p_{x,x+1}(\Delta t) = \lambda(x)\Delta t + o(\Delta t). \tag{7.6.2}$$

(iii)　　If Δt is small, the probability of 2 or more arrivals (the count increases by more than 1) in the time interval $[t, t + \Delta t]$ is negligible. Equivalently, this may be stated as: $p_{xy} = 0$ for $y > x+1$ (since $p_{xy}(\Delta t) = p_{xy}\Delta t + o(\Delta t) = o(\Delta t)$, when $p_{xy} = 0$), i.e.,

$$P\{X(t+\Delta t) > x+1 | X(t) = x\} = o(\Delta t). \tag{7.6.3}$$

(iv)   The count remains the same or increases over time. The count will never decrease, i.e.,

$$P\{X(t+\Delta t) < x | X(t) = x\} = 0. \tag{7.6.4}$$

Thus, a pure birth process experiences arrivals but no departures. The particle counter is set to i at the start of the experiment and the count can only move upward over time. Property (iii) implies that, in a small enough time interval, the probability of counting more than 1 particle is negligible. This assumption indicates, among other things, that particles are assumed to arrive at the counter one at a time.

Note that as a consequence of (i) through (iv), for the probability of staying put we have:

$$P\{X(t+\Delta t) = x | X(t) = x\} = p_{x,x}(\Delta t) = 1 - [\lambda(x)\Delta t + o(\Delta t)] - o(\Delta t)$$
$$= 1 - \lambda\Delta t - 2\,o(\Delta t) = 1 - \lambda(x)\Delta t + o(\Delta t). \tag{7.6.5}$$

The transition rate diagram for a general pure birth process is shown in Figure 7.6.1. The transition rates $\cdots, \lambda(x-1), \lambda(x), \lambda(x+1), \cdots$ need not be equal.

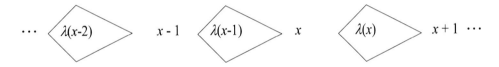

Figure 7.6.1. General Pure Birth Transition Rate Diagram.

Example 7.6.1. (Poisson Process)

As an example of a pure birth process, we will discuss the homogeneous Poisson process.

Definition 7.6.2.

A pure birth process $\{X(t), t \ge 0\}$ is called a *Poisson process with parameter* λ, $\lambda > 0$, provided that all four conditions of a pure birth in Definition 7.6.1 holds with the transition rate constant λ, rather than a function of x, and the initial state 0 instead of m. Hence, relations (7.6.1), (7.6.2.) and (7.6.5) for a Poison process will, respectively, change to:

$$P\{X(0) = 0\} = 1. \tag{7.6.6}$$

$$P\{X(t+\Delta t) = x+1 | X(t) = x\} = p_{x,x+1}(\Delta t) = \lambda\Delta t + o(\Delta t). \tag{7.6.7}$$

$$P\{X(t+\Delta t)=x\,|\,X(t)=x\}=p_{x,x}(\Delta t)=1-[\lambda\Delta t+o(\Delta t)]-o(\Delta t)$$
$$=1-\lambda\Delta t-2\,o(\Delta t)=1-\lambda\Delta t+o(\Delta t). \tag{7.6.8}$$

We will now derive a set of differential difference equations (DDE), whose solutions are the probability mass functions (*pdf*) $p_x(t)\equiv P\{X(t)=x\}$, for $t\geq 0$ and $x=0,1,2,\cdots$.

Theorem 7.6.1.

In a Poisson process with transition rate λ, the DDE are

$$\begin{cases} p_0'(t)=-\lambda p_0(t), \\ p_x'(t)=-\lambda p_x(t)+\lambda p_{x-1}(t), \quad x=1,2,\cdots, \end{cases} \tag{7.6.9}$$

with $p_0(0)=\delta_{i0}$, where δ_{i0} is called the *Kronecker delta* and is defined as

$$\delta_{i0}=\begin{cases} 1, & i=0, \\ 0, & i\neq 0. \end{cases} \tag{7.6.10}$$

Proof

Assume $x\geq 1$. By the law of total probability we have

$$p_x(t+\Delta t)=P\{X(t+\Delta t)=x\}$$
$$=P\{X(t+\Delta t)=x\,|\,X(t)=x-1\}P\{X(t)=x-1\}$$
$$+P\{X(t+\Delta t)=x\,|\,X(t)=x\}P\{X(t)=x\}$$
$$+P\{X(t+\Delta t)=x\,|\,X(t)<x-1\}P\{X(t)<x-1\}$$
$$+P\{X(t+\Delta t)=x\,|\,X(t)>x\}P\{X(t)>x\}$$

$$=[\lambda\Delta t+o(\Delta t)]p_{x-1}(t)+[1-\lambda\Delta t+o(\Delta t)]p_x(t)$$
$$+\sum_{k=0}^{x-2}P\{X(t+\Delta t)=x\,|\,X(t)=k\}P\{X(t)=k\}$$
$$+\sum_{k=x+1}^{\infty}P\{X(t+\Delta t)=x\,|\,X(t)=k\}P\{X(t)=k\}$$

$$= \lambda \Delta t p_{x-1}(t) + o(\Delta t) p_{x-1}(t) + p_x(t) - \lambda \Delta t p_x(t) + o(\Delta t) p_x(t)$$

$$+ \sum_{k=0}^{x-2} o(\Delta t) p_k(t) + \sum_{k=x+1}^{\infty} (0) p_{x\backslash k}(t).$$

In summary, we have:

$$p_x(t + \Delta t) - p_x(t) = \lambda \Delta t p_{x-1}(t) + o(\Delta t) p_{x-1}(t) - \lambda \Delta t p_x(t) + o(\Delta t) p_x(t)$$

$$+ \sum_{k=0}^{x-2} o(\Delta t) p_k(t). \qquad (7.6.11)$$

Dividing (7.6.11) through by Δt and passing to the limit as $\Delta t \to 0$ on both sides (noting that Δt is constant relative to t) we will have:

$$\lim_{\Delta t \to 0} \frac{p_x(t + \Delta t) - p_x(t)}{\Delta t} = \lambda p_{x-1}(t) - \lambda p_x(t) + \lim_{\Delta t \to 0} \left[\frac{o(\Delta t)}{\Delta t} \right] p_{x-1}(t) + \lim_{\Delta t \to 0} \left[\frac{o(\Delta t)}{\Delta t} \right] p_x(t)$$

$$+ \lim_{\Delta t \to 0} \left[\frac{o(\Delta t)}{\Delta t} \right] \sum_{k=0}^{x-2} p_k(t),$$

or

$$p_x'(t) = \lambda p_{x-1}(t) - \lambda p_x(t), \quad x = 1, 2, \cdots. \qquad (7.6.12)$$

This proves the second equation of (7.6.9). In the case of $x = 0$ we have

$$p_0(t + \Delta t) = P\{X(t + \Delta t) = 0\}$$

$$= P\{X(t + \Delta t) = x | X(t) = 0\} P\{X(t) = 0\}$$

$$+ P\{X(t + \Delta t) = 0 | X(t) > 0\} P\{X(t) > 0\}$$

$$= [1 \quad \lambda \Delta t + o(\Delta t)] p_0(t) + \sum_{k=1}^{\infty} (0) p_0(t)$$

$$= p_0(t) - \lambda \Delta t p_0(t) + o(\Delta t),$$

or equivalently,

$$p_0(t + \Delta t) - p_0(t) = -\lambda \Delta t p_0(t) + o(\Delta t) p_0(t). \qquad (7.6.13)$$

Again, dividing (7.6.13) through by Δt and passing to the limit as $\Delta t \to 0$ on both sides gives

$$p_0'(t) = -\lambda p_0(t). \tag{7.6.14}$$

This proves the first equation of (7.6.9) and completes the proof of the theorem.

The following figures illustrate DDE for Poisson Process. The system (7.6.9) may be generated from these illustrations.

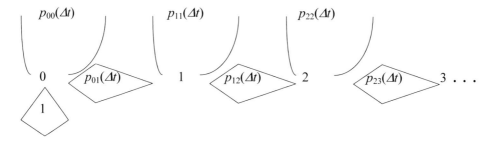

Figure 7.6.2(a). Infinitesimal Transition Probability Diagram/ Poisson Process.

As with random walks, the arrows contain the transition probabilities and self-loop probabilities. The arrow under state 1 indicates that the (initial) probability of beginning in state 0 at time 0 is 1, that is, $p_0(0) = 1$. Transition probabilities with value zero or value $o(h)$ are suppressed, as well as initial probabilities with value zero.

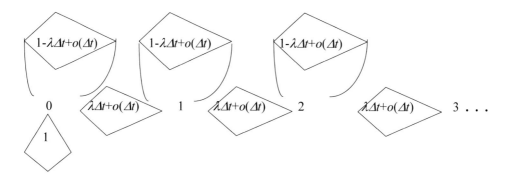

Figure 7.6.2(b). Infinitesimal Transition Probability Diagram/Poisson Process.

Although informative, this diagram is cumbersome to draw and actually contains no more information than the diagram below.

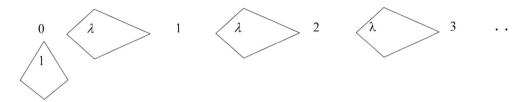

Figure 7.6.3. Transition Rate Diagram/Poisson Process.

The simple diagram above contains all information needed to generate the Poisson DDE's. It is this diagram we shall refer to as the Transition Rate Diagram. In the next figure, we illustrate how to generate the particular DDE: $d/dt\, p_3(t) = \lambda\, p_2(t) - \lambda p_3(t)$. With an eye to the diagram we may think of $\lambda p_2(t)$ as the "flow into state 3", $\lambda p_3(t)$ as the "flow out of state 3", and the DDE above as $d/dt\, p_3(t) = \lambda p_2(t) - \lambda\, p_3(t) =$ "flow in" - "flow out":

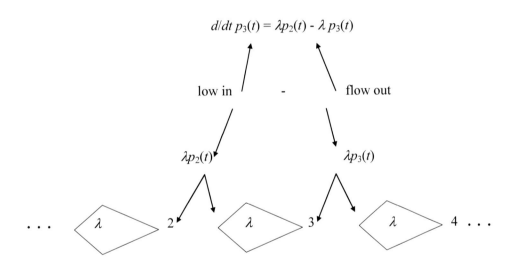

Figure 7.6.4(a). DDE Generation.

The DDE for state 0, $d/dt\, p_0(t) = -\,\lambda\, p_0(t)$, is generated in the same way:

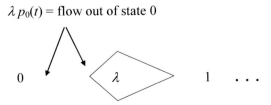

Figure 7.6.4(b). DDE Generation.

There is zero flow into state 0 and thus flow in - flow out = 0 - $\lambda\, p_0(t)$ and the DDE is simply $d/dt\, p_0(t) = -\lambda\, p_0(t)$. Similarly, for any state x, $x \geq 1$, we have:

$$\text{Net flow} = \text{Flow in} - \text{Flow out}$$
$$\text{DDE (Poisson):}\ \ d/dt\, p_x(t) = \lambda p_{x-1}(t) - \lambda\, p_x(t) = \text{flow into } x - \text{flow out of } x$$

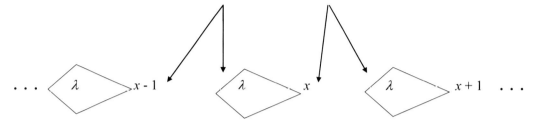

Figure 7.6.4(c). DDE Generation.

This diagrammatic process can be used to generate DBE's for all of the birth-death processes we will encounter.

Different methods may be used to solve the DDE (7.6.1) for $p_x(t)$, $x = 0,1,2,\cdots$, the probability mass function over $x = 0,1,2,\cdots$, at any fixed but arbitrary time t. Here we solve it by iteration beginning with $p_0(t)$. That is, having solved the first few DDE's $p_0(t)$, $p_1(t)$, . . ., we endeavor to detect a pattern and use the pattern to write the remaining solutions without solving explicitly.

We begin with state 0, for which the equation is (7.6.6), $p_0'(t) = -\lambda p_0(t)$. This is a homogeneous linear first order differential equation with constant coefficient and initial value (7.6.2). Thus, the solution is

$$p_0(t) = p_0(0)e^{-\lambda t} = e^{-\lambda t}. \tag{7.6.15}$$

For state 1, from (7.6.12), the DDE is $p_1'(t) = \lambda p_0(t) - \lambda p_1(t)$, which is a non-homogeneous linear first order differential equation with constant coefficient and initial value (7.6.10), i.e., $p_1(0) = 0$. Thus, using the variation of parameters method, the solution is

$$p_1(t) = e^{-\lambda t}\int_0^t \lambda e^{\lambda u} p_0(u)du = \lambda e^{-\lambda t}\int_0^t e^{\lambda u}(e^{-\lambda u})du = \lambda e^{-\lambda t}\int_0^t (du) = \lambda t e^{-\lambda t} = \frac{(\lambda t)^1}{1!}e^{-\lambda t}.$$

The DDE for state 2 has the same form as that of state 1 with $p_2(0) = 0$. Hence, the solution is

$$p_2(t) = e^{-\lambda t}\int_0^t \lambda e^{\lambda u} p_1(u)du = \lambda e^{-\lambda t}\int_0^t e^{\lambda u}(\lambda u e^{-\lambda u})du = \lambda^2 e^{-\lambda t}\int_0^t u\,du = \lambda^2 t^2 e^{-\lambda t} = \frac{(\lambda t)^2}{2!}e^{-\lambda t}.$$

A pattern is seen and, thus, we can state the following theorem:

Theorem 7.6.2.

In a Poisson process with transition rate λt, the probability mass function $p_x(t)$ is Poisson with parameter λt, i.e.,

$$p_x(t) = \frac{(\lambda t)^x}{x!} e^{-\lambda t}, \quad x = 0, 1, 2, \cdots . \tag{7.6.16}$$

Proof

Left as an exercise.

According to Theorem 7.6.2, we see that on the average, there are λt occurrences (changes, arrivals, birth, etc.) in a time interval of length t. That is, λ is the mean number of arrivals in a unit of time.

Example 7.6.2.

In Example 7.5.2, the particle counting experiment for 100 picograms of radium 226, the half-life of 1600 years means the decay is very slow. Thus, it is reasonable to assume that over any "practical" period of time measurement, the mean arrival rate of $\lambda = 3.7$ alpha particles per second remains essentially constant. So the Poisson model seems appropriate for this experiment. Assuming the model fits this experiment, let us find

(1) Probability that 8 particles are counted by $t = 2$ seconds.
(2) Expected value of particles counted by $t = 2$.
(3) Standard deviation of particles counted by $t = 2$.

Solution
(1) $P(\text{8 particles are counted by } t = 2 \text{ seconds})$

$$= P(X(2) = 8) = p_8(2) = \frac{(3.7 \times 2)^8 e^{-(3.7)(2)}}{8!} = 0.136.$$

(2) The expected number of particles counted by $t = 2$ seconds $= (3.7)(2) = 7.4$.
(3) The variance of particles counted by $t = 2 = (3.7)(2) = 7.4$. Thus, the standard deviation is $\sqrt{7.4} = 2.72$ alpha particles.

A Poisson process is an example of a birth-death process, which takes place over continuous time. Unlike a random walk, which concerns itself only with which state is entered on the next trial, a birth-death process considers the state of the process at every instant of time t, $t \geq 0$. The state space of a birth-death process is the set of non-negative integers $\{0,1,2,\cdots\}$ or some ordered subset thereof. Intuitively, we may think of a birth-death process as a function of continuous time, which changes state in a random fashion after some random amount of waiting time in the current state. In the diagram below we assume a Poisson changed states (only) at the times indicated: $\tau_1, \tau_2, \cdots$. In diagram 7.6.7 the process is in state 0 over the time interval $[0, \tau_1)$, that is, $X(t) = 0$, for $0 \leq t < \tau_1$. In other words, each $X(t)$ "chooses" state 0 while $0 \leq t < \tau_1$. Then, at τ_1 the process makes a transition to state 1 and remains there a random amount of time, $\tau_2 - \tau_1$, (until time τ_2). For $\tau_1 < t < \tau_2$, each $X(t)$ "chooses" state 1. At τ_2 another transition moves the process to state 2 and the process remains there until time τ_3, etc. .

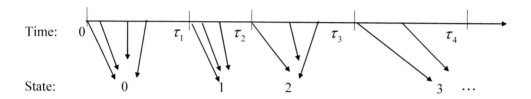

Figure 7.6.5. Poisson Process Time Line.

As we mentioned before, a process with the Markov properties is "*forgetful*". Suppose we notice at time τ_1 that such a process is in a state x. Then, at a later time τ_2, we see that the process is *still* in state x. It is interesting to note that, under the Markov assumptions, the process is no more likely to make a transition to a state y at $\tau_2 + \Delta t$ than it was at $\tau_1 + \Delta t$, since $P(X(\tau_1 + \Delta t) = y \mid X(\tau_1) = x) = P(X(\tau_2 + \Delta t) = y \mid X(\tau_2) = x)$. The process ignores or "forgets" the time it has already spent in state x, namely $\tau_2 - \tau_1$; it is as if transition probabilities are determined "from scratch" at each instant of time.

Example 7.6.3.

A biologist is interested in studying a colony of bacteria generated from a single bacterial cell. The reproductive method for this bacterial species is simple fission wherein one cell becomes 2 identical cells. For counting purposes it is convenient to think of the 2 resultant cells as the parent cell itself and one new arrival (a "daughter" cell) with a net gain in count of 1. Past studies indicate that under optimal conditions the mean arrival rate for an individual cell is 3 new arrivals per day so that, on average, a cell divides every 8 hours (but not specifically eight hours from the last division).

The diagrams below indicate 2 of many possible scenarios, where 1 parent cell gives rise to 3 new offspring (the letter N_i indicates the ith new arrival):

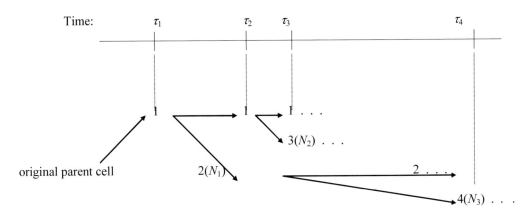

Figure 7.6.2. 3 Births in Example 7.6.3 (Scenario 1). New cells 2, 3 and 4 come into being at times τ_2, τ_3 and τ_4 respectively.

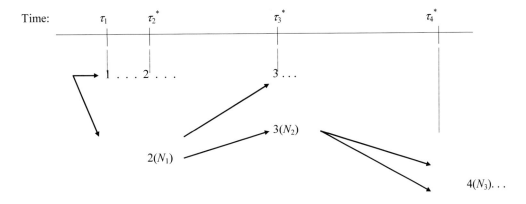

Figure 7.6.3. 3 Births in Example 7.6.3 (Scenario 2). Successive new arrivals 2, 3 and 4 occur at times τ_2^*, τ_3^* and τ_4^*, respectively.

Note that other arrivals could occur within these same time frames (from 1 and/ or 3 beyond τ_3 in Figure 7.6.2, and from 1 beyond τ_2^* and/ or 2 beyond τ_3^* in Figure 7.6.3), but the diagrams are truncated for simplicity.

Below we illustrate another scenario for 3 arrivals in a continuation of Figure 7.6.3.

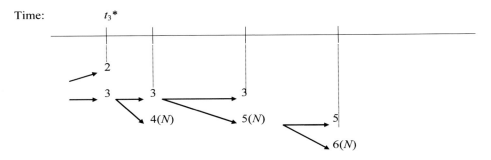

Figure 7.6.4. 3 Births in Example 7.6.3 (Scenario 3). Here, cell 3 gives rise to new arrivals 4, 5 and 6.

In all three illustrations one cell gives rise to 3 new cells (on average per day).

We now formulize this model. Let $X(t)$ denote the number in the population at time t with an initial count of $X(0) = 1$ parent cell. Since it is very unlikely that any 2 cells divide at exactly the same instant, it is reasonable to assume that the population size increases (never decreasing) in one-step transitions in the fashion $1 \to 2 \to 3 \to \cdots$. Further, we assume the probability of 2 or more arrivals in a small time interval [t, t + Δt] is negligible. Under the assumptions we have made, a pure birth process seems a likely model for this experiment. All that remains is to show that the transition probabilities $p_{x,x+1}(\Delta t)$ are approximately proportional to Δt for each x (as in property (ii) for a pure birth process).

To this end, we use the diagram in Figure 7.6.5 below to help examine the transition probabilities in this experiment.

Explanation of Figure 7.6.5: The diagram represents one of several possible sequences of cell divisions (and hence arrivals) over time. The tree in the upper portion of the diagram maps the cell divisions and shows the resulting cells after each. The graph in the lower portion of the diagram shows the population size between successive divisions and the set of existing cells between division times. We begin with A, a single cell. At the first division, A "gives rise" to A (itself) and B and thus $X(\tau_1) = 2$. At the next division, A gives rise to A and C so that $X(\tau_2) = 3$ (at this point the existing individuals are A, B and C). This is followed in time by B giving rise to B and D (at which point we have cells A, B, C and D with $X(\tau_3) = 4$).

Note that from an initial population of size $X(0) = 1$ through the first transition, where the population size becomes 2, this process is indistinguishable from the Poisson process as far as mean arrival rate is concerned. From state 1 to state 2 the mean arrival rate is a constant $\lambda = \lambda_1 = 3$ arrivals per day. Treating this transition as a Poisson type transition, it follows that

$$p_{12}(\Delta t) = P(X(t + \Delta t) = 2) \mid X(t) = 1) = 3\Delta t + o(\Delta t). \qquad (7.6.17)$$

Now consider $P[X(t) = 3 \mid X(t) = 2] = p_{23}(\Delta t)$. With an eye to the diagram above we see that

$p_{23}(\Delta t) = P(\text{transition from 2 cells to 3 cells in } \Delta t)$
   $= P(A \text{ divides in } \Delta t \text{ and } B \text{ does not OR } B \text{ divides in } \Delta t \text{ and } A \text{ does not})$
    (we are assuming a negligible probability of 2 or more arrivals in a small Δt,
    so that not both A and B divide in Δt)
   $= P(AB^C) + P(A^CB)$ (abbreviating and noting mutual exclusivity)
   $= P(A) P(B^C) + P(A^C) P(B)$ (assuming cells are independent) .

Since the process is forgetful (Markovian), we may treat the division of A as a Poisson transition "as if from scratch". We do the same for B. Hence,

$p_{23}(\Delta t) = [3\Delta t + o(\Delta t)][1 - 3\Delta t + o(\Delta t)] + [3\Delta t + o(\Delta t)][1 - 3\Delta t + o(\Delta t)]$
  $= 2[3\Delta t - 3^2(\Delta t)^2 + o(\Delta t) + (o(\Delta t))^2]$ (noting that $(\Delta t)^2$ and $(o(\Delta t))^2$ are both $o(\Delta t)$)
  $= 2[3\Delta t + o(\Delta t)] = (2)(3)\Delta t + o(\Delta t) = 6\Delta t + o(\Delta t).$

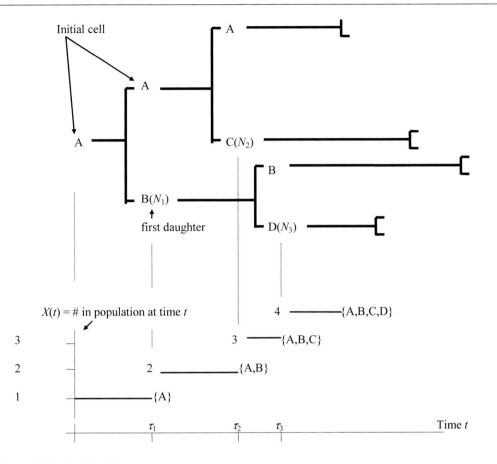

Figure 7.6.5. Cell Division.

Thus,

$$p_{23}(\Delta t) = (2)(3)\Delta t + o(\Delta t).\qquad(7.6.18)$$

Next we consider $p_{34}(\Delta t) = P(X(t + \Delta t) = 4 \mid X(t) = 3)$. For convenience, assume that the conditioning event $X(t) = 3$ is represented by the individuals A, B and C at time t.

Note that individual C could be the daughter of A or the daughter of B, but this is not important since we "start from scratch" all over again.

In similar fashion to the calculations above, we have

$$\begin{aligned}
p_{34}(\Delta t) &= P(AB^C C^C) + P(A^C BC^C) + P(A^C B^C C)\\
&= 3(3\Delta t + o(\Delta t))(1 - 3\Delta t + o(\Delta t))^2\\
&= 3(3\Delta t + o(\Delta t))[1 - 6\Delta t - 6\Delta t(o(\Delta t)) + o(\Delta t)]\\
&= 3[3\Delta t - 3\Delta t(o(\Delta t)) - 6\Delta t(o(\Delta t))^2 + o(\Delta t)]\\
&= 3(3\Delta t + o(\Delta t)) = (3)(3)\Delta t + o(\Delta t) = 9\Delta t + o(\Delta t).
\end{aligned}$$

Hence,

$$p_{34}(\Delta t) = (3)(3)\Delta t + o(\Delta t). \tag{7.6.19}$$

At this point, the pattern is clear: $p_{x,x+1}(\Delta t) = (x)(3)\Delta t + o(\Delta t)$ and it can be shown by mathematical induction that, in general,

$$p_{x,x+1}(\Delta t) = P(X(t + \Delta t) = x + 1 \mid X(t) = x) = 3x\Delta t + o(\Delta t).$$

Since $p_{x,x+1}(\Delta t)$ has the form $\lambda_x \Delta t + o(\Delta t)$, with rates $\lambda_x = \lambda(x) = 3x$ in particular, we see that the pure birth model fits this bacterial growth process. The appropriate transition rate diagram is shown below.

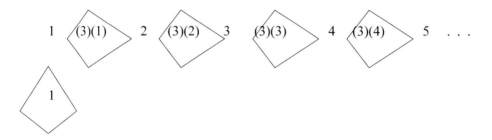

or equivalently:

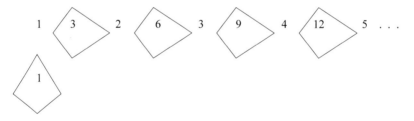

Figure 7.6.6. Transition Rate Diagram for Bacterial Growth/Example 7.6.4.
The (pure birth) DDE's for the bacterial growth process are, therefore, as follows:

State	Initial Probability	DBE (Net flow = Flow in - Flow out)
$m = 1$	$p_1(0) = 1$	$d/dt\, p_1(t) = -3p_1(t)$
2	$p_2(0) = 0$	$d/dt\, p_2(t) = 3p_1(t) - (3)(2)p_2(t)$
3	$p_3(0) = 0$	$d/dt\, p_3(t) = (3)(2)p_2(t) - (3)(3)p_3(t)$
...		
k	$p_k(0) = 0$	$d/dt\, p_k(t) = 3(k-1)p_{k-1}(t) - 3kp_k(t).$

We leave it as an exercise to solve this system sequentially in order to get $p_1(t) = e^{-3t}$, then $p_2(t) = e^{-3t}(1 - e^{-3t})$, then $p_3(t) = e^{-3t}(1 - e^{-3t})^2$, etc., and show by mathematical induction that, in general,

$$p_x(t) = e^{-3t}(1 - e^{-3t})^{x-1}, \qquad x = 1, 2, \cdots. \tag{7.6.19}$$

The reader will recognize this probability mass function is the geometric *pmf* for the random variable $X(t)$, where $P(X(t) = x) = p(1 - p)^{x-1}$, $x = 1, 2, 3, \ldots$, with parameter $p = e^{-3t}$. Thus, we have $E(X(t)) = e^{3t}$ and $Var(X(t)) = (1 - e^{-3t})(e^{6t}) = e^{6t} - e^{3t}$. Hence, for example, the expected size of the bacterial population at $t = 2.5$ days is $E(X(2.5)) = e^{3(2.5)} \doteq 1{,}808$ individuals and the variance of $X(2.5)$ is $e^{6(2.5)} - e^{3(2.5)} \doteq 3{,}267{,}209$, i.e., the standard deviation of $1{,}807$. The probability of exactly $1{,}808$ individuals after 2.5 days is $p_{1808}(2.5) = e^{-3(2.5)}(1 - e^{-3(2.5)})^{1807} \doteq 0.000204$, while the probability of at least $1{,}808$ individuals after 2.5 days is $P(X(2.5) \geq 1{,}808) = P(X(2.5) > 1{,}807) = (1 - e^{-3(2.5)})^{1807} \doteq 0.368$.

Definition 7.6.3.

A pure birth process $\{X(t), t \geq 0\}$ is called the *Yule-Furry process* if $\lambda_x = \lambda x$, for all x. In other words, the Yule-Furry process is a pure birth process with linear transition rate.

Assuming $p_x(0) = 1$, it can be shown that (see Example 7.6.4) the DDE for the Yule-Furry process is:

$$p_x'(t) = -x\lambda p_x(t) + (x-1)\lambda p_{x-1}(t), \quad x = 1, 2, \cdots. \tag{7.6.20}$$

It is left as an exercise to show that the solution of (7.6.20) is

$$p_x(t) = e^{-\lambda t}(1 - e^{-\lambda t})^{x-1}, \qquad x = 1, 2, \cdots. \tag{7.6.21}$$

Therefore, in the Yule process, the random variable $X(t)$ (= the population size at time t) has a geometric distribution with parameter $p = e^{-\lambda t}$.

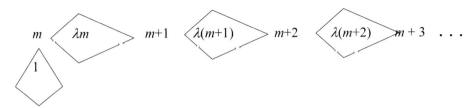

Figure 7.6.7. Transition Rate Diagram for the Yule Process with initial m.

Example 7.6.7.

In this example we investigate what happens in the Yule process when the initial population is m, $m \geq 1$.

Solution

The transition rate diagram for the Yule process initialized with $m \geq 1$ members is shown below. λ is constant and denotes the mean birth rate for an individual in the population.

Let $Y(t)$ be the number of individuals in the population at time t, so that $Y(0) = m$. $Y(t)$ may be considered as the sum of m distinct population sizes, where each of the m sub-populations begins as a Yule process with one member (m "parallel" Yule processes). Define $X_i(t)$ to be the population size, at time t, generated by individual i for $i = 1, 2, 3, \ldots, m$. $Y(t) = \sum_{i=1}^{m} X_i(t)$. We assume that sub-populations grow independently of each other. Thus, $X_1(t)$, $X_2(t)$, ..., $X_m(t)$ are independent, identically distributed geometric random variables, with common state space $\{1, 2, 3, \ldots\}$ and *pmf* as in (7.6.21), i.e.,

$$p_{X_i}(t) = P\{X_i(t) = x\} = e^{-\lambda t}\left(1 - e^{-\lambda t}\right)^{x-1}. \qquad (7.6.22)$$

It is known that the sum, Y, of m independent, identically distributed geometric random variables, $X_1, X_2, \ldots, X_m$, each with *pmf* as in (7.6.22), has a negative binomial distribution with *pmf*

$$P(Y = y) = \binom{y-1}{m-1} p^m (1-p)^{y-m}, \quad m, m+1, m+2, \cdots, \qquad (7.6.23)$$

and

$$E(Y) = m/p \text{ and } Var(Y) = [m(1 - p)]/p^2. \qquad (7.6.24)$$

Hence, in our case for the Yule process under consideration, we have

$$P(Y = y) = \binom{y-1}{m-1} e^{-\lambda t m}(1 - e^{-\lambda t})^{y-m}, \quad m, m+1, m+2, \cdots, \qquad (7.6.25)$$

and

$$E(Y(t)) = me^{\lambda t} \text{ and } Var(Y(t)) = m(1 - e^{-\lambda t})e^{2\lambda t}. \qquad (7.6.26)$$

So, now suppose a Yule process, with mean birth rate for an individual of $\lambda = 2$ new arrivals per hour, begins at $t = 0$ with 5 members. The probability that the population size is 145 after 3 hours is $p_{145}(3) = P(Y(3) = 145) = \binom{145-1}{5-1} e^{-(2)(3)(5)}[1 - e^{-(2)(3)}]^{145-5} \doteq 0.00000114$.

The expected size of the population at $t = 3$ is $E(Y(3)) = 5e^{(2)(3)} \doteq 2{,}017$.

7.7. INTRODUCTION OF QUEUEING PROCESS AND ITS HISTORICAL DEVELOPMENT

It seems the word "*queue*" takes on different meanings in different books and literature. In Pablo Picasso's book entitled *Le Desire Attrape par la Queue*, "*queue*" means a *Desire Caught by the Tail*. This is in line with Hall (1991) explanation that the original meaning of "*queue*" is "*tail of a beast*". This, however, is not the intended meaning in this book. For us, a *queue* is a *waiting line*, where tasks wait to be served.

Queueing theory has played as one of the most dominated theories of stochastic models. Newer areas of application of the queueing theory are continuously emerging in industry and business. Some of the classical examples of queues can be found in cases associated with the transportation industry, machine interference, system design, facility utilization, inventory fluctuation, finance, and conveyor theory. For instance, in manufacturing economics, a case is how to assign the correct ratio of man to machine. This is because, if the man is assigned too few machines, he often may be idle and unproductive. On the other hand, if he is assign too many machines, he may get busy with some and others may be waiting too long to obtain service and thus the manufacturing time will be lost. It is important to incorporate cost factor in the study of queues, although not many studies do so.

There are all sorts of possibilities to have a queue in the real life. As a simple example for a queueing system we can consider a post office which receives an average of 5 local mail items per minute and has five mail delivery persons to distribute them the next day. The electronic posting system is no different than the traditional one when it comes to the queueing modeling of the system.

Another less obvious example: Although all of us human beings experience a queue as we are born, seldom we notice that we get in line for passing. The mean waiting time in this line is the mean life span of us.

The notion of *task*, *job*, *unit*, or *customer* may have many interpretations: a computer program in line awaiting execution, a broken device waiting in line for repair, a task waiting in a bank teller's line, a suit waiting to be dry-cleaned, thousands of people waiting (and even dieing) to receive organ implants to save their lives, goods waiting to be conveyed from a distribution center to retail stores of a company, planes waiting their turns to take off in a busy airport, etc. Although every minute that a task spends waiting to be served translates into lost business and lost revenue in most instances, sometimes long waiting times and waiting lines are indications of importance of high quality and may make for good publicity for certain businesses.

Humans waiting for the dole and the unemployed waiting for a bowl of soup are examples of queues history books mention, thus, showing that the existence of queue historically. However, the study of queues is just a century old. At the beginning of the twentieth century the practical situations such as the examples below gave rise to develop the theory of queues and started with the work of A. K. Erlang, 1878-1929, a famous Danish scientist of Copenhagen Telephone Company, who was among the first pioneers in the development of the theory of queues [see Brockmeyer et al. (1948)]. His basic research in this field dates from 1908 to 1922. His most noted paper is the one of 1917 [see Erlang (1917)]. This introduction attracted the interest of more and more mathematicians, as well as of engineers and even of economists in the kinds of problems formulated by Erlang. It turns out that problems arising in telephone traffic are also relevant to some other fields of research including engineering, economics, transport, military problems, and organization of industry.

But it took thirteen years before the next result in the field appeared, and that was due to F. Pollaczek (1930). Two years later the famous work of A. Y. Khintchine (1932) was published. Practical demands made many new problems to be formulated in the field of queueing theory. For these problems to be applied in practice and for physical conditions to be approximated, a thorough investigation of these problems is necessary. Moreover, such investigations have to be instructive for developing research methods and creating a systematic theory. Though such a theory does not yet exist, there have been many attempts to create one. Among the most important attempts so far is the theory of stochastic processes and particularly important is the Markov processes and their generalizations. It took more than a decade before theoretical analysis of queueing systems started to grow considerably with the advent of operations research in the late 1940's and early 1950's. The first textbook on this subject entitled *Queues, Inventories, and Maintenance* was authored in 1958 by Morse. Not much later, Saaty wrote his famous book entitled *Elements of Queueing Theory with Application* in 1961. Today, over 40 books have been written on queues, over 1000 papers have been published and several journals publish papers on queues. Haghighi and Mishev, two authors of this book, (2008) is one of the most recent books on queue entitled *Queueing Models in Industry and Business* that was published in June 2008 by Nova Publishers, Inc. in New York.

The theory of queues is generally concerned with how physical systems behave. In a sense, the theory of queues deals with the stochastic behavior of processes arising in connection with mass servicing in cases when random fluctuation occur, such as telephone traffic, particle counting, processes of automatic ticket machines, stock processes, etc. The purpose of developing this theory has been to predict fluctuating demands from observational data which will ultimately enable an enterprise to provide adequate service with tolerable waiting times for its tasks.

The queueing theory is less concerned with optimization than it tries to explore and understand various queueing situations. For instance, suppose that the facility is a teller in a bank. If the waiting is too long, the tasks may become impatient and leave or they may not join the line at all, thus, causing a loss of transactions for the bank. The bank's manager may decide that having another teller is worthwhile because cost will be offset by the profits from the transactions of the impatient tasks more of whom remain to be served. This illustrates how optimization can be used in the theory of queues.

There is a concern about the theory of queues that it is too mathematical rather than being scientific. In other words, the concern is that operation researchers should address more behavior of the systems than abstraction of them. In fact, the best way to solve a queueing problem is to understand its behavior. Sophisticated mathematical involvement in the analysis of queueing models has caused some researchers approach simulation. However, set back is that theoretical discussion from simulation is not possible.

Arrivals to a queue may be referred to as *items*, *units*, *jobs*, *customers*, or *tasks*. Arrivals may come from without, as with tasks entering a retail store, or from within, as with broken machines lining up for repair and reuse in a shop with a fixed number of machines.

Definition 7.7.1.

A queueing system that allows arrivals from without, i.e., the source is infinite, will be called an *open* queue; otherwise the queue is termed a *closed* queue. The number of tasks in the system is referred to as *queue size*. The *state* of a queue at time *t* is the number of customers in the system at time *t*.

The input may arrive deterministically or probabilistically according to a probabilistic law (distribution). We in this text consider the latter. The probabilistic law may be *Poisson*, *Erlang*, *Palm*, *Recurrent*, *Yule*, etc.

Definition 7.7.2.

A task may arrive at the system, but does not attend due to variety of reasons. In this case we say that the task *balked*. On the other hand, a task may attend the system, wait a while and leave the system before receiving service. In this case, we say the task is *reneging*.

In the examples above, those waiting for organs that cannot receive them and die are examples of tasks *renege*. Note that both concepts of reneging and balking could be considered the same since both are leaving the system, but one immediate at the arrival and one after a while. Because of this commonality, some authors use reneging for both concepts. We distinguish in this book.

A queueing system may be modeled as a birth-death process. Arrivals to a system may want to be served, as in B-D process. In that case, they may have to wait for service. The place where arrivals await for service is called *buffer* or *waiting room*. The buffer's capacity may be finite (a finite number of waiting spaces) or infinite (where no customers are turned away). We refer to the location of a server as *counter* or *service station*. A counter may contain one or more *servers* (a repairman, grocery checkout clerks, bank tellers, etc.).

Definition 7.7.3.

A queueing system with one server is called a *single-server* and with more than one server is called *multichannel* or *multi-server*. If no arrival has to wait before starting its service, then the system is called *infinite-server queue*. In case of a multi-server queue, a task may switch between the server lines. In this case we say that the task *jockeys*. We assume *service times* as random variables following some given probabilistic law such as constant, negative exponential, geometric. The duration of time in which one or more servers is busy is called a *busy period*.

Definition 7.7.4.

A queueing system may have more than one service station. These stations may be set parallel. In that case, each station may have its own buffer or a common buffer for all. In the first case, an arrival may chose a waiting line, in the latter case, however, all arrivals must line up before the same buffer. Each station may have one or several servers. Such a system is called a *parallel queueing system*. On the other hand, the stations may be set in series with the same types of servers. In this case an output of a station may have to join the next station. The system as such is called a *tandem queueing system*.

Definition 7.7.5.

Service of tasks will be according to a certain principle. The principle sometimes is referred to as the *queue discipline*. The discipline could be such as service in order of arrival, i.e., *first-come-first-served* (*FCFS*) or *first-in-first-out* (*FIFO*), *random service*, *priority service*, *last-come-first-served* (*LCFS*), *batch* (*bulk*) *service*, etc. After being served, a served tasks or *outputs* may *exit* from the system or move on to other activities such as *feedbacking* the service stations, going for *splitting*.

Definition 7.7.6.

Suppose tasks (such as calls to a telephone exchange) arrive according to some probabilistic law. If an arriving task (call) finds the system empty (a free line) at its arrival time, its service will start (the call will be connected) and it will take as long as necessary (the conversation will last as long as necessary). That is, the service times (holding times) are random variables. However, if an arriving task (call) finds all servers (lines), which there are a fixed number of them, busy, then it will be sent either to a waiting facility (buffer) in which case we will speak of a *waiting system*, or it will be refused, in which case we will speak of a *loss system*. Of course, the system may be a combination of these two kinds in which case we will call it a *combined waiting and loss system*.

Example 7.7.1. (Servicing of Machine)

Suppose there are a fixed number of identical machines subject to occasional breakdowns. Suppose also that breakdowns occur according to some probabilistic law. Assume that there is a *server* who is the repairman, and the service times of machines are random variables. If a machine breaks down and the repairman is idle, then its service will start immediately, otherwise queues until its turn. The order of service may vary from one case to another according to set regulations. This system is a closed system.

Example 7.7.2. (Particle Counting)

Suppose particles arrive at a counting device according to some stochastic law. Not all the particles will be registered. This is due to the inertia of the device. The processes of registration constitute a queue.

Definition 7.7.7.

An *M/M/1 queueing system* is a single-server queueing system with Markovian arrival and Markovian service processes with infinite source of arrivals (open system), in which we assume the following:

(i) The interarrival times of customers are exponentially distributed (i.e., Poisson input) with mean arrival rate (birth rate) of λ customers per unit time into the system. On average, there are $1/\lambda$ time units between arrivals. This exponential distribution is a consequence of the Markov property. Hence, the first M in the $M/M/1$ designates this "Markovian" property.

(ii) The service times are exponentially distributed with a mean service rate (death rate) of μ services per unit time. μ is the average service rate of customers or departure out of the service facility. That is, on average, the service time is $1/\mu$ time units per customer. The second M in $M/M/1$ designates this "Markovian" property.

(iii) The system has 1 server. Hence, the "1" in $M/M/1$ represents this fact.

(iv) There is no limit on the system's capacity. The buffer (waiting room) has no size limit and no job will be turned away. This fact sometimes is denoted as $M/M/1/\infty$.

(v) The queue discipline is FIFO, that is, jobs are serviced in the same order in which they arrive.

We have seen that exponentially distributed interarrival times tasks with mean $1/\lambda$ are equivalent (over a time interval of length t) to a Poisson arrival pattern with mean rate λ.

Definition 7.7.8.

An *M/M/1* system is often referred to as a *single-server, infinite capacity queueing systems with Poisson input and exponential service times.*

Example 7.7.3.

Consider an *M/M/*1 queueing process. The states $0, 1, 2, 3, \cdots$ of this process represent the possible number of customers in the queue. In the simplest case, the arrival and service rates are both assumed to be constant (and, for instance, not state dependent). That is, in a birth-death terminology, for each $k \geq 0$, $\lambda_k = \lambda$ and for each $k \geq 1$, $\mu_k = \mu$. The system of DDE's is as follows:

State	Initial Probability	DDE
0	$p_0(0)$	$p_0'(t) = -\lambda p_0(t) + \mu p_1(t)$
1	$p_1(0)$	$p_1'(t) = -(\lambda + \mu)p_1(t) + \lambda p_0(t) + \mu p_2(t)$
$\vdots$	$\vdots$	$\vdots$
k	$p_k(0)$	$p_k'(t) = -(\lambda + \mu)p_k(t) + \lambda p_{k-1}(t) + \mu p_{k+1}(t)$
$\vdots$	$\vdots$	$\vdots$

We summarize this DDE for this *M/M/*1 system as:

$$\begin{cases} p_0'(t) = -\lambda p_0(t) + \mu p_1(t), \\ p_k'(t) = -(\lambda + \mu)p_k(t) + \lambda p_{k-1}(t) + \mu p_{k+1}(t), \quad k = 1, 2, \cdots, \end{cases} \quad (7.7.1)$$

with $p_0(0) = \delta_{i0}$, where δ_{i0} is called the *Kronecker delta* and is defined as

$$\delta_{i0} = \begin{cases} 1, & i = 0, \\ 0, & i \neq 0. \end{cases} \quad (7.7.2)$$

To solve (7.7.1), we may first apply the Laplace transform and then generating function or vice versa. Going with the latter, after some algebra we need to invert the Laplace transform and then find the coefficients of Taylor expansion of the generating function. Due to complexity of the solution process, we refer the reader to Haghighi and Mishev (2008, p. 161).

Definition 7.7.9.

The *M/M/*1/*N* system is an open queueing system with defining criteria identical to those of the *M/M/*1 system discussed in Definition 7.7.7, with the exception that system capacity is limited to *N* customers, $N \geq 1$. Due to the capacity limitation, we assume that new arrivals are turned away when the system is full.

Figures 7.7.1 and 7.7.2 describe the job traffic flow and transition diagram, respectively, for the case *M/M/*1/*N*.

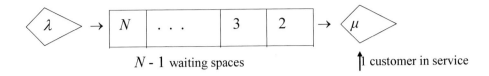

Figure 7.7.1. Traffic Flow Diagram for the *M/M/*1*/N* Queue.

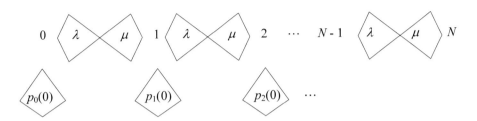

Figure 7.7.2. Transition Rate Diagram for the M/M/1/N Queue.

For the *M/M/*1*/N* system, the states of the process are the possible queue sizes $0, 1, 2, 3, \cdots, N$. Again, the arrival and departure rates are constant. That is, for $k = 0, 1, 2, \cdots, N - 1, \lambda_k = \lambda$ and for $k = 1, 2, 3, \cdots, N$, $\mu_k = \mu$. Using the transition rate diagram, Figure 7.7.2, we may generate the DDE system for the *M/M/*1*/N* queue:

State	Initial Probability	DDE
0	$p_0(0)$	$p_0'(t) = -\lambda p_0(t) + \mu p_1(t)$
1	$p_1(0)$	$p_1'(t) = -(\lambda + \mu)p_1(t) + \lambda p_0(t) + \mu p_2(t)$
$\vdots$	$\vdots$	$\vdots$
k	$p_k(0)$	$p_k'(t) = -(\lambda + \mu)p_k(t) + \lambda p_{k-1}(t) + \mu p_{k+1}(t)$
$\vdots$	$\vdots$	$\vdots$
N	$p_N(0)$	$p_N'(t) = -\mu p_N(t) + \lambda p_{N-1}(t)$

Again we summarize this DDE system as:

$$\begin{cases} p_0'(t) = -\lambda p_0(t) + \mu p_1(t), \\ p_k'(t) = -(\lambda + \mu)p_k(t) + \lambda p_{k-1}(t) + \mu p_{k+1}(t), \quad 1 \le k \le N-1, \\ p_N'(t) = -\mu p_N(t) + \lambda p_{N-1}(t), \end{cases} \tag{7.7.3}$$

with $p_0(0) = \delta_{i0}$, where δ_{i0} is called the Kronecker delta.

Note that in both *M/M/*1 and *M/M/*1*/N* systems described above the arrival and departure patterns are independent of the state of the system.

Definition 7.7.10.

The $C/M/1/N$ system is a closed queueing system (the "C" is indicative of this fact) with defining criteria identical to those of the $M/M/1/N$ system discussed in the Definition 7.7.9 with the exception that the arrival source is finite. Unlike $M/M/1/N$ and $M/M/1$ systems, arrivals in this case come from within, i.e., this queue allows no arrival from outside the system. Within this system machines that are in need of repair are awaiting in the repair queue.

Example 7.7.4. (N Machine Failure-Repair)

In the $C/M/1/N$ system, we assume there are N (≥ 1) identical machines, each operating independently of the others. Each machine has an average individual failure (death) rate of μ failures per unit time. Thus, at any time that k (of the N total) machines are "*up*" (*in good working condition*). These k "*up*" machines have a "*collective mean failure rate*" of $k\mu$. For example, if there are $N = 5$ machines in the system, each with a mean failure rate of $\mu = 2$ failures per month, then we can expect an overall rate of $5\mu = 10$ failures per month for the 5 machines. Associated with the N machines is a single repair facility (server) which restores machines to good working condition at a mean rate of λ machines per unit time. The mean repair (birth) rate is a constant λ, regardless of the number of "*down*" machines (*in not working condition*).

Treating this system as a birth-death process, it may be modeled with the transition rate diagram, Figure 7.7.3.

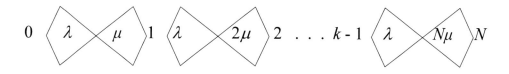

Figure 7.7.3. Transition Rate Diagram for the $C/M/1/N$ Failure-Repair System.

This queue may assume the states $0, 1, 2, \cdots, N$. The DDE's for the $C/M/1/N$ system are:

State	Initial Probability	DDE
0	$p_0(0)$	$p_0'(t) = -\lambda p_0(t) + \mu p_1(t)$
1	$p_1(0)$	$p_1'(t) = -(\lambda + \mu) p_1(t) + \lambda p_0(t) + 2\mu p_2(t)$
$\vdots$	$\vdots$	$\vdots$
k	$p_k(0)$	$p_k'(t) = -(\lambda + k\mu) p_k(t) + \lambda p_{k-1}(t) + (k+1)\mu p_{k+1}(t)$
$\vdots$	$\vdots$	$\vdots$
N	$p_N(0)$	$p_N'(t) = -k\mu p_N(t) + \lambda p_{N-1}(t)$

As with the $M/M/1/N$ system, we will summarize the DDE system for the $C/M/1/N$ system as follows:

$$\begin{cases} p_0'(t) = -\lambda p_0(t) + \mu p_1(t), \\ p_k'(t) = -(\lambda + k\mu)p_k(t) + \lambda p_{k-1}(t) + (k+1)\mu p_{k+1}(t), \quad 1 \le k \le N-1, \\ p_N'(t) = -k\mu p_N(t) + \lambda p_{N-1}(t), \end{cases} \qquad (7.7.4)$$

with $p_0(0) = \delta_{i0}$, where δ_{i0} is called the Kronecker delta.

7.8. STATIONARY DISTRIBUTIONS

All states in the $M/M/1$, $M/M/1/N$ and $C/M/1/N$ systems are recurrent. Thus, we now focus on the study of such systems after they have been in operation for a "long time". That is, we want to study the system in the stationary (steady-state or equilibrium) case. In other words, we will study the time independent distribution denoted by π_k, i.e.,

$$\pi_k = \lim_{t \to \infty} p_k(t), \qquad (7.8.1)$$

for each state k (assuming these limits exist). If the process has run long enough, we consider π_k to be, simply, the probability of being in state k, without regard to time.

Definition 7.8.1.

In a stationary queueing system, the ratio of the average number of arrivals, λ, per the average number of services, μ, denoted by, ρ, i.e., $\rho = \lambda / \mu$, is called the *traffic intensity* (or *utilization factor*) of the system, i.e., the probability that the server is busy.

Example 7.8.1. (Stationary M/M/1).

Let X be a random variable representing the number of tasks in the $M/M/1$ system. We assume X is time independent, i.e., if the time-dependent random variable $X(t)$ represents the number of tasks at time t, then its time-independent form is $X = X(\infty)$. The stationary probability distribution of X is as follows:

k	0	1	2	. . .
$P(X=k) = \pi_k$	π_0	π_1	π_2	. . .

The states $0, 1, 2, \cdots$ represent the number of jobs awaiting service, one of which may be receiving service. The random variable X being time-independent, derivatives on the left of 7.8.1 will be zero and that leads to the *difference balance equations*, denoted by *DBE's*, (or *limiting balance equations*, denoted by *LBE's*). hence, DBE's for the $M/M/1$ is:

$$\lambda \pi_0 = \mu \pi_1,$$
$$(\lambda + \mu) \pi_k = \lambda \pi_{k-1} + \mu \pi_{k+1}, \qquad k = 1, 2, \cdots, \tag{7.8.2}$$

with $\sum_{k=0}^{\infty} \pi_k = 1$.

Solving the system (7.8.3) iteratively, for the system $M/M/1$ we will have:

$$\pi_k = (1 - \rho) \rho^k, \quad k = 0, 1, 2, \cdots, \tag{7.8.3}$$

where ρ is the traffic intensity, i.e., $\rho = \dfrac{\lambda}{\mu}$, and $\rho < 1$.

Note that the probability π_0 represents the proportion of time the server is idle, i.e., there is no tasks in the system. In other words, $\pi_0 = P(X = 0)$. Thus, the utilization factor, ρ, is $1 - \pi_0$. This fact can also be seen from (7.8.4) since $\pi_0 = 1 - \rho$ and $1 - \pi_0 = \rho$. Note also that the interest in more or less value of π_0 varies depending upon the case being described by the model. For instance, if we are paying a server, such as a bank teller, it may be desirable to have a small π_0.

Stationary expected queue length in this case is

$$E(X) = \sum_{k=0}^{\infty} k \pi_k = \sum_{k=0}^{\infty} k \rho^k (1 - \rho) = (1 - \rho) \rho \sum_{k=0}^{\infty} k \rho^{k-1}. \tag{7.8.4}$$

However, since

$$\frac{1}{1 - \rho} = \sum_{k=0}^{\infty} \rho^k \text{ and } \sum_{k=1}^{\infty} k \rho^{k-1} = \frac{d}{d\rho} \left(\frac{1}{1 - \rho} \right) = \frac{1}{(1 - \rho)^2}, \tag{7.8.5}$$

from (7.8.5) we will have

$$E(X) = (1 - \rho) \rho \sum_{k=0}^{\infty} k \rho^{k-1} = \frac{(1 - \rho) \rho}{(1 - \rho)^2}.$$

Therefore, the expected number of jobs in the system $M/M/1$ is the

$$E(X) = \frac{\rho}{1 - \rho}. \tag{7.8.6}$$

As a specific example, suppose a large computer at an engineering laboratory receives incoming programs for execution at a mean rate of 4 programs per minute. The computer has a mean service time of 12 seconds per task (the mean service rate = 5 tasks per minute). Storage capacity is sufficiently large so that all submitted programs can be stored while awaiting processing. This system can be modeled as an *M/M/*1 queueing process with the transition rate diagram below.

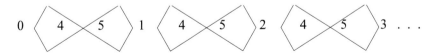

Figure 7.8.1. Transition Rate Diagram for an M/M/1 Queue with Parameters $\lambda = 4$, $\mu = 5$.

The states $0, 1, 2, \cdots$ represent the number of programs awaiting execution plus any currently in execution process. The DBE's for this example are:

Flow out = Flow in
$$4\pi_0 = 5\pi_1$$
$$4\pi_1 + 5\pi_1 = 4\pi_0 + 5\pi_2 \qquad\qquad\qquad (7.8.7)$$
$$4\pi_2 + 5\pi_2 = 4\pi_1 + 5\pi_3$$
$$\vdots$$

Going through detail solving the system (7.8.7), we will have

$$\pi_1 = (4/5)\pi_0$$
$$\pi_2 = (4/5)^2 \pi_0$$
$$\pi_3 = (4/5)^3 \pi_0$$
$$\vdots$$
$$\pi_k = (4/5)^k \pi_0$$
$$\vdots$$

In order to have a proper stationary distribution, we must include the *pmf* condition $\sum_{k=0}^{\infty} \pi_k - 1$. Thus,

$$\sum_{k=0}^{\infty} (4/5)^k \pi_0 = \pi_0 \sum_{k=0}^{\infty} (4/5)^k = 1. \qquad\qquad\qquad (7.8.8)$$

The series in (7.8.8) is geometric, with ratio 4/5 < 1, and, thus, it converges. Hence,

$$\pi_0 \sum_{k=0}^{\infty} \left(\frac{4}{5}\right)^k = \pi_0 \frac{1}{1 - \dfrac{4}{5}} = 5\pi_0 = 1$$

implies that $\pi_0 = 1/5$. Consequently,

$$\pi_1 = (4/5)(1/5) = 4/25, \; \pi_2 = (4/5)^2(1/5) = 16/125, \; \cdots, \; \pi_k = (4/5)^k(1/5), \; \cdots.$$

The expected number of tasks in this $M/M/1$ system is

$$E(X) = \sum_{k=0}^{\infty} k\pi_k = \sum_{k=0}^{\infty} k\left(\frac{4}{5}\right)^k\left(\frac{1}{5}\right) = \frac{\frac{4}{5}}{1-\frac{4}{5}} = 4 \; tasks.$$

Example 7.8.2. (Stationary M/M/1/N)

The steady-state distribution of X in this case is:

k	0	1	2	. . .	N
$P(X=k) = \pi_k$	π_0	π_1	π_2	. . .	π_N

Hence, $E(X) = \sum_{k=0}^{N} k\pi_k$.

Note that in this case π_N is the proportion of times that customers are being turned away or the fraction of lost customers. If turning jobs away is considered undesirable, then small values of π_N are preferred. On the other hand, $1 - \pi_N$ is the proportion of time that the system is not full and customers are not being turned away. From a retail sales perspective, small values of $1 - \pi_k$ might be preferable.

If customers arrive at a mean rate of λ, but some balk for some reason (the long length of the queue, for example) while the system is not full, the actual rate of arrival will not be λ, but some other number, say λ_e. We define λ_e as the *effective arrival rate*. This rate may be looked at a fraction of λ, namely, $\lambda_e = (1 - \pi_N)\lambda$. If $\pi_N \neq 0$, the arrival rate for jobs, which actually enter the system is λ adjusted by the proportion factor $(1 - \pi_N)$. In the $M/M/1$ case, $\lambda_e = \lambda$ since that system does not turn jobs away.

The DBE's for the $M/M/1/N$ system are:

Flow out = Flow in
$$\lambda\pi_0 = \mu\pi_1$$
$$\lambda\pi_1 + \mu\pi_1 = \lambda\pi_0 + \mu\pi_2$$
$$\vdots$$
$$\lambda\pi_{N-1} + \mu\pi_{k-1} = \lambda\pi_{k-2} + \mu\pi_N$$
$$\mu\pi_N = \lambda_{N-1},$$

(7.8.9)

together with the *pmf* condition $\pi_0 + \pi_1 + \pi_2 + \cdots + \pi_N = 1$. This system is a truncated version of the one for $M/M/1$, (7.8.7). Thus, we can summarize the DBE's as:

$$\lambda\pi_0 = \mu\pi_1,$$
$$(\lambda + \mu)\pi_k = \lambda\pi_{k-1} + \mu\pi_{k+1}, \qquad k = 1, 2, \cdots, N-1, \qquad (7.8.10)$$
$$\mu\pi_N = \lambda\pi_{N-1},$$

with $\sum_{k=0}^{\infty} \pi_k = 1$.

Solving the system (7.8.10) iteratively, for the system $M/M/1/N$ we will have:

$$\pi_k = \begin{cases} \dfrac{(1-\rho)\rho^k}{1-\rho^{N+1}}, & k = 0, 1, 2, \cdots, N, \quad \rho \neq 1, \\[4mm] \dfrac{1}{N+1}, & \rho = 1, \end{cases} \qquad (7.8.11)$$

where $\rho = \dfrac{\lambda}{\mu}$.

Stationary expected queue length in this case is

$$E(X) = \sum_{k=0}^{N} k\pi_k = \begin{cases} \sum_{k=0}^{N} k \dfrac{(1-\rho)\rho^k}{1-\rho^{N+1}}, & \rho \neq 1, \\[4mm] \sum_{k=0}^{N} k \dfrac{1}{1-N}, & \rho = 1, \end{cases} \qquad (7.8.12)$$

which can easily be shown to be

$$E(X) = \begin{cases} \dfrac{\rho(1-\rho^N)}{(1-\rho)(1-\rho^{N+1})} - \dfrac{N\rho^N}{1 \quad \rho^{N+1}}, & \rho \neq 1, \\[4mm] \dfrac{N}{2}, & \rho = 1. \end{cases} \qquad (7.8.13)$$

Suppose, specifically, the system is defined by $N = 3$, jobs arrive at a mean rate of $\lambda = 2$ jobs per minute, and a mean rate of $\mu = 3$ jobs per minute. Hence, $\rho = 2/3$. Then, the system is an $M/M/1/3$, modeled by the transition rate diagram, Figure 7.8.2.

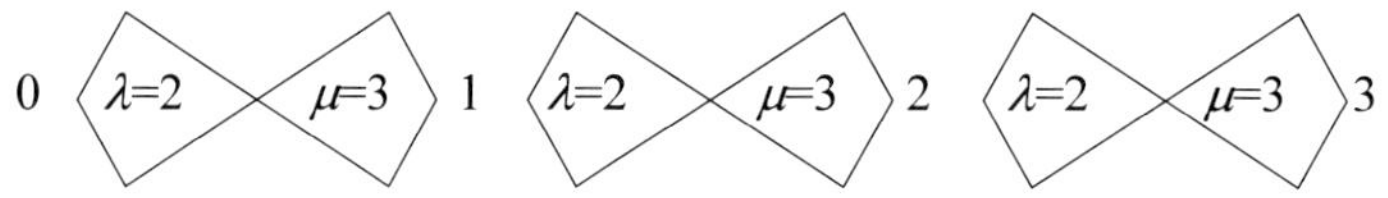

Figure 7.8.2. Transition Rate Diagram for the $M/M/1/3$ Queueing System.

Thus, from (7.8.11) we have:

$$\pi_0 = \frac{1-\dfrac{2}{3}}{1-\left(\dfrac{2}{3}\right)^4} = \frac{\dfrac{1}{3}}{\dfrac{65}{81}} = \frac{27}{65}, \quad \pi_1 = \frac{\left(1-\dfrac{2}{3}\right)\left(\dfrac{2}{3}\right)}{1-\left(\dfrac{2}{3}\right)^4} = \frac{\dfrac{2}{9}}{\dfrac{65}{81}} = \frac{18}{65},$$

$$\pi_2 = \frac{\left(1-\dfrac{2}{3}\right)\left(\dfrac{2}{3}\right)^2}{1-\left(\dfrac{2}{3}\right)^4} = \frac{12}{65}, \text{ and } \pi_3 = \frac{\left(1-\dfrac{2}{3}\right)\left(\dfrac{2}{3}\right)^3}{1-\left(\dfrac{2}{3}\right)^4} = \frac{8}{65}.$$

So, the proportion of times this queue is empty is 27/65, while of the times turning away jobs is 8/65. The expected number of jobs in the system is

$$E(X) = 0(27/65) + 1(18/65) + 2(12/65) + 3(8/65) = 1\frac{1}{65}.$$

So, on the average, there are $1\dfrac{1}{65}$ jobs in the system (1 in service and $\dfrac{1}{65}$ in the waiting line). The effective arrival rate in this system is

$$\lambda_e = (1 - \pi_3)\lambda = 1\frac{49}{65} \text{ jobs per minute. Hence, only } 1\frac{49}{65} \text{ of arrivals actually enters the}$$

system per minute.

Note that a large value of π_3 might be considered desirable (a store that is often full of customers). On the other hand, a large π_3 means that potential customers are often being turned away. This might indicate a need to increase the capacity of the system. A large value of π_0, on the other hand, would perhaps indicate the possibility of using a somewhat less capable server (perhaps less expensive and/or a slower service capability). From the server's point of view, a small π_0 may not be preferable. For instance, a mechanical server might wear out more quickly if π_0 is small and the server stays very busy.

Example 7.8.3. (Stationary C/M/1/N)

Suppose there are N machines working in an office. At any time, some of these machines may be in working "up" or not working "down" condition. We let X be a random variable as defined in Examples 7.8.1 and 7.8.2.

The transition rate diagram for the $C/M/1/N$ system is shown in Figure 7.8.3.

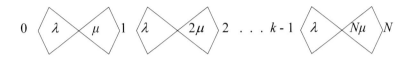

Figure 7.8.3. Transition Rate Diagram for the $C/M/1/N$ Failure-Repair Process

The DBE's for this general $C/M/1/N$ system are:

Flow out = Flow in
$$\lambda \pi_0 = \mu \pi_1$$
$$\lambda \pi_1 + \mu \pi_1 = \lambda \pi_0 + 2 \mu \pi_2$$
$$\lambda \pi_2 + 2\mu \pi_2 = \lambda \pi_1 + 3 \mu \pi_3 \qquad (7.8.14)$$
$$\vdots$$
$$N \mu \pi_N = \lambda_{N-1}$$

As before, solving (7.8.14) iteratively, we obtain:

$$\pi_1 = (\lambda / \mu) \pi_0$$
$$\pi_2 = (1/2)(\lambda / \mu)^2 \pi_0$$
$$\pi_3 = (1/[(2)(3)])(\lambda / \mu)^3 \pi_0$$
$$\vdots$$
$$\pi_k = (1/k!)(\lambda / \mu)^k \pi_0$$
$$\vdots$$
$$\pi_N = (1/N!)(\lambda / \mu)^k \pi_0$$

Using the *pmf* condition $\pi_0 + \pi_1 + \pi_2 + \cdots + \pi_N = 1$, we will have

$$\sum_{k=0}^{N} \pi_k = \sum_{k=0}^{N} \frac{1}{k!} \left(\frac{\lambda}{\mu} \right)^k \pi_0 = \pi_0 \sum_{k=0}^{N} \frac{1}{k!} \left(\frac{\lambda}{\mu} \right)^k = 1 . \qquad (7.8.15)$$

Hence, $\pi_0 = \left[\sum_{k=0}^{N} \frac{1}{k!} \left(\frac{\lambda}{\mu} \right)^k \right]^{-1}$ and, therefore, we have

$$\pi_k = \left(\frac{1}{k!}\right)\left(\frac{\lambda}{\mu}\right)^k \left[\sum_{k=0}^{N}\frac{1}{k!}\left(\frac{\lambda}{\mu}\right)^k\right]^{-1}, \quad k = 0,1,2,\cdots,N. \tag{7.8.16}$$

Again, letting $\rho \equiv \dfrac{\lambda}{\mu}$, (7.8.16) can be rewritten as

$$\pi_k = \left(\frac{\rho^k}{k!}\right)\left[\sum_{k=0}^{N}\frac{\rho^k}{k!}\right]^{-1}, \quad k = 0,1,2,\cdots,N. \tag{7.8.17}$$

The expected value of X in this case will be

$$E(X) = \sum_{k=0}^{N} k\pi_k = \sum_{k=0}^{N} k\left(\frac{\rho^k}{k!}\right)\left[\sum_{k=0}^{N}\frac{\rho^k}{k!}\right]^{-1} = \left[\sum_{k=0}^{N}\frac{\rho^k}{k!}\right]^{-1}\sum_{k=0}^{N} k\left(\frac{\rho^k}{k!}\right). \tag{7.8.18}$$

Specifically, suppose $N = 3$. Any one of these machines has a mean failure rate of $\mu = 5$ failures per month, while the associated repair facility fixes these machines at a mean rate of $\lambda = 4$ machines per month. The transition rate diagram for this process is:

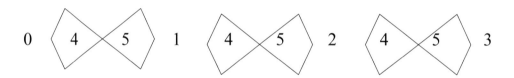

Figure 7.8.4. Transition Rate Diagram for a $C/M/1/3$.

Hence, the states 0, 1, 2 and 3 represent the number of machines "up". The corresponding DBE's are:

Flow out = Flow in
$4\pi_0 = 5\pi_1$
$5\pi_1 + 4\pi_1 = 4\pi_0 + 10\pi_2$
$10\pi_2 + 4\pi_2 = 4\pi_1 + 15\pi_3 \tag{7.8.19}$
$15\pi_3 = 4\pi_2,$

and the *pmf* condition is $\pi_0 + \pi_1 + \pi_2 + \pi_3 = 1$. Thus, from (7.8.17), solutions to the system (7.8.19) are

$\pi_0 = 375/827$, $\pi_1 = 300/827$, $\pi_2 = 120/827$ and $\pi_3 = 32/827$.

Hence, all 3 machines are "down" with probability $\pi_0 = 375/827$, while all 3 machines are "up" only $\pi_3 = 32/827$ of the time.

From (7.8.18), the expected value of X is:

$$E(X) = \frac{375}{827}\left[0\cdot 1 + 1\cdot\frac{4}{5} + 2\cdot\frac{8}{25} + 3\cdot\frac{32}{375}\right] = 96\frac{540}{827}.$$

7.9. WAITING TIME

As we mentioned at the beginning of the section, another quantity of interest in a queueing systems is the waiting time in the queue. Hence, in a steady-state case, we let the random variables X and X_Q represent the number of customers in the system (in waiting line and in service) and in the waiting line alone, respectively. Thus, for a single-server queue such as $M/M/1$ and $M/M/1/N$, for instance, we will have

$$X_Q = \begin{cases} X-1, & \text{if } X \geq 1, \\ 0, & \text{if } X = 0. \end{cases} \tag{7.9.1}$$

Note that for the waiting line to be empty, the system must be empty or only one customer in service, i.e.,

$$P(X_Q = 0) = P(X = 0) + P(X = 1) = \pi_0 + \pi_1. \tag{7.9.2}$$

For a single-server queue, in general, the event of having $k > 0$ customers waiting in line is equivalent to the event of having $k + 1$ in the system.

$$P(X_Q = k) = P(X = k + 1) = \pi_{k+1} \text{ for } k \geq 1.$$

In summary, the stationary distribution for X_Q in case $M/M/1$ system is shown below (and by truncating it we will have for $M/M/1/N$).

$X_Q = k$	0	1	2	. . .	$k - 1$	. . .
$P(X_Q = k)$	$\pi_0 + \pi_1$	π_2	π_3	. . .	π_k	. . .

Now from (7.8.3) and (7.8.6) the expected number of jobs in the waiting line will be

$$E(X_Q) = \sum_{k=1}^{\infty}(k-1)\pi_k = \sum_{k=1}^{\infty}k\pi_k - \sum_{k=1}^{\infty}\pi_k$$

$$= E(X) + \pi_0 - \sum_{k=0}^{\infty}\pi_k = \frac{\rho}{1-\rho} + (1-\rho) - \sum_{k=0}^{\infty}(1-\rho)\rho^k$$

$$= \frac{\rho}{1-\rho} + (1-\rho) - (1-\rho)\frac{1}{1-\rho} = \frac{\rho}{1-\rho} - \rho = \frac{\rho - \rho(1-\rho)}{1-\rho}$$

or

$$E(X_Q) = \frac{\rho^2}{1-\rho}. \tag{7.9.3}$$

Example 7.9.1.

In Example 7.8.1 we considered an $M/M/1$ system with mean arrival rate of $\lambda = 4$ computer programs per minute and a mean execution rate of $\mu = 5$ programs per minute. We found that $\pi_k = (4/5)^k(1/5)$ for $k = 0, 1, 2, \cdots$ and $E(X) = 4$ programs in the system. Thus, the expected number of programs in the system, waiting to be executed, is

$$E\left(X_Q\right) = 0\left(\pi_0 + \pi_1\right) + \sum_{k=1}^{\infty}k\pi_{k+1}$$

$$= \sum_{k=1}^{\infty}k\left(\frac{4}{5}\right)^{k+1}\left(\frac{1}{5}\right) = \frac{16}{5} = 3\frac{1}{5} \text{ programs.}$$

Example 7.9.2.

In Example 7.8.2, for the case $M/M/1/3$ queueing system with a mean arrival rate of $\lambda = 2$ customers per minute and mean service rate of $\mu = 3$ customers per minute, we found the distribution of the number of customers in the system as:

$X = k$	0	1	2	3
$P(X = k)$	27/65	18/65	12/65	8/65

So, the distribution of X_Q will be

$X_Q = k$	0	1	2
$P(X_Q = k)$	45/65	12/65	8/65

Thus,

$E(X_Q) = 0(45/65) + 1(12/65) + 2(8/65) = 28/65$ customers.

We now define the following random variables:

(i) T_Q = the time a customer spends in the waiting line,
(ii) T_S = the time a customer spends in service, and
(iii) T = the time a customer spends in the system (= $T_Q + T_S$, the so called *sojourn time*),

with their expected values as W_Q, W_S, and W, respectively. That is,

$$E(T_Q) = W_Q,\ E(T_S) = W_S,\ \text{and}\ E(T) = W. \tag{7.9.4}$$

We also denote the expected values of X and X_Q as L and L_Q, respectively. That is,

$$E(X_Q) = L_Q\ \text{and}\ E(X) = L. \tag{7.9.5}$$

In the queueing literature, it is well known, under some restriction, that for a stationary $M/M/1$ with arrival rate λ,

$$L = \lambda W. \tag{7.9.6}$$

Relation (7.9.4) is called *Little's Formulas*, named after J. D. C. Little (1961), who was the first to present a formal proof. It says that the mean sojourn time is equals to the ratio of the mean number in the system to the mean arrivals.

A similar relation to (7.9.6) can be derived for the waiting line as

$$L_Q = \lambda W_Q. \tag{7.9.7}$$

Thus, from (7.8.7) and (7.9.6) we will have:

$$W = \frac{L}{\lambda} = \frac{\dfrac{\rho}{1-\rho}}{\lambda} = \frac{\dfrac{\lambda/\mu}{1-\lambda/\mu}}{\lambda} = \frac{\dfrac{\lambda}{\mu-\lambda}}{\lambda}$$

or

$$W = \frac{1}{\mu-\lambda}. \tag{7.9.8}$$

Also, from (7.9.3) and (7.9.7) we will have:

$$W_Q = \frac{L_Q}{\lambda} = \frac{\frac{\rho^2}{1-\rho}}{\lambda} = \frac{\frac{(\lambda/\mu)^2}{1-\lambda/\mu}}{\lambda} = \frac{\frac{\lambda^2/\mu}{\mu-\lambda}}{\lambda}$$

or

$$W_Q = \frac{\lambda}{\mu(\mu-\lambda)}.$$ (7.9.9)

We note that from (7.9.4) since $W = W_Q + W_S$ and since the average duration of a service is $1/\mu$, from (7.9.9) we will have $W = \frac{\lambda}{\mu(\mu-\lambda)} + \frac{1}{\mu} = \frac{\lambda + (\mu-\lambda)}{\mu(\mu-\lambda)} = \frac{1}{\mu-\lambda}$ that confirms (7.9.8).

Example 7.9.3.

In Example 7.9.1, given $\lambda = 4$ and $\mu = 5$, we found $L = 4$ and $L_Q = 16/5$. Now, from (7.9.8) and (7.9.9) we will have $W = 1/(5-4) = 1$ minute and $W_Q = 4/(5(5-4)) = 4/5$ minute.

7.10. SINGLE-SERVER QUEUE WITH FEEDBACK

Let us consider a stationary single-server queueing system as will be described. Suppose after departure from the service, a job needs to return to the end of the waiting line for further service. Let p, $0 < p < 1$, be the probability of return. Hence, a job will exit the system with probability of $1 - p = q$. Eventually, after all service is performed, the completed job leaves the system. Hence, the *effective job service-rate* (the rate at which jobs actually leave the system and a change in state occurs, i.e., departure rte) is $q\mu$. The rate at which jobs are fed back into the system for further work (with no change in state) is $p\mu$. Therefore, we have developed a new single-server queue that is called an *M/M/1 with feedback*, and, more generally a *single-sever queue with feedback*. Figure 7.10.1 illustrates such a system.

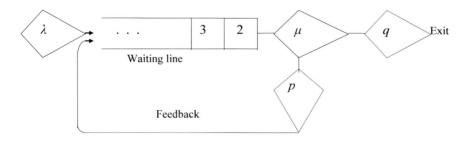

Figure 7.10.1. Traffic Flow Diagram for *M/M/*1 Feedback.

We can find similar performance measure as for $M/M/1$ for this new model, $M/M/1$ feedback. So, let us suppose a certain type of job (perhaps a computer program in execution) arrives with an average of λ per minute. We assume that on any one trip through the queue, the probability that a job leaves the system after exiting the service facility is $q = 1/10$. The expected proportion of the time that the job is fed back through the system is, then, $p = 1 - q = 9/10$. If the expected service rate is $\mu = 7$ jobs per minute, then, on the average, the effective service rate (departure) is $q\mu = 7/10$ jobs per minute. Figure 7.10.2 shows the activities and the number of jobs in the system in a stationary single-server feedback queue.

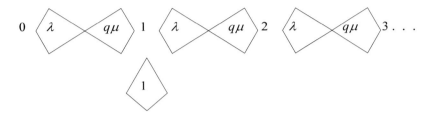

Figure 7.10.2. Transition Rate Diagram for the stationary single-server feedback queue.

7.10.1. M/M/1/N with Feedback

Suppose in the time interval $(0, \infty)$ tasks arrive according to a Poisson distribution of density λ into a waiting room of capacity $N - 1$. If an arriving task finds the system in full capacity it will go away. The service times are assumed to be independent identically distributed random variables with a common negative exponential distribution function of parameter μ and independent of the input process. After being served, a task may leave the system with probability q or it may immediately return to the waiting room with probability p. The event that a task returns is independent of any other event involved in the system and, in particular, independent of the number of its previous returns. Note, however, that as soon as a task gets into the system, it can obtain service as many times as it wishes as long as it returns immediately each time.

If we denote $\xi(t)$ as the queue size at time t, then $\{\xi(t)\}$ will be a homogeneous Markov process with state space $S = \{0, 1, 2, \cdots, N\}$. Let the transition probabilities, $p_{i,k}(t)$, be defined by

$$p_{i,k}(t) = P\{\xi(t) = k | \xi(0) = i\}, \qquad k, i = 0, 1, 2, ..., N.$$

Then, $\lim_{t \to \infty} p_{ik}(t) = p_k$, $k = 0, 1, \cdots, N$, exists and is independent of the initial distribution. p_k can be obtained using the usual method of standard Markov processes.

Hence, letting $\rho_1 = \dfrac{\lambda}{q\mu}$, the DBE for this *finite single-server queue with feedback* model is:

$$\lambda p_0 = q\mu p_1$$
$$(\lambda + q\mu)p_1 = \lambda p_0 + q\mu p_2 \qquad (7.10.1)$$
$$\vdots$$
$$(\lambda + q\mu)p_{N-1} = \lambda p_{N-2} + q\mu p_N.$$

It is easy to see that the solution to the system (7.10.1) is:

$$p_1 = \rho_1 p_0$$
$$p_2 = \rho_1^2 p_0$$
$$p_3 = \rho_1^3 p_0$$
$$\vdots$$
$$p_N = \rho_1^N p_0.$$

Since $\sum_{k=0}^{N} p_k = 1$, we obtain

$$p_0 + \rho_1 p_0 + \rho_1^2 p_0 + ... + \rho_1^N p_0 = p_0\left(1 + \rho_1 + \rho_1^2 + ... + \rho_1^N\right)$$

$$= \begin{cases} \dfrac{1-\rho_1}{N+1}\rho_1^N, & if \quad \rho_1 \neq 1 \\ p_0(N+1), & if \quad \rho_1 = 1 \end{cases}.$$

Hence, we will have:

$$p_k = \begin{cases} \dfrac{1-\rho_1}{N+1}, & if \rho_1 \neq 1, k = 1, 2, \cdots, N \\ \dfrac{1}{N+1}, & if \rho_1 = 1, k = 0. \end{cases} \qquad (7.10.2)$$

7.10.2. Infinite Single-Server Queue with Feedback

Now suppose that in the time interval $(0, \infty)$, tasks arrive to a service station according to a Poisson process of density λ. We assume that the service times are independent identically distributed random variables with a common distribution function $H(x)$ and independent of the input process. Tasks are being served by a single server on the basis of first-come-first-served. The server will be idle if, and only if, there is no task in the system. After being served each task either leaves the system with probability q or immediately joins the queue again with probability p, where $p + q = 1$. We also assume that the return of a task is an event independent of any other event involved in the system and in particular independent of the number of its previous returns. This is the Takács type $[F(x), H(x), p]$ model. If we assume $H(x)$ to be a negative exponential distribution function of parameter μ and there is no buffer limit, then we have an *infinite Markovian single-server queue with feedback*. If $p = 0$, then there is no feedback.

Let us denote by τ_n, $n = 0, 1, 2, \cdots$, the epoch of arrivals with $\tau_0 = 0$. Then, since the arrival process is Poisson of parameter λ, it follows that $\tau_{n+1} - \tau_n$ $(n = 0, 1, 2, ...)$ are mutually independent, identically distributed random variables with a common distribution function

$$F(x) = \begin{cases} 1 - e^{-\lambda x}, & \text{if } x \geq 0 \\ 0, & \text{if } x < 0 \end{cases}.$$

$(7.10.3)$

Now let us denote by $\xi_n(t)$ the queue size at time t, i.e., the number of tasks in the system (including the one at the service station). Let us also denote by $\xi(t)$ the queue size immediately before the arrival of the nth task, i.e., $\xi(t) = \xi(\tau - 0)$. Finally, let us denote by θ_n the *sojourn time* (*total time spent in the system*) of the nth task, i.e., total time the nth task spends in the system.

We would like to determine the stationary distribution of $\xi(t)$ and of θ_n, i.e., the distribution of $\xi(t)$ when it is the same distribution for all $t \geq 0$, and distribution of θ_n when it is the same for every $n = 0, 1, 2, \cdots$.

It is not too difficult to prove the following two lemmas:

Lemma 7.10.1.

The limiting distribution of $\xi(t)$, i.e., $\lim_{t \to \infty} P\{\xi(t) = k\}$, $k = 0, 1, 2, \cdots$, exists and is independent of the initial state if and only if $\xi(t)$ has a stationary distribution and the limiting distribution is identical with the stationary distribution.

Proof

Left as exercise.

Lemma 7.10.2.

The limiting distribution $\lim_{t \to \infty} P\{\theta_n \le x\}$ exists and is independent of the initial state if and only if θ_n has a stationary distribution and the limiting distribution is identical with the stationary distribution.

Proof

Left as exercise.

Notations:

1. $\Psi(s)$ the *Laplace-Stieltjes transform* of $H(x)$ defined by

$$,$$
(7.10.4)

2. α_r, the rth moment of $H(x)$ define by

$$\alpha_r = \int_0^\infty x^r dH(x), \quad r = 0,1,2,\cdots.$$
(7.10.5)

3. $\alpha = $ the average service time defined by $\alpha = \alpha_1 = \int_0^\infty x dH(x)$.
(7.10.6)

4. $H^*(x)$ be the distribution function of the total service time of a task,

5. $H_k(x)$, the kth iterated convolution of $H(x)$ with itself,

6. $\psi^*(s)$, the Laplace-Stieltjes transform of $H(x)$, i.e.,

$$\psi^*(s) = \int_0^\infty e^{-sx} dH^*(x), \quad \text{Re}(s) \ge 0.$$
(7.10.7)

7. α_r^*, the rth moment of $H^*(x)$ defined by

$$\alpha_r^* = \int_0^\infty x^r dH^*(x), \quad r = 0,1,\cdots.$$
(7.10.8)

Note that, clearly, $\alpha^* = \alpha_1^*$.
(7.10.9)

Remark 7.10.2.

$$H^*(x) = q \sum_{k=1}^{\infty} p^{k-1} H_k(x) \tag{7.10.10}$$

Proof

The probability that a task joins the queue exactly k times is qp^{k-1}, $k = 0, 1, 2, \cdots$, because once it joins the queue for the first time it would have to return $k - 1$ times before it departs permanently. Each return having a probability p and permanent departure having probability q gives us the probability of joining the queue k times to be qp^{k-1}. Now if it does join the queue k times, then its total service times is equal to the sum of k mutually independent random variables each of which has the distribution function $H(x)$. The distribution of the total service times will be the convolution of $H(x)$ with itself multiplied by the probability of joining the queue k times, and summed over all k's. This proves (7.10.10).

Remark 7.10.3.

Since Laplace-Stieltjes transform of the distribution of the sum of mutually independent random variables is the product of the Laplace-Stieltjes transform of distribution of each one of the random variable, (7.10.10) implies that

$$\psi^*(s) = q \sum_{k=1}^{\infty} p^{k-1} (\psi(s))^k.$$

The sum being a geometric series with ratio $p\psi(s)$, we obtain

$$\psi^*(s) = \frac{q\psi(s)}{1 - p\psi(s)} \qquad |p\psi(s)| < 1. \tag{7.10.11}$$

Remark 7.10.4.

$$\alpha^* = \frac{\alpha}{q}. \tag{7.10.12}$$

Proof

From (7.10.4) we have $\psi(0) = 1$ since $H(x)$ is a distribution function. Now by (7.10.4), (7.10.6), (7.10.7), (7.10.8) and (7.10.9) we have $\alpha = -\psi'(0)$ and $\alpha^* = -\psi^*{}'(0)$ which proves (7.10.12).

Remark 7.10.5.

$$\psi_2^* = \frac{\alpha_2}{q} + \frac{2p\alpha_1^2}{q^2}.$$ (7.10.13)

Proof

$$\frac{d^2\psi(s)}{ds^2} = \frac{q\dfrac{d^2\psi(s)}{ds^2}\left(1 - p\psi(s)\right) + 2pq\left(\dfrac{d\psi(s)}{ds}\right)^2}{\left(1 - p\psi(s)\right)^2}.$$

By (7.10.4) - (7.10.8) we have

$$\psi(0) = 1, \quad \left.\frac{d\psi(s)}{ds}\right|_{s=0} = -\alpha_1 = -\alpha, \quad \left.\frac{d^2\psi(s)}{ds^2}\right|_{s=0} = \alpha_2, \text{ and } \alpha_2^* = \left.\frac{d^2\psi^*(s)}{ds^2}\right|_{s=0}.$$

Thus,

$$\alpha_2^* = \frac{q^2\alpha_2 + 2qp\alpha^2}{q^3} = \frac{\alpha_2}{q} + \frac{2p\alpha_1^2}{q^2},$$

which proves (7.10.13).

Remark 7.10.6.

$$\alpha_r^* = \alpha_r + \frac{p}{q}\sum_{j=1}^{r}\binom{r}{j}\alpha_j\alpha_{r-j}^*.$$ (7.10.14)

Proof

Writing (7.10.11) in the form $\psi^*(s)(1 - p\psi(s)) = q\psi(s)$ and then taking derivatives, evaluating at $s = 0$ we obtain the recurrence formula for α_r^* given by (7.10.14).

7.10.3. The Stationary Distribution of the Feedback Queue Size

Now, as far as the distribution of the queue size is concerned, we can consider an equivalent process to the process $[F(x), H(x), p]$ such that tasks supposedly joining the queue only once are served in one stretch. In this case the service time of the tasks will be equal to their total service time if they were served in the original manner. The process obtained is, according to Takács, of type $[F(x), H^*(x), 0]$. Thus, the distribution of the queue size for the two processes $[F(x), H(x), p]$ and $[F(x), H^*(x), 0]$, are identical. Takács (1962a, 1962b) showed that for the process $[F(x), H^*(x), 0]$ a stationary distribution $\{p_j^*\}$ exists if, and only if, $\lambda \alpha^* < 1$ and is given by *A. Y. Khintchin's formula*:

$$U^*(z) = \sum_{j=0}^{\infty} p_j^* z^j = \frac{(1 - \lambda \alpha^*)(1 - z)\psi^*[\lambda](1 - z)}{\psi^*[\lambda(1 - z)] - z}, \qquad (7.10.15)$$

where $U^*(z)$ is the generating function of the stationary probability distribution of the queue size denoted by p_j^*. From (7.10.12) and the above discussion we obtain the following theorem of Takács.

Theorem 7.10.1.

If $\lambda \alpha < q$, then the process $\{\xi(t), 0 \le t < \infty\}$ has a unique distribution $P\{\xi(t) = j\} = p_j^*$, $j = 0,1,\cdots$, and for $|z| \le 1$ we have

$$U^*(z) = \sum_{j=0}^{\infty} p_j^* z^j = \left(1 - \frac{\lambda \alpha}{q}\right) \frac{q(1 - z)\, \psi[\lambda(1 - z)]}{(q + pz)\, \psi[\lambda(1 - z)] - z}. \qquad (7.10.16)$$

If $\lambda \alpha \ge q$, then a stationary distribution does not exists.

Now let us denote by x_n the time needed to complete the current service (if any) immediately before the arrival of the nth task, i.e. at the time instant $t = \tau - 0$. Of course, if $\xi(0) = 0$, then the server is idle at the time $\tau_n = 0$. The vector sequence

$\{\xi_n, x_n, n = 1, 2, \cdots\}$ forms a Markov sequence. For a stationary sequence $\{\xi_n, x_n\}$ we define the following:

$$p_j = P\{\xi_n = j\}, \quad j = 0, 1, 2, \cdots \tag{7.10.17}$$

$$p_j(x) = P\{x_n \le x, \xi = j\}, \quad j = 1, 2, \cdots \tag{7.10.18}$$

$$\prod_j (s) = \int_0^\infty e^{-st} dp_j(x), \quad j = 1, 2, \cdots \tag{7.10.19}$$

$$U(s, z) = \sum_{j=1}^\infty \prod_j (s) z^j \tag{7.10.20}$$

Takács (1963b) proved that for the process $\left[F(x), H^*(x), 0 \right]$ the sequence $\{\xi_n, x_n\}$ has a unique stationary distribution if and only if $\lambda \alpha < q$. This distribution is $P\{\xi_n = j\} = P^*, j = 1, 2, \cdots$, given by (7.10.14) and, thus, the generating function corresponding to (7.10.20) will be

$$U^*(s, z) = \frac{(1 - \lambda \alpha^*) \lambda z (1 - z) \{\psi^*(s) - \psi^*[\lambda(1-z)]\}}{\{z - \psi^*[\lambda(1-z)]\}[s - \lambda(1-z)]}. \tag{7.10.21}$$

We have already mentioned that the distribution function of the two processes $[F(x), H(x), p]$ and $[F(x), H^*(x), 0]$ are identical. In fact, the only distinction between these two processes is that, in the process $[F(x), H^*(x), 0]$ the remaining length of the current service at the time of arrival of the nth task, x, is replaced by n, the remaining part of the total service time of the task just being served at the time of arrival of the nth task. But the additional time for the process $[F(x), H^*(x), 0]$, i.e., the time added to X_n is independent of the queue size. Time has the distribution function

$$\hat{H}(x) = q \sum_{k=0}^\infty p^k H(x). \tag{7.10.22}$$

This is because after a task arrives, the probability that it returns k more times is, as previously explained, qp^k. The additional time, being the sum of k independent identically distributed random variables having a common distribution function $H(x)$, has distribution function $H_k(x)$ with $H_0(x) = 1$, if $x \ge 0$ and $H_0(x) = 0$, if $x < 0$. Thus, the Laplace-

Stieltjes transform of $H(x)$ in (7.10.21) is $\hat{\psi}(s) = \int_0^\infty e^{-sx} d\hat{H}(x) = q \sum_{k=0}^\infty p^k \left[\psi(s) \right]^k$. The sum

being a geometric series, we have

$$\hat{\psi}(s) = \frac{q}{1 - p\psi(s)} . \tag{7.10.23}$$

Thus,

$$\psi^*(s) = \psi(s)\hat{\psi}(x). \tag{7.10.24}$$

From (7.10.21) and (7.10.24) we have

$$U^*(x,z) = \frac{\left(1 - \dfrac{\lambda\alpha}{q}\right)\lambda z(1-z)\{\psi(s)\hat{\psi}(s) - \psi[\lambda(1-z)]\,\hat{\psi}[\lambda(1-z)]\}}{\{z - \psi[\lambda(1-z)]\,\hat{\psi}[\lambda(1-z)]\}\,[s - \lambda(1-z)]}$$

$$= \frac{\left(1 - \dfrac{\lambda\alpha}{q}\right)\lambda z(1-z)\hat{\psi}(s)\psi(s) - \psi[\lambda(1-z)]\left(\dfrac{\hat{\psi}[\lambda(1-z)]}{\psi(s)}\right)}{s - \lambda(1-z)z - \psi[\lambda(1-z)]\,\hat{\psi}[\lambda(1-z)]}$$

$$= \hat{\psi}(z)\frac{\left(1 - \dfrac{\lambda\alpha}{q}\right)\lambda z(1-z)}{\{z - (q + pz)\psi[\lambda(1-z)]\}[s - \lambda(1-z)]}$$

Hence,

$$U^*(s,z) = \hat{\psi}(z)U(s,z). \tag{7.10.25}$$

Remark 7.10.6.

We could also derive (7.10.25) in the following way:

Let the additional service time discussed above be denoted by θ. Then, distribution function of θ is given by (7.10.22). x_n and θ, being independent, imply that

$$U^*(s,z) = E\left\{e^{-s(x_n + \theta)}\, z^{\xi_n}\right\} = E\left\{e^{-sx_n} z^{\xi_n}\right\}.$$

$$E\left\{e^{-s\theta}\right\} = U(s,z)\hat{\psi}(s).$$

The above discussion proves the following theorem:

Theorem 7.10.2.

If $\lambda\alpha < q$, then the vector process $\{\xi_n, x_n ; n = 1, 2, \cdots\}$ has a unique stationary distribution, for which $p_j = p_j^*$ (7.10.16) and

$$U(s,z) = \sum_{j=1}^{\infty} \prod_j (s) z^j = \left(1 - \frac{\lambda\alpha}{q}\right) \frac{\lambda z(1-z)\{\psi(s) - \psi[\lambda(1-z)]\}}{\{z - (q+pz)\psi[\lambda(1-z)]\}[s - \lambda(1-z)]}. \qquad (7.10.26)$$

Remark 7.10.7.

Let $X(t)$ denote the time needed to complete the current service (if any) at time t. $X(t) = 0$ if $\xi(t) = 0$. Then, Takács remarked that the vector process $\{\xi(t), X(t), 0 \le t < \infty\}$ is a Markov process. The sequence $\{\xi(t), x(t)\}$ has a stationary distribution, if, and only if, $\lambda\alpha < q$, and coincides with the stationary distribution of $\{\xi_n, x_n\}$. We will show this for the case of the one-dimensional sequences $\{\xi(t)\}$ and $\{\xi_n\}$.

7.10.4. The Stationary Distribution of Sojourn Time of the nth Customer

To find the distribution of sojourn time θ_n of the nth task we need the joint distribution of ξ_n and X_n. Since ξ_n and X_n determine θ_n, if $\{\xi_n, X_n\}$ has a stationary distribution, every θ_n $n = 1, 2, \cdots$, has the same distribution. Now for a stationary process let

$$E\{e^{-s\theta} n\} = \Phi(s), \qquad \mathrm{Re}(s) \ge 0. \qquad (7.10.27)$$

Then, we obtain the third theorem of Takács (1963).

Theorem 7.10.3.

If $\lambda\alpha < q$, then θ_n has a unique stationary distribution $P\{\theta_n \le X\}$. The Laplace-Stieltjes transform of this distribution is given by

$$\Phi(s) = q \sum_{k=1}^{\infty} P^{k-1} U_k(s,1), \ Re(s) \geq 0, \tag{7.10.28}$$

where

$$U_1(s,z) = p_0 \psi[s + \lambda(1-z)] + U\{s + (1-z), (q+pz)\psi[s+\lambda(1-z)]\} \tag{7.10.29}$$

for $Re(s) \geq 0$ and $|z \leq 1|$, $p_0 = 1 - \dfrac{\lambda\alpha}{q}$, $U(s,z)$ is defined by (7.10.26), and

$$U_{k+1}(s,z) = \psi(s + \lambda[1-z])U_k\{s, (q+pz)\psi[s+\lambda(1-z)]\}, \ k = 1,2,\cdots \ . \tag{7.10.30}$$

Proof

Note that $U_k(s,1)$ will determine the stationary distribution of θ_n . We have already mentioned, on several occasions before, that the probability that a task joins the queue exactly k times (including the original arrival) is qp^{k-1}, $k = 1,2,\cdots$. Now suppose that the total time spent by the nth task until its kth departure from the service station is $\theta_n^{(k)}$ (if it joins the queue k times). Let us also denote by $\xi_n^{(k)}$ the number of tasks in the queue immediately after the departure of the nth task from the service station for the nth time. Finally let

$$U_k(s,z) = E\left\{e^{-s\theta_n^{(k)}} z^{\xi_n^{(k)}}\right\}, \ k = 1,2,\cdots . \tag{7.10.31}$$

Then, for a stationary sequence $\{\xi_n, X_n\}$, we have

$$U_1(s,z) = p_0 \psi[s + \lambda(1-z)] + \sum_{j=1}^{\infty} \prod_j [s + \lambda(1-z)](q+pz)^j \{\psi[s+\lambda(1-z)]\}^j$$

$$= p_0 \psi[s + \lambda(1-z)] + U\{s + \lambda(1-z), (q+pz)\psi[s+\lambda(1-z)]\}. \tag{7.10.32}$$

Relation (7.10.32) follows from the conditioning on the expectation of the queue size and the total service time and then, dropping the condition (using the theorem of total expectation) (see the Appendix).

Note:

1) $P_0 = 1 - \dfrac{\lambda\alpha}{q}$ and $U(s, z)$ is defined by (7.10.26);

2) we may obtain the Laplace-Stieltjes transform of $\theta_n^{(k)}$, if in (7.10.31) we let $z = 1$;

3) we may also obtain the generating function of $\xi_n^{(k)}$, if we let $s = 0$;

4) $e^{\lambda(1-z)x}$ is the generating function of the arrival process;

5) $(q + pz)^j$ is the generating function of the distribution of the number of returned tasks if the nth task leaves j tasks behind when it departs from the station for the kth time; and

6) $U_1(s,1) = (1 - \dfrac{\lambda\alpha}{q})[\psi(s) + U(s, \psi(s))]$, where $U(s, z)$ is defined by (7.10.26).

To prove (7.10.30), let us suppose the following:

1) the queue size immediately after the kth departure of the nth task from the station is $\xi_n^{(k)} = j$;

2) the total time spent by the nth task until immediately after his k^{th} departure from the station is $\theta_n^{(k)} = x$; and

3) the nth task joins the queue again after its kth departure from the station.

Under the above conditions the difference $\theta_n^{(k+1)} - \theta_n^{(k)}$, denoted by y, will be the length of time required for $j+1$ services. Service, being mutually independent, the distribution of y is $H_{j+1}(x)$. Now suppose υ_1 is the number of tasks arrived during the time of $j+1$ services, i.e., during the time interval y, and suppose that υ_2 is the number of tasks returned during the time interval y. Then, by assumption, υ_1 has a Poisson distribution of density λy with the generating function $e^{-\lambda(1-z)y}$ and υ_2 has a Bernoulli distribution of parameters j and p with the generating function $(q + pz)^j$. υ_1 and υ_2 are independent and $\xi_n^{(k+1)}$ will be the sum of these two independent random variables. We, thus, have:

$$E\left\{e^{-s\theta_n^{(k+1)}} z^{\xi_n^{(k+1)}} \Big| \xi_n^{(k)} = j, \theta_n^{(k)} = x\right\} = e^{-sx}(q + pz)^j \int_0^\infty e^{-sy - \lambda(1-z)y} dH_{j+1}(y)$$

$$= e^{-sx}(q + pz)^j \left\{\psi[s + \lambda(1-z)]\right\}^{j+1}$$

or

$$E\left\{e^{-s\theta_n^{(k+1)}}\, z^{\xi_n^{(k+1)}}\,\Big|\, \xi_n^{(k)}, \theta_n^{(k)}\right\} = \psi\left[s+\lambda(1-z)\right]^{-s\theta_n^{(k)}} \left\{(q+pz)\,\psi\left[s+\lambda(1-z)\right]\right\}^{\xi_n^{(k)}}$$

If we now drop the condition, we have

$$U_{k+1}(s,z) = \psi[s+\lambda(1-z)]\, U_k\left\{[s,(q+pz)\,\psi[s+\lambda(1-z)]\right\} \qquad (7.10.33)$$

as was to prove. From (7.10.23) and (7.10.33) we finally obtain

$$E\left\{e^{-s\theta_n}\right\} = q\sum_{k=1}^{\infty} p^{k-1} E\left\{e^{-s\theta_n^{(k)}}\right\} = q\sum_{k=1}^{\infty} p^{k-1} U_k(s,1),$$

which proves the theorem.

7.10.5. The Moments of Sojourn Time of the nth Customer

In this sub-section we will obtain explicit expressions for the moments of the sojourn time of a task. Let

$$\Phi(s,z) = q\sum_{k=1}^{\infty} p^{k-1} U_k(s,z), \qquad (7.10.34)$$

where $U_k(s,z)$ is given by (7.10.29) and (7.10.30).

Note that if $z = 1$, then

$$\Phi(s,z) = q\sum_{k=1}^{\infty} p^{k-1} U_k(s,z) = q\sum_{k=0}^{\infty} p^k U_{k+1}(s,z)$$

$$= qU_1(s,z) + q\sum_{k=1}^{\infty} p^k \psi\left\{s+\lambda(1-z)U_k\left[s,(q+pz)\psi[s+\lambda(1-z)]\right]\right\}$$

$$= qU_1(s,z) + p\psi[s+\lambda(1-z)]q\sum_{k=1}^{\infty} p^{k-1} U_k\left\{s,(q+pz)\psi[s+\lambda(1-z)]\right\}.$$

giving:

$$\Phi(s,z) = qU_1(s,z) + p\psi[s+\lambda(1-z)]\Phi\left\{s,(q+pz)\psi[s+\lambda(1-z)]\right\}. \qquad (7.10.35)$$

To obtain the moments, we let

$$\Phi_{ij} = \left(\frac{\partial^{i+j} \Phi(s,z)}{\partial s^i \, \partial z^j} \right)_{s=0, \, z=1}. \qquad (7.10.36)$$

The partial derivatives in (7.10.36) leads to a system of $r+1$ linear equations with $r+1$ unknowns yielding Φ_i, where $i + j = r$, $i = 0, 1, 2, \ldots, r$. This system can be solved step by step for $r = 1, 2, \cdots$. Also from (7.10.28) we see that

$$E\{\theta_n^r\} = (-1)^r \Phi_{r0}, \quad r = 0, 1, 2, \cdots, \qquad (7.10.37)$$

for a stationary process.

It is not too difficult to find Φ_{ij} for $r = 1$, and, hence, the expectation of θ_n. However, finding Φ_{ij} for $r = 2$ may be a little more manually formidable. With the assistance of algebraic programming called ALPAK using IBM 7090, however, Brown (1963) was able to obtain the second moment. We are, thus, led to the fourth theorem of Takács.

Theorem 7.10.4.

If $\lambda \alpha < q$ and θ_n has a stationary distribution, then assuming α_2 is finite, we have

$$E\{\theta_n\} = \frac{\lambda \alpha_2 + 2\alpha_1 (1 - \lambda \alpha_1)}{2(q - \lambda \alpha_1)} \qquad (7.10.38)$$

and assuming α_3 is finite, we have

$$E\{\theta_n^2\} = \frac{q^2 - 2q}{\left[\begin{array}{l} 6(q - \lambda \alpha_1)^2 \left[q^2 - q(2 + \lambda \alpha_1) + \lambda \alpha_1 \right] \\ \cdot \left\{ \begin{array}{l} 2q \left[6\lambda \alpha_1^3 - 6\alpha_1^2 - 6\lambda \alpha_1 \alpha_2 + 3\alpha_2 + \lambda \alpha_3 \right] \\ - \left[12\lambda \alpha_1^3 - 12\alpha_1^2 - 6\lambda \alpha_1 \alpha_2 + 2\lambda^2 \alpha_1 \alpha_2 - 3\lambda^2 \alpha_2^2 \right] \end{array} \right\} \end{array} \right]}. \qquad (7.10.39)$$

Note:

1. Brown (1963) assumes all α_r's to be finite for convenience, though, in order to calculate the moments, finiteness of α_{r+1} would be sufficient.

2. In his method, Brown replaces z by $t = 1 - z$ and U_1 by $W = \dfrac{qU_1}{q - \lambda\alpha_1}$ and, thus,

writes the rth moment of θ_n as $\beta_r = (-1)^r \Phi_{r0}$, where

$$\Phi_{ij} = \left[\left(\frac{\partial}{\partial s}\right)^i \left(\frac{\partial}{\partial t}\right)^j \Phi(s,t) \right]_{s=1, t=0} . \text{ He then finds}$$

$$\Phi(s,t) = (q - \lambda\alpha_1)\, W(s,t) + p\psi(s, \lambda t)\, \Phi[s, W(s,t)],$$

where

$$W(s,t) = 1 - (1 - pt)\, \Psi(s + \lambda t),$$
$$W(s,t) = \psi(s, \lambda t) + S\{s + \lambda t, \lambda W(s,t)\, T[W(s,t)]\},$$
$$S(x,y) = \frac{\Psi(x) - \psi(y)}{x - y}, \text{ and } T(w) = \frac{\lambda W(1 - W)}{1 - W - (1 - pW)\,\Psi(\lambda W)}.$$

And that

$$\beta_0 = 1,$$

$$\beta_1 = \alpha_1\left(\frac{1 - \alpha_1}{q - \alpha_1}\right) + \frac{\alpha_2}{2(q - \alpha_{1)}},$$

and

$$\beta_2 = \frac{(2qF - G)(q^2 - 2q)}{6(q - \alpha_1)^2\left[q^2(\alpha_1 + 2) + \alpha_1\right]},$$

where

$$F = 6\alpha_1^3 - 6\alpha_1^2 - 6\alpha_1\alpha_2 + 3\alpha_2 + \alpha_3,$$

and

$$G = 12\alpha_1^2 - 12\alpha_1^2 - 6\alpha_1\alpha_2 + 2\alpha_1\alpha_2 - 3\alpha_2^2.$$

Example 7.10.7. (Exponential Service Single-server Infinite Queue with Feedback)

We now consider the single-server queue mentioned earlier in this chapter. We assume that $H(x)$ is a negative exponential distribution function with parameter μ i.e.,

$$H(x) = \begin{cases} 1 - e^{-\mu x}, & \text{for } x \geq 0 \\ 0, & x < 0. \end{cases} \tag{7.10.40}$$

Under this assumption, we will have

$$H^*(x) = \begin{cases} \{1 - e^{-\mu q x}, & \text{for } x \geq 0 \\ 0, & \text{for } x < 0. \end{cases} \tag{7.10.41}$$

since a task leaves the system with probability q.

The corresponding Laplace-Stieltjes transforms of (7.10.40) and (7.10.41) are, respectively, $\psi(s) = \dfrac{u}{u+s}$, and $\psi^*(s) = \dfrac{\mu q}{q+s}$. Thus, equations (7.10.26) and (7.10.27) for this particular case become:

$$U_{k+1}(s,z) = \frac{u}{u+s+\lambda(1-z)} U_k\left(s, \frac{u(q+pz)}{u+s+\lambda(1-z)}\right), \quad k = 0,1,\ldots, \tag{7.10.42}$$

where

$$U_0(s,z) = \frac{1 - \dfrac{\lambda}{q\mu}}{1 - \dfrac{\lambda z}{q\mu}}. \tag{7.10.43}$$

Takács expresses $U_k(s,1)$ from (7.10.42) as follows

$$U_k(s,1) = \frac{(1 - \dfrac{\lambda}{q\mu})}{a_k(s) - b_k(s)}, \tag{7.10.44}$$

where $a_k(s)$ and $b_k(s), k = 0,1,\ldots$, are given by the following matrix

$$\begin{pmatrix} a_k(s) \\ b_k(s) \end{pmatrix} = \begin{pmatrix} \dfrac{\lambda + \mu + s}{\mu} & -q \\ \dfrac{\lambda}{\mu} & p \end{pmatrix}^k \begin{pmatrix} 1 \\ \dfrac{\lambda}{q\mu} \end{pmatrix}.$$

Thus, using (7.10.34), we obtain the Laplace-Stieltjes transform of the distribution of θ_n as

$$\Phi(s) = (1 - \frac{\lambda}{q\mu}) \sum_{k=1}^{\infty} \frac{qp^{k-1}}{a_k(s) - b_k(s)}. \tag{7.10.45}$$

Hence, taking the first and the second derivatives of (7.10.45) and evaluating them at $s = 0$ with $\alpha_1 = \alpha = \dfrac{1}{\mu}$ and $\alpha_2 = \dfrac{2}{\mu^2}$ we obtain, for this particular case, the first and the second moments as:

$$E(\theta_n) = \frac{1}{q\mu - \lambda} \tag{7.10.46}$$

$$E(\theta_n^2) = \frac{2(2q - q^2)\mu}{(q\mu - \lambda)^2 \left[(2q - q^2)\mu - (1 - q)\lambda \right]}. \tag{7.10.47}$$

EXERCISES

7.1.   Introduction

7.1.1.   Find mean and variance of the empirical distribution defined in Example 7.1.5.

7.2.   Markov Chain and Markov Process

7.2.1.   Prove that if $\mathbf{P}(t)$ is a stochastic matrix, then $\mathbf{P}^n(t)$ is also a stochastic matrix for any positive integer n.

7.2.2.   Suppose that the transition probability matrix for a discrete-time Markov chain is

$$P = \begin{bmatrix} \dfrac{1}{2} & \dfrac{1}{2} \\ \dfrac{3}{4} & \dfrac{1}{4} \end{bmatrix}.$$

(a) Find $\left[\mathbf{I} - z\mathbf{P}\right]^{-1}$.

(b) Find the general form of $\mathbf{P}^n$.

7.2.3.   Consider a Markov chain $\{X_n, n = 0, 1, 2, \cdots\}$ with state space $S = \{0, 1, 2\}$ and transition matrix

$$P = \begin{bmatrix} 0.5 & 0.0 & 0.5 \\ 0.4 & 0.2 & 0.4 \\ 0 & 0.4 & 0.6 \end{bmatrix}.$$

(a) Find $P\left(X_2 = 2 \mid X_1 = 0, X_0 = 1\right)$.

(b) Find $P\left(X_2 = 2, X_1 = 0 \mid X_0 = 1\right)$.

(c) Find $P\left(X_2 = 2, X_1 = 0 \mid X_0 = 0\right)$.

(d) For n > 1, find $P\left(X_{n+1} = 2, X_n = 0 \mid X_{n-1} = 0\right)$.

7.2.4. Consider a Markov chain $\{X_n, n = 0, 1, 2, \cdots\}$ with state space $S = \{0, 1, 2\}$ and transition matrix

$$P = \begin{bmatrix} 0.2 & 0.3 & 0.5 \\ 0.8 & 0.2 & 0 \\ 0.6 & 0 & 0.4 \end{bmatrix}.$$

(a) Find the matrix of 2-transition probabilities.

(b) Given   the   initial   distribution $P\left(X_0 = i\right) = 1/3, i = 0, 1, 2,$ find
$P\left(X_2 = 0\right)$ and $P\left(X_0 = 0, X_1 = 1, X_2 = 2\right)$.

7.3.   Classification of States of a Markov Chain

7.3.1.   Prove properties $1 - 3$ for $x \leftrightarrow y$ defined in Definition 7.3.4.

7.3.2.   Suppose a Markov chain has a state space $S = \{0, 1, 2, 3\}$ and transition matrix

$$P = \begin{bmatrix} 0 & 0 & 0 & 0 \\ 1 & 0 & 0 & 0 \\ 0.4 & 0.6 & 0 & 0 \\ 0.1 & 0.4 & 0.2 & 0.3 \end{bmatrix}.$$

Determine all classes for this chain.

7.3.3.   Suppose a Markov chain has a state space $S = \{0, 1, 2, 3, 4\}$ and transition matrix

$$P = \begin{bmatrix} 0 & 0.2 & 0.8 & 0 & 0 \\ 0 & 0 & 0 & 0.9 & 0.1 \\ 0 & 0 & 0 & 0.1 & 0.9 \\ 1 & 0 & 0 & 0 & 0 \\ 1 & 0 & 0 & 0 & 0 \end{bmatrix}.$$

(a) Verify that the Markov chain is irreducible with period 3.

(b) Find the stationary distribution.

7.3.4.   Two irreducible Markov chains 1 and 2 with their respective transition probabilities are given as:

	Chain 1	Chain 2	i
$p_{i,0}$	$\dfrac{i+1}{i+2}$	$\dfrac{1}{i+2}$	$0, 1, 2, \cdots$
$p_{i,i+1}$	$\dfrac{1}{i+2}$	$\dfrac{i+1}{i+2}$	$0, 1, 2, \cdots$

Determine whether each of these chains is transient, null recurrent, or positive recurrent.

7.4.   Random Walk Process

7.4.1.   Answer the following questions about general random walks:

(i)   What is E(number of visits to the state x by trial n)?
(ii)   What is E(total number of visits to the state x over all trials)?
(iii)   What is E(proportion of visits to the state x by trial n)?
(iv) E(proportion of visits to the state x over all trials)?
(v)   If it exists, what is E(total trials)?
(vi)   What is P(the trials never end)?

7.4.2.   *Fortune Dependent Gambling.* Consider the Gambler's Ruin problem (Example 7.4.2). Here, G1 begins with \$1 and G2 with \$3.
Find $p_x(n)$ for $x = 0, 1, 2, 3, 4$ and $n > 0$.

(i)   What is P(G1 loses all his money) $= P$(eventual absorption into state 0)?
(ii)   What is P(G2 loses all his money) $= P$(eventual absorption into state 4)?
(iii)   What is P(neither player ever wins all) $= P$(the game never ends)?
(iv)   What is P(G1 has \$3 after the 5th trial) $= P$(G2 has \$1 after the 5th trial)
        $= P$(the walk is in state 3 after the 5th trial)?
(v)   What is P(G2 has \$3 after 19 trials)?
(vi)   How many times can we expect G1 to have exactly \$3? In other words, how many times can we expect G1 to "*visit*" state 3?
(vii)   What fraction of the trials can we expect to result in G1's having exactly \$2? In other words, what proportion of the trials can we expect G1 to "*visit*" state 2?
(viii)   Let N denote number of trials until G1 or G2 is ruined. Then, what is E(number of trials until G1 or G2 is ruined) $= E$(number of gambles) $= E$(number of trials until absorption)?

7.4.3.   *A Failure-Repair model (FR).* In a certain office the photocopy machine has a probability, q, of failure (total breakdown) each time (trial) it is used. Let $1 - q = p$ be the probability that the machine functions without breakdown on a given trial. Each time the machine fails a repairman is called and no use is made of the machine until it is completely repaired. Assume the process begins functioning.

What proportion of the time can we expect the machine to be "up" or, restated, what is E(proportion of trials resulting in state 1)?

7.4.4.   Consider a recurrent random walks.

(a) For each state x, what proportion of the trials can we expect to spend in state x? In other words, what is E(proportion of visits to state x)?

(b) After a "large" number (as $n \to \infty$) of trials, what is the probability of being in a particular state?

7.5. Birth-Death (B-D) Processes

7.5.1. Consider a birth-death process with transition probabilities

$$\lambda_k = \lambda, \text{ and } \mu_k = k\mu, k \geq 0.$$

(a) Find DDE for $p_k(t) = P(k \text{ customers in system at time } t)$, for all k, $k \geq 0$.

(b) Define a generating function $G(z, t)$ for $p_k(t)$ and find the partial differential equation that $G(z, t)$ should satisfy.

(c) Assume that $p_0(0) = 1$. Show that the solution to the partial differential equation in part (a) is $G(z,t) = e^{\frac{\lambda}{\mu}\left(1-e^{-\mu t}\right)(z-1)}$.

7.5.2. Consider a birth-death process with transition probabilities

$$\lambda_k = \begin{cases} (N-k)\lambda, & k \leq N, \\ 0, & k \geq N. \end{cases}, \text{ and } \mu_k = \begin{cases} k\mu, & k \leq N, \\ 0, & k \geq N. \end{cases}$$

Find DDE for $p_k(t) = P(k \text{ customers in system at time } t)$, for all k, $0 \leq k \leq N$.

7.5.3. Suppose the transition probabilities of a birth-death process is given by

$$\lambda_0 = 1, \ \lambda_i = \frac{1}{1+\left(\dfrac{i}{i+1}\right)^2}, \ \mu_i = 1 - \lambda_i, i = 1, 2, \cdots.$$

Show that this is a transient process.

7.6. Pure Birth Process – Poisson Process

7.6.1. Prove Theorem 7.6.2.

7.6.2. Suppose that the number of automobiles passing a traffic light daily within a rush hours, 8:00 am – 10:00 am, follows a homogeneous Poisson process with mean $\lambda = 40$ per half an hour. Among these automobiles, 80% disregard the red light. What is the probability that at least one automobile disregards the red light between 8:00 am and 9:00 am?

7.6.3.   A Geiger counter is struck by radioactive particles according to a homogeneous Poisson process with parameter $\lambda = 1$ per 12 seconds. On the average, the counter only records 8 out of 10 particles.

   (a) What is the probability that the Geiger counter records at least 2 particles per minutes?

   (b) What are the mean and variance (in minutes) of the random variable Y, representing time, between the occurrences of two successively recorded particles?

7.7.   Introduction of Queueing Process and its Historical Development

7.7.1.   Explain the difference between balking and reneging queueing models.

7.7.2.   (*Project*) Solve the system of DDE (7.7.1) with restriction (7.7.2).

7.8.   Stationary Distributions

7.8.1.   Consider a stationary $M/M/1$ queue with arrival and service rates λ and μ, respectively. Suppose a customer balks (depart without service) within an interval of length Δt with probability $\alpha \Delta t + o(\Delta t)$, where "o" is the little "o".

   (a) Express p_{n+1}, the probability that there are $n + 1$ in the system, in terms of p_n.

   (b) Solve DBE found in part (a) for p_n, $n = 0, 1, 2, \cdots$, when $\alpha = \mu$.

7.8.2.   Consider a stationary $M/M/1/N$ queue with arrival and service rates λ and μ, respectively. Express p_{n+1}, the probability that there are $n + 1$ in the system, $n = 0, 1, 2, \cdots, N$, in terms of p_n.

7.8.3.   Consider a machine that it may be up and running perfectly, up and running un-efficiently, or it is down (not operating). Assume that state changes can only occur after a fixed time unit of length 1. The machine may stay in the same state for while, i.e., it may work perfectly for while, stay down for while until it is repaired, or work for while, but not efficiently. Let X_n be a random variable denoting the states of the machine at time n. Also assume that the sequence $\{X_n, n = 0,1,2,\cdots\}$ forms a Markov chain with transition matrix

$$P = \begin{bmatrix} 0.8 & 0.1 & 0.1 \\ 0 & 0.6 & 0.4 \\ 0.8 & 0 & 0.2 \end{bmatrix}.$$

(a) Find the stationary state probabilities for each state.

(b) Suppose if the machine is in perfect working condition, it will generate $1,000 profit per unit time. If it is in un-efficient working condition, it will generate $600 profit per unit time. However, if it is in the "down" condition it will generate $100 loss. What is the expected profit per unit time, if the machine is to let work for a sufficiently long time?

7.9.   Waiting Time

7.9.1.   In Example 7.9.1, use $\lambda = 8$ and $\mu = 10$ in $M/M/1$. Find the expected number of programs in the system, waiting to be executed

7.9.2.   In Exercise 7.9.1, find W and W_Q.

7.9.3.   Find the waiting time distribution for stationary $C/M/1/N$.

7.9.4.   For Example 7.9.3 write the DBE system and solve.

7.10.   Single-server Queue with Feedback

7.10.1. Prove Lemma 7.10.1.

7.10.2. Prove Lemma 7.10.2.

7.10.3. Find corresponding formula similar to (7.10.15), when the inter-arrival distribution is exponential with parameter λ and the service distribution is also exponential but with parameter μ.

7.10.4. (*Project*) Write the DBE system for $M/M/1$ with feedback and solve.

APPENDICES (DISTRIBUTION TABLES)

Table 1. The Binomial Probability Distribution

$$F(x) = P(X \le x) = \sum_{k=0}^{x} \binom{n}{k} p^k (1-p)^{n-k}$$

n	x	0.05	0.10	0.15	0.20	0.25	0.30	0.35	0.40	0.45	0.50
							p				
2	0	0.902500	0.810000	0.722500	0.640000	0.562500	0.490000	0.422500	0.360000	0.302500	0.250000
	1	0.997500	0.990000	0.977500	0.960000	0.937500	0.910000	0.877500	0.840000	0.797500	0.750000
	2	1.000000	1.000000	1.000000	1.000000	1.000000	1.000000	1.000000	1.000000	1.000000	1.000000
3	0	0.857375	0.729000	0.614125	0.512000	0.421875	0.343000	0.274625	0.216000	0.166375	0.125000
	1	0.992750	0.972000	0.939250	0.896000	0.843750	0.784000	0.718250	0.648000	0.574750	0.500000
	2	0.999875	0.999000	0.996625	0.992000	0.984375	0.973000	0.957125	0.936000	0.908875	0.875000
	3	1.000000	1.000000	1.000000	1.000000	1.000000	1.000000	1.000000	1.000000	1.000000	1.000000
4	0	0.814506	0.656100	0.522006	0.409600	0.316406	0.240100	0.178506	0.129600	0.091506	0.062500
	1	0.985981	0.947700	0.890481	0.819200	0.738281	0.651700	0.562981	0.475200	0.390981	0.312500
	2	0.999519	0.996300	0.988019	0.972800	0.949219	0.916300	0.873519	0.820800	0.758519	0.687500
	3	0.999994	0.999900	0.999494	0.998400	0.996094	0.991900	0.984994	0.974400	0.958994	0.937500
	4	1.000000	1.000000	1.000000	1.000000	1.000000	1.000000	1.000000	1.000000	1.000000	1.000000
5	0	0.773781	0.590490	0.443705	0.327680	0.237305	0.168070	0.116029	0.077760	0.050328	0.031250
	1	0.977408	0.918540	0.835210	0.737280	0.632812	0.528220	0.428415	0.336960	0.256218	0.187500
	2	0.998842	0.991440	0.973388	0.942080	0.896484	0.836920	0.764831	0.682560	0.593127	0.500000
	3	0.999970	0.999540	0.997772	0.993280	0.984375	0.969220	0.945978	0.912960	0.868780	0.812500
	4	1.000000	0.999990	0.999924	0.999680	0.999023	0.997570	0.994748	0.989760	0.981547	0.968750
	5	1.000000	1.000000	1.000000	1.000000	1.000000	1.000000	1.000000	1.000000	1.000000	1.000000
6	0	0.735092	0.531441	0.377150	0.262144	0.177979	0.117649	0.075419	0.046656	0.027681	0.015625
	1	0.967226	0.885735	0.776484	0.655360	0.533936	0.420175	0.319080	0.233280	0.163567	0.109375
	2	0.997770	0.984150	0.952661	0.901120	0.830566	0.744310	0.647085	0.544320	0.441518	0.343750
	3	0.999914	0.998730	0.994115	0.983040	0.962402	0.929530	0.882576	0.820800	0.744736	0.656250
	4	0.999998	0.999945	0.999601	0.998400	0.995361	0.989065	0.977678	0.959040	0.930802	0.890625
	5	1.000000	0.999999	0.999989	0.999936	0.999756	0.999271	0.998162	0.995904	0.991696	0.984375
	6	1.000000	1.000000	1.000000	1.000000	1.000000	1.000000	1.000000	1.000000	1.000000	1.000000
7	0	0.698337	0.478297	0.320577	0.209715	0.133484	0.082354	0.049022	0.027994	0.015224	0.007812
	1	0.955619	0.850306	0.716584	0.576717	0.444946	0.329417	0.233799	0.158630	0.102418	0.062500
	2	0.996243	0.974308	0.926235	0.851968	0.756409	0.647070	0.532283	0.419904	0.316440	0.226562
	3	0.999806	0.997272	0.987897	0.966656	0.929443	0.873964	0.800154	0.710208	0.608288	0.500000
	4	0.999994	0.999824	0.998778	0.995328	0.987122	0.971204	0.944392	0.903744	0.847072	0.773438
	5	1.000000	0.999994	0.999931	0.999629	0.998657	0.996209	0.990992	0.981158	0.964294	0.937500
	6	1.000000	1.000000	0.999998	0.999987	0.999939	0.999781	0.999357	0.998362	0.996263	0.992188
	7	1.000000	1.000000	1.000000	1.000000	1.000000	1.000000	1.000000	1.000000	1.000000	1.000000

Table 1. The Binomial Probability Distribution (Continued)

n	x	0.05	0.10	0.15	0.20	0.25	0.30	0.35	0.40	0.45	0.50
8	0	0.663420	0.430467	0.272491	0.167772	0.100113	0.057648	0.031864	0.016796	0.008373	0.003906
	1	0.942755	0.813105	0.657183	0.503316	0.367081	0.255298	0.169127	0.106376	0.063181	0.035156
	2	0.994212	0.961908	0.894787	0.796918	0.678543	0.551774	0.427814	0.315395	0.220130	0.144531
	3	0.999628	0.994976	0.978648	0.943718	0.886185	0.805896	0.706399	0.594086	0.476956	0.363281
	4	0.999985	0.999568	0.997146	0.989594	0.972702	0.942032	0.893909	0.826330	0.739619	0.636719
	5	1.000000	0.999977	0.999758	0.998769	0.995773	0.988708	0.974682	0.950193	0.911544	0.855469
	6	1.000000	0.999999	0.999988	0.999916	0.999619	0.998710	0.996429	0.991480	0.981877	0.964844
	7	1.000000	1.000000	1.000000	0.999997	0.999985	0.999934	0.999775	0.999345	0.998318	0.996094
	8	1.000000	1.000000	1.000000	1.000000	1.000000	1.000000	1.000000	1.000000	1.000000	1.000000
9	0	0.630249	0.387420	0.231617	0.134218	0.075085	0.040354	0.020712	0.010078	0.004605	0.001953
	1	0.928789	0.774841	0.599479	0.436208	0.300339	0.196003	0.121085	0.070544	0.038518	0.019531
	2	0.991639	0.947028	0.859147	0.738198	0.600677	0.462831	0.337273	0.231787	0.149503	0.089844
	3	0.999357	0.991669	0.966068	0.914358	0.834274	0.729659	0.608894	0.482610	0.361385	0.253906
	4	0.999967	0.999109	0.994371	0.980419	0.951073	0.901191	0.828281	0.733432	0.621421	0.500000
	5	0.999999	0.999936	0.999366	0.996934	0.990005	0.974705	0.946412	0.900647	0.834178	0.746094
	6	1.000000	0.999997	0.999954	0.999686	0.998657	0.995709	0.988818	0.974965	0.950227	0.910156
	7	1.000000	1.000000	0.999998	0.999981	0.999893	0.999567	0.998604	0.996199	0.990920	0.980469
	8	1.000000	1.000000	1.000000	0.999999	0.999996	0.999980	0.999921	0.999738	0.999243	0.998047
	9	1.000000	1.000000	1.000000	1.000000	1.000000	1.000000	1.000000	1.000000	1.000000	1.000000
10	0	0.598737	0.348678	0.196874	0.107374	0.056314	0.028248	0.013463	0.006047	0.002533	0.000977
	1	0.913862	0.736099	0.544300	0.375810	0.244025	0.149308	0.085954	0.046357	0.023257	0.010742
	2	0.988496	0.929809	0.820196	0.677800	0.525593	0.382783	0.261607	0.167290	0.099560	0.054688
	3	0.998972	0.987205	0.950030	0.879126	0.775875	0.649611	0.513827	0.382281	0.266038	0.171875
	4	0.999936	0.998365	0.990126	0.967206	0.921873	0.849732	0.751496	0.633103	0.504405	0.376953
	5	0.999997	0.999853	0.998617	0.993631	0.980272	0.952651	0.905066	0.833761	0.738437	0.623047
	6	1.000000	0.999991	0.999865	0.999136	0.996494	0.989408	0.973976	0.945238	0.898005	0.828125
	7	1.000000	1.000000	0.999991	0.999922	0.999584	0.998410	0.995179	0.987705	0.972608	0.945312
	8	1.000000	1.000000	1.000000	0.999996	0.999970	0.999856	0.999460	0.998322	0.995498	0.989258
	9	1.000000	1.000000	1.000000	1.000000	0.999999	0.999994	0.999972	0.999895	0.999659	0.999023
	10	1.000000	1.000000	1.000000	1.000000	1.000000	1.000000	1.000000	1.000000	1.000000	1.000000
11	0	0.568800	0.313811	0.167343	0.085899	0.042235	0.019773	0.008751	0.003628	0.001393	0.000488
	1	0.898105	0.697357	0.492186	0.322123	0.197097	0.112990	0.060582	0.030233	0.013931	0.005859
	2	0.984765	0.910438	0.778812	0.617402	0.455201	0.312740	0.200129	0.118917	0.065224	0.032715
	3	0.998448	0.981465	0.930555	0.838861	0.713305	0.569562	0.425550	0.296284	0.191123	0.113281
	4	0.999888	0.997249	0.984112	0.949590	0.885374	0.789695	0.668312	0.532774	0.397140	0.274414
	5	0.999994	0.999704	0.997343	0.988346	0.965672	0.921775	0.851316	0.753498	0.633123	0.500000
	6	1.000000	0.999977	0.999678	0.998035	0.992439	0.978381	0.949857	0.900647	0.826200	0.725586
	7	1.000000	0.999999	0.999972	0.999765	0.998812	0.995709	0.987758	0.970719	0.939037	0.886719
	8	1.000000	1.000000	0.999998	0.999981	0.999874	0.999422	0.997962	0.994076	0.985197	0.967285
	9	1.000000	1.000000	1.000000	0.999999	0.999992	0.999953	0.999793	0.999266	0.997787	0.994141
	10	1.000000	1.000000	1.000000	1.000000	1.000000	0.999998	0.999990	0.999958	0.999847	0.999512
	11	1.000000	1.000000	1.000000	1.000000	1.000000	1.000000	1.000000	1.000000	1.000000	1.000000

Table 1. The Binomial Probability Distribution (Continued)

n	X	p									
		0.05	0.10	0.15	0.20	0.25	0.30	0.35	0.40	0.45	0.50
12	0	0.540360	0.282430	0.142242	0.068719	0.031676	0.013841	0.005688	0.002177	0.000766	0.000244
	1	0.881640	0.659002	0.443460	0.274878	0.158382	0.085025	0.042441	0.019591	0.008289	0.003174
	2	0.980432	0.889130	0.735818	0.558346	0.390675	0.252815	0.151288	0.083443	0.042142	0.019287
	3	0.997764	0.974363	0.907794	0.794569	0.648779	0.492516	0.346653	0.225337	0.134468	0.072998
	4	0.999816	0.995671	0.976078	0.927444	0.842356	0.723655	0.583345	0.438178	0.304432	0.193848
	5	0.999989	0.999459	0.995358	0.980595	0.945598	0.882151	0.787265	0.665209	0.526930	0.387207
	6	1.000000	0.999950	0.999328	0.996097	0.985747	0.961399	0.915368	0.841788	0.739315	0.612793
	7	1.000000	0.999997	0.999928	0.999419	0.997218	0.990511	0.974493	0.942690	0.888260	0.806152
	8	1.000000	1.000000	0.999995	0.999938	0.999608	0.998308	0.994390	0.984733	0.964425	0.927002
	9	1.000000	1.000000	1.000000	0.999995	0.999962	0.999794	0.999152	0.997190	0.992122	0.980713
	10	1.000000	1.000000	1.000000	1.000000	0.999998	0.999985	0.999921	0.999681	0.998920	0.996826
	11	1.000000	1.000000	1.000000	1.000000	1.000000	0.999999	0.999997	0.999983	0.999931	0.999756
	12	1.000000	1.000000	1.000000	1.000000	1.000000	1.000000	1.000000	1.000000	1.000000	1.000000
13	0	0.513342	0.254187	0.120905	0.054976	0.023757	0.009689	0.003697	0.001306	0.000421	0.000122
	1	0.864576	0.621345	0.398277	0.233646	0.126705	0.063670	0.029578	0.012625	0.004904	0.001709
	2	0.975492	0.866117	0.691964	0.501652	0.332602	0.202478	0.113191	0.057902	0.026908	0.011230
	3	0.996897	0.965839	0.881997	0.747324	0.584253	0.420606	0.278275	0.168580	0.092921	0.046143
	4	0.999713	0.993540	0.965835	0.900869	0.793962	0.654314	0.500503	0.353042	0.227948	0.133423
	5	0.999980	0.999080	0.992466	0.969965	0.919787	0.834603	0.715893	0.574396	0.426806	0.290527
	6	0.999999	0.999901	0.998732	0.992996	0.975710	0.937625	0.870532	0.771156	0.643742	0.500000
	7	1.000000	0.999992	0.999838	0.998754	0.994351	0.981777	0.953799	0.902329	0.821235	0.709473
	8	1.000000	1.000000	0.999985	0.999834	0.999011	0.995969	0.987426	0.967916	0.930151	0.866577
	9	1.000000	1.000000	0.999999	0.999984	0.999874	0.999348	0.997485	0.992207	0.979658	0.953857
	10	1.000000	1.000000	1.000000	0.999999	0.999989	0.999927	0.999652	0.998685	0.995861	0.988770
	11	1.000000	1.000000	1.000000	1.000000	0.999999	0.999995	0.999970	0.999862	0.999476	0.998291
	12	1.000000	1.000000	1.000000	1.000000	1.000000	1.000000	0.999999	0.999993	0.999969	0.999878
	13	1.000000	1.000000	1.000000	1.000000	1.000000	1.000000	1.000000	1.000000	1.000000	1.000000
14	0	0.487675	0.228768	0.102770	0.043980	0.017818	0.006782	0.002403	0.000784	0.000232	0.000061
	1	0.847014	0.584629	0.356671	0.197912	0.100968	0.047476	0.020519	0.008098	0.002887	0.000916
	2	0.969946	0.841640	0.647911	0.448051	0.281128	0.160836	0.083927	0.039792	0.017006	0.006470
	3	0.995827	0.955867	0.853492	0.698190	0.521340	0.355167	0.220496	0.124309	0.063215	0.028687
	4	0.999573	0.990770	0.953260	0.870160	0.741535	0.584201	0.422723	0.279257	0.167186	0.089783
	5	0.999967	0.998526	0.988472	0.956146	0.888331	0.780516	0.640506	0.485855	0.337320	0.211975
	6	0.999998	0.999819	0.997793	0.988390	0.961729	0.906718	0.816408	0.692452	0.546121	0.395264
	7	1.000000	0.999983	0.999672	0.997603	0.989690	0.968531	0.975657	0.941681	0.881139	0.604736
	8	1.000000	0.999999	0.999963	0.999618	0.997846	0.991711	0.975657	0.941681	0.881139	0.788025
	9	1.000000	1.000000	0.999997	0.999954	0.999658	0.998334	0.993965	0.982490	0.957380	0.910217
	10	1.000000	1.000000	1.000000	0.999996	0.999960	0.999754	0.998894	0.996094	0.988569	0.971313
	11	1.000000	1.000000	1.000000	1.000000	0.999997	0.999975	0.999859	0.999391	0.997849	0.993530
	12	1.000000	1.000000	1.000000	1.000000	1.000000	0.999998	0.999989	0.999941	0.999747	0.999084
	13	1.000000	1.000000	1.000000	1.000000	1.000000	1.000000	1.000000	0.999997	0.999986	0.999939
	14	1.000000	1.000000	1.000000	1.000000	1.000000	1.000000	1.000000	1.000000	1.000000	1.000000

Table 1. The Binomial Probability Distribution (Continued)

n	x	p 0.05	0.10	0.15	0.20	0.25	0.30	0.35	0.40	0.45	0.50
15	0	0.463291	0.205891	0.087354	0.035184	0.013363	0.004748	0.001562	0.000470	0.000127	0.000031
	1	0.829047	0.549043	0.318586	0.167126	0.080181	0.035268	0.014179	0.005172	0.001692	0.000488
	2	0.963800	0.815939	0.604225	0.398023	0.236088	0.126828	0.061734	0.027114	0.010652	0.003693
	3	0.994533	0.944444	0.822655	0.648162	0.461287	0.296868	0.172696	0.090502	0.042421	0.017578
	4	0.999385	0.987280	0.938295	0.835766	0.686486	0.515491	0.351943	0.217278	0.120399	0.059235
	5	0.999947	0.997750	0.983190	0.938949	0.851632	0.721621	0.564282	0.403216	0.260760	0.150879
	6	0.999996	0.999689	0.996394	0.981941	0.943380	0.868857	0.754842	0.609813	0.452160	0.303619
	7	1.000000	0.999966	0.999390	0.995760	0.982700	0.949987	0.886769	0.786897	0.653504	0.500000
	8	1.000000	0.999997	0.999919	0.999215	0.995807	0.984757	0.957806	0.904953	0.818240	0.696381
	9	1.000000	1.000000	0.999992	0.999887	0.999205	0.996347	0.987557	0.966167	0.923071	0.849121
	10	1.000000	1.000000	0.999999	0.999988	0.999885	0.999328	0.997169	0.990652	0.974534	0.940765
	11	1.000000	1.000000	1.000000	0.999999	0.999988	0.999908	0.999521	0.998072	0.993673	0.982422
	12	1.000000	1.000000	1.000000	1.000000	0.999999	0.999991	0.999943	0.999721	0.998893	0.996307
	13	1.000000	1.000000	1.000000	1.000000	1.000000	0.999999	0.999996	0.999975	0.999879	0.999512
	14	1.000000	1.000000	1.000000	1.000000	1.000000	1.000000	1.000000	0.999999	0.999994	0.999969
	15	1.000000	1.000000	1.000000	1.000000	1.000000	1.000000	1.000000	1.000000	1.000000	1.000000
16	0	0.440127	0.185302	0.074251	0.028147	0.010023	0.003323	0.001015	0.000282	0.000070	0.000015
	1	0.810760	0.514728	0.283901	0.140737	0.063476	0.026112	0.009763	0.003291	0.000988	0.000259
	2	0.957062	0.789249	0.561379	0.351844	0.197111	0.099360	0.045090	0.018337	0.006620	0.002090
	3	0.992996	0.931594	0.789891	0.598134	0.404987	0.245856	0.133860	0.065147	0.028125	0.010635
	4	0.999143	0.982996	0.920949	0.798245	0.630186	0.449904	0.289207	0.166567	0.085309	0.038406
	5	0.999919	0.996703	0.976456	0.918312	0.810345	0.659782	0.489964	0.328840	0.197598	0.105057
	6	0.999994	0.999495	0.994414	0.973343	0.920443	0.824687	0.688146	0.527174	0.366030	0.227249
	7	1.000000	0.999939	0.998941	0.992996	0.972870	0.925648	0.840595	0.716063	0.562899	0.401810
	8	1.000000	0.999994	0.999840	0.998524	0.992530	0.974326	0.932943	0.857730	0.744109	0.598190
	9	1.000000	1.000000	0.999981	0.999752	0.998356	0.992870	0.977144	0.941681	0.875897	0.772751
	10	1.000000	1.000000	0.999998	0.999967	0.999715	0.998434	0.993804	0.980858	0.951376	0.894943
	11	1.000000	1.000000	1.000000	0.999997	0.999962	0.999734	0.998698	0.995104	0.985061	0.961594
	12	1.000000	1.000000	1.000000	1.000000	0.999996	0.999966	0.999796	0.999062	0.996544	0.989365
	13	1.000000	1.000000	1.000000	1.000000	1.000000	0.999997	0.999977	0.999873	0.999435	0.997910
	14	1.000000	1.000000	1.000000	1.000000	1.000000	1.000000	0.999998	0.999989	0.999942	0.999741
	15	1.000000	1.000000	1.000000	1.000000	1.000000	1.000000	1.000000	1.000000	0.999997	0.999985
	16	1.000000	1.000000	1.000000	1.000000	1.000000	1.000000	1.000000	1.000000	1.000000	1.000000
17	0	0.418120	0.166772	0.063113	0.022518	0.007517	0.002326	0.000660	0.000169	0.000039	0.000008
	1	0.792228	0.481785	0.252454	0.118219	0.050113	0.019275	0.006701	0.002088	0.000575	0.000137
	2	0.949747	0.761797	0.519758	0.309622	0.163702	0.077385	0.032725	0.012319	0.004086	0.001175
	3	0.991199	0.917359	0.755614	0.548876	0.353018	0.201907	0.102790	0.046423	0.018448	0.006363
	4	0.998835	0.977856	0.901290	0.758223	0.573886	0.388690	0.234835	0.125999	0.059576	0.024521
	5	0.999880	0.995333	0.968130	0.894299	0.765306	0.596819	0.419699	0.263931	0.147068	0.071732
	6	0.999990	0.999216	0.991720	0.962337	0.892918	0.775215	0.618782	0.447841	0.290235	0.166153
	7	0.999999	0.999894	0.998262	0.989066	0.959763	0.895360	0.787238	0.640508	0.474308	0.314529
	8	1.000000	0.999989	0.999705	0.997419	0.987615	0.959723	0.900621	0.801064	0.662564	0.500000
	9	1.000000	0.999999	0.999960	0.999507	0.996899	0.987307	0.961674	0.908101	0.816592	0.685471
	10	1.000000	1.000000	0.999996	0.999924	0.999375	0.996765	0.987973	0.965187	0.917410	0.833847
	11	1.000000	1.000000	1.000000	0.999991	0.999900	0.999344	0.996985	0.989406	0.969902	0.928268
	12	1.000000	1.000000	1.000000	0.999999	0.999988	0.999897	0.999411	0.997479	0.991377	0.975479
	13	1.000000	1.000000	1.000000	1.000000	0.999999	0.999988	0.999914	0.999549	0.998134	0.993637
	14	1.000000	1.000000	1.000000	1.000000	1.000000	0.999999	0.999991	0.999943	0.999714	0.998825
	15	1.000000	1.000000	1.000000	1.000000	1.000000	1.000000	0.999999	0.999995	0.999972	0.999863
	16	1.000000	1.000000	1.000000	1.000000	1.000000	1.000000	1.000000	1.000000	0.999999	0.999992
	17	1.000000	1.000000	1.000000	1.000000	1.000000	1.000000	1.000000	1.000000	1.000000	1.000000

Table 1. The Binomial Probability Distribution (Continued)

n	x	p									
		0.05	0.10	0.15	0.20	0.25	0.30	0.35	0.40	0.45	0.50
18	0	0.397214	0.150095	0.053646	0.018014	0.005638	0.001628	0.000429	0.000102	0.000021	0.000004
	1	0.773523	0.450284	0.224053	0.099079	0.039464	0.014190	0.004587	0.001320	0.000334	0.000072
	2	0.941871	0.733796	0.479662	0.271342	0.135305	0.059952	0.023617	0.008226	0.002506	0.000656
	3	0.989127	0.901803	0.720236	0.501025	0.305689	0.164550	0.078267	0.032781	0.011985	0.003769
	4	0.998454	0.971806	0.879439	0.716354	0.518669	0.332655	0.188620	0.094169	0.041069	0.015442
	5	0.999828	0.993585	0.958104	0.867084	0.717451	0.534380	0.354997	0.208758	0.107697	0.048126
	6	0.999985	0.998828	0.988181	0.948729	0.861015	0.721696	0.549103	0.374277	0.225810	0.118942
	7	0.999999	0.999827	0.997281	0.983720	0.943052	0.859317	0.728278	0.563441	0.391475	0.240341
	8	1.000000	0.999979	0.999489	0.995748	0.980652	0.940414	0.860937	0.736841	0.577849	0.407265
	9	1.000000	0.999998	0.999921	0.999089	0.994578	0.979032	0.940305	0.865286	0.747280	0.592735
	10	1.000000	1.000000	0.999990	0.999841	0.998756	0.993927	0.978768	0.942353	0.872042	0.759659
	11	1.000000	1.000000	0.999999	0.999978	0.999769	0.998570	0.993831	0.979718	0.946281	0.881058
	12	1.000000	1.000000	1.000000	0.999997	0.999966	0.999731	0.998562	0.994250	0.981713	0.951874
	13	1.000000	1.000000	1.000000	1.000000	0.999996	0.999960	0.999738	0.998721	0.995093	0.984558
	14	1.000000	1.000000	1.000000	1.000000	1.000000	0.999996	0.999964	0.999785	0.999003	0.996231
	15	1.000000	1.000000	1.000000	1.000000	1.000000	1.000000	0.999996	0.999974	0.999856	0.999344
	16	1.000000	1.000000	1.000000	1.000000	1.000000	1.000000	1.000000	0.999998	0.999987	0.999928
	17	1.000000	1.000000	1.000000	1.000000	1.000000	1.000000	1.000000	1.000000	0.999999	0.999996
	18	1.000000	1.000000	1.000000	1.000000	1.000000	1.000000	1.000000	1.000000	1.000000	1.000000
19	0	0.377354	0.135085	0.045599	0.014412	0.004228	0.001140	0.000279	0.000061	0.000012	0.000002
	1	0.754707	0.420265	0.198492	0.082866	0.031007	0.010422	0.003132	0.000833	0.000193	0.000038
	2	0.933454	0.705445	0.441321	0.236889	0.111345	0.046224	0.016956	0.005464	0.001528	0.000364
	3	0.986764	0.885002	0.684150	0.455089	0.263093	0.133171	0.059140	0.022959	0.007719	0.002213
	4	0.997987	0.964806	0.855558	0.673288	0.465424	0.282224	0.149996	0.069614	0.027981	0.009605
	5	0.999759	0.991407	0.946304	0.836938	0.667755	0.473863	0.296765	0.162922	0.077714	0.031784
	6	0.999977	0.998304	0.983670	0.932400	0.825124	0.665502	0.481166	0.308069	0.172659	0.083534
	7	0.999998	0.999727	0.995916	0.976722	0.922543	0.818030	0.665567	0.487775	0.316926	0.179642
	8	1.000000	0.999964	0.999157	0.993342	0.971252	0.916085	0.814506	0.667481	0.493981	0.323803
	9	1.000000	0.999996	0.999856	0.998421	0.991097	0.967447	0.912526	0.813908	0.671036	0.500000
	10	1.000000	1.000000	0.999980	0.999691	0.997712	0.989459	0.965306	0.911526	0.815899	0.676197
	11	1.000000	1.000000	0.999998	0.999950	0.999516	0.997177	0.988559	0.964772	0.912874	0.820358
	12	1.000000	1.000000	1.000000	0.999993	0.999917	0.999383	0.996906	0.988437	0.965769	0.916466
	13	1.000000	1.000000	1.000000	0.999999	0.999988	0.999892	0.999326	0.996932	0.989072	0.968216
	14	1.000000	1.000000	1.000000	1.000000	0.999999	0.999985	0.999885	0.999359	0.997244	0.990395
	15	1.000000	1.000000	1.000000	1.000000	1.000000	0.999998	0.999985	0.999899	0.999472	0.997787
	16	1.000000	1.000000	1.000000	1.000000	1.000000	1.000000	0.999999	0.999989	0.999928	0.999636
	17	1.000000	1.000000	1.000000	1.000000	1.000000	1.000000	1.000000	0.999999	0.999994	0.999962
	18	1.000000	1.000000	1.000000	1.000000	1.000000	1.000000	1.000000	1.000000	1.000000	0.999998
	19	1.000000	1.000000	1.000000	1.000000	1.000000	1.000000	1.000000	1.000000	1.000000	1.000000
20	0	0.358486	0.121577	0.038760	0.011529	0.003171	0.000798	0.000181	0.000037	0.000006	0.000001
	1	0.735840	0.391747	0.175558	0.069175	0.024313	0.007637	0.002133	0.000524	0.000111	0.000020
	2	0.924516	0.676927	0.404896	0.206085	0.091260	0.035483	0.012118	0.003611	0.000927	0.000201
	3	0.984098	0.867047	0.647725	0.411449	0.225156	0.107087	0.044376	0.015961	0.004933	0.001288
	4	0.997426	0.956826	0.829847	0.629648	0.414842	0.237508	0.118197	0.050952	0.018863	0.005909
	5	0.999671	0.988747	0.932692	0.804208	0.617173	0.416371	0.245396	0.125599	0.055334	0.020695
	6	0.999966	0.997614	0.978065	0.913307	0.785782	0.608010	0.416625	0.250011	0.129934	0.057659
	7	0.999997	0.999584	0.994079	0.967857	0.898188	0.772272	0.601027	0.415893	0.252006	0.131588
	8	1.000000	0.999940	0.998671	0.990018	0.959075	0.886669	0.762378	0.595599	0.414306	0.251722
	9	1.000000	0.999993	0.999752	0.997405	0.986136	0.952038	0.878219	0.755337	0.591361	0.411901
	10	1.000000	0.999999	0.999961	0.999437	0.996058	0.982855	0.946833	0.872479	0.750711	0.588099
	11	1.000000	1.000000	0.999995	0.999898	0.999065	0.994862	0.980421	0.943474	0.869235	0.748278
	12	1.000000	1.000000	0.999999	0.999985	0.999816	0.998721	0.993985	0.978971	0.941966	0.868412

Table 1. The Binomial Probability Distribution (Continued)

n	x	p									
		0.05	0.10	0.15	0.20	0.25	0.30	0.35	0.40	0.45	0.50
	13	1.000000	1.000000	1.000000	0.999998	0.999970	0.999739	0.998479	0.993534	0.978586	0.942341
	14	1.000000	1.000000	1.000000	1.000000	0.999996	0.999957	0.999689	0.998388	0.993566	0.979305
	15	1.000000	1.000000	1.000000	1.000000	1.000000	0.999994	0.999950	0.999683	0.998469	0.994091
	16	1.000000	1.000000	1.000000	1.000000	1.000000	0.999999	0.999994	0.999953	0.999723	0.998712
	17	1.000000	1.000000	1.000000	1.000000	1.000000	1.000000	0.999999	0.999995	0.999964	0.999799
	18	1.000000	1.000000	1.000000	1.000000	1.000000	1.000000	1.000000	1.000000	0.999997	0.999980
	19	1.000000	1.000000	1.000000	1.000000	1.000000	1.000000	1.000000	1.000000	1.000000	0.999999
	20	1.000000	1.000000	1.000000	1.000000	1.000000	1.000000	1.000000	1.000000	1.000000	1.000000
21	0	0.340562	0.109419	0.032946	0.009223	0.002378	0.000559	0.000118	0.000022	0.000004	0.000000
	1	0.716972	0.364730	0.155038	0.057646	0.019027	0.005585	0.001450	0.000329	0.000064	0.000010
	2	0.915082	0.648409	0.370496	0.178703	0.074523	0.027129	0.008623	0.002376	0.000560	0.000111
	3	0.981119	0.848035	0.611301	0.370376	0.191682	0.085606	0.033085	0.011021	0.003131	0.000745
	4	0.996760	0.947848	0.802529	0.586008	0.367420	0.198381	0.092359	0.036956	0.012595	0.003599
	5	0.999558	0.985555	0.917265	0.769296	0.566590	0.362712	0.200876	0.095740	0.038922	0.013302
	6	0.999951	0.996727	0.971259	0.891488	0.743630	0.550518	0.356695	0.200246	0.096364	0.039177
	7	0.999996	0.999387	0.991677	0.956947	0.870087	0.722993	0.536486	0.349540	0.197073	0.094624
	8	1.000000	0.999905	0.997982	0.985586	0.943853	0.852350	0.705905	0.523716	0.341271	0.191655
	9	1.000000	0.999988	0.999590	0.995928	0.979370	0.932427	0.837675	0.691442	0.511686	0.331812
	10	1.000000	0.999999	0.999930	0.999030	0.993577	0.973610	0.922818	0.825622	0.679003	0.500000
	11	1.000000	1.000000	0.999990	0.999806	0.998313	0.991260	0.968665	0.915076	0.815899	0.668188
	12	1.000000	1.000000	0.999999	0.999968	0.999628	0.997563	0.989237	0.964772	0.909237	0.808345
	13	1.000000	1.000000	1.000000	0.999995	0.999932	0.999434	0.996906	0.987709	0.962107	0.905376
	14	1.000000	1.000000	1.000000	0.999999	0.999990	0.999892	0.999266	0.996447	0.986825	0.960823
	15	1.000000	1.000000	1.000000	1.000000	0.999999	0.999983	0.999859	0.999165	0.996263	0.986698
	16	1.000000	1.000000	1.000000	1.000000	1.000000	0.999998	0.999979	0.999845	0.999159	0.996401
	17	1.000000	1.000000	1.000000	1.000000	1.000000	1.000000	0.999998	0.999978	0.999856	0.999255
	18	1.000000	1.000000	1.000000	1.000000	1.000000	1.000000	1.000000	0.999998	0.999982	0.999889
	19	1.000000	1.000000	1.000000	1.000000	1.000000	1.000000	1.000000	1.000000	0.999999	0.999990
	20	1.000000	1.000000	1.000000	1.000000	1.000000	1.000000	1.000000	1.000000	1.000000	1.000000
	21	1.000000	1.000000	1.000000	1.000000	1.000000	1.000000	1.000000	1.000000	1.000000	1.000000
22	0	0.323534	0.098477	0.028004	0.007379	0.001784	0.000391	0.000077	0.000013	0.000002	0.000000
	1	0.698151	0.339199	0.136724	0.047962	0.014865	0.004077	0.000984	0.000206	0.000037	0.000005
	2	0.905177	0.620041	0.338177	0.154491	0.060649	0.020666	0.006112	0.001558	0.000337	0.000061
	3	0.977818	0.828072	0.575180	0.332041	0.162392	0.068063	0.024524	0.007563	0.001974	0.000428
	4	0.995978	0.937866	0.773844	0.542882	0.323486	0.164549	0.071613	0.026582	0.008336	0.002172
	5	0.999419	0.981784	0.900055	0.732638	0.516797	0.313413	0.162895	0.072226	0.027075	0.008450
	6	0.999932	0.995610	0.963160	0.867049	0.699370	0.494176	0.302158	0.158444	0.070515	0.026239
	7	0.999993	0.999121	0.988614	0.943855	0.838472	0.671251	0.473559	0.289822	0.151754	0.066900
	8	0.999999	0.999853	0.997036	0.979858	0.925412	0.813543	0.646608	0.454046	0.276382	0.143139
	9	1.000000	0.999979	0.999348	0.993859	0.970491	0.908404	0.791555	0.624352	0.435000	0.261734
	10	1.000000	0.999998	0.999879	0.998410	0.990026	0.961255	0.893018	0.771950	0.603711	0.415906
	11	1.000000	1.000000	0.999981	0.999651	0.997129	0.985965	0.952619	0.879294	0.754296	0.584094
	12	1.000000	1.000000	0.999997	0.999935	0.999300	0.995672	0.982037	0.944894	0.867235	0.738266
	13	1.000000	1.000000	1.000000	0.999990	0.999856	0.998873	0.994222	0.978534	0.938315	0.856861
	14	1.000000	1.000000	1.000000	0.999999	0.999975	0.999754	0.998440	0.992952	0.975702	0.933100
	15	1.000000	1.000000	1.000000	1.000000	0.999996	0.999956	0.999651	0.998078	0.992016	0.973761
	16	1.000000	1.000000	1.000000	1.000000	1.000000	0.999994	0.999937	0.999573	0.997856	0.991550
	17	1.000000	1.000000	1.000000	1.000000	1.000000	0.999999	0.999991	0.999925	0.999542	0.997828
	18	1.000000	1.000000	1.000000	1.000000	1.000000	1.000000	0.999999	0.999990	0.999925	0.999572
	19	1.000000	1.000000	1.000000	1.000000	1.000000	1.000000	1.000000	0.999999	0.999991	0.999939
	20	1.000000	1.000000	1.000000	1.000000	1.000000	1.000000	1.000000	1.000000	0.999999	0.999995
	21	1.000000	1.000000	1.000000	1.000000	1.000000	1.000000	1.000000	1.000000	1.000000	1.000000
	22	1.000000	1.000000	1.000000	1.000000	1.000000	1.000000	1.000000	1.000000	1.000000	1.000000

Table 1. The Binomial Probability Distribution (Continued)

n	x	p 0.05	0.10	0.15	0.20	0.25	0.30	0.35	0.40	0.45	0.50
	0	0.307357	0.088629	0.023803	0.005903	0.001338	0.000274	0.000050	0.000008	0.000001	0.000000
	1	0.679420	0.315127	0.120416	0.039845	0.011595	0.002971	0.000666	0.000129	0.000021	0.000003
	2	0.894826	0.591957	0.307959	0.133185	0.049203	0.015690	0.004317	0.001017	0.000202	0.000033
	3	0.974185	0.807269	0.539630	0.296531	0.136957	0.053844	0.018080	0.005161	0.001237	0.000244
	4	0.995070	0.926887	0.744045	0.500714	0.283212	0.135603	0.055132	0.018974	0.005473	0.001300
	5	0.999246	0.977392	0.881123	0.694687	0.468469	0.268754	0.130947	0.053969	0.018642	0.005311
	6	0.999906	0.994227	0.953694	0.840167	0.653727	0.439947	0.253416	0.123957	0.050967	0.017345
	7	0.999990	0.998770	0.984796	0.928494	0.803697	0.618128	0.413569	0.237271	0.115197	0.046570
	8	0.999999	0.999780	0.995773	0.972658	0.903677	0.770855	0.586041	0.388356	0.220300	0.105020
	9	1.000000	0.999967	0.999002	0.991059	0.959221	0.879946	0.740824	0.556229	0.363622	0.202436
	10	1.000000	0.999996	0.999799	0.997500	0.985142	0.945400	0.857506	0.712911	0.527791	0.338820
	11	1.000000	1.000000	0.999966	0.999403	0.995353	0.978552	0.931759	0.836357	0.686533	0.500000
23	12	1.000000	1.000000	0.999995	0.999878	0.998757	0.992760	0.971741	0.918654	0.816412	0.661180
	13	1.000000	1.000000	0.999999	0.999979	0.999717	0.997912	0.989957	0.965078	0.906329	0.797564
	14	1.000000	1.000000	1.000000	0.999997	0.999945	0.999490	0.996964	0.987185	0.958878	0.894980
	15	1.000000	1.000000	1.000000	1.000000	0.999991	0.999895	0.999227	0.996027	0.984675	0.953430
	16	1.000000	1.000000	1.000000	1.000000	0.999999	0.999982	0.999837	0.998975	0.995228	0.982655
	17	1.000000	1.000000	1.000000	1.000000	1.000000	0.999998	0.999972	0.999784	0.998783	0.994689
	18	1.000000	1.000000	1.000000	1.000000	1.000000	1.000000	0.999996	0.999964	0.999753	0.998700
	19	1.000000	1.000000	1.000000	1.000000	1.000000	1.000000	1.000000	0.999995	0.999962	0.999756
	20	1.000000	1.000000	1.000000	1.000000	1.000000	1.000000	1.000000	1.000000	0.999996	0.999967
	21	1.000000	1.000000	1.000000	1.000000	1.000000	1.000000	1.000000	1.000000	1.000000	0.999997
	22	1.000000	1.000000	1.000000	1.000000	1.000000	1.000000	1.000000	1.000000	1.000000	1.000000
	23	1.000000	1.000000	1.000000	1.000000	1.000000	1.000000	1.000000	1.000000	1.000000	1.000000
	0	0.291989	0.079766	0.020233	0.004722	0.001003	0.000192	0.000032	0.000005	0.000001	0.000000
	1	0.660817	0.292477	0.105924	0.033057	0.009031	0.002162	0.000450	0.000081	0.000012	0.000001
	2	0.884055	0.564274	0.279828	0.114517	0.039801	0.011874	0.003040	0.000662	0.000121	0.000018
	3	0.970218	0.785738	0.504879	0.263862	0.115018	0.042398	0.013263	0.003503	0.000771	0.000139
	4	0.994025	0.914925	0.713382	0.459877	0.246648	0.111075	0.042164	0.013449	0.003567	0.000772
	5	0.999038	0.972342	0.860561	0.655892	0.422155	0.228808	0.104411	0.039971	0.012716	0.003305
	6	0.999873	0.992544	0.942808	0.811071	0.607412	0.388589	0.210552	0.095961	0.036421	0.011328
	7	0.999986	0.998316	0.980131	0.910829	0.766204	0.564674	0.357516	0.191945	0.086293	0.031957
	8	0.999999	0.999679	0.994127	0.963825	0.878682	0.725037	0.525676	0.327922	0.173003	0.075795
	9	1.000000	0.999948	0.998517	0.987379	0.945335	0.847218	0.686650	0.489080	0.299127	0.153728
	10	1.000000	0.999993	0.999680	0.996212	0.978662	0.925764	0.816667	0.650238	0.453915	0.270628
	11	1.000000	0.999999	0.999941	0.999022	0.992800	0.968606	0.905770	0.786978	0.615099	0.419410
24	12	1.000000	1.000000	0.999991	0.999783	0.997906	0.988498	0.957747	0.885735	0.757967	0.580590
	13	1.000000	1.000000	0.999999	0.999959	0.999477	0.996367	0.983581	0.946508	0.865867	0.729372
	14	1.000000	1.000000	1.000000	0.999993	0.999888	0.999017	0.994511	0.978342	0.935231	0.846272
	15	1.000000	1.000000	1.000000	0.999999	0.999980	0.999774	0.998435	0.992490	0.973066	0.924205
	16	1.000000	1.000000	1.000000	1.000000	0.999997	0.999956	0.999623	0.997796	0.990479	0.968043
	17	1.000000	1.000000	1.000000	1.000000	1.000000	0.999993	0.999925	0.999460	0.997183	0.988672
	18	1.000000	1.000000	1.000000	1.000000	1.000000	0.999999	0.999988	0.999892	0.999316	0.996695
	19	1.000000	1.000000	1.000000	1.000000	1.000000	1.000000	0.999998	0.999983	0.999868	0.999228
	20	1.000000	1.000000	1.000000	1.000000	1.000000	1.000000	1.000000	0.999998	0.999980	0.999861
	21	1.000000	1.000000	1.000000	1.000000	1.000000	1.000000	1.000000	1.000000	0.999998	0.999982
	22	1.000000	1.000000	1.000000	1.000000	1.000000	1.000000	1.000000	1.000000	1.000000	0.999998
	23	1.000000	1.000000	1.000000	1.000000	1.000000	1.000000	1.000000	1.000000	1.000000	1.000000
	24	1.000000	1.000000	1.000000	1.000000	1.000000	1.000000	1.000000	1.000000	1.000000	1.000000

Table 1. The Binomial Probability Distribution (Continued)

n	x	\multicolumn{10}{c}{p}
		0.05	0.10	0.15	0.20	0.25	0.30	0.35	0.40	0.45	0.50
25	0	0.277390	0.071790	0.017198	0.003778	0.000753	0.000134	0.000021	0.000003	0.000000	0.000000
	1	0.642376	0.271206	0.093070	0.027390	0.007024	0.001571	0.000304	0.000050	0.000007	0.000001
	2	0.872894	0.537094	0.253742	0.098225	0.032109	0.008961	0.002133	0.000429	0.000072	0.000010
	3	0.965909	0.763591	0.471121	0.233993	0.096214	0.033241	0.009685	0.002367	0.000479	0.000078
	4	0.992835	0.902006	0.682107	0.420674	0.213741	0.090472	0.032048	0.009471	0.002309	0.000455
	5	0.998787	0.966600	0.838485	0.616689	0.378279	0.193488	0.082625	0.029362	0.008599	0.002039
	6	0.999831	0.990524	0.930471	0.780035	0.561098	0.340655	0.173403	0.073565	0.025754	0.007317
	7	0.999980	0.997739	0.974532	0.890877	0.726506	0.511849	0.306078	0.153552	0.063851	0.021643
	8	0.999998	0.999542	0.992027	0.953226	0.850562	0.676928	0.466820	0.273531	0.133984	0.053876
	9	1.000000	0.999921	0.997859	0.982668	0.928672	0.810564	0.630309	0.424617	0.242371	0.114761
	10	1.000000	0.999988	0.999505	0.994445	0.970330	0.902200	0.771161	0.585775	0.384260	0.212178
	11	1.000000	0.999999	0.999902	0.998460	0.989266	0.955754	0.874584	0.732282	0.542566	0.345019
	12	1.000000	1.000000	0.999983	0.999631	0.996630	0.982530	0.939555	0.846232	0.693676	0.500000
	13	1.000000	1.000000	0.999998	0.999924	0.999084	0.994006	0.974539	0.922199	0.817312	0.654981
	14	1.000000	1.000000	1.000000	0.999986	0.999785	0.998222	0.990686	0.965608	0.904017	0.787822
	15	1.000000	1.000000	1.000000	0.999998	0.999957	0.999546	0.997062	0.986831	0.956040	0.885239
	16	1.000000	1.000000	1.000000	1.000000	0.999993	0.999901	0.999208	0.995674	0.982643	0.946124
	17	1.000000	1.000000	1.000000	1.000000	0.999999	0.999982	0.999819	0.998795	0.994166	0.978357
	18	1.000000	1.000000	1.000000	1.000000	1.000000	0.999997	0.999966	0.999719	0.998356	0.992683
	19	1.000000	1.000000	1.000000	1.000000	1.000000	1.000000	0.999995	0.999946	0.999620	0.997961
	20	1.000000	1.000000	1.000000	1.000000	1.000000	1.000000	0.999999	0.999992	0.999930	0.999545
	21	1.000000	1.000000	1.000000	1.000000	1.000000	1.000000	1.000000	0.999999	0.999990	0.999922
	22	1.000000	1.000000	1.000000	1.000000	1.000000	1.000000	1.000000	1.000000	0.999999	0.999990
	23	1.000000	1.000000	1.000000	1.000000	1.000000	1.000000	1.000000	1.000000	1.000000	0.999999
	24	1.000000	1.000000	1.000000	1.000000	1.000000	1.000000	1.000000	1.000000	1.000000	1.000000
	25	1.000000	1.000000	1.000000	1.000000	1.000000	1.000000	1.000000	1.000000	1.000000	1.000000

Table 2. The Poisson Probability Distribution

$$F(x) = P(X \le x) = \sum_{k=0}^{x} \frac{\lambda^k e^{-\lambda}}{k!}$$

$$\lambda = E(x)$$

	0.1	0.2	0.3	0.4	0.5	0.6	0.7	0.8	0.9	1.0
0	0.904837	0.818731	0.740818	0.670320	0.606531	0.548812	0.496585	0.449329	0.406570	0.367879
1	0.995321	0.982477	0.963064	0.938448	0.909796	0.878099	0.844195	0.808792	0.772482	0.735759
2	0.999845	0.998852	0.996400	0.992074	0.985612	0.976885	0.965858	0.952577	0.937143	0.919699
3	0.999996	0.999943	0.999734	0.999224	0.998248	0.996642	0.994247	0.990920	0.986541	0.981012
4	1.000000	0.999998	0.999984	0.999939	0.999828	0.999606	0.999214	0.998589	0.997656	0.996340
5	1.000000	1.000000	0.999999	0.999996	0.999986	0.999961	0.999910	0.999816	0.999656	0.999406
6	1.000000	1.000000	1.000000	1.000000	0.999999	0.999997	0.999991	0.999979	0.999957	0.999917

	1.1	1.2	1.3	1.4	1.5	1.6	1.7	1.8	1.9	2.0
0	0.332871	0.301194	0.272532	0.246597	0.223130	0.201897	0.182684	0.165299	0.149569	0.135335
1	0.699029	0.662627	0.626823	0.591833	0.557825	0.524931	0.493246	0.462837	0.433749	0.406006
2	0.900416	0.879487	0.857112	0.833498	0.808847	0.783358	0.757223	0.730621	0.703720	0.676676
3	0.974258	0.966231	0.956905	0.946275	0.934358	0.921187	0.906811	0.891292	0.874702	0.857123
4	0.994565	0.992254	0.989337	0.985747	0.981424	0.976318	0.970385	0.963593	0.955919	0.947347
5	0.999032	0.998500	0.997769	0.996799	0.995544	0.993960	0.992001	0.989622	0.986781	0.983436
6	0.999851	0.999749	0.999596	0.999378	0.999074	0.998664	0.998125	0.997431	0.996554	0.995466
7	0.999980	0.999963	0.999936	0.999893	0.999830	0.999740	0.999612	0.999438	0.999207	0.998903
8	0.999998	0.999995	0.999991	0.999984	0.999972	0.999955	0.999928	0.999890	0.999837	0.999763

	2.2	2.4	2.6	2.8	3.0	3.2	3.4	3.6	3.8	4.0
0	0.110803	0.090718	0.074274	0.060810	0.049787	0.040762	0.033373	0.027324	0.022371	0.018316
1	0.354570	0.308441	0.267385	0.231078	0.199148	0.171201	0.146842	0.125689	0.107380	0.091578
2	0.622714	0.569709	0.518430	0.469454	0.423190	0.379904	0.339740	0.302747	0.268897	0.238103
3	0.819352	0.778723	0.736002	0.691937	0.647232	0.602520	0.558357	0.515216	0.473485	0.433470
4	0.927504	0.904131	0.877423	0.847676	0.815263	0.780613	0.744182	0.706438	0.667844	0.628837
5	0.975090	0.964327	0.950963	0.934890	0.916082	0.894592	0.870542	0.844119	0.815556	0.785130
6	0.992539	0.988406	0.982830	0.975589	0.966491	0.955381	0.942147	0.926727	0.909108	0.889326
7	0.998022	0.996661	0.994666	0.991869	0.988095	0.983170	0.976926	0.969211	0.959893	0.948866
8	0.999530	0.999138	0.998513	0.997567	0.996197	0.994286	0.991707	0.988329	0.984016	0.978637
9	0.999899	0.999798	0.999624	0.999340	0.998898	0.998238	0.997291	0.995976	0.994201	0.991868
10	0.999980	0.999957	0.999913	0.999836	0.999708	0.999503	0.999190	0.998729	0.998071	0.997160
11	0.999996	0.999992	0.999982	0.999963	0.999929	0.999871	0.999777	0.999630	0.999408	0.999085
12	0.999999	0.999998	0.999996	0.999992	0.999984	0.999969	0.999943	0.999900	0.999832	0.999726

	4.2	4.4	4.6	4.8	5.0	5.2	5.4	5.6	5.8	6.0
0	0.014996	0.012277	0.010052	0.008230	0.006738	0.005517	0.004517	0.003698	0.003028	0.002479
1	0.077977	0.066298	0.056290	0.047733	0.040428	0.034203	0.028906	0.024406	0.020587	0.017351
2	0.210238	0.185142	0.162639	0.142539	0.124652	0.108787	0.094758	0.082388	0.071511	0.061969
3	0.395403	0.359448	0.325706	0.294230	0.265026	0.238065	0.213291	0.190622	0.169963	0.151204
4	0.589827	0.551184	0.513234	0.476259	0.440493	0.406128	0.373311	0.342150	0.312718	0.285056
5	0.753143	0.719912	0.685760	0.651006	0.615961	0.580913	0.546132	0.511861	0.478315	0.445680
6	0.867464	0.843645	0.818029	0.790805	0.762183	0.732393	0.701671	0.670258	0.638391	0.606303
7	0.936057	0.921421	0.904949	0.886666	0.866628	0.844922	0.821659	0.796975	0.771026	0.743980
8	0.972068	0.964197	0.954928	0.944183	0.931906	0.918065	0.902650	0.885678	0.867186	0.847237
9	0.988873	0.985110	0.980473	0.974859	0.968172	0.960326	0.951245	0.940870	0.929156	0.916076
10	0.995931	0.994312	0.992223	0.989583	0.986305	0.982301	0.977486	0.971778	0.965099	0.957379
11	0.998626	0.997992	0.997137	0.996008	0.994547	0.992690	0.990368	0.987513	0.984050	0.979908
12	0.999569	0.999342	0.999021	0.998578	0.997981	0.997191	0.996165	0.994856	0.993210	0.991173
13	0.999874	0.999799	0.999688	0.999527	0.999302	0.998992	0.998573	0.998019	0.997297	0.996372
14	0.999966	0.999942	0.999907	0.999853	0.999774	0.999661	0.999502	0.999284	0.998990	0.998600
15	0.999991	0.999984	0.999974	0.999957	0.999931	0.999892	0.999836	0.999756	0.999644	0.999491
16	0.999998	0.999996	0.999993	0.999988	0.999980	0.999968	0.999949	0.999922	0.999882	0.999825

Table 2. The Poisson Probability Distribution (Continued)

	$\lambda = E(x)$									
	6.5	7.0	7.5	8.0	8.5	9.0	9.5	10.0	10.5	11.0
0	0.001503	0.000912	0.000553	0.000335	0.000203	0.000123	0.000075	0.000045	0.000028	0.000017
1	0.011276	0.007295	0.004701	0.003019	0.001933	0.001234	0.000786	0.000499	0.000317	0.000200
2	0.043036	0.029636	0.020257	0.013754	0.009283	0.006232	0.004164	0.002769	0.001835	0.001211
3	0.111850	0.081765	0.059145	0.042380	0.030109	0.021226	0.014860	0.010336	0.007147	0.004916
4	0.223672	0.172992	0.132062	0.099632	0.074364	0.054964	0.040263	0.029253	0.021094	0.015105
5	0.369041	0.300708	0.241436	0.191236	0.149597	0.115691	0.088528	0.067086	0.050380	0.037520
6	0.526524	0.449711	0.378155	0.313374	0.256178	0.206781	0.164949	0.130141	0.101632	0.078614
7	0.672758	0.598714	0.524639	0.452961	0.385597	0.323897	0.268663	0.220221	0.178511	0.143192
8	0.791573	0.729091	0.661967	0.592547	0.523105	0.455653	0.391823	0.332820	0.279413	0.231985
9	0.877384	0.830496	0.776408	0.716624	0.652974	0.587408	0.521826	0.457930	0.397133	0.340511
10	0.933161	0.901479	0.862238	0.815886	0.763362	0.705988	0.645328	0.583040	0.520738	0.459889
11	0.966120	0.946650	0.920759	0.888076	0.848662	0.803008	0.751990	0.696776	0.638725	0.579267
12	0.983973	0.973000	0.957334	0.936203	0.909083	0.875773	0.836430	0.791556	0.741964	0.688697
13	0.992900	0.987189	0.978435	0.965819	0.948589	0.926149	0.898136	0.864464	0.825349	0.781291
14	0.997044	0.994283	0.989740	0.982743	0.972575	0.958534	0.940008	0.916542	0.887888	0.854044
15	0.998840	0.997593	0.995392	0.991769	0.986167	0.977964	0.966527	0.951260	0.931665	0.907396
16	0.999570	0.999042	0.998041	0.996282	0.993387	0.988894	0.982273	0.972958	0.960394	0.944076
17	0.999849	0.999638	0.999210	0.998406	0.996998	0.994680	0.991072	0.985722	0.978138	0.967809
18	0.999949	0.999870	0.999697	0.999350	0.998703	0.997574	0.995716	0.992813	0.988489	0.982313
19	0.999984	0.999956	0.999889	0.999747	0.999465	0.998944	0.998038	0.996546	0.994209	0.990711
20	0.999995	0.999986	0.999961	0.999906	0.999789	0.999561	0.999141	0.998412	0.997212	0.995329
21	0.999999	0.999995	0.999987	0.999967	0.999921	0.999825	0.999639	0.999300	0.998714	0.997748
22	1.000000	0.999999	0.999996	0.999989	0.999971	0.999933	0.999855	0.999704	0.999430	0.998958
23	1.000000	1.000000	0.999999	0.999996	0.999990	0.999975	0.999944	0.999880	0.999758	0.999536
	11.5	12.0	12.5	13.0	13.5	14.0	14.5	15.0	15.5	16.0
0	0.000010	0.000006	0.000004	0.000002	0.000001	0.000001	0.000000	0.000000	0.000000	0.000000
1	0.000127	0.000080	0.000050	0.000032	0.000020	0.000012	0.000008	0.000005	0.000003	0.000002
2	0.000796	0.000522	0.000341	0.000223	0.000145	0.000094	0.000061	0.000039	0.000025	0.000016
3	0.003364	0.002292	0.001555	0.001050	0.000707	0.000474	0.000317	0.000211	0.000140	0.000093
4	0.010747	0.007600	0.005346	0.003740	0.002604	0.001805	0.001246	0.000857	0.000587	0.000400
5	0.027726	0.020341	0.014823	0.010734	0.007727	0.005532	0.003940	0.002792	0.001970	0.001384
6	0.060270	0.045822	0.034567	0.025887	0.019254	0.014228	0.010450	0.007632	0.005544	0.004006
7	0.113734	0.089504	0.069825	0.054028	0.041483	0.031620	0.023936	0.018002	0.013456	0.010000
8	0.190590	0.155028	0.124916	0.099758	0.078995	0.062055	0.048379	0.037446	0.028787	0.021987
9	0.288795	0.242392	0.201431	0.165812	0.135264	0.109399	0.087759	0.069854	0.055190	0.043298
10	0.401730	0.347229	0.297075	0.251682	0.211226	0.175681	0.144861	0.118464	0.096116	0.077396
11	0.519798	0.461597	0.405761	0.353165	0.304453	0.260040	0.220131	0.184752	0.153783	0.126993
12	0.632947	0.575965	0.518975	0.463105	0.409333	0.358458	0.311082	0.267611	0.228269	0.193122
13	0.733040	0.681536	0.627835	0.573045	0.518247	0.464448	0.412528	0.363218	0.317081	0.274511
14	0.815260	0.772025	0.725032	0.675132	0.623271	0.570437	0.517597	0.465654	0.415407	0.367527
15	0.878295	0.844416	0.806029	0.763607	0.717793	0.669360	0.619163	0.568090	0.517011	0.466745
16	0.923601	0.898709	0.869308	0.835493	0.797545	0.755918	0.711208	0.664123	0.615440	0.565962
17	0.954250	0.937034	0.915837	0.890465	0.860878	0.827201	0.789716	0.748859	0.705184	0.659344
18	0.973831	0.962584	0.948148	0.930167	0.908378	0.882643	0.852960	0.819472	0.782464	0.742349
19	0.985682	0.978720	0.969406	0.957331	0.942128	0.923495	0.901224	0.875219	0.845508	0.812249
20	0.992497	0.988402	0.982692	0.974988	0.964909	0.952092	0.936216	0.917029	0.894367	0.868168
21	0.996229	0.993935	0.990600	0.985919	0.979554	0.971156	0.960377	0.946894	0.930430	0.910773
22	0.998179	0.996953	0.995094	0.992378	0.988541	0.983288	0.976301	0.967256	0.955837	0.941759
23	0.999155	0.998527	0.997536	0.996028	0.993816	0.990672	0.986340	0.980535	0.972960	0.963314
24	0.999622	0.999314	0.998808	0.998006	0.996783	0.994980	0.992406	0.988835	0.984018	0.977685
25	0.999837	0.999692	0.999444	0.999034	0.998385	0.997392	0.995923	0.993815	0.990874	0.986881
26	0.999932	0.999867	0.999749	0.999548	0.999217	0.998691	0.997885	0.996688	0.994962	0.992541

Table 2. The Poisson Probability Distribution (Continued)

	\multicolumn{10}{c}{$\lambda = E(x)$}								
	11.5	12.0	12.5	13.0	13.5	14.0	14.5	15.0	15.5	16.0
27	0.999973	0.999944	0.999891	0.999796	0.999633	0.999365	0.998939	0.998284	0.997308	0.995895
28	0.999989	0.999977	0.999954	0.999911	0.999833	0.999702	0.999485	0.999139	0.998607	0.997811
29	0.999996	0.999991	0.999981	0.999962	0.999927	0.999864	0.999757	0.999582	0.999301	0.998869
30	0.999999	0.999997	0.999993	0.999984	0.999969	0.999940	0.999889	0.999803	0.999660	0.999433
31	0.999999	0.999999	0.999997	0.999994	0.999987	0.999974	0.999951	0.999910	0.999839	0.999724
32	1.000000	1.000000	0.999999	0.999998	0.999995	0.999989	0.999979	0.999960	0.999926	0.999869

Table 3. The Chi-Square Distribution

$$F(x) = P(X \le x) = \int_0^x \frac{w^{r/2-1}e^{-w/2}}{\Gamma(r/2)2^{r/2}} \, dw$$

$$0.005 \le P(X \le x) \le 0.50$$

r	$P(X \le x)$								
	0.005	0.01	0.025	0.05	0.10	0.20	0.30	0.40	0.50
1	0.000039	0.000157	0.000982	0.003932	0.015791	0.064185	0.148472	0.274996	0.454936
2	0.010025	0.020101	0.050636	0.102587	0.210721	0.446287	0.713350	1.021651	1.386294
3	0.071722	0.114832	0.215795	0.351846	0.584374	1.005174	1.423652	1.869168	2.365974
4	0.206989	0.297109	0.484419	0.710723	1.063623	1.648777	2.194698	2.752843	3.356694
5	0.411742	0.554298	0.831212	1.145476	1.610308	2.342534	2.999908	3.655500	4.351460
6	0.675727	0.872090	1.237344	1.635383	2.204131	3.070088	3.827552	4.570154	5.348121
7	0.989256	1.239042	1.689869	2.167350	2.833107	3.822322	4.671330	5.493235	6.345811
8	1.344413	1.646497	2.179731	2.732637	3.489539	4.593574	5.527422	6.422646	7.344121
9	1.734933	2.087901	2.700390	3.325113	4.168159	5.380053	6.393306	7.357034	8.342833
10	2.155856	2.558212	3.246973	3.940299	4.865182	6.179079	7.267218	8.295472	9.341818
11	2.603222	3.053484	3.815748	4.574813	5.577785	6.988674	8.147868	9.237285	10.340998
12	3.073824	3.570569	4.403788	5.226029	6.303796	7.807328	9.034277	10.181971	11.340322
13	3.565035	4.106915	5.008751	5.891864	7.041505	8.633861	9.925682	11.129140	12.339756
14	4.074675	4.660425	5.628726	6.570631	7.789534	9.467328	10.821478	12.078482	13.339274
15	4.600916	5.229349	6.262138	7.260944	8.546756	10.306959	11.721169	13.029750	14.338860
16	5.142205	5.812212	6.907664	7.961646	9.312236	11.152116	12.624349	13.982736	15.338499
17	5.697217	6.407760	7.564186	8.671760	10.085186	12.002266	13.530676	14.937272	16.338182
18	6.264805	7.014911	8.230746	9.390455	10.864936	12.856953	14.439862	15.893212	17.337902
19	6.843971	7.632730	8.906516	10.117013	11.650910	13.715790	15.351660	16.850433	18.337653
20	7.433844	8.260398	9.590777	10.850811	12.442609	14.578439	16.265856	17.808829	19.337429
21	8.033653	8.897198	10.282898	11.591305	13.239598	15.444608	17.182265	18.768309	20.337228
22	8.642716	9.542492	10.982321	12.338015	14.041493	16.314040	18.100723	19.728791	21.337045
23	9.260425	10.195716	11.688552	13.090514	14.847956	17.186506	19.021087	20.690204	22.336878
24	9.886234	10.856361	12.401150	13.848425	15.658684	18.061804	19.943229	21.652486	23.336726
25	10.519652	11.523975	13.119720	14.611408	16.473408	18.939754	20.867034	22.615579	24.336587
26	11.160237	12.198147	13.843905	15.379157	17.291885	19.820194	21.792401	23.579434	25.336458
27	11.807587	12.878504	14.573383	16.151396	18.113896	20.702976	22.719236	24.544005	26.336339
28	12.461336	13.564710	15.307861	16.927875	18.939242	21.587969	23.647457	25.509251	27.336229
29	13.121149	14.256455	16.047072	17.708366	19.767744	22.475052	24.576988	26.475134	28.336127
30	13.786720	14.953457	16.790772	18.492661	20.599235	23.364115	25.507759	27.441622	29.336032
35	17.191820	18.508926	20.569377	22.465015	24.796655	27.835874	30.178172	32.282116	34.335638
40	20.706535	22.164261	24.433039	26.509303	29.050523	32.344953	34.871939	37.133959	39.335345
45	24.311014	25.901269	28.366152	30.612259	33.350381	36.884407	39.584701	41.995025	44.335118
50	27.990749	29.706683	32.357364	34.764252	37.688648	41.449211	44.313307	46.863776	49.334937
60	35.534491	37.484852	40.481748	43.187958	46.458888	50.640618	53.809126	56.619995	59.334666
70	43.275180	45.441717	48.757565	51.739278	55.328940	59.897809	63.346024	66.396114	69.334474
80	51.171932	53.540077	57.153173	60.391478	64.277844	69.206939	72.915342	76.187932	79.334330
90	59.196304	61.754079	65.646618	69.126030	73.291090	78.558432	82.511097	85.992545	89.334218
100	67.327563	70.064895	74.221927	77.929465	82.358136	87.945336	92.128944	95.807848	99.334129
120	83.851572	86.923280	91.572642	95.704637	100.623631	106.805606	111.418574	115.464544	119.333996

Table 3. The Chi-Square Distribution (Continued)

$$0.60 \leq P(X \leq x) \leq 0.9995$$

r	$P(X \leq x)$								
	0.60	0.70	0.80	0.90	0.95	0.975	0.99	0.995	0.9995
1	0.708326	1.074194	1.642374	2.705543	3.841459	5.023886	6.634897	7.879439	12.115665
2	1.832581	2.407946	3.218876	4.605170	5.991465	7.377759	9.210340	10.596635	15.201805
3	2.946166	3.664871	4.641628	6.251389	7.814728	9.348404	11.344867	12.838156	17.729996
4	4.044626	4.878433	5.988617	7.779440	9.487729	11.143287	13.276704	14.860259	19.997355
5	5.131867	6.064430	7.289276	9.236357	11.070498	12.832502	15.086272	16.749602	22.105327
6	6.210757	7.231135	8.558060	10.644641	12.591587	14.449375	16.811894	18.547584	24.102799
7	7.283208	8.383431	9.803250	12.017037	14.067140	16.012764	18.475307	20.277740	26.017768
8	8.350525	9.524458	11.030091	13.361566	15.507313	17.534546	20.090235	21.954955	27.868046
9	9.413640	10.656372	12.242145	14.683657	16.918978	19.022768	21.665994	23.589351	29.665808
10	10.473236	11.780723	13.441958	15.987179	18.307038	20.483177	23.209251	25.188180	31.419813
11	11.529834	12.898668	14.631421	17.275009	19.675138	21.920049	24.724970	26.756849	33.136615
12	12.583838	14.011100	15.811986	18.549348	21.026070	23.336664	26.216967	28.299519	34.821275
13	13.635571	15.118722	16.984797	19.811929	22.362032	24.735605	27.688250	29.819471	36.477794
14	14.685294	16.222099	18.150771	21.064144	23.684791	26.118948	29.141238	31.319350	38.109404
15	15.733223	17.321694	19.310657	22.307130	24.995790	27.488393	30.577914	32.801321	39.718760
16	16.779537	18.417894	20.465079	23.541829	26.296228	28.845351	31.999927	34.267187	41.308074
17	17.824387	19.511022	21.614561	24.769035	27.587112	30.191009	33.408664	35.718466	42.879213
18	18.867904	20.601354	22.759546	25.989423	28.869299	31.526378	34.805306	37.156451	44.433771
19	19.910199	21.689127	23.900417	27.203571	30.143527	32.852327	36.190869	38.582257	45.973120
20	20.951368	22.774545	25.037506	28.411981	31.410433	34.169607	37.566235	39.996846	47.498452
21	21.991497	23.857789	26.171100	29.615089	32.670573	35.478876	38.932173	41.401065	49.010812
22	23.030661	24.939016	27.301454	30.813282	33.924438	36.780712	40.289360	42.795655	50.511119
23	24.068925	26.018365	28.428793	32.006900	35.172462	38.075627	41.638398	44.181275	52.000189
24	25.106348	27.095961	29.553315	33.196244	36.415028	39.364077	42.979820	45.558512	53.478751
25	26.142984	28.171915	30.675201	34.381587	37.652484	40.646469	44.314105	46.927890	54.947455
26	27.178880	29.246327	31.794610	35.563171	38.885139	41.923170	45.641683	48.289882	56.406890
27	28.214078	30.319286	32.911688	36.741217	40.113272	43.194511	46.962942	49.644915	57.857586
28	29.248618	31.390875	34.026565	37.915923	41.337138	44.460792	48.278236	50.993376	59.300025
29	30.282536	32.461168	35.139362	39.087470	42.556968	45.722286	49.587884	52.335618	60.734647
30	31.315863	33.530233	36.250187	40.256024	43.772972	46.979242	50.892181	53.671962	62.161853
35	36.474606	38.859140	41.777963	46.058788	49.801850	53.203349	57.342073	60.274771	69.198558
40	41.622193	44.164867	47.268538	51.805057	55.758479	59.341707	63.690740	66.765962	76.094602
45	46.760687	49.451713	52.728815	57.505305	61.656233	65.410159	69.956832	73.166061	82.875687
50	51.891584	54.722794	58.163797	63.167121	67.504807	71.420195	76.153891	79.489978	89.560519
60	62.134840	65.226507	68.972069	74.397006	79.081944	83.297675	88.379419	91.951698	102.694755
70	72.358347	75.689277	79.714650	85.527043	90.531225	95.023184	100.425184	104.214899	115.577584
80	82.566250	86.119710	90.405349	96.578204	101.879474	106.628568	112.328792	116.321056	128.261312
90	92.761420	96.523762	101.053723	107.565008	113.145270	118.135893	124.116319	128.298944	140.782281
100	102.945944	106.905761	111.666713	118.498004	124.342113	129.561197	135.806723	140.169489	153.166955
120	123.288988	127.615901	132.806284	140.232569	146.567358	152.211403	158.950166	163.648184	177.602904

Table 4. The Standard Normal Probability Distribution

$$F(z) = P(X \leq z) = \frac{1}{\sqrt{2\pi}} \int_{-\infty}^{z} e^{-t^2/2} \, dt$$

z	0.00	0.01	0.02	0.03	0.04	0.05	0.06	0.07	0.08	0.09
-3.4	0.000337	0.000325	0.000313	0.000302	0.000291	0.000280	0.000270	0.000260	0.000251	0.000242
-3.3	0.000483	0.000466	0.000450	0.000434	0.000419	0.000404	0.000390	0.000376	0.000362	0.000349
-3.2	0.000687	0.000664	0.000641	0.000619	0.000598	0.000577	0.000557	0.000538	0.000519	0.000501
-3.1	0.000968	0.000935	0.000904	0.000874	0.000845	0.000816	0.000789	0.000762	0.000736	0.000711
-3.0	0.001350	0.001306	0.001264	0.001223	0.001183	0.001144	0.001107	0.001070	0.001035	0.001001
-2.9	0.001866	0.001807	0.001750	0.001695	0.001641	0.001589	0.001538	0.001489	0.001441	0.001395
-2.8	0.002555	0.002477	0.002401	0.002327	0.002256	0.002186	0.002118	0.002052	0.001988	0.001926
-2.7	0.003467	0.003364	0.003264	0.003167	0.003072	0.002980	0.002890	0.002803	0.002718	0.002635
-2.6	0.004661	0.004527	0.004396	0.004269	0.004145	0.004025	0.003907	0.003793	0.003681	0.003573
-2.5	0.006210	0.006037	0.005868	0.005703	0.005543	0.005386	0.005234	0.005085	0.004940	0.004799
-2.4	0.008198	0.007976	0.007760	0.007549	0.007344	0.007143	0.006947	0.006756	0.006569	0.006387
-2.3	0.010724	0.010444	0.010170	0.009903	0.009642	0.009387	0.009137	0.008894	0.008656	0.008424
-2.2	0.013903	0.013553	0.013209	0.012874	0.012545	0.012224	0.011911	0.011604	0.011304	0.011011
-2.1	0.017864	0.017429	0.017003	0.016586	0.016177	0.015778	0.015386	0.015003	0.014629	0.014262
-2.0	0.022750	0.022216	0.021692	0.021178	0.020675	0.020182	0.019699	0.019226	0.018763	0.018309
-1.9	0.028717	0.028067	0.027429	0.026803	0.026190	0.025588	0.024998	0.024419	0.023852	0.023295
-1.8	0.035930	0.035148	0.034380	0.033625	0.032884	0.032157	0.031443	0.030742	0.030054	0.029379
-1.7	0.044565	0.043633	0.042716	0.041815	0.040930	0.040059	0.039204	0.038364	0.037538	0.036727
-1.6	0.054799	0.053699	0.052616	0.051551	0.050503	0.049471	0.048457	0.047460	0.046479	0.045514
-1.5	0.066807	0.065522	0.064255	0.063008	0.061780	0.060571	0.059380	0.058208	0.057053	0.055917
-1.4	0.080757	0.079270	0.077804	0.076358	0.074934	0.073529	0.072145	0.070781	0.069437	0.068112
-1.3	0.096800	0.095098	0.093418	0.091759	0.090123	0.088508	0.086915	0.085343	0.083793	0.082264
-1.2	0.115070	0.113139	0.111232	0.109349	0.107488	0.105650	0.103835	0.102042	0.100273	0.098525
-1.1	0.135666	0.133500	0.131357	0.129238	0.127143	0.125072	0.123024	0.121000	0.119000	0.117023
-1.0	0.158655	0.156248	0.153864	0.151505	0.149170	0.146859	0.144572	0.142310	0.140071	0.137857
-0.9	0.184060	0.181411	0.178786	0.176186	0.173609	0.171056	0.168528	0.166023	0.163543	0.161087
-0.8	0.211855	0.208970	0.206108	0.203269	0.200454	0.197663	0.194895	0.192150	0.189430	0.186733
-0.7	0.241964	0.238852	0.235762	0.232695	0.229650	0.226627	0.223627	0.220650	0.217695	0.214764
-0.6	0.274253	0.270931	0.267629	0.264347	0.261086	0.257846	0.254627	0.251429	0.248252	0.245097
-0.5	0.308538	0.305026	0.301532	0.298056	0.294599	0.291160	0.287740	0.284339	0.280957	0.277595
-0.4	0.344578	0.340903	0.337243	0.333598	0.329969	0.326355	0.322758	0.319178	0.315614	0.312067
-0.3	0.382089	0.378280	0.374484	0.370700	0.366928	0.363169	0.359424	0.355691	0.351973	0.348268
-0.2	0.420740	0.416834	0.412936	0.409046	0.405165	0.401294	0.397432	0.393580	0.389739	0.385908
-0.1	0.460172	0.456205	0.452242	0.448283	0.444330	0.440382	0.436441	0.432505	0.428576	0.424655
-0.0	0.500000	0.496011	0.492022	0.488034	0.484047	0.480061	0.476078	0.472097	0.468119	0.464144

Table 4. The Standard Normal Probability Distribution (Continued)

z	0.00	0.01	0.02	0.03	0.04	0.05	0.06	0.07	0.08	0.09
0.0	0.500000	0.503989	0.507978	0.511966	0.515953	0.519939	0.523922	0.527903	0.531881	0.535856
0.1	0.539828	0.543795	0.547758	0.551717	0.555670	0.559618	0.563559	0.567495	0.571424	0.575345
0.2	0.579260	0.583166	0.587064	0.590954	0.594835	0.598706	0.602568	0.606420	0.610261	0.614092
0.3	0.617911	0.621720	0.625516	0.629300	0.633072	0.636831	0.640576	0.644309	0.648027	0.651732
0.4	0.655422	0.659097	0.662757	0.666402	0.670031	0.673645	0.677242	0.680822	0.684386	0.687933
0.5	0.691462	0.694974	0.698468	0.701944	0.705401	0.708840	0.712260	0.715661	0.719043	0.722405
0.6	0.725747	0.729069	0.732371	0.735653	0.738914	0.742154	0.745373	0.748571	0.751748	0.754903
0.7	0.758036	0.761148	0.764238	0.767305	0.770350	0.773373	0.776373	0.779350	0.782305	0.785236
0.8	0.788145	0.791030	0.793892	0.796731	0.799546	0.802337	0.805105	0.807850	0.810570	0.813267
0.9	0.815940	0.818589	0.821214	0.823814	0.826391	0.828944	0.831472	0.833977	0.836457	0.838913
1.0	0.841345	0.843752	0.846136	0.848495	0.850830	0.853141	0.855428	0.857690	0.859929	0.862143
1.1	0.864334	0.866500	0.868643	0.870762	0.872857	0.874928	0.876976	0.879000	0.881000	0.882977
1.2	0.884930	0.886861	0.888768	0.890651	0.892512	0.894350	0.896165	0.897958	0.899727	0.901475
1.3	0.903200	0.904902	0.906582	0.908241	0.909877	0.911492	0.913085	0.914657	0.916207	0.917736
1.4	0.919243	0.920730	0.922196	0.923641	0.925066	0.926471	0.927855	0.929219	0.930563	0.931888
1.5	0.933193	0.934478	0.935745	0.936992	0.938220	0.939429	0.940620	0.941792	0.942947	0.944083
1.6	0.945201	0.946301	0.947384	0.948449	0.949497	0.950529	0.951543	0.952540	0.953521	0.954486
1.7	0.955435	0.956367	0.957284	0.958185	0.959070	0.959941	0.960796	0.961636	0.962462	0.963273
1.8	0.964070	0.964852	0.965620	0.966375	0.967116	0.967843	0.968557	0.969258	0.969946	0.970621
1.9	0.971283	0.971933	0.972571	0.973197	0.973810	0.974412	0.975002	0.975581	0.976148	0.976705
2.0	0.977250	0.977784	0.978308	0.978822	0.979325	0.979818	0.980301	0.980774	0.981237	0.981691
2.1	0.982136	0.982571	0.982997	0.983414	0.983823	0.984222	0.984614	0.984997	0.985371	0.985738
2.2	0.986097	0.986447	0.986791	0.987126	0.987455	0.987776	0.988089	0.988396	0.988696	0.988989
2.3	0.989276	0.989556	0.989830	0.990097	0.990358	0.990613	0.990863	0.991106	0.991344	0.991576
2.4	0.991802	0.992024	0.992240	0.992451	0.992656	0.992857	0.993053	0.993244	0.993431	0.993613
2.5	0.993790	0.993963	0.994132	0.994297	0.994457	0.994614	0.994766	0.994915	0.995060	0.995201
2.6	0.995339	0.995473	0.995604	0.995731	0.995855	0.995975	0.996093	0.996207	0.996319	0.996427
2.7	0.996533	0.996636	0.996736	0.996833	0.996928	0.997020	0.997110	0.997197	0.997282	0.997365
2.8	0.997445	0.997523	0.997599	0.997673	0.997744	0.997814	0.997882	0.997948	0.998012	0.998074
2.9	0.998134	0.998193	0.998250	0.998305	0.998359	0.998411	0.998462	0.998511	0.998559	0.998605
3.0	0.998650	0.998694	0.998736	0.998777	0.998817	0.998856	0.998893	0.998930	0.998965	0.998999
3.1	0.999032	0.999065	0.999096	0.999126	0.999155	0.999184	0.999211	0.999238	0.999264	0.999289
3.2	0.999313	0.999336	0.999359	0.999381	0.999402	0.999423	0.999443	0.999462	0.999481	0.999499
3.3	0.999517	0.999534	0.999550	0.999566	0.999581	0.999596	0.999610	0.999624	0.999638	0.999651
3.4	0.999663	0.999675	0.999687	0.999698	0.999709	0.999720	0.999730	0.999740	0.999749	0.999758

Table 5. The (Student) t Probability Distribution

$$F(x) = P(X \leq x) = \frac{\Gamma\left(\frac{d+1}{2}\right)}{\Gamma\left(\frac{d}{2}\right)\sqrt{d\pi}} \int_{-\infty}^{x} \left(1 + \frac{t^2}{d}\right)^{-(d+1)/2} dt$$

$$0.0005 \leq P(X \leq x) \leq 0.30$$

d	$P(X \leq x)$								
	0.30	0.20	0.10	0.05	0.025	0.01	0.005	0.001	0.0005
1	0.726543	1.376382	3.077684	6.313752	12.706205	31.820516	63.656741	318.308839	636.619249
2	0.617213	1.060660	1.885618	2.919986	4.302653	6.964557	9.924843	22.327125	31.599055
3	0.584390	0.978472	1.637744	2.353363	3.182446	4.540703	5.840909	10.214532	12.923979
4	0.568649	0.940965	1.533206	2.131847	2.776445	3.746947	4.604095	7.173182	8.610302
5	0.559430	0.919544	1.475884	2.015048	2.570582	3.364930	4.032143	5.893430	6.868827
6	0.553381	0.905703	1.439756	1.943180	2.446912	3.142668	3.707428	5.207626	5.958816
7	0.549110	0.896030	1.414924	1.894579	2.364624	2.997952	3.499483	4.785290	5.407883
8	0.545934	0.888890	1.396815	1.859548	2.306004	2.896459	3.355387	4.500791	5.041305
9	0.543480	0.883404	1.383029	1.833113	2.262157	2.821438	3.249836	4.296806	4.780913
10	0.541528	0.879058	1.372184	1.812461	2.228139	2.763769	3.169273	4.143700	4.586894
11	0.539938	0.875530	1.363430	1.795885	2.200985	2.718079	3.105807	4.024701	4.436979
12	0.538618	0.872609	1.356217	1.782288	2.178813	2.680998	3.054540	3.929633	4.317791
13	0.537504	0.870152	1.350171	1.770933	2.160369	2.650309	3.012276	3.851982	4.220832
14	0.536552	0.868055	1.345030	1.761310	2.144787	2.624494	2.976843	3.787390	4.140454
15	0.535729	0.866245	1.340606	1.753050	2.131450	2.602480	2.946713	3.732834	4.072765
16	0.535010	0.864667	1.336757	1.745884	2.119905	2.583487	2.920782	3.686155	4.014996
17	0.534377	0.863279	1.333379	1.739607	2.109816	2.566934	2.898231	3.645767	3.965126
18	0.533816	0.862049	1.330391	1.734064	2.100922	2.552380	2.878440	3.610485	3.921646
19	0.533314	0.860951	1.327728	1.729133	2.093024	2.539483	2.860935	3.579400	3.883406
20	0.532863	0.859964	1.325341	1.724718	2.085963	2.527977	2.845340	3.551808	3.849516
21	0.532455	0.859074	1.323188	1.720743	2.079614	2.517648	2.831360	3.527154	3.819277
22	0.532085	0.858266	1.321237	1.717144	2.073873	2.508325	2.818756	3.504992	3.792131
23	0.531747	0.857530	1.319460	1.713872	2.068658	2.499867	2.807336	3.484964	3.767627
24	0.531438	0.856855	1.317836	1.710882	2.063899	2.492159	2.796940	3.466777	3.745399
25	0.531154	0.856236	1.316345	1.708141	2.059539	2.485107	2.787436	3.450189	3.725144
26	0.530892	0.855665	1.314972	1.705618	2.055529	2.478630	2.778715	3.434997	3.706612
27	0.530649	0.855137	1.313703	1.703288	2.051831	2.472660	2.770683	3.421034	3.689592
28	0.530424	0.854647	1.312527	1.701131	2.048407	2.467140	2.763262	3.408155	3.673906
29	0.530214	0.854192	1.311434	1.699127	2.045230	2.462021	2.756386	3.396240	3.659405
30	0.530019	0.853767	1.310415	1.697261	2.042272	2.457262	2.749996	3.385185	3.645959
32	0.529665	0.852998	1.308573	1.693889	2.036933	2.448678	2.738481	3.365306	3.621802
34	0.529353	0.852321	1.306952	1.690924	2.032244	2.441150	2.728394	3.347934	3.600716
36	0.529076	0.851720	1.305514	1.688298	2.028094	2.434494	2.719485	3.332624	3.582150
38	0.528828	0.851183	1.304230	1.685954	2.024394	2.428568	2.711558	3.319030	3.565678
40	0.528606	0.850700	1.303077	1.683851	2.021075	2.423257	2.704459	3.306878	3.550966
50	0.527760	0.848869	1.298714	1.675905	2.008559	2.403272	2.677793	3.261409	3.496013
60	0.527198	0.847653	1.295821	1.670649	2.000298	2.390119	2.660283	3.231709	3.460200
120	0.526797	0.846786	1.293763	1.666914	1.994437	2.380807	2.647905	3.210789	3.435015

Table 6. The F-Distribution

$$F(x) = P(X \le x) = \left(\frac{\nu_1}{\nu_2}\right)^{\nu_1/2} \frac{\Gamma\left(\dfrac{\nu_1+\nu_2}{2}\right)}{\Gamma\left(\dfrac{\nu_1}{2}\right)\Gamma\left(\dfrac{\nu_2}{2}\right)} \int_0^x t^{\nu_1/2-1}\left(1+\frac{\nu_1 t}{\nu_2}\right)^{-(\nu_1+\nu_2)/2} dt$$

$P(X \le x)$	ν_2	ν_1 1	2	3	4	5	6	7	8	9	10	15	25	60	129	∞
0.90	1	39.86	49.50	53.59	55.83	57.24	58.20	58.91	59.44	59.86	60.19	61.22	62.05	62.79	63.08	63.32
0.925		71.38	88.39	95.62	99.58	102.07	103.78	105.02	105.96	106.70	107.30	109.11	110.59	111.90	112.40	112.83
0.95		161.45	199.50	215.71	224.58	230.16	233.99	236.77	238.88	240.54	241.88	245.95	249.26	252.20	253.33	254.29
0.975		647.79	799.50	864.16	899.58	921.85	937.11	948.22	956.66	963.28	968.63	984.87	998.08	1009.80	1014.32	1018.16
0.99		4052.18	4999.50	5403.35	5624.58	5763.65	5858.99	5928.36	5981.07	6022.47	6055.85	6157.28	6239.83	6313.03	6341.23	6365.23
0.90	2	8.53	9.00	9.16	9.24	9.29	9.33	9.35	9.37	9.38	9.39	9.42	9.45	9.47	9.48	9.49
0.925		11.85	12.33	12.50	12.58	12.63	12.66	12.68	12.70	12.72	12.73	12.76	12.79	12.81	12.82	12.83
0.95		18.51	19.00	19.16	19.25	19.30	19.33	19.35	19.37	19.38	19.40	19.43	19.46	19.48	19.49	19.50
0.975		38.51	39.00	39.17	39.25	39.30	39.33	39.36	39.37	39.39	39.40	39.43	39.46	39.48	39.49	39.50
0.99		98.50	99.00	99.17	99.25	99.30	99.33	99.36	99.37	99.39	99.40	99.43	99.46	99.48	99.49	99.50
0.90	3	5.54	5.46	5.39	5.34	5.31	5.28	5.27	5.25	5.24	5.23	5.20	5.17	5.15	5.14	5.13
0.925		7.19	6.93	6.79	6.70	6.64	6.60	6.57	6.55	6.53	6.51	6.46	6.42	6.38	6.37	6.36
0.95		10.13	9.55	9.28	9.12	9.01	8.94	8.89	8.85	8.81	8.79	8.70	8.63	8.57	8.55	8.53
0.975		17.44	16.04	15.44	15.10	14.88	14.73	14.62	14.54	14.47	14.42	14.25	14.12	13.99	13.94	13.90
0.99		34.12	30.82	29.46	28.71	28.24	27.91	27.67	27.49	27.35	27.23	26.87	26.58	26.32	26.21	26.13
0.90	4	4.54	4.32	4.19	4.11	4.05	4.01	3.98	3.95	3.94	3.92	3.87	3.83	3.79	3.77	3.76
0.925		5.72	5.30	5.09	4.96	4.88	4.82	4.77	4.74	4.71	4.68	4.61	4.55	4.50	4.48	4.46
0.95		7.71	6.94	6.59	6.39	6.26	6.16	6.09	6.04	6.00	5.96	5.86	5.77	5.69	5.66	5.63
0.975		12.22	10.65	9.98	9.60	9.36	9.20	9.07	8.98	8.90	8.84	8.66	8.50	8.36	8.31	8.26
0.99		21.20	18.00	16.69	15.98	15.52	15.21	14.98	14.80	14.66	14.55	14.20	13.91	13.65	13.55	13.47

Table 6. The F-Distribution (Continued)

$P(X \le x)$	ν_2	ν_1 1	2	3	4	5	6	7	8	9	10	15	25	60	129	∞
0.90	5	4.06	3.78	3.62	3.52	3.45	3.40	3.37	3.34	3.32	3.30	3.24	3.19	3.14	3.12	3.11
0.925		5.03	4.55	4.30	4.16	4.06	4.00	3.95	3.91	3.87	3.85	3.77	3.70	3.63	3.61	3.59
0.95		6.61	5.79	5.41	5.19	5.05	4.95	4.88	4.82	4.77	4.74	4.62	4.52	4.43	4.40	4.37
0.975		10.01	8.43	7.76	7.39	7.15	6.98	6.85	6.76	6.68	6.62	6.43	6.27	6.12	6.07	6.02
0.99		16.26	13.27	12.06	11.39	10.97	10.67	10.46	10.29	10.16	10.05	9.72	9.45	9.20	9.11	9.02
0.90	6	3.78	3.46	3.29	3.18	3.11	3.05	3.01	2.98	2.96	2.94	2.87	2.81	2.76	2.74	2.72
0.925		4.63	4.11	3.86	3.71	3.60	3.53	3.48	3.44	3.40	3.37	3.29	3.21	3.14	3.11	3.09
0.95		5.99	5.14	4.76	4.53	4.39	4.28	4.21	4.15	4.10	4.06	3.94	3.83	3.74	3.70	3.67
0.975		8.81	7.26	6.60	6.23	5.99	5.82	5.70	5.60	5.52	5.46	5.27	5.11	4.96	4.90	4.85
0.99		13.75	10.92	9.78	9.15	8.75	8.47	8.26	8.10	7.98	7.87	7.56	7.30	7.06	6.96	6.88
0.90	7	3.59	3.26	3.07	2.96	2.88	2.83	2.78	2.75	2.72	2.70	2.63	2.57	2.51	2.49	2.47
0.925		4.37	3.84	3.57	3.42	3.31	3.24	3.18	3.13	3.10	3.07	2.98	2.90	2.83	2.80	2.77
0.95		5.59	4.74	4.35	4.12	3.97	3.87	3.79	3.73	3.68	3.64	3.51	3.40	3.30	3.26	3.23
0.975		8.07	6.54	5.89	5.52	5.29	5.12	4.99	4.90	4.82	4.76	4.57	4.40	4.25	4.19	4.14
0.99		12.25	9.55	8.45	7.85	7.46	7.19	6.99	6.84	6.72	6.62	6.31	6.06	5.82	5.73	5.65
0.90	8	3.46	3.11	2.92	2.81	2.73	2.67	2.62	2.59	2.56	2.54	2.46	2.40	2.34	2.31	2.29
0.925		4.19	3.64	3.38	3.21	3.11	3.03	2.97	2.93	2.89	2.86	2.76	2.68	2.60	2.57	2.55
0.95		5.32	4.46	4.07	3.84	3.69	3.58	3.50	3.44	3.39	3.35	3.22	3.11	3.01	2.96	2.93
0.975		7.57	6.06	5.42	5.05	4.82	4.65	4.53	4.43	4.36	4.30	4.10	3.94	3.78	3.72	3.67
0.99		11.26	8.65	7.59	7.01	6.63	6.37	6.18	6.03	5.91	5.81	5.52	5.26	5.03	4.94	4.86
0.90	9	3.36	3.01	2.81	2.69	2.61	2.55	2.51	2.47	2.44	2.42	2.34	2.27	2.21	2.18	2.16
0.925		4.05	3.50	3.23	3.07	2.96	2.88	2.82	2.77	2.74	2.70	2.61	2.52	2.44	2.41	2.38
0.95		5.12	4.26	3.86	3.63	3.48	3.37	3.29	3.23	3.18	3.14	3.01	2.89	2.79	2.74	2.71
0.975		7.21	5.71	5.08	4.72	4.48	4.32	4.20	4.10	4.03	3.96	3.77	3.60	3.45	3.39	3.33
0.99		10.56	8.02	6.99	6.42	6.06	5.80	5.61	5.47	5.35	5.26	4.96	4.71	4.48	4.39	4.31
0.90	10	3.29	2.92	2.73	2.61	2.52	2.46	2.41	2.38	2.35	2.32	2.24	2.17	2.11	2.08	2.06
0.925		3.95	3.39	3.12	2.95	2.84	2.76	2.70	2.66	2.62	2.59	2.49	2.40	2.31	2.28	2.25
0.95		4.96	4.10	3.71	3.48	3.33	3.22	3.14	3.07	3.02	2.98	2.85	2.73	2.62	2.58	2.54
0.975		6.94	5.46	4.83	4.47	4.24	4.07	3.95	3.85	3.78	3.72	3.52	3.35	3.20	3.14	3.08
0.99		10.04	7.56	6.55	5.99	5.64	5.39	5.20	5.06	4.94	4.85	4.56	4.31	4.08	3.99	3.91

Table 6. The *F*-Distribution (Continued)

$P(X \leq x)$	ν_2	ν_1 1	2	3	4	5	6	7	8	9	10	15	25	60	129	∞
0.90	15	3.07	2.70	2.49	2.36	2.27	2.21	2.16	2.12	2.09	2.06	1.97	1.89	1.82	1.78	1.76
0.925		3.66	3.09	2.81	2.64	2.53	2.45	2.38	2.33	2.29	2.26	2.15	2.05	1.96	1.92	1.88
0.95		4.54	3.68	3.29	3.06	2.90	2.79	2.71	2.64	2.59	2.54	2.40	2.28	2.16	2.11	2.07
0.975		6.20	4.77	4.15	3.80	3.58	3.41	3.29	3.20	3.12	3.06	2.86	2.69	2.52	2.46	2.40
0.99		8.68	6.36	5.42	4.89	4.56	4.32	4.14	4.00	3.89	3.80	3.52	3.28	3.05	2.95	2.87
0.90	25	2.92	2.53	2.32	2.18	2.09	2.02	1.97	1.93	1.89	1.87	1.77	1.68	1.59	1.55	1.52
0.925		3.45	2.88	2.59	2.42	2.30	2.22	2.15	2.10	2.06	2.02	1.90	1.80	1.69	1.64	1.60
0.95		4.24	3.39	2.99	2.76	2.60	2.49	2.40	2.34	2.28	2.24	2.09	1.96	1.82	1.76	1.71
0.975		5.69	4.29	3.69	3.35	3.13	2.97	2.85	2.75	2.68	2.61	2.41	2.23	2.05	1.98	1.91
0.99		7.77	5.57	4.68	4.18	3.85	3.63	3.46	3.32	3.22	3.13	2.85	2.60	2.36	2.26	2.17
0.90	60	2.79	2.39	2.18	2.04	1.95	1.87	1.82	1.77	1.74	1.71	1.60	1.50	1.40	1.34	1.29
0.925		3.28	2.71	2.42	2.24	2.12	2.03	1.96	1.91	1.86	1.83	1.70	1.58	1.45	1.39	1.33
0.95		4.00	3.15	2.76	2.53	2.37	2.25	2.17	2.10	2.04	1.99	1.84	1.69	1.53	1.46	1.39
0.975		5.29	3.93	3.34	3.01	2.79	2.63	2.51	2.41	2.33	2.27	2.06	1.87	1.67	1.57	1.48
0.99		7.08	4.98	4.13	3.65	3.34	3.12	2.95	2.82	2.72	2.63	2.35	2.10	1.84	1.72	1.60
0.90	129	2.74	2.34	2.13	1.99	1.89	1.82	1.76	1.72	1.68	1.65	1.54	1.44	1.31	1.25	1.19
0.925		3.22	2.64	2.35	2.18	2.06	1.97	1.90	1.84	1.80	1.76	1.63	1.50	1.36	1.29	1.21
0.95		3.91	3.07	2.67	2.44	2.28	2.17	2.08	2.01	1.95	1.90	1.74	1.59	1.42	1.34	1.25
0.975		5.14	3.80	3.22	2.89	2.67	2.51	2.39	2.29	2.21	2.15	1.94	1.74	1.52	1.41	1.30
0.99		6.84	4.77	3.94	3.47	3.16	2.94	2.78	2.65	2.55	2.46	2.18	1.92	1.64	1.51	1.37
0.90	∞	2.71	2.30	2.08	1.95	1.85	1.78	1.72	1.67	1.63	1.60	1.49	1.38	1.24	1.16	1.04
0.925		3.17	2.59	2.30	2.13	2.00	1.91	1.84	1.78	1.74	1.70	1.56	1.43	1.28	1.19	1.04
0.95		3.84	3.00	2.61	2.37	2.22	2.10	2.01	1.94	1.88	1.83	1.67	1.51	1.32	1.22	1.05
0.975		5.03	3.69	3.12	2.79	2.57	2.41	2.29	2.19	2.12	2.05	1.83	1.63	1.39	1.26	1.05
0.99		6.64	4.61	3.78	3.32	3.02	2.80	2.64	2.51	2.41	2.32	2.04	1.78	1.48	1.32	1.06

REFERENCES

Allen, L. J. S. (2003). An Introduction to Stochastic Processes with Biology Applications, Prentice Hall, Upper Saddle River, NJ.

Ash, R. (1970). Basic Probability Theory, John Wiley & Sons.

Awad, A. M. and Gharraf, M.K. (1986). Estimation of P $(Y < X)$ in the Burr case: a Comparative Study, Communications in Statistics, Simulation and Computation 15, 389–403.

Baklizi, A, and Abu Dayyeh, W. (2003). Shrinkage estimation of $P(Y < X)$ in the exponential case. Communications in Statistics, Simulation and Computation 32, 31–42.

Barlow, R. E., Bartholomew, D.J., Bremner, J.M., and Brunk, H. D. (1972). Statistical Inference under Order Restrictions, John Wiley & Sons, New York.

Beichelt, Frank (2006). Stochastic Processes in Science, Engineering and Finance, Chapman & Hall/CRC, Taylor & Francis Group.

Bergland, G. D. (1969). A guided tour of the fast Fourier transform, IEEE Spectrum 6, 41–52.

Bhat, U. Narayan (2008). An Introduction to Queueing Theory: Modeling and Analysis in Applications, Birkhäuser, Boston.

Birnbaum, Z. W. (1956). On a use of the Mann-Whitney statistic, Proc., Third Berkeley Symp. Math. Statist. Probab, Vol. 1, Univ. of California Press, 13-17.

Blahut, R. E. (1984). Fast Algorithms for Digital Signal Processing, Addison-Wesley, New York, NY.

Bojadjiev, L. and Kamenov, O. (1999). Higher mathematics, Vol. IV, Ciela, Sofia, Bulgaria.

Bojadjiev, L. and Kamenov, O. (2000). Higher mathematics, Vol. III, Ciela, Sofia, Bulgaria.

Bownik, M. (2001). On characterizations of multiwavelets in $L_2(R^n)$, in: Proceeding of AMS 11, 3265–3274.

Brigham, E. O. (1988). The Fast Fourier Transform and Applications, Prentice Hall, Englewood Cliffs, NJ.

Bronson, R. (1982). Schaum's Outline of Theory and Problems of Operations Research, McGraw-Hill.

Bronson, R. (1994). Schaums's Outline of Theory and Problems of Differential Equations, 2nd edition, McGraw-Hill.

Brownlee, K.A., Hodges, J.L., and Rosenblatt, M. (1953). The up-and-down method with small samples, Journal of the American Statistical Association 48, 262–277.

Casella, George, and Berger, Roger (1990). Statistical Inference, Wadsworth publishing Co., Duxbury Press.

Chernoff, H. (1979). Sequential Analysis and Optimal Design, Philadelphia: Society for Industrial and Applied Mathematics.

Chu, E. and A. George, (2000). Inside the FFT Black Box: Serial and Parallel Fast Fourier Transform Algorithms, CRC Press, Boca Raton, FL.

Chui, C.K. (1992). An Introduction to Wavelets, Academic Press, San Diego, CA.

Chui, C.K. and Lian, J.-A. (1995). Construction of compactly supported symmetric and antisymmetric orthonormal wavelets with scales = 3, Applied and Computational Harmonic Analysis 2, 21–51.

Chui, C.K. and Lian, J.-A. (1996). A study of orthonormal multi-wavelets, Applied Numerical Math. 20, 273–298.

Chui, C. K. and Lian, J.-A. (2002). Nonstationary wavelets and refinement sequences of nonuniform B-splines, in: Approximation Theory X, Wavelets, Splines, and Applications, C. K. Chui, L. L. Schumaker, and J. Sto¨eckler (eds.), Vanderbilt University Press, Nashville, TN, 207–229.

Chui, C. K., and Lian, J.-A. (2006). Construction of orthonormal multi-wavelets with additional vanishing moments, Advanced in Computational Mathematics 24, 239–262.

Church, J.D. and Harris, B. (1970). The estimation of reliability from stress-strength relationships, Technometrics, 12, 49–54.

Clogg, C. C., Rubin, D. B., Schenker, N., Schultz, B. and Weidman, L. (1991). Multiple imputation of industry and occupation codes in census public-use samples using Bayesian logistic regression. Journal of the American Statistical Association 86, 68–78.

Cohen, A., Daubechies, I., and Feauveau, J.C., (1992). Biorthogonal bases of compactly supported wavelets, Comm. Pure & Appl. Math. 45, 485–560.

Constantine, K, Karson, M., and Tse, H. (1986). Estimation of $P(Y<X)$ in the gamma case, Commun. Statist.-Simula 15, 365-88.

Cooley, J. W. and Tukey, O. W. (1965). An algorithm for the machine calculation of complex Fourier series, Math. Comput. 19, 297–301.

Daubechies, I. (1988). Orthonormal bases of compactly supported wavelets, Comm. Pure & Appl. Math. 41, 909–996.

Daubechies, I. (1992). Ten Lectures on Wavelets, CBMS-NSF Lecture Notes no. 61, SIAM.

Davis, T.A. and Sigmon, K. (2005). MATLAB Primer, 7th ed., Chapman & Hall/CRC.

Delprat, N., Escudié, B., Guillemain, P., Kronland-Martinet, R., Tchamitchian, P., and Torrésani, B. (1992). Asymptotic wavelet and Gabor analysis: extraction of instantaneous frequencies. IEEE Trans. Inf. Th. 38, 644–664.

Devore, J.L. (1999). Probability and Statistics for Engineering and the Sciences, Duxbury.

Dixon, W. J., and Mood, A.M. (1948). A method for obtaining and analyzing sensitivity data. Journal of the American Statistical Association 43, 109–126.

Downton, F. (1973). The estimation of P (Y < X) in the normal case, Technometrics 15, 551–558.

Duhamel, P. and Vetterli, M. (1990). Fast Fourier transforms: A tutorial review, Signal Processing 19, 259–299.

Durham, S., Haghighi, A. Montazer and Goddard, P. (1991). Differential Markov Chain: An Introduction to Applied probability. Lecture Notes, Department of Statistics, University of South Carolina.

Durham, S.D., and Flournoy, N. (1993). Convergence results for an adaptive ordinal urn design, Journal of the Theory of Probability and Its Applications 37, 14–17.

Durham, S.D., and Flournoy, N. (1995). Up-and-down designs I. Stationary treatment distributions, In Adaptive Designs, N. Flournoy and W. F. Rosenberger (eds.), 139–157. Hayward, California: Institute of Mathematical Statistics.

Durham, S.D., Flournoy, N., and Rosenberger, W.F. (1997). A random walk rule for phase I clinical trials. Biometrics 53, 745–760.

Durham, Stephen; Flournoy, Nancy, Goddard, Charles, and Haghighi, Aliakbar Montazer (1995). Lecture Notes, Department of Statistics, University of South Carolina.

Enis, P., and Geisser, S. (1971). "Estimation of the probability that $Y < X$", JASA 66, 162–168.

Finizio, N. and Ladas, G. (1982). An Introduction to Differential Equations with Difference Equations, Fourier Series, and Partial Differential Equations, Wadsworth Publishing Company, Belmont, CA.

Finney, R.L., Weir, M.D., and Giordano, F.R. (2001). Thomas' Calculus—Early Transcendentals, 10th edition, Addison Wesley Longman, Boston, MA.

Gezmu, M. (1996). The Geometric Up-and-Down Design for Allocating Dosage Levels. Dissertation, American University.

Giovagnoli, A., and Pintacuda, N. (1998). Properties of frequency distributions induced by general `up-and-down' methods for estimating quantiles. Journal of Statistical Planning and Inference 74, 51–63.

Goddard, Charles Roberts (1994). Random Walks and Queues: A Visual Approach, submitted as a partial fulfillment of the requirements for the Degree of Master of Science in the Department of Statistics, University of South Carolina, Columbia, SC, USA.

Goldberg, Samuel (1961). Introduction to Difference Equations, Science Editions, Inc. (originally by John Wiley and Sons, Inc., 1958).

Goupillaud, P., A. Grossmann, and J. Morlet (1984-1985). Cycle-octave and related transforms in seismic signal analysis, Geoexploration 23, 85–102.

Gross, D. and Harris, C.M. (1998). Fundamentals of Queueing Theory, 3rd edition, John Wiley & Sons.

Haar, Alfred (1910). Zur Theorie der orthogonalen Funktionensysteme, (German) Mathematische Annalen 69, 331–371.

Haghighi, A. Montazer and Shayib, M. A. (2007). Preferred Distribution Function for Estimating $P(Y < X)$, Proceedings, American Statistics Association.

Haghighi, A. Montazer and Shayib, M. A. (To appear in 2009). Shrinkage estimators for calculating reliability, Weibull case, Journal of Applied Statistical Science, Nova Science Publishers, NY.

Haghighi, Aliakbar Montazer and Mishev, Dimitar P. (2008). Queueing Models in Industry and Business, Nova Science Publishers, Inc., New York.

Higgins J.J. and Keller-McNulty, S. (1995). Concepts in Probability and Stochastic Modeling, Duxbury Press.

Hogg, R. V., and Tanis, E. A. (1988). Probability and Statistical Inference, 3rd Edition, MacMillan.

Hogg, R.V. and Tanis, E.A. (1993). Probability and Statistical Inference, 4th edition, Macmillan Publishing Company, NY.

Ivanova, Anastasia, Montazer-Haghighi, Aliakbar, Mohanty, Sri Gopal, and Durham, Steven D. (2003). Improved up-and-down designs for phase I trials, Statistics in Medicine, 22:69-82.

James, R.C. (1966), Advanced Calculus, Belmont, CA, Wadsworth.

Johnson, R. A. (1988). "Stress-Strength models for reliability" in Handbook of Statistics, P. R. Krismaiah and C. R., Rao, eds. Elsevier, 27–54.

Johnson, N. L. and Kotz, S. (1970). Continuous Univariate Distributions – 1, John Wiley & Sons, Inc.

Johnson, N. L., Kotz, S. and Balakrishnan, N. Continuous Univariate Distributions: Vol. 1, 2nd Edition, John Wiley, New York, 1994.

Johnson, R.A. (2000). Miller and Freund's Probability and Statistics for Engineers, 6th edition, Prentice Hall.

Kao, Edward P.C. (1997). An introduction to Stochastic Processes. Duxbury Press.

Karlin, S. and Taylor, H. (1975). A First Course in Stochastic Processes, second edition, Academic Press, New York.

Kelley, W.G. and Peterson, A. (2001). Difference Equations: An Introduction with Applications, Academic Press.

Kemeny, J. G. and J. L. Snell (1960). Finite Markov Chains. Van Nostrand, Princeton, NJ.

Kemeny, J.G., and Snell, J.L. (1960). Finite Markov Chains. Van Nostrand, Princeton.

Kleinrock, L. and Gail, R. (1996). Queueing Systems, Problems and Solutions, John Wiley & Sons.

Korn, E. L., Midthune, D., Chen, T. T., Rubinstein, L. V., Christian M. C., and Simon, R. M. (1994). A comparison of two phase I trial designs, Statistics in Medicine 13, 1799 - 1806.

Kotz, S. and Nadarajah, S. (2000). Extreme Value Distributions, Theory and Applications, Imperial College Press.

Kotz, S., Lumelskii, Y., and Pensky, M. (2003). The Stress-Strength Model and its Generalizations, Theory and Applications, World Scientific.

Kreyszig, E. (1978). Introductory Functional Analysis with Applications, John Wiley & Sons, Inc.

Kreyszig, E. (2006). Advanced Engineering Mathematics, 9th Ed., John Wiley & Sons, Inc.

Lehmann, E. L. (1983). Theory of Point Estimation, John Wiley & Sons, Inc.

Leon-Garsia, A. (2008). Probability and Random Processes for Electrical Engineering, 3rd edition, Pearson Prentice Hall.

Levy, H. and Lessman, F. (1992). Finite Difference Equations, Dover Publishing, (originally published: New York: Macmillan, 1961).

Lian, J.-A. (1998). Orthogonality criteria for multi-scaling functions, Applied and Computational Harmonic Analysis 5, 277-311.

Lian, J.-A. (1999). On intra- and inter-orthogonal scaling and wavelet vectors, in: Approximation Theory IX, Vol. 2: Computational Aspects, Vanderbilt University Press, Nashville, TN, 169–178.

Lian, J.-A. (2002). Polynomial identities of Bezout type, in: Trends in Approximation Theory, K. Kopotun, T. Lyche, and M. Neamtu (eds.), Vanderbilt University Press, Nashville, TN, 243–252.

Lian, J.-A. (2005). Armlets and balanced multiwavelets: Flipping filter construction, IEEE Transaction in Signal Processing, Vol. 53, no. 5, 1754–1767.

Lian, J.-A. (2006). On bivariate scaling functions and wavelets, in: Wavelets and Splines: Athens 2005, Guanrong Chen and Ming-Jun Lai (eds.), Nashboro Press, Brentwood, TN, 328–345.

Lian, J.-A. (2007). Bidimensional PR QMF with FIR filters, Applications and Applied Mathematics 2, 66–78.

Lian, J.-A. (2007). A new family of PR two channel filter banks, Applications and Applied Mathematics 2, 136–143.

Lindley, D. V. (1969). Introduction to Probability and Statistics from a Bayesian Viewpoint. Vol 1, Cambridge University Press.

Lipsky, L. (2009). Queueing Theory, 2nd edition, Springer.

Lipson, J. D. (1981). Elements of Algebra and Algebraic Computing, Reading, MA: Addison-Wesley.

Little, J.D.C. (1961). A proof of a queueing formula: $L = \lambda W$, Oper. Res. 9, 383–387.

Mallat, S.G. (1999). A Wavelet Tour of Signal Processing, Academic Press, San Diego, CA.

Mats, V.A., Rosenberger, W.F., and Flournoy, N. (1998). Restricted optimality for phase I clinical trials. In New Developments and Applications in Experimental Design (eds. Flournoy, N., Rosenberger, W.F., and Wong, W.K.), 50–61.

McCool, J. I. (1991). Inference on $P(Y < X)$ in the Weibull case, Communications in Statistics, Simulation and Computation 20(1): 129-148.

Mendenhall, W., Beaver, R.J. and Beaver B.M. (2009). Introduction to Probability and Statistics, 13th edition, Brooks/Cole, Cengage Learning.

Meyer, Y. (1992). Wavelets and Operators, Cambridge University Press.

Meyer, Y. (1993). Wavelets, Algorithms & Applications, SIAM, Philadelphia.

Mickens, R.E. (1987). Difference Equations: Theory and Applications, 2nd edition, Van Nostrand Reinhold, NY.

Miller, K. (1968). Linear Difference Equations, W.A. Benjamin, Inc.

Nadarajah, S. (2003). Reliability for Extreme Value Distributions, Mathematical and Computer Modeling, 37, 915 – 922.

Narayana, T. V. (1953). Sequential Procedures in the Probit Analysis, Dissertation, University of North Carolina, Chapel Hill.

Navidi, W. (2010). Principles of Statistics for Engineers and Scientists, McGraw Hill.

Navidi, W. (2010). Principles of Statistics for Engineers and Scientist, McGraw-Hill, Higher Education.

Navidi, William Cyrus (2010). 1st ed. Principles of Statistics for Engineers and Scientists, The McGraw-Hill Companies, Inc.

Neter, J., Kutner M., Nachtsheim, C. and Wasserman, W. (1996). Applied Linear Statistical Models, 4th edition, Times Mirror Higher Education.

Neuts, M.F. (1973). Probability, Allyn and Bacon, Boston, MA.

Norris, J. R. (1997). Markov Chains, Cambridge Series in Statistical and Probabilistic mathematics, Cambridge University Press, Cambridge.

Nussbaumer, H. J. (1982). Fast Fourier Transform and Convolution Algorithms, 2nd ed. Springer-Verlag, New York.

Owen, D.B., Craswell, K.J., and Hanson, D.L. (1964). "Nonparametric upper confidence bounds for $P(Y < X)$ when X and Y are normal," JASA 50, 906–924.

Papoulis, A. (1962). The Fourier Integral and its Applications, McGraw-Hill, New York.

Prabhu, Narahari U. (2007). Stochastic Processes, World Scientific.

Press, W, Teukolsky, S., Vetterling, W., Flannery, B. (1992). Numerical Recipes in Fortran, 2nd Edition, Cambridge University Press.

Press, W. H., B.P. Flannery, S.A. Teukolsky, and W.T. Vetterling (1992). Fast Fourier transform, Ch. 12, in: Numerical Recipes in FORTRAN: The Art of Scientific Computing, 2nd ed., Cambridge University Press, Cambridge, England, 490–529.

Ramirez, R. W. (1985). The FFT: Fundamentals and Concepts, Prentice-Hall, Englewood Cliffs, NJ.

Robertson, T., Wright, F.T., and Dykstra, R.L. (1988). Ordered Restricted Statistical Inference. John Wiley & Sons, New York.

Ross, S. (1982). Introduction to Probability Models, 3rd Edition, Academic Press, 1985.

Ross, Sheldon M. (1996). Introductory Statistics, The McGraw-Hill Companies, Inc.

Ryan, T. (1989). Statistical Methods for Quality Improvement, John Wiley & Sons.

Schork, M. Anthony, and Remington, Richard D. (2000). Statistics with Applications to the Biological and Health Science, third edition, Prentice Hall.

Shayib, M. A. (2005). Effects of Parameters and Sample Size on the Estimation of P ($Y < X$), A Simulation Study, Part I, JSM Proceedings, 142–144.

Spiegel, M. (1971). Theory and Problems of Calculus of Finite Differences and Difference Equations, McGraw-Hill.

Spiegel, M. (1965). Schaums's Outline of Theory and Problems of Laplace Transforms, McGraw-Hill.

Spiegel, M. (1974). Schaums's Outline of Theory and Problems of Fourier Analysis with Applications to Boundary Value Problems, McGraw-Hill.

Stidham, S. (2009). Optimal Design of Queueing System, Taylor and Francis Group.

Stoer, J. and Bulirsch, R. (1980). Introduction to Numerical Analysis, Springer-Verlag, New York.

Storer, B.E. (1989). Design and analysis of Phase I clinical trials, Biometrics 45, 925–937.

Strang, G. (1993). Wavelet transforms versus Fourier transforms, Bull. Amer. Math. Soc. 28, 288–305.

Suhov, Y. and Kelbert, M. (2008). Probability and Statistics by Example: II Markov Chains: a Primer in Random Processes and their Applications, Cambridge University Press.

Surles, J.G. and Padgett, W.J. (1998). Inference for P ($Y < X$) in the Burr type X Model, Journal of Applied Statistical Science 7, 225-238.

Takacs, Lajos (1962). Introduction to the Theory of Queues, Oxford University Press.

Thompson, J. (1968). Some shrinkage techniques for estimating the mean, JASA 63, 113–122.

Trench, W.F. (2000). Elementary Differential Equations, Brooks/Cole, Thomson Learning.

Tse, S. and Tso, G. (1996). Shrinkage estimation of reliability for exponentially distributed lifetimes, Communications in Statistics, Simulation and Computation 25(2): 415-430.

Tsutakawa, R. K. (1967). Random walk design in bio-assay, Journal of the American Statistical Association 62, 842–856.

Van Loan, C. (1992). Computational Frameworks for the Fast Fourier Transform, SIAM, Philadelphia, PA.

Vetterli, M. and J. Kovačević (1995). Wavelets and Subband Coding, Prentice Hall PTR, Englewood Cliffs, NJ.

Wackerly, D.D., Mendenhall, W. and Scheaffer, R.L. (2008). Mathematical Statistics with Applications, Thomson Brooks/Cole.

Wagon, S. (1991). Mathematica in Action, W. H. Freeman, New York.

Walker, J. S. (1996). Fast Fourier Transform, 2nd ed., CRC Press, Boca Raton, FL.

Wetherill, G.B, and Glazebrook, K.D. (1986). Sequential Methods in Statistics, Chapman and Hall, London, New York.

Whittaker, E. T. & Watson, G. N. (1965). A course in Modern Analysis, Cambridge at The University Press, 4th Edition, Reprinted.

Zill, D.G. (2009). A First Course in Differential Equations with Modeling Applications, 9th ed., Brooks/Cole, Belmont, CA.

ANSWERS TO SELECTED EXERCISES

1.1.1.	$D = \{(x, y) : \ y \geq 3x\}$	1.2.9.	$\frac{\partial f}{\partial x} = yzx^{yz-1}, \ \frac{\partial f}{\partial y} = zx^{yz}\ln x, \ \frac{\partial f}{\partial z} = yx^{yz}\ln x$
1.1.3.	$D = \{(x, y) : \ y \geq -x^2\}$	1.2.11.	$f_{xx} = 2y, \ f_{xy} = 2x + \frac{1}{2\sqrt{y}}, \ f_{yy} = -\frac{x}{4y\sqrt{y}}$
1.1.4.	$D = \{(x, y) : \ (x, y) \neq (0, 0)\}$	1.2.14.	$\mathrm{grad}(f)(1, 1) = (1, 1)$
1.1.5.	$D = \{(x, y) : \ (x, y) \neq (0, 0)\}$	1.2.15.	21
1.1.6.	$D = \mathbb{R}^2$	1.3.2.	$\mathrm{Re}\, g = \frac{2x}{x^2 + y^2}, \ \mathrm{Im}\, g = -\frac{2y}{x^2 + y^2}$
1.1.8.	$D = \{(x, y) : \ xy < 1\}$	1.3.5.	$-3 + 4i$
1.1.9.	$D = \{(x, y) : \ x > 0, \ y > 0\}$	1.3.6.	$8i$
1.1.10.	$D = \{(x, y) : \ xy > 0\}$	1.3.10.	$\ln \sqrt{2} + i\frac{\pi}{4}$
1.1.11.	-84	1.3.12.	$\frac{5}{3}\cos 3 + i\frac{4}{3}\sin 3$
1.1.12.	$\frac{12}{5}$	1.3.22.	$\{1, i, -1, i\}$
1.2.1.	$\frac{\partial f}{\partial x}(1, 3) = -10, \ \frac{\partial f}{\partial y}(1, 3) = -6$	1.4.5.	Convergent, Sum= 2
1.2.2.	$f_x(1, 0) = 1, \ f_y(1, 0) = 2$	1.4.6.	Convergent, Sum= $-\frac{3}{8}$
1.2.3.	$\frac{\partial z}{\partial x}(1, 2) = -\frac{3}{25}, \ \frac{\partial z}{\partial y}(1, 2) = \frac{24}{25}$	1.4.7.	Convergent, Sum = $\frac{12}{5}$
1.2.4.	$\frac{\partial z}{\partial x} = -\frac{z + y}{x + y}, \ \frac{\partial z}{\partial y} = -\frac{z + x}{x + y}$	1.4.8.	Convergent, Sum= $\frac{1}{\pi^3 - 1}$
1.2.5.	$\frac{\partial R}{\partial R_i} = \left(\frac{R}{R_i}\right)^2, \ i = 1, 2, 3$	1.4.10.	Convergent
1.2.6.	$\frac{\partial z}{\partial x} = \frac{1}{\sqrt{x^2 + y^2}}, \ \frac{\partial z}{\partial y} = \frac{y}{x\sqrt{x^2 + y^2} + x^2 + y^2}$	1.4.11.	Divergent

2.2.5.	$f(x) = \dfrac{4}{\pi}\left(\sin x + \dfrac{1}{3}\sin 3x + \dfrac{1}{5}\sin 5x + \cdots\right)$		
2.3.3.	$\sin x = \dfrac{2}{\pi} - \dfrac{4}{\pi}\left(\dfrac{\cos 2x}{2^2-1} + \dfrac{\cos 4x}{4^2-1} + \dfrac{\cos 6x}{6^2-1} + \cdots\right)$		
2.3.4.	(a) $\cos x = \dfrac{8}{\pi}\displaystyle\sum_{n=1}^{\infty} \dfrac{n\sin 2nx}{4n^2-1}$	2.3.4.	(b) $f(0) = f(\pi) = 0$
2.3.8.	(a) $x(\pi-x) = \dfrac{8}{\pi}\left(\dfrac{\sin x}{1^3} + \dfrac{\sin 3x}{3^3} + \dfrac{\sin 5x}{5^3} + \cdots\right)$. (b) $x(\pi-x) = \dfrac{\pi^2}{6} - \left(\dfrac{\cos 2x}{1^2} + \dfrac{\cos 4x}{2^2} + \dfrac{\cos 6x}{3^2} + \cdots\right)$		
2.4.4.	$u(x,t) = 4 - \dfrac{24}{\pi^2}\displaystyle\sum_{k=1}^{\infty} \dfrac{1}{(2k-1)^2}\, e^{-\left(\frac{(2k-1)\pi}{6}\right)^2 t}\cos\left(\dfrac{(2k-1)\pi}{6}x\right)$		

3.1.1.	$\dfrac{2}{s}\left(1 - e^{-2s}\right)$	3.2.2.	$\dfrac{s-5}{(s-5)^2+9}$
3.1.2.	$\dfrac{1}{s} - \dfrac{2}{s+2}$	3.2.3.	$\dfrac{3}{(s-5)^2+9}$
3.1.4.	$\dfrac{1}{2}\left(\dfrac{1}{s} + \dfrac{s}{s^2+36}\right)$	3.2.4.	$\dfrac{3}{(s-4)^2-9}$
3.1.6.	$\dfrac{2}{s^2+16}$	3.2.5.	$\dfrac{s-1}{(s-1)^2-81}$
3.1.7.	$\dfrac{2}{s^3}$	3.2.6.	$e^{-t}\left(2\cos\sqrt{3}t - \dfrac{\sqrt{3}}{3}\sin\sqrt{3}t\right)$
3.1.9.	$\dfrac{1}{2}\left(\dfrac{1}{s} - \dfrac{s}{s^2+36}\right)$	3.2.8.	$\dfrac{1}{2}e^{t}\sin 2t$
3.1.11.	$2e^{-t}$	3.2.10.	$\dfrac{\sqrt{3}}{3}e^{-t}\sin\sqrt{3}t$
3.1.13.	$\dfrac{1}{3}t^3 e^{3t}$	3.3.1.	$\dfrac{5}{s-5}$
3.1.15.	$\dfrac{1}{2}t^2 e^{-2t}$	3.3.2.	$-\dfrac{s+2}{s+7}$
3.1.17.	$\dfrac{5}{6}\left(e^{3t} - e^{-3t}\right)$	3.3.3.	$\dfrac{3s+21}{s^2}$

3.2.1.	$\dfrac{6}{(s-7)^4}$	3.3.4.	$\dfrac{4s^2+25}{s^2-25}$
3.4.1.	$e^{-t}\left(2\cos 2t + \dfrac{1}{2}\sin 2t\right)$	3.4.5.	$\dfrac{11}{5}\cos 2t - \dfrac{1}{5}\cos 3t + \sin 2t$
3.5.1.	$\dfrac{1}{s^2}\left(1 - 2e^{-3s} + e^{-6s}\right)$	3.5.3.	$\dfrac{2}{s} + \dfrac{4e^{-3s}}{s} - \dfrac{5e^{-6s}}{s}$

3.5.5.	$\dfrac{1}{s^2+1}\left(1 - e^{-4\pi s}\right)$
3.5.8.	$\dfrac{1}{3}\sin(3t-3)\,u(t-1) = \begin{cases} 0, & \text{if } 0 \le t < 1 \\ \dfrac{1}{3}\sin(3t-3), & \text{if } t \ge 1 \end{cases}$
3.5.9.	$(t-2)e^{-3(t-2)}\,u(t-2) = \begin{cases} 0, & \text{if } 0 \le t < 2 \\ (t-2)e^{-3(t-2)}, & \text{if } t \ge 2 \end{cases}$
3.6.1.	$y(t) = 2(1 - \cos t) - 2u(t-\pi)(1 + \cos t)$
3.6.2.	$y(t) = -\dfrac{2}{3}t + \dfrac{10}{9} - \dfrac{1}{9}e^{3t} + \left[-\dfrac{2}{3}t + \dfrac{10}{9} + \dfrac{2}{9}e^{3(t-2)}\right]u(t-2)$

4.1.6.	B	4.2.7.	A	4.2.20.	a) 1/3 b) 2/3 c) 1/3	4.2.33.	C	4.2.34.	B
4.2.35.	D	4.4.3.	C	4.4.4.	A	4.4.5.	D	4.4.6.	B	4.4.7.	A
4.4.8.	B	4.7.1.		4.7.2.	B	4.7.3.	B	4.7.4.	C	4.7.5.	C
4.7.6.	E	4.7.7.	A	4.7.10.	D	4.10.3.	A		
4.10.15. (a)	$E(X) = 2.93,\;\; E(Y) = -0.22,\;\; E(X-Y) = 3.15,\;\; E(2X-3Y) = 6.52$							
4.10.15. (b)	No	4.11.1.	C	4.11.2.	B	4.11.3.	A	4.11.4.	B
4.11.5.	A		4.11.6.	B					

5.1.1.	C	5.1.2.	E	5.1.3.	B	5.1.8.	D	5.1.9.	D	5.1.10.	C
5.1.11.	B					5.3.4.	$(0.83374, 0.96626)$				
5.4.6.	$power = 0.2920$					5.5.1.	$(-7.555, -2.005)$				
5.7.1.	$\widehat{\beta}_1 = -0.99390,\;\; \widehat{\beta}_0 = 8.54268,\;\; r^2 = 0.80201,\;\; r = -0.89555$										

6.1.1.	$y_k = C_1 + C_2(-1)^k,\ k = 1, 2, \cdots$
6.1.3.	$y_k = (C_1 + C_2 k)(-1)^k,\ k = 1, 2, \cdots$
6.1.13.	$Y_k = \dfrac{1}{17} \left(-7 \sin \dfrac{k\pi}{2} + 6 \cos \dfrac{k\pi}{2} \right)$
6.2.1.	$y_k = 2(k + 1)$
6.2.3.	$y_k = \dfrac{3^{k+1} + 1}{2}$
6.4.3.	$\dfrac{2}{(1 - z)^3}$
6.4.7.	$a_0 = a_1 = 0,\ a_k = 1,\ \text{if } k \geq 2$
6.4.9.	$P_{k,n} = \dfrac{1}{2^k} \left(1 + \dfrac{k}{2} + \dfrac{k(k+1)}{2^2 2!} + \cdots + \dfrac{k(k+1)\cdots(k+n-2)}{2^{n-1}(n-1)!} \right)$

INDEX

Mcgawnty

→ May Hall   120

10a-12p
1:30-3:30